Principles of Renewable Energy Engineering with Worked Examples

Nihal E Wijeysundera

World Scientific

NEW JERSEY • LONDON • SINGAPORE • BEIJING • SHANGHAI • HONG KONG • TAIPEI • CHENNAI • TOKYO

Published by

World Scientific Publishing Co. Pte. Ltd.
5 Toh Tuck Link, Singapore 596224
USA office: 27 Warren Street, Suite 401-402, Hackensack, NJ 07601
UK office: 57 Shelton Street, Covent Garden, London WC2H 9HE

British Library Cataloguing-in-Publication Data
A catalogue record for this book is available from the British Library.

PRINCIPLES OF RENEWABLE ENERGY ENGINEERING WITH WORKED EXAMPLES

ISBN 978-981-125-114-6 (hardcover)
ISBN 978-981-125-115-3 (ebook for institutions)
ISBN 978-981-125-116-0 (ebook for individuals)

For any available supplementary material, please visit
https://www.worldscientific.com/worldscibooks/10.1142/12691#t=suppl

This book is dedicated to the memory of

My father *T.B. Wijeysundera*
and
My mother *G.M.E. Wijeysundera*

Preface

Over the past two decades commercial applications of renewable energy sources, particularly, solar thermal power, solar photovoltaic power, wind energy, and biomass energy have been growing at a rapid pace. Concerns on global warming, manufacturing cost reduction, and improvements in device-efficiency have been the main reasons for this development.

Universities and other educational institutions now offer courses on renewable energy on a regular basis. There are numerous popular books, handbooks, and textbooks on individual renewable energy sources. However, textbooks that include the engineering principles of all the different renewable energy systems are few. Moreover, there is a need for textbooks on renewable energy that provide detailed worked examples and problems as in other engineering textbooks.

Physical principles that underpin the design and operation of renewable energy conversion devices are covered in different engineering courses. The main objective of this textbook is to bring these principles together in a single volume, tailored for a basic course on renewable energy, including detailed worked examples and problems. It is hoped that the use of this book would avoid the need to refer to several textbooks on individual energy sources to develop a course on renewable energy. Students using the book would benefit in a similar way.

A distinguishing feature of the book is the inclusion of 20-30 worked examples and problems in each chapter, carefully chosen to illustrate the various concepts covered in the chapter. Students would be able to develop their problem-solving skills by following worked examples carefully.

Moreover, for instructors the book should provide a useful source of problems, for illustration in courses, and inclusion in tutorials, quizzes, and examinations. The inclusion of comprehensive worked examples is a novel feature of the book, but it is not done at the expense of the subject matter covered, but more to complement the latter.

The present book covers the engineering principles of the following renewable energy applications: (i) novel systems, currently being deployed on a commercial scale (solar, wind, biomass) (ii) systems operating mainly as pilot projects (wave, ocean-thermal), and (iii) systems well-established over a long period (tidal, geothermal, hydroelectric).

The first chapter incudes a review of the current patterns of energy use and the growth of renewable sources in recent decades. Estimation of solar radiation, which is necessary for the design of all solar-based energy systems, is covered in chapter 2. Solar heating and power generation systems are included in chapters 3 and 4. Chapters 5 and 6 deal with solar-PV systems, and wind energy systems respectively.

Energy storage which is necessary to deal with the intermittent nature of renewable sources is covered in chapter 7. Wave, tidal and OTEC, systems are included in chapter 8. Hydroelectric and geothermal plants, and biomass energy systems are covered in chapters 9 and 10 respectively.

I was fortunate to have had the opportunity to teach several courses on energy systems, including energy conversion, solar energy systems, and energy engineering at the National University of Singapore (NUS). The notes developed for these courses provided the framework and much of the material for this book. I am thankful to my colleagues in the energy and bio-thermal division at NUS, with whom I shared the teaching of these courses, for valuable discussions on energy systems.

I wish to thank Dr. Jerome Ellepola who willingly reviewed early drafts of the two chapters on wind energy systems, and solar PV power. I benefitted much from his insightful comments and suggestions on the content, and presentation of the subject matter.

I am thankful to Dr. Kanishka Kobbekaduwa for fruitful discussions on the physical aspects of solar cells, and for reviewing an early draft of the chapter on solar PV power.

Thanks are due to my sons Duminda and Harindra, and my daughters-in-law Sindhu and Sophia, for their constant encouragement.

Finally, my heartfelt thanks are given to my wife Kamani for encouragement and generous support towards the completion of this project.

Nihal E. Wijeysundera

Contents

Chapter 1

Introduction to Energy Sources and Utilization

1.1 Background

Since the dawn of civilization, humans have relied on various forms of energy to improve their quality of life. Animals were first employed to replace human effort for lifting and moving loads. Draft animals were particularly useful in agriculture and transportation. Water wheels driven by flowing streams, and simple windmills were used to generate mechanical energy for processes, such as, grinding grain and other food products. Sunlight provided thermal energy for heating and drying. Cooking food relied largely on firewood, animal dung and charcoal. These energy resources were renewable because they occurred naturally in the environment. Over time, there were gradual improvements in the technologies used to harness these natural energy sources [9].

The industrial revolution that happened in the 1800s marked a historic turning point in the use of energy for human purposes. The steam engine, invented by Thomas Newcomen in 1712, was first used for pumping water from mines. Later, more advanced versions of the steam engine, developed by Stephenson and Watt in 1814, were adopted for rail transportation [15]. Coal was the fuel of choice for these steam engines.

The first commercially successful, coal gas-powered, internal combustion engine was patented in 1860 by the Belgian engineer Jean Lenoir. At about the same time the German inventor named Nikolaus Otto developed his internal combustion engine, which operated on the four-stroke cycle, and used liquid hydrocarbon fuels [16]. The rapid industrialization of a number of western countries, that followed the invention of these prime movers, increased the demand for fossil fuels, such as, coal, oil and gas. This trend has continued to the present day. Fossil fuels, which include crude oil and natural gas, are buried combustible geologic deposits. These were formed over millions of years from decayed plants and animals exposed to extreme heat and pressure of the earth's crust [9].

The discovery of nuclear fission [7] by Hahn, Strassman and Meitner in 1939, and the development of the first experimental nuclear reactor by a group of scientists led by Enrico Fermi in 1942, ushered the atomic - age. The first commercial gas-cooled nuclear reactor to generate electricity was developed in the United Kingdom in 1956. This was followed by the introduction of the first commercial pressurized-water-reactor (PWR) in the USA in 1960. The nuclear fuel used in these reactors was uranium, which has been present as ore since the formation of the earth.

Due to the abundant supply of relatively inexpensive fossil fuels, and their high energy density, they became the main fuel for transportation and electricity generation over the last two centuries. Consequently, the development of renewable energy sources like wind and solar received much less attention. However, hydro-electric power generation projects, which required the construction of dams and reservoirs, continued to be developed due to their indirect contribution to water management in irrigation, and flood control.

The origin of all renewable energy sources like solar, wind and wave energy, and fossil fuels like coal, crude-oil, and natural gas can be traced to solar radiation received on earth, either recently or millions of years ago. Natural processes responsible for the conversion of solar radiation reaching the earth to other energy sources, are shown schematically in Fig. 1.1.

The numbers in Fig. 1.1, represent the different quantities of various energy forms available for utilization. Numerical data was obtained from Ref.[8], which presents a detailed discussion on the abundance of different energy sources. Of the short-wave solar radiation striking the earth about 30 percent is reflected to space by the atmosphere, clouds, and surfaces on earth. About 47 percent of solar radiation is absorbed by the earth's surface, and is converted directly to heat. About 23 percent

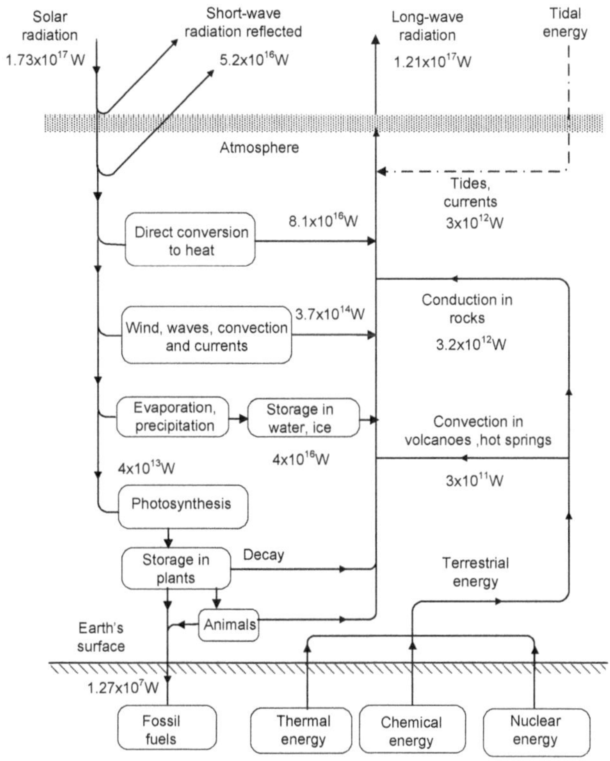

Energy flow diagram (data from Ref.[8])

Fig. 1.1 Conversion of solar radiation to different energy forms (Data from Ref.[8])

of incident solar energy is absorbed by the atmosphere. The heating of the land and bodies of water like oceans and seas causes winds, convection currents and waves, which are available in the form of wind energy and wave energy. These sources are harnessed using wind turbines and wave-devices respectively. The evaporation from water surfaces and precipitation at higher elevations leads to the formation of rivers and lakes, with the potential to produce hydro-electric power, using water turbines and electric generators.

Solar radiation is absorbed by plants for photosynthesis, which is a process used by plants and other organisms to convert solar energy into chemical energy that can be later released to fuel the organism's activities. This energy can be utilized as heat by directly burning the plant, or by converting the plant to a biofuel like ethanol. The latter conversion requires several additional thermo-chemical processes. The decaying of plants, and animals that consume plants as food, release a fraction the original energy in the form of heat. Once buried in the earth for millions of years, under intense heat and pressure of the earth's crust, the decayed plants and animals form fossil fuels such as coal, oil and gas [9].

The earth also has thermal energy available in the form of geothermal heat that comes from radioactive decay of minerals, and continual heat loss from the earth's original formation [9,22]. Some of the terrestrial energy from the earth is released by convection in volcanoes and hot springs, and conduction in rocks. Tidal energy is an energy form that originates from periodic variations in gravitational attraction exerted by celestial bodies, particularly the sun and the moon. These changing forces cause temporary variations in water levels of oceans that can be used to generate mechanical energy and electricity [5,23].

The different energy forms mentioned above eventually manifest as thermal radiation, a fraction of which is reradiated back to earth by the atmosphere, and the rest is radiated to space. Ideally, the balance between the net solar radiation absorbed by the earth and thermal energy radiated back to space maintains the average temperature of the earth at an equilibrium value [8].

A renewable energy resource is one that is replenished again after use to avoid depletion. However, this process should occur on a human time scale. Therefore, fossil fuels and nuclear fuels are non-renewable energy sources because it takes thousands of millions of years to create them. On the other hand, solar, wind, wave, hydro, tidal, biomass and geothermal resources are renewable because they are being replenished continuously.

1.2 Patterns of Global Energy Use

Data on global energy supply and consumption are compiled, analyzed and published periodically by institutions such as the International

Energy Agency (IEA), U.S. Energy Information Administration (EIA), and the European Environment Agency (EEA). The British Petroleum (BP) Statistical Review of World Energy Report [10-14], published annually, provides a comprehensive survey of the global energy situation. The information in these sources reveals trends that are useful for formulation of energy policy.

An important parameter that captures global energy consumption is the world total primary energy consumption (WTPEC) [17], which has deep implications in the areas of economics and politics. The WTPEC, measured per year, is the total energy from all sources, used for human purposes in all countries. A related energy consumption metric is the total primary energy supply (TPES) [17]. This is the sum of the global energy supply less the changes in net energy storage.

The conversion efficiency from the primary form of an energy resource to a useable form is not considered in estimating the TPES. For example, crude oil extracted from the ground has to be refined, processed and delivered to gas stations before it could be used in a vehicle by consumers. Significant amount of the energy in crude oil is used for converting it to gasoline, which is the directly usable form.

Table 1.1 World total energy supply and consumption (Data from Ref.[17])

Year	TPES (TWh)	WTPEC (TWh)	Electricity	WTPEC/TPES	Elect./WTPEC
2010	147899	100914	21431	0.682	0.212
2011	152504	103914	22126	0.681	0.213
2012	155505	104426	22668	0.672	0.217
2013	157482	108171	23322	0.687	0.216
2014	155481	109613	23816	0.705	0.217

The fraction of primary energy required for processing and delivery for use by consumers is different for different energy sources. There is also a qualitative difference between energy in the form of heat and electricity. The conversion of heat to work or electricity is subject to the Carnot limit [6], which is about 30 to 40 percent for present day power plants. This implies that it takes about 3 to 4 units of heat energy to produce a unit of electricity. The rest of the heat is usually discharged to the environment, unless it is used as process-heat, as in combined heat and power (CHP) systems [9].

The TPES and WTPEC figures for 5 years are listed in Table 1.1[17]. The ratio of total energy consumption to total energy supply, which is a measure of the average efficiency of conversion from primary energy forms to useable forms, shows little variation. The fraction of total primary energy supply used to generate electricity on a large scale is given in the last column of the table. The average value here is about 22 percent.

The variation of total global consumption of different primary energy sources for a 50-year period, beginning in 1965, is depicted graphically in Fig.1.2 [10-14]. The graphs show yearly fluctuations that can be attributed to economic and socio-political factors. Oil whose applications are mainly in the transport and industrial sectors is the dominant primary energy source. During the period from 1965 to 1980 the average yearly oil consumption increased more rapidly than from 1980 to 2015.

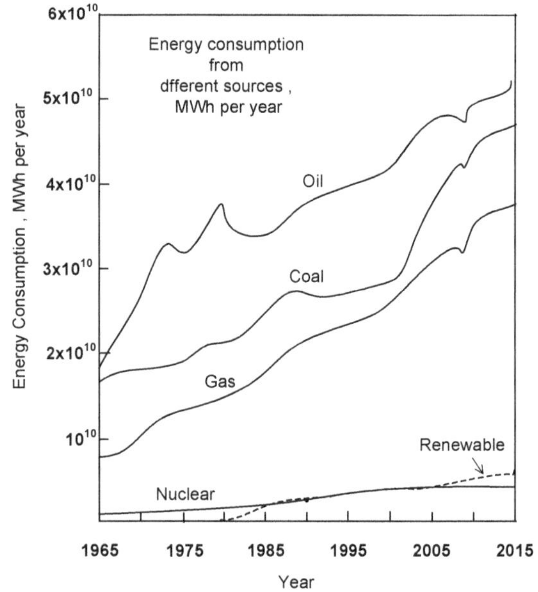

Fig. 1.2 Yearly variation of global energy consumption (Data from Ref.[14])

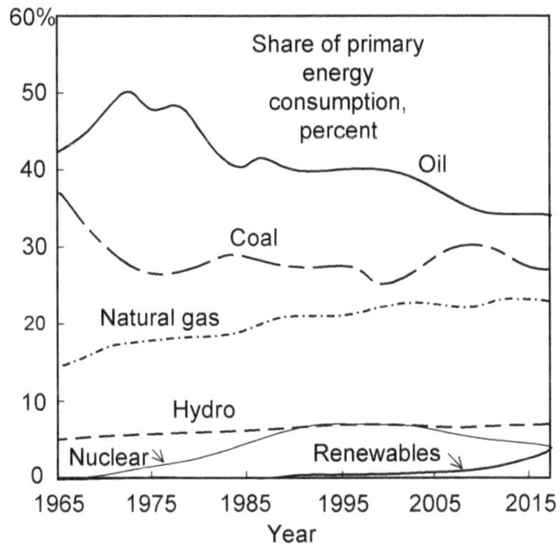

Fig. 1.3 Share of primary energy consumption by source (Data from Ref.[14])

Coal, which is used mainly to generate electricity, displays an opposite trend. The increase of average yearly consumption of coal from 2000 to 2015 was significantly more than that between the period from 1965 to 2000. In the case of natural gas, used mainly for industrial and commercial applications, the average yearly consumption has increased steadily.

Nuclear energy consumption shows a slight decrease since 2007. This is partly due to continuing public misgivings over nuclear power plant safety, following the accidents at the Three Mile Island plant in the USA, the Chernobyl plant in Russia, and the Fukushima Daiichi plant in Japan [7]. However, in recent years nuclear generation capacity has been increasing in some countries like China and France, but several countries including Germany and Japan have phased-out the development of nuclear power.

The contribution of renewable energy sources, although small in comparison to oil, coal and natural gas, has been increasing steadily since 1990. The factors responsible for the resurgence of renewable energy are discussed in the next section.

The percentage contribution of different primary energy sources to the total world energy consumption, during a period of 52 years from

1965 to 2017, is shown graphically in Fig. 1.3 [10-14]. The share of oil peaked at 50 percent in 1973, and has since decreased to about 35 percent. The contribution of coal has decreased from 37 percent in 1965 to about 28 percent in 2017. The fraction of natural gas has increased steadily. The share of hydro-power has remained at about 7 percent, while the share of nuclear power has been decreasing gradually since about 2000. The contribution of renewable sources such as solar, wind, tidal and geothermal heat, although small in relative terms, has increased steadily since about 2000.

1.3 Development of Renewable Energy

Fossil fuels, which include wood, coal, oil and natural gas have become the primary energy source worldwide. These fuels are used mainly for heating, electricity generation, and transportation. Although the impact of fossil fuel combustion on global warming and the local environment, are of much concern, they are likely to remain an important energy source for many years to come. Therefore, the depletion of fossil fuel resources, and any interruption of their supply could have dramatic effects on the world economy. In view of this situation, efforts are underway to develop advanced combustion systems and power plants that use fuels more efficiently. Such power systems include combined cycle power plants and combined heat and power systems. Techniques are also being developed to mitigate the harmful environmental effects of fossil fuel combustion.

1.3.1 *Energy insecurity*

Several factors have contributed to the growth in deployment of renewable energy sources. The sudden and unexpected price increase of crude oil in 1973 and 1979 came as a warning to the world's industrial countries on the vulnerability of their economies to energy supply interruptions [1]. Oil prices first tripled due to an Arab oil embargo in 1973, and then nearly doubled in 1979, due to political turmoil in Iran. These events generated considerable interest in renewable energy research, which was accompanied by generous research funding from various sources, especially in the USA.

However, as the political tensions responsible for the above oil price hikes eased, and as more fossil fuel resources were discovered in the North Sea of Europe and the North Slope of Alaska [1], interest in renewables, and the research funding available, dwindled. Moreover, the cost of renewables like solar and wind were too high at the time to compete with cheap oil, of which there was an abundant supply. Fortunately, the interest generated in the efficient use of energy, or commercial and industrial energy conservation, as it is usually called, continued.

1.3.2 *Fossil fuel resource depletion*

A second factor that spurred the development of renewable energy sources is concerns on depletion of fossil fuel resources. Although it took millions of years for these energy reserves to be formed, humans have managed to consume a large fraction of it in a short period of 200 years. As the graphs in Fig. 2 indicate, global fossil fuel consumption is on the rise, and new reserves are becoming harder to find [1]. This situation has prompted many people to ask how long it will be before fossil fuels run out. From the published literature, it is clear that the answer to this question is not straightforward, because estimates vary widely.

An estimate of US fossil fuel reserves was first made by M. King Hubbert [1,9], in 1950. He demonstrated that fossil fuel production would follow a bell-shaped curve, with the production curve first increasing following the discovery of new resources, and improved extraction methods. The curve peaks and then declines as resources become depleted. His predictions agreed with the actual oil production trends of the USA at the time.

Years of extraction left for oil, coal and natural gas are commonly reported as the reserve-to-production (R/P) ratio, which is defined as the number of years of production left, based on known reserves, and the current annual production levels. This value, however, can change with time due to the discovery of new reserves, and changes in annual production levels, which are affected by economic conditions. The (R/P) ratio is published annually in the British Petroleum (BP) Statistical Review of World Energy Report [10-14]. The number of years of resource left estimated for different years, listed in Table 1.2, were obtained from this publication. Coal resources are estimated to last for about 100 years while crude oil and natural gas supplies would be exhausted in about 50 years.

Table 1.2 Years of extraction left for oil, gas and coal (Data from Ref.[10-14])

Year/Resource	Oil	Natural gas	Coal
2014	53.3	55.1	113
2015	52.5	54	110
2016	50.7	52.8	114
2018	50.2	52.6	134

1.3.3 *Environmental effects*

Combustion of fossil fuels, coal in particular, produces a number of profoundly harmful environmental impacts, and pollutants that harm public health. Air pollution and global warming are two of the most serious. This is the third and more urgent factor that is driving the search for alternative forms of energy.

A simplified model [9,24] that captures the main energy exchange processes responsible for global warming are shown schematically in Fig. 1.4. Incoming solar radiation is partially reflected by the atmosphere and the earth's surface, and the rest is absorbed by the earth. The absorbed radiation is later emitted as long-wave heat radiation by the earth's surface. Gases in the atmosphere such as carbon dioxide and methane tend to absorb this heat radiation, and later reradiate it in all directions including the earth's surface. These gases are called greenhouse gases (GHG) because they behave like the transparent roof of a typical greenhouse used to grow plants. The roof of the greenhouse lets in solar radiation almost completely, but absorbs much of heat radiation emitted by the plants and other items inside the greenhouse.

As more man-made greenhouse gases (GHG) like carbon dioxide and methane accumulate in the atmosphere, the heat energy reradiated toward the earth's surface increases. This additional heat causes ice sheets to melt and oceans to become warmer. Both these effects have the potential to cause water levels in oceans to rise, which could have dramatic consequences for people living in low lying areas of the world. Quantitative predictions of the effects of global warming can be made only by using complex computer models known as general circulation models [9].

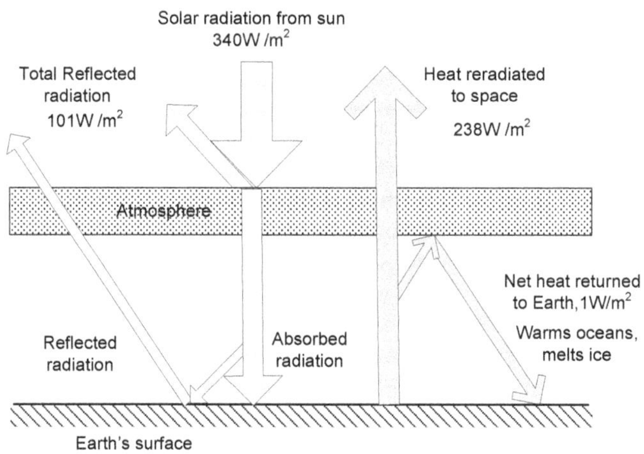

Fig. 1.4 Simulated planetary energy imbalance (Data from Ref.[3])

The quantities of energy, indicated in Fig. 1.4, have been obtained by such computer simulations. The net heat energy reradiated back to earth by the atmosphere, which is called the planetary energy imbalance [3], is about 1 W per m^2. This quantity is a critical metric, in that it predicts future global warming already taking place. A more detailed schematic of the energy exchange between the earth, the atmosphere and the sun is presented in worked example 1.15. Two simplified models to estimate the average temperature of the earth are given in worked examples 1.16 and 1.17.

Lack of avenues to verify computer-generated results for global temperature increase by direct measurement has led to an on-going debate on the extent of global temperature rise, and the steps that need to be taken now to mitigate the detrimental effects of global warming.

Fortunately, there are serious worldwide actions undertaken to reverse global temperature rise before it is too late to avoid the catastrophic consequences forecast by experts. The estimation of the carbon dioxide production in typical fossil fuel combustion systems is illustrated in worked examples 1.13 and 1.14.

The most prominent example of worldwide action on climate change has been the Kyoto Protocol, an international agreement drawn up in 1997 in the city of Kyoto in Japan [3,9]. The member states of the European Union, Canada, Japan, Russia, and Australia are signatory nations to the Kyoto Protocol. By 2008 some of the signatory countries achieved significant reductions in GHG emissions compared to those in 1990, which is used as the baseline for comparison. Despite disagreements between countries on some aspects of the Kyoto Protocol, especially, between industrial and emerging countries, there is broad consensus on the need to reduce GHG emissions.

In addition, the United Nations established an international body in 1988 called 'The Intergovernmental Panel on Climate Change (IPCC)' [3,9]. The mission of the IPCC is to provide a scientific view of climate change, and its impact on the natural environment, the economy and politics of the world, and possible responses. The IPCC provides an internationally accepted authority of climate change, publishing reports which have the agreement of leading climate scientists. Thousands of scientists and other experts contribute on a voluntary basis to writing and reviewing these reports, that are then reviewed by governments.

The findings on global warming and its impact on climate change, published in IPCC reports, have helped generate public awareness of the situation, and initiate action at every level of society for reducing the so-called 'carbon footprint'. The gradual replacement of fossil fuels with a mix of alternative energy sources has been identified as an important option in the strategies for reducing carbon dioxide emissions.

1.3.4 *Development of solar, wind and biomass*

The rapid growth of wind, solar and biofuels during the last decade, depicted in Figs. 1.5, 1.6 and 1.7, is partly due to technological developments that have made them more efficient and affordable. In

2018, the contribution of renewable energy to power generation (excluding hydro) increased by 14%, slightly below the 10-year average growth (16%). China accounted for 45% of global growth, and its consumption has increased 20-fold in the last 10 years [14]. Wind power accounted for about 50% of renewable power generation in the last few years, and solar has constantly increased its share, and represents about 24% now. The growth in production of biofuels averaged 9.7% in 2018, the highest growth since 2010 [14].

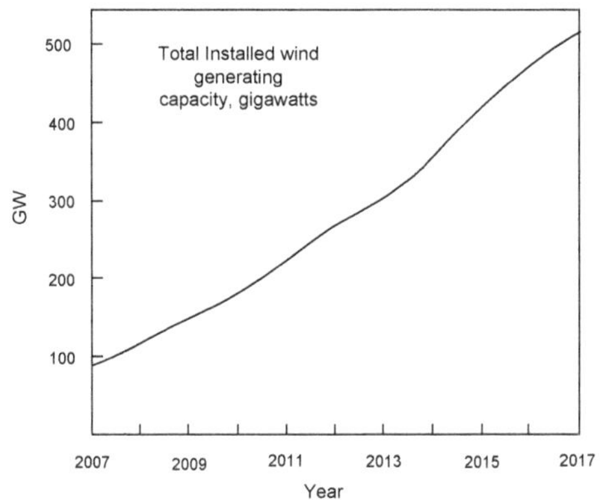

Fig. 1.5 Growth of the total installed capacity of wind power (Data from Ref.[14])

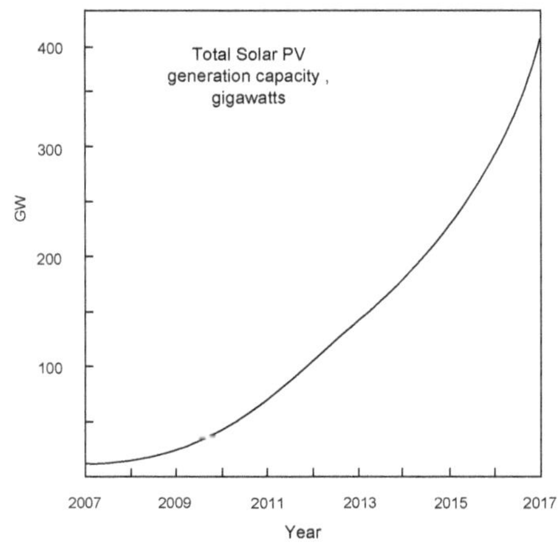

Fig. 1.6 Growth of the total installed capacity of PV power (Data from Ref.[14])

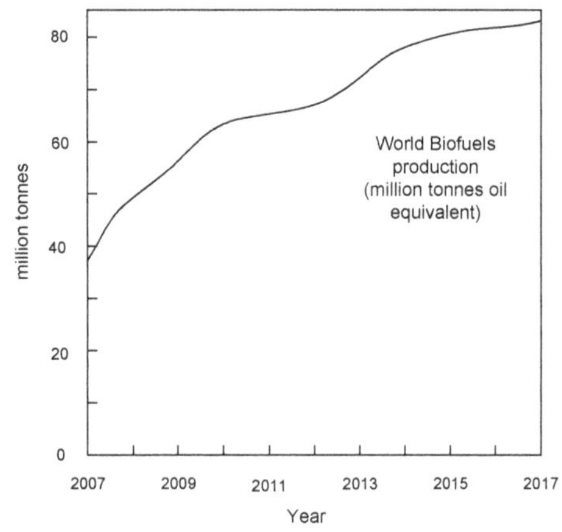

Fig. 1.7 Growth of the total world biofuel production (Data from Ref.[14])

1.4 Renewable Energy Conversion Processes

In this section we shall briefly describe the different forms of renewable energy [5,6] that are currently developed on a commercial scale. The energy conversion processes will be illustrated using energy flow diagrams.

1.4.1 *Solar energy conversion to heat and power*

The energy flow diagram in Fig. 1.8 shows two direct applications of solar radiation, which are process heating and electricity generation.

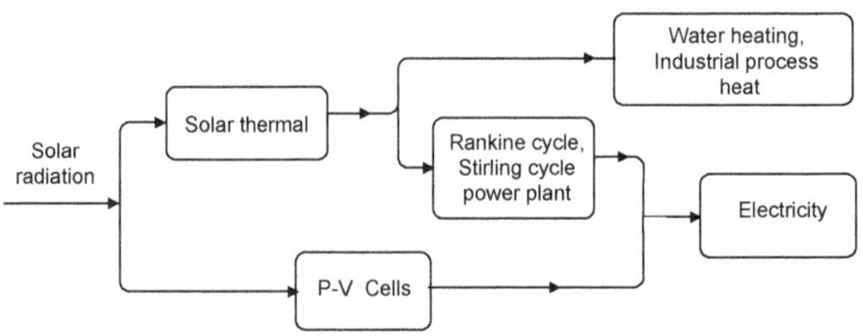

Fig. 1.8 Energy-flow diagram for solar energy conversion

Using solar energy for producing hot water for domestic, commercial and industrial applications is well established in many parts of the world. The required water temperatures are about 50° C to 60° C, which can be achieved with flat-plate solar collectors [2,6]. Solar radiation is also used as the heat source in Rankine-cycle power plants and Stirling-cycle engines to produce electricity [2,6]. In this case the temperatures required are much higher and therefore concentrator-type solar collectors must be employed. Photovoltaic (PV) cells [2,5,18] are devices that convert solar radiation directly to electricity. These process will be considered in detail in later chapters of the book.

1.4.2 *Wind and wave energy conversion*

The movement of air across the surface of the earth, caused by areas of high pressure and low pressure is called wind. Incoming solar radiation (Fig. 1.1) interacts with the atmosphere to create a thermal engine, absorbing heat at a higher temperature, and releasing heat at a lower temperature. This temperature difference drives the winds, maintaining circulation of the atmosphere against frictional dissipation [2,5,19]. Waves in water bodies like lakes, seas and oceans are most caused by wind. Friction between the wind and the water surface creates wind-driven waves or surface waves [5,20].

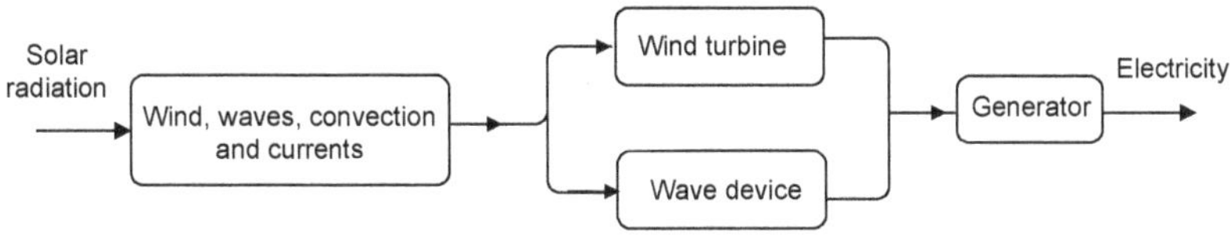

Fig. 1.9 Energy-flow diagram for wind and wave energy conversion

The kinetic energy of wind and waves can be converted to useful mechanical energy with aid of wind turbines and wave energy devices respectively, as depicted in the energy-flow diagram in Fig. 1.9. These devices, in turn, can drive a generator to produce electricity.

1.4.3 *Ocean thermal energy conversion (OTEC)*

This is an indirect method of utilizing solar energy. At certain locations in the tropics, the temperature difference between the surface water and deep water in the ocean is about 20 to 25° C. In an OTEC power plant [6,21] warm surface water is used as the heat source to vaporize a low boiling working fluid such as Ammonia. (see Fig. E1.11) Vapor expands in a turbine to produce mechanical work that drives a generator. Cold water from deep water is used to condense the vapor returning from the turbine. The condensed liquid is then pumped to the evaporator to complete the cycle. Typical overall efficiency of OTEC plants is about 3%.

1.4.4 *Biomass energy conversion*

Biomass is a form of renewable energy that comes from plants and animals [5,9]. It is renewable because the necessary raw materials are replaced on a human time scale. Conversion of solar radiation to biomass occurs by photosynthesis as depicted in Fig. 1.1. The sunlight-to-biomass conversion efficiency is about 0.1% for typical plants, and 1-2% for crop plants. The direct combustion of biomass releases its chemical energy to produce heat. Alternatively, biomass can be converted to liquid biofuels or biogas that can be burned as a fuel. The energy flow diagram in Fig. 1.10 shows the main steps in the production of biofuels, such as ethanol using feed stock like corn or sugar cane.

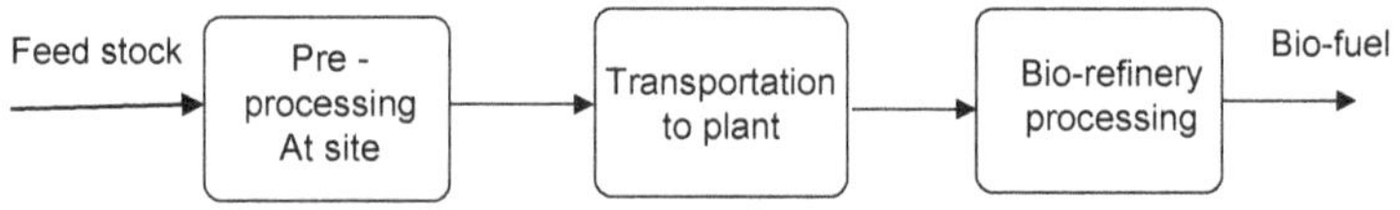

Fig. 1.10 Energy-flow diagram for biomass energy conversion

1.4.5 *Hydroelectric energy conversion*

The basic energy flow diagram for a hydroelectric power generation system [5,6] is shown in Fig. 1.11. Solar energy falling on bodies of water causes evaporation. The resulting cloud formation, and subsequent rain lead to the formation of water sources at high elevations. The potential energy of this water is used to operate water turbines that drive generators to produce electricity. Hydroelectric power generation is a well-established renewable energy conversion process that contributes significantly to global electricity production.

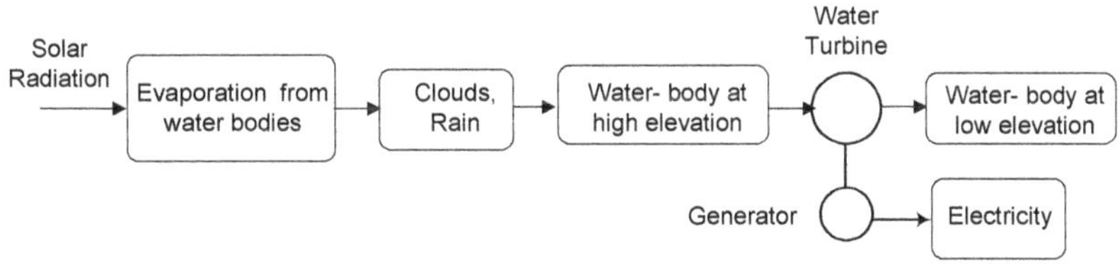

Fig. 1.11 Energy-flow diagram for hydroelectric energy conversion

1.4.6 *Geothermal energy conversion*

The production of energy using the internal heat of the earth's crust is referred to as geothermal energy [6,22]. The source of this heat energy is the radioactive decay of minerals and the heat flow from the original formation of the earth. In order to extract geothermal energy, wells are drilled into the earth's crust to a depth of 3 to 10 km. The most common method of extracting heat is by using water and steam. The steam produced is used as the working fluid in Rankine cycle power plants to generate electricity, and also as process heat to heat buildings. Geothermal energy can be extracted only in locations with specific geological conditions. These tend to be in regions that are volcanically and tectonically active, such as Iceland, Indonesia, New Zealand, Hawaii and California.

1.4.7 *Tidal energy conversion*

Tidal energy is extracted from the periodic change of heights of the earth's oceanic tides [23]. Tides are caused by the variation of the gravitational attraction exerted on the earth by celestial bodies, particularly the moon and the sun, and the earth's rotation. Water in oceans tend to bulge when these attractive forces are large. As the earth rotates, this bulge of the ocean water meets the shallow water close to the shoreline, thus creating a tide. The kinetic energy of the moving water stream is used to operate turbines, which in turn spin generators to produce electricity. An alternative arrangement of energy extraction is the use of tidal barrages which use the difference in water height or potential energy between high and low tides to operate turbines [5,23].

We shall discuss these different renewable energy production systems in greater detail in later chapters of this book. Some applications of the concepts of mechanical, thermal, and chemical energy are illustrated in the worked examples to follow in this chapter.

1.5 Cost Analysis of Energy Systems

From the brief description above it is clear that the basic input energy to a renewable energy conversion system is free. This energy input could be solar radiation, wind and wave kinetic energy, or geothermal heat. However, the equipment required to convert the basic energy form to a usable form, such as, electricity or process heat is not free.

Moreover, the total usable energy delivered by a renewable energy system over its lifetime depends on the availability of the resource. Solar radiation, wind and wave energy are variable on all time scales, and this needs to be accounted for in estimating the unit cost of energy delivered by the system over its lifetime. In this section we shall outline basic principles of engineering cost analysis needed to calculate the various economic figures-of-merit for renewable energy systems.

1.5.1 *Life-cycle costing*

Life-cycle costing is a concept that includes both the initial cost of an energy system as well as the yearly operating cost during the lifetime of the system. The following are the main components of the life-cycle cost: (i) capital cost of equipment, (ii) installation cost, (iii) operating cost, (iv) interest charges, if money is borrowed, (v) maintenance cost, (vi) taxes, which usually vary with the country, and (vii) salvage value.

Some of the above cost components vary from year to year during the lifetime of the system. This is accounted for by using the time value of money.

1.5.2 *Time value of money*

It is a common experience that the purchasing power of money in most economies is deceasing all the time. The yearly rate at which this decline occurs is called the *annual inflation rate.* Governments track the inflation rate by calculating the average price of a representative collection of goods. Moreover, investors whose money is invested in projects or in a bank expect an yearly rate of return which is more than the rate of inflation. The change in the value of money due to the above factors is called the *time value of money*.

The expectation of investors to obtain a return on their investment in incorporated in economic evaluations by using the concept of *discounting*. The discount rates are also used to determine the present value of future incomes and expenditures (cash flows) of a firm. In the case of money deposited and maintained in a bank, the yearly *interest rate* paid by the bank may be thought of as the discount rate.

1.5.3 *Discounted cash flow analysis*

We shall first consider discounted cash flow *without* accounting for inflation. The *present worth* or present value of a future *cash flow* is the value of that cash flow with the time value of money included. As an example, consider the future value F, after n years, of a sum of money A invested at an annual interest rate of r dollars per dollar per year. With the interest compounded at rate r per year, the initial sum A appreciates by a factor of $(1 + r)$ each year, disregarding inflation. Hence after n years we have

$$F = A(1 + r)^n \tag{1.1}$$

$$A = \frac{F}{(1+r)^n} \tag{1.2}$$

Equation (1.2) indicates that the *present worth* of a future sum of money is discounted, in constant dollars, by a factor $(1+r)^{-1}$ for each year in the future. For this context, r is generally called the *discount rate*. The factor $(1+r)^{-n}$ is defined as the present-worth factor, $PWF(r,n)$.

The present-worth is the key concept used in discounted cash flow analysis. It allows us to perform all economic calculations involving different time periods using the present value of money. The discount rate is a measure of an investor's or a firm's preference for the time value of money.

1.5.4 *Effect of inflation*

The effect of inflation, as was mentioned earlier, is to reduce the future value of money by a factor, $(1+s)$ per year, where s is the yearly rate of inflation. When inflation is included, the future value, F in Eq. (1.1) becomes

$$F = A\left[\frac{(1+r)}{(1+s)}\right]^n \tag{1.3}$$

Hence the present-worth A of a future sum F, including the effect of inflation, is given by

$$A = \frac{F}{(1+r_e)^n} \tag{1.4}$$

where r_e is the *effective interest rate* and $PWF(r_e,n)$ is the corresponding present-worth factor.

The effective interest rate is given by [2]

$$r_e = \frac{(1+r)}{(1+s)} - 1 \tag{1.5}$$

The present-worth factor, defined in Eq. (1.4), takes into account the effect of inflation.

1.5.5 *Capital-recovery factor*

Consider a series of annual payments of equal amount, P_a made over period of n years. The present-worth of the total payment, S is obtained by applying Eq. (1.2) to each yearly payment. Therefore

$$S = \frac{P_a}{(1+r)} + \frac{P_a}{(1+r)^2} + \frac{P_a}{(1+r)^3} + \cdots + \frac{P_a}{(1+r)^n} \tag{1.6}$$

The RHS of Eq. (1.6) is a geometric series with a ratio of $(1 + r)^{-1}$. Using the expression for the sum of the geometric series [2] we obtain

$$S = P_a \left[\frac{1-(1+r)^{-n}}{r}\right] \tag{1.7}$$

The constant yearly payments, P_a over n years required to form a total present amount, S is obtained as

$$P_a = S \left[\frac{r}{1-(1+r)^{-n}}\right] \tag{1.8}$$

The term in the square brackets is called the capital-recovery factor, $CRF(r, n)$, given by

$$CRF(r, n) = \frac{r}{1-(1+r)^{-n}} \tag{1.9}$$

The term P_a is Eq. (1.8) is the yearly interest and principal payment on a self-amortizing loan of present value S. Due to the large balance, the initial payments are mostly interest, while later payments are mainly principal payment.

1.5.6 *Economic figures of merit*

Simple payback period

A common figure of merit used to evaluate the viability of a renewable energy system or an energy conservation measure is the *simple payback period* (PP). It is determined by comparing the total initial cost of the system with the cost of the annual energy production or the energy saving per year.

The simple payback period is given by

$$PP = \frac{C_o}{a_i c_{ei}} \tag{1.10}$$

where C_o is the initial cost, a_i is the average yearly rate of energy delivery or energy saving (kWh), and c_{ei} is the unit cost of energy delivered per kWh.

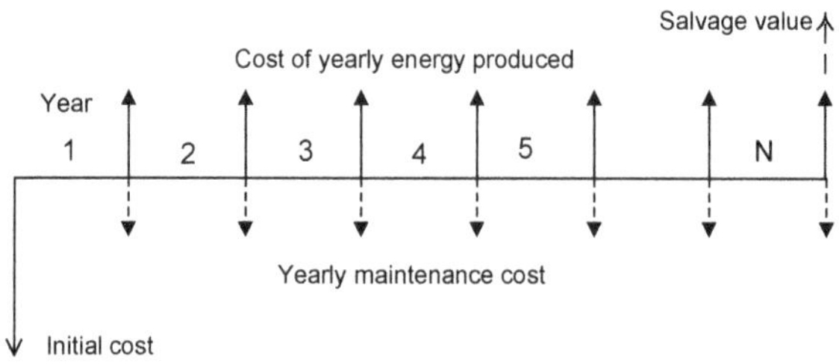

Fig. 1.12 Cash-flow diagram

Net present worth

The net present worth (PW) is a convenient method to compare different investment options. Consider an investment in an energy project that has a series of future yearly cash flows, A_n representing the different cost items. These are shown schematically in the yearly-cash-flow diagram in Fig. 1.12. Some of the cash flows like the initial cost and the maintenance cost are negative while cash flows such as the cost of energy produced or energy saved, and the salvage value are positive. The present-worth of the algebraic sum of the cash flows over the lifetime, N of the project is given by

$$PW = \sum_{n=1}^{N} \frac{A_n}{(1+r)^n} \tag{1.11}$$

If the yearly cost item, A_n is constant, then the geometric series can be summed to obtain

$$PW = A_c \frac{[1-(1+r)^{-N}]}{r} \tag{1.12}$$

For a project the *net present value* of all the cash flow streams should be *positive*. However, this occurs under an assumed value of the discount rate or rate of return, r. The *minimum attractive rate of return* (MARR) is defined as the discount rate required for the returns (PW) from the project to be financially attractive [9]. The value of the MARR is set by the firm or agency making the investment in the project. Using the MARR as the discount rate all cash flows are discounted to the present. If the present-worth thus obtained is positive, then the project is financially viable.

Internal rate of return

The *internal rate of return* (IRR) is the value of the discount rate r that gives a zero present-worth PW. Computation of the IRR requires a numerical procedure due to the non-linear equations involved. This information can be used to compare the economic attractiveness of different projects.

Benefit-cost ratio method

Another approach used to evaluate the financial viability of a project is the benefit-cost ratio (B/C) method. This approach allows the inclusion of all benefits accruing from a project, some of which may not be directly related to the main objective of the project. For instance, in a hydro-electric power generation scheme, the reservoir that stores water may be used for boating, fishing, flood control and as source of water for irrigation. These additional benefits could be estimated and included in the sum of the benefits [2].

For a project to be financially attractive the ratio of benefits to costs should be greater than 1. There are two definitions of the B/C ratio. These are:

Conventional B/C:

$$[B/C]_{con} = \frac{Total\ Benefits}{Initial\ cost + operating\ cost} \tag{1.13}$$

Modified B/C:

$$[B/C]_{mod} = \frac{Total\ Benefits - Operating\ cost}{Initial\ cost} \tag{1.14}$$

The typical cost items are shown in Fig. 1.12. The above ratios can be evaluated using either the present-worth of all the benefits and the costs or the corresponding annualized values.

1.5.7 *Levelized cost of energy*

A widely used metric for comparing different energy production systems is the *levelized cost of energy* (LCOE). For instance, we could use the LCOE to compare the cost of electrical energy from a fossil fuel plant with that from a wind energy system. The LCOE is the average cost of a unit of energy ($ per kWh), delivered by the energy conversion system during its lifetime. The computation of the LCOE can be done using either the present worth of all the costs and the total energy delivered, or the annualized costs and the average yearly energy production. These two approaches yield identical results.

For purposes of illustration consider a fossil fuel plant operating on the Rankine cycle using coal as the fuel. There are three main cost components, which are:

(i) Capital cost: The main items of the capital cost are – the purchase of land, the power generation system including the steam generators, the turbines, the condensing equipment, the pumps. and other mechanical devices. Any costs associated with debt payment is also included.

(ii) Fuel cost: The fuel cost is estimated using the annual output of electrical energy and the efficiencies of the various conversion equipment such as the steam generator, the turbine, the pumps and the generator.

(iii) Balance of cost: This is a cost component that includes items, like the wages, operation costs, maintenance costs and other costs that are specific to the plant.

Using *annualized costs* the LCOE may be written as [9]:

$$[LCOE]_{an} = \frac{Annual\ capital\ cost + Fuel\ cost + Balance\ of\ cost}{Annual\ energy\ production\ (kWh)} \tag{1.15}$$

An expression for the LCOE in terms of the *present worth* of the various costs is given by

$$\sum_1^N LOCE \times (1+r)^{-n} \times E_n = I + \sum_1^N C_n (1+r)^{-n} - S(1+r)^{-N} \tag{1.16}$$

Note that in Eq. (1.16) the yearly values of the LCOE are discounted to obtain the present worth. I is the initial capital cost, C_n includes the yearly fuel cost, operation and maintenance cost and others, S is the salvage value, r is the discount rate, and N is the lifetime of the project. The annual energy production rate is E_n.

Hence we have

$$[LCOE]_{pw} = \frac{I+\sum_1^N C_n(1+r)^{-n} - S(1+r)^{-N}}{\sum_1^N E_n \times (1+r)^{-n}} \quad (1.17)$$

The application of the different economic figures-of-merit listed above to typical energy projects will be illustrated in the worked examples to follow in this chapter.

1.6 Worked Examples

The worked examples presented in this section illustrate the applications of mechanical energy, thermal energy, chemical energy, global energy exchange, and cost evaluation of energy systems.

Mechanical energy

Example 1.1 An aircraft is cruising steadily at a velocity of 800 km per hour at an altitude of 8 km. The total mass of the aircraft is 150×10^3 kg and the thrust provided by the engines is 300 kN. Find (i) the kinetic energy, (ii) the gravitational potential energy, and (iii) the power used to propel the aircraft.

Solution The speed of the aircraft, v is: 800 kmh^{-1} = 222 ms^{-1}

Let the mass of the aircraft be m, the altitude be h, and the gravitational acceleration be g. The kinetic energy is given by:

$$E = \frac{mv^2}{2} = \frac{150 \times 10^3 \times 222^2}{2} = 3696 \text{ MJ}$$

The potential energy is given by:

$$P = mgh = 150 \times 10^3 \times 9.81 \times 8000 = 11772 \text{ MJ}$$

The thrust force, T produced by the engines propels the aircraft forward. Under steady conditions it is equal to the drag force on the aircraft. The power used to propel the aircraft is equal to the rate of work done, W by the thrust force, which is given by:

$$W = Tv = 300 \times 10^3 \times 222 = 66 \times 10^6 = 66.6 \text{ MW}$$

Example 1.2 A train of mass 300×10^3 kg is pulled by a locomotive up a slope of 1 in 120. The steady speed of the train is 40 km per hour, and the frictional resistance is 0.04 *N per kg*. Find (i) the traction force of the locomotive, (ii) the power delivered by the locomotive, (iii) the kinetic energy of the train, and (iv) the rate of increase of potential energy of the train.

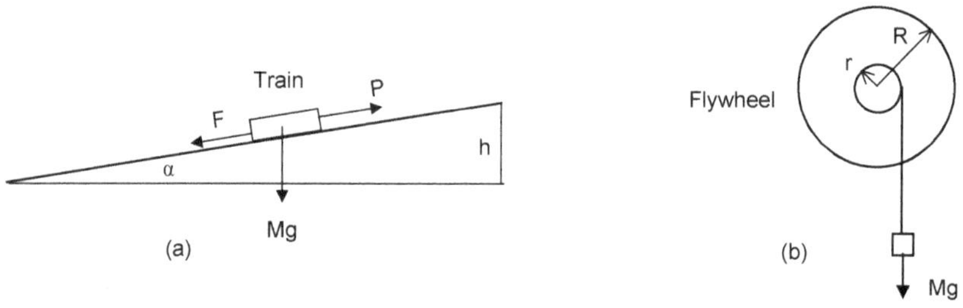

Fig. E1.2 (a) Kinetic and potential energy of train (b) Rotational kinetic energy of flywheel

Solution The forces acting on the train are shown in Fig. E1.2(a). Since the speed is constant, the traction force, P is balanced by the frictional resistance and the component of the weight parallel to the slope. Therefore we have

$$P = F + Mgsin\alpha$$

$$P = 0.04 \times 300 \times 10^3 + 300 \times 10^3 \times 9.81 \times (1/120) = 36.5 \text{ kN}$$

The steady speed of the train is:

$$v = 40\frac{km}{h} = \frac{40\times10^3}{3600} = 11.1 \text{ ms}^{-1}$$

The power delivered by the locomotive is given by

$$W = Pv$$

$$W = 36.5 \times 10^3 \times 11.1 = 405.6 = 405.6 \text{ kJ}$$

The kinetic energy of the train is

$$E = \frac{mv^2}{2} = \frac{300{\times}10^3{\times}11.1^2}{2} = 18.48 \text{ MJ}$$

The rate of increase of potential energy is

$$PE = mg(dh/dt) = mgvsin\alpha$$

$$PE = 300 \times 10^3 \times 9.81 \times 11.1 \times (1/120) = 272.2 \text{ kW}$$

Example 1.3 A flywheel of mass 500 kg and diameter 0.65 m is attached to an axle of mass 50 kg and diameter 0.1 m. A thin rope is wound round the axle carries a mass of 9 kg at the end. The mass is released from rest and it descends 4.5 m before it reaches the ground. Calculate (i) the moment of inertia of the flywheel and axle, (ii) the initial potential energy of the mass, (iii) the kinetic energy of the wheel and axle when the mass touches the ground, and (iv) the energy lost when the mass comes to rest on the ground.

Solution The flywheel and weight are shown schematically in Fig. E1.2(b). The moment of inertia of a disc about its center is given by

$$I = \frac{1}{2}MR^2$$

where M is the total mass and R is the radius.

Applying the above formula to the flywheel and the axle we obtain the total moment inertia as:

$$I_t = \frac{500}{2} \times \frac{0.65^2}{4} + \frac{50}{2} \times \frac{0.1^2}{4} = 26.46 \text{ kgm}^2$$

The initial potential energy of the mass is

$$E_p = mgh = 9 \times 9.81 \times 4.5 = 397.3 \text{ J}$$

Let the rotational speed of the flywheel and axle be ω rad.s^{-1}. Then the speed of the mass is given by

$$v = r_a\omega = 0.05\omega$$

The total kinetic energy of the flywheel, axle, and the mass is

$$E_k = \frac{1}{2}I_t\omega^2 + \frac{1}{2}mv^2 \qquad \text{(E1.3.1)}$$

$$E_k = \frac{1}{2}\times 26.46\omega^2 + \frac{1}{2}\times 9\times(0.05\omega)^2 = 13.24\omega^2$$

From mechanical energy conservation, the total kinetic energy and potential energy is constant. Therefore

$$397.3 = 13.24\omega^2$$

Hence we get the final angular speed as 5.47 rad.s^{-1}.

Substituting in Eq. (E1.3.1) we obtain the kinetic energy of the flywheel and axle, and the mass, just before the mass touches the ground, as 395.8 J and 0.337 J respectively. The inelastic impact of the mass with ground converts its kinetic energy completely to heat.

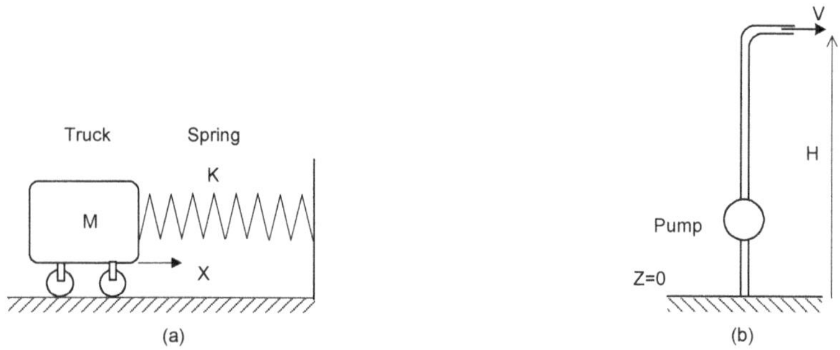

Fig. E1.4 (a) Strain energy of spring (b) Potential and kinetic energy of water stream

Example 1.4 A mass of 4 kg placed on a spring will compress it by 2.5 cm. The spring is used as a buffer, and a model truck of mass 0.25 kg runs into it with a velocity of 100 cm s^{-1}. (i) How far will the spring be compressed before the truck comes to rest. (ii) Calculate the maximum strain energy stored in the spring.

Solution The truck and spring are shown schematically in Fig. E1.4(a). If the spring constant is k, then for static balance

$$kx = mg$$

$$k = \frac{4\times 9.81}{0.025} = 1569.6 \text{ Nm}^{-1}$$

When the truck comes to rest its initial kinetic energy is converted to strain energy of the spring. Hence by energy conservation we have

$$\frac{1}{2}mv^2 = \frac{1}{2}k(\Delta x)^2 \tag{E1.4.1}$$

where Δx is the change in length of the spring due to compression.

Substituting in Eq. (E1.4.1) we have

$$\frac{1}{2}\times 0.25\times 1.0^2 = \frac{1}{2}\times 1569.6(\Delta x)^2$$

From the above equation we obtain the compression of the spring as 1.26 cm. The maximum strain energy of the spring is 0.125 J.

Example 1.5 A water pump driven by a motor is used to raise water from a depth of 16.5 m at the rate of 0.075 $m^3 s^{-1}$. The water is discharged with a velocity of 13 ms^{-1}. Calculate (i) the kinetic energy and potential energy of the water per kg at delivery, and (ii) the power input to the pump if 1/3 of it is used to overcome frictional losses in the pumping system.

Solution The water pumping system is shown in Fig. E1.4(b). The energy input to the motor is used by the pump to lift the water and impart kinetic energy to the water. The kinetic energy *per unit mass* of water is

$$\frac{1}{2}v^2 = \frac{1}{2}(13)^2 = 84.5 \text{ Jkg}^{-1}$$

The potential energy *per unit mass* of water is

$$gh = 9.81\times 16.5 = 161.86 \text{ Jkg}^{-1}$$

Therefore the total energy of water per unit mass at delivery is 246.36 Jkg^{-1}.

The mass flow rate of water at delivery is

$$\dot{m} = \dot{Q}\rho = 0.075 \times 1000 = 75 \text{ kgs}^{-1}$$

where $\dot{Q}$ is the volume flow rate of water in m^3s^{-1}.

The energy transferred to the water by the pump per unit time is

$$E = \dot{m} \times \left(\frac{1}{2}v^2 + gh\right) = 75 \times 246.36 = 18.48 \text{ kW}$$

One third of the energy input to the pump is used to overcome frictional losses in the system. Therefore the total energy input to the pump is

$$(1.5 \times 18.48) = 27.7 \text{ kW}$$

Thermal energy

Example 1.6 A flywheel of mass 200 kg and radius 30 cm is rotating with a steady angular speed of 50 rpm. It is brought to rest by activating a frictional braking device that applies a constant braking torque of 5 Nm on the flywheel. The thermal capacity of the braking device is 20 JK^{-1}. Calculate (i) the initial kinetic energy of the flywheel, (ii) the number of revolutions required for the flywheel to come to rest, (iii) the frictional work done by the brake, and (iv) the temperature rise in the brake if no heat is lost to the ambient.

Solution The moment of inertia of the flywheel is

$$I = \frac{1}{2}MR^2 = \frac{1}{2} \times 200 \times 0.3^2 = 9 \text{ kgm}^2$$

The initial kinetic energy of the flywheel is

$$E_k = \frac{1}{2}I\omega^2 = \frac{1}{2} \times 9 \times \left(\frac{50 \times 2\pi}{60}\right)^2 = 123.3 \text{ J}$$

The frictional work done by the constant braking torque is equal to the initial kinetic energy of the flywheel.

Therefore

$$T\theta = \frac{1}{2}I\omega^2$$

$$5\theta = 123.3$$

The number of revolutions is 3.9.

The work done by the braking torque is converted to heat which results in a rise in temperature of the brake. The increase in thermal energy of the brake is equal to the loss in kinetic energy of the flywheel since no heat is lost to the surroundings. Hence we have

$$mc\Delta t - \frac{1}{2}I\omega^2 - 123.3$$

$$20\Delta t = 123.3$$

The rise in temperature of the brake is 6.15°C.

Example 1.7 An electric water heater consists of a cylindrical tank holding 50 gallons of water. The heating element has a resistance of 15 ohms and draws a steady current of 16 amps. Calculate the time required to heat the water from 20°C to 70°C. Neglect heat losses from the tank and the heat capacity of the tank.

Solution The mass of water in the tank is

$$m = \left(\frac{50}{264.2}\right) \times 1000 = 189 \text{ kg}$$

The rate of heat production in the electrical heating coil is

$$H = i^2R = 16^2 \times 15 = 3.84 \text{ kW}$$

The energy input required to raise the temperature of water from 20°C to 70°C is

$$Q = mc\Delta T = 189 \times 4.2 \times (70 - 20) = 39690 \text{ kJ}$$

From energy balance we have

$$H(3600\tau) = mc\Delta T$$

where τ is the number of hours required to heat the water. From the above equation we obtain

$$\tau = \frac{39690}{3600\times3.84} = 2.87\text{h}$$

Example 1.8 A hot water storage tank contains 0.5 m^3 of water at 70°C. The water looses heat to the ambient across an area of 1.8 m^2. The ambient temperature is 20°C. The rate of heat loss per unit area per unit temperature difference (the heat transfer coefficient) is 20 Wm^{-2} per °C. Calculate the mean temperature of the water after 5 hours.

Solution We shall first obtain the solution using an approximate method. Let the temperatures at the beginning and end of the cooling period of 5 hours be T_1 and T_2 respectively. Then the mean temperature is given by, $T_m = 0.5(T_1 + T_2)$. The total heat loss during the cooling period of τ hours is equal to the decrease in thermal energy of the water in the tank. Hence we have

$$UA[0.5(T_1 + T_2) - T_a]\tau \times 3600 = mc(T_1 - T_2)$$

where U is the heat transfer coefficient and A is the area of heat loss. The time τ is in hours.

$$20 \times 10^{-3} \times 1.8[0.5(70 + T_2) - 20]\tau \times 3600 = 500 \times 4.2(70 - T_2)$$

$$(T_2 + 30) = 6.48(70 - T_2)$$

The temperature of the water after 5 hours is 56.6°C. This procedure is approximate because the water temperature decreases continuously in a nonlinear manner. Therefore using the average water temperature to compute the heat loss is not accurate.

The exact procedure requires the solution of a differential equation which gives the temperature of the water at any instant. We write the energy balance at any instant as follows:

$$mc\frac{dT}{dt} = -AU(T - T_a)$$

$$500 \times 4.2 \times \frac{dT}{dt} = -1.8 \times 20 \times 10^{-3}(T - 20)$$

$$\frac{dT}{d\tau} = -\left(\frac{3600\times1.8\times20\times10^{-3}}{500\times4.2}\right)(T-20) = -0.0617(T-20)$$

The time τ is in hours. The solution of the above equation gives the water temperature after 5 hours as 56.7°C. The two methods give close results.

Example 1.9 A fixed quantity of air of mass 0.2 kg is contained in a cylinder behind a piston. The initial pressure and temperature are 0.6 bar and 32°C. The air is compressed to a final pressure of 1.2 bar according to the polytropic law is $PV^{1.25} = C$, where P is the pressure and V is the volume. Calculate (i) the initial and final volumes of air, (ii) the final temperature, (iii) the work done by the air, (iv) the change in internal energy, and (v) the heat transfer. Assume that air is an ideal gas with constant specific heat capacities, $c_v = 0.718$ kJkg^{-1}K^{-1} and $c_p = 1.005$ kJkg^{-1}K^{-1}.

Solution For an ideal gas

$$R = c_p - c_v = 1.005 - 0.718 = 0.287 \text{ kJkg}^{-1}\text{K}^{-1}$$

Apply the ideal gas equation to the initial state 1

$$P_1V_1 = mRT_1$$

$$0.6\times100\times V_1 = 0.2\times0.287\times305$$

Hence the initial volume is 0.291 m^3.

Apply the polytropic law to the process 1-2

$$P_1{V_1}^{1.25} = P_2{V_2}^{1.25}$$

$$0.6\times0.291^{1.25} = 1.2{V_2}^{1.25}$$

The final volume is 0.167 m^3.

Apply the ideal gas equation to the final state 2

$$P_2V_2 = mRT_2$$

$$1.2\times100\times0.167 = 0.2\times0.287\times T_2$$

The final temperature is 349.13 K.

The work done *on* air during the polytropic process is given by

$$W_{12} = \frac{P_2V_2 - P_1V_1}{n-1}$$

Applying the ideal gas equation to the above equation we get

$$W_{12} = \frac{mRT_2 - mRT_1}{n-1}$$

$$W_{12} = \frac{0.2 \times 0.287(349.13 - 305)}{1.25 - 1} = 10.13 \text{ kJ}$$

The *increase* in internal energy of air is given by

$$U_2 - U_1 = mc_v(T_2 - T_1)$$

$$U_2 - U_1 = 0.2 \times 0.718 \times (349.13 - 305) = 6.34 \text{ kJ}$$

Applying the first law of thermodynamics to the process 1-2 we have

$$Q_{in} = U_2 - U_1 + W_{out}$$

$$Q_{in} = 6.34 - 10.34 = -4 \text{ kJ}$$

Hence the heat output from air is 4 kJ.

Example 1.10 Air flows through a horizontal converging nozzle whose area of cross-section varies from 3×10^{-3} m^2 at the entrance to 2×10^{-3} m^2 at the exit. The pressure, temperature and air speed at the entrance are 100 kPa, 27°C and 60 ms^{-1} respectively. At the exit of the nozzle, the pressure is 80 kPa, and the temperature is 50°C. Assume that air is an ideal gas with $c_v = 0.718$ kJkg^{-1}K^{-1} and $R = 0.287$ kJkg^{-1}K^{-1}. Calculate (i) the mass flow rate of air, (ii) the air speed at the exit, and (iii) the rate of heat transfer between the air and the surroundings.

Solution The density of air at the entrance section 1 is obtained from the ideal gas equation as

$$\rho_1 = \frac{m_1}{V_1} = \frac{P_1}{RT_1} = \frac{100}{0.287 \times 300} = 1.161 \text{ kgm}^{-3}$$

The air mass flow rate at the entrance is given by

$$m = A_1\rho_1 V_1 = 3 \times 10^3 \times 1.161 \times 60 = 209 \text{ kgs}^{-1}$$

The density of air at the exit section 2 is obtained from the ideal gas equation as

$$\rho_2 = \frac{m_2}{V_2} = \frac{P_2}{RT_2} = \frac{80}{0.287 \times 323} = 0.863 \text{ kgm}^{-3}$$

Air flows steadily through the nozzle. Hence the mass flow rate at the exit section is 209 kgs^{-1}.

$$m = A_2\rho_2 V_2 = 2 \times 10^3 \times 0.863 \times V_2 = 209 \text{ kgs}^{-1}$$

Air velocity at the exit is 121.1 ms^{-1}.

Applying the first law of thermodynamics (steady flow energy equation) to the nozzle we have

$$m\left(h_1 + \frac{V_1{}^2}{2} + gz_1\right) + Q_{in} = m\left(h_2 + \frac{V_2{}^2}{2} + gz_2\right) + W_{out}$$

The nozzle is horizontal and therefore there is no change in gravitational potential energy from 1 to 2. The work output of the nozzle is zero. Apply the above conditions to the energy equation. Hence we get

$$m\left(h_1 + \frac{V_1{}^2}{2}\right) + Q_{in} = m\left(h_2 + \frac{V_2{}^2}{2}\right)$$

The specific enthalpy of air is given by $h = c_p T$. For an ideal gas

$$c_p = c_v + R = 0.716 + 0.287 = 1.003 \text{ kJkg}^{-1}\text{K}^{-1}$$

Substituting numerical values in the above equation we get

$$Q_{in} = 0.209\left(1.003 \times 323 + \frac{121.12^2}{2 \times 10^3}\right) - 0.209\left(1.003 \times 300 + \frac{60^2}{2 \times 10^3}\right)$$

Hence the heat input rate to the nozzle is 5.98 kW.

Thermal energy conversion

Example 1.11 A Carnot heat engine receives 9 kJ of heat from a source at 300°C, and performs 5.2 kJ of work per cycle. Calculate (i) the thermal efficiency, (ii) the temperature of the cold body, and (iii) the heat rejected.

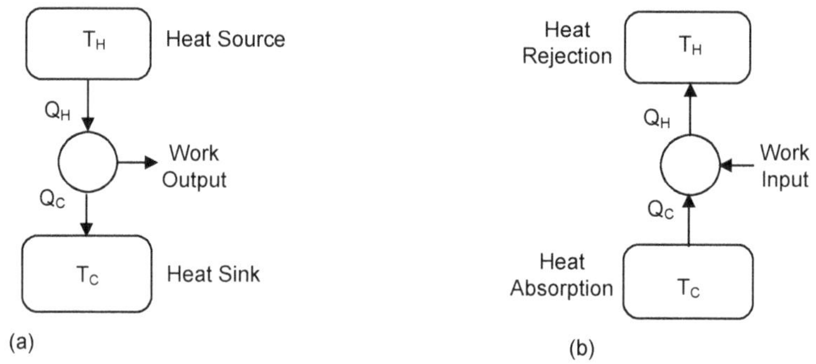

Fig. E1.11 (a) Carnot heat engine (b) Reversed Carnot heat engine

Solution A Carnot heat engine is shown schematically in Fig. E1.11(a). The efficiency of the engine is given by

$$\eta_C = \frac{W}{Q_H} = \frac{5.2}{9} = 57.8\%$$

The efficiency of a Carnot engine [6] may be expressed as:

$$\eta_c = 1 - \frac{T_C}{T_H} = 1 - \frac{T_C}{(273+300)} = 0.578$$

Hence the temperature of the cold body is 241.9 K or -31.2°C.

The heat rejected is given by

$$Q_c = Q_H - W = 9 - 5.2 = 3.8 \text{ kJ}$$

Example 1.12 A reversed Carnot engine absorbs 420 kJ/min from a source at a temperature of -10°C. Heat is rejected to a body at 32°C. (a) Calculate the COP and work input of the system if it functions as a refrigerator. (b) What is the COP, if it is operating as a heat pump?.

Solution A reversed Carnot engine is shown in Fig. E1.11(b). The coefficient of performance (COP) of the refrigerator is given by

$$(COP)_{ref} = \frac{Q_C}{W} = \frac{Q_C}{Q_H - Q_C} = \frac{T_C}{T_H - T_C}$$

$$(COP)_{ref} = \frac{(-10+273)}{(32+273)-(-10+273)} = 6.26$$

Work input is

$$W = \frac{Q_C}{(COP)_r} = \frac{420}{6.26} = 67.07 \text{ kJ/min}$$

COP of the heat pump is

$$(COP)_{hp} = \frac{Q_H}{W} = \frac{Q_C + W}{W}$$

$$(COP)_{hp} = \frac{420+67.07}{67.07} = 7.26$$

Example 1.13 For the complete combustion of the hydrocarbon fuel C_8H_{18}, calculate (i) the air-fuel ratio, (ii) the specific carbon content of the fuel, and (iii) the carbon dioxide produces per kg of fuel.

Solution The complete combustion of the fuel may be represented by the chemical equation:

$$C_8H_{18} + \left(\frac{25}{2}\right)O_2 + \left(\frac{25}{2} \times 3.76\right)N_2 \rightarrow 8CO_2 + 9H_2O + (25/2 \times 3.76)N_2$$

Note that $(25/2 \times 3.76)$ kmols of nitrogen are included on both sides of the equation because air used in combustion has oxygen and nitrogen in the mole ratio of 1:3.76. Also nitrogen does not take part in the combustion reaction.

Mass of fuel: $12 \times 8 + 18 \times 1 = 114$ kg
Mass of air: $(25/2) \times (1 + 3.76) \times 29 = 1725.5$ kg

The air-fuel ratio by mass:

$$(A/F)_{mass} = \frac{1725.5}{114} = 15.1 \text{ kg per kg fuel}$$

Specific carbon content of the fuel:

$$\frac{M_{carbon}}{M_{fuel}} = \frac{8\times12}{114} = 0.84 \text{ kg per kg}$$

Carbon dioxide produced per unit mass of fuel:

$$\frac{M_{CO_2}}{M_{fuel}} = \frac{8\times44}{114} = 3.08 \text{ kg per kg}$$

Example 1.14 A coal-fired steam power plant with an electrical output of 400 MW, operates on the Rankine cycle with an ideal thermal efficiency of 36%. The efficiency of the actual cycle is 76% of the ideal Rankine efficiency. The efficiencies of the boiler, and the generator are 83% and 96% respectively. The capacity factor based on the number of hours of operation per year is 0.7. The energy density of coal used is 18 MJ per kg. The specific carbon content of coal used is 0.4 kg of carbon per kg of coal. Calculate (i) the coal consumption per year, and (ii) the mass of carbon dioxide emitted per year.

Solution The electrical energy output per year based on the number of hours of operation per year (capacity factor) is given by

$$E_{elet} = 400 \times 3600 \times 8760 \times 0.7 = 8.83 \times 10^6 \text{ GJ}$$

The overall efficiency of the plant is

$$\eta_o = 0.36 \times 0.76 \times 0.83 \times 0.96 = 0.218$$

$$\eta_o = \frac{E_{elet}}{E_{fuel}} = \frac{8.83\times10^6}{E_{fuel}} = 0.218$$

Therefore the energy input from coal is

$$E_{fuel} = 40.5 \times 10^6 \text{GJ}$$

The mass of coal required per year is

$$M_{coal} = \frac{40.5\times10^6\times10^3}{18} = 2.25 \times 10^9 \text{ kg}$$

The combustion of carbon may be represented by the chemical equation:

$$C + O_2 = CO_2$$

Therefore 12 kg of carbon produces 44 kg of carbon dioxide. The coal used has 0.4 kg of carbon per kg coal. Hence the mass of carbon dioxide produced per kg of coal is

$$0.4 \times (44/12) = 1.47 \text{ kg}$$

The total mass of carbon dioxide produced per year is

$$Mass - co_2 = 1.47 \times 2.25 \times 10^9 = 3.31 \times 10^9 \text{ kg}$$

Global energy exchange

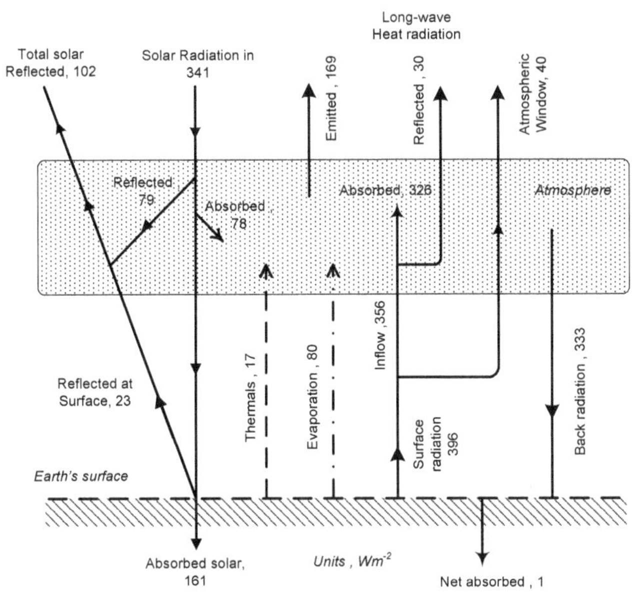

Fig. E 1.15 Global energy exchange (Data from Ref.[24])

Example 1.15 Using data from published sources show schematically the various energy exchange processes between the earth's surface, the atmosphere and the sun.

Solution Using the numerical data in Ref.[24] we have drawn Fig. E1.15 to depict schematically the energy exchange between the sun, the earth's atmosphere, and the earth's surface. The interactions of solar radiation are represented on the left of the figure, while the long-wave thermal radiation exchanges are given on the right. The units of the energy quantities indicated for the various processes are in watts per square meter.

The incoming solar radiation (341) is partially reflected (79) and partially absorbed (78) by the atmosphere, and the rest is transmitted (184) to the earth's surface. The surface reflects a fraction of this radiation (23) and absorbs the rest (161). The energy absorbed by the earth warms the earth's surface.

Due to its temperature, the earth emits long-wave radiation (396). The convection currents or thermals (17), and evaporation from water bodies (80) absorb a portion of the earth's energy. Of the thermal radiation emitted a fraction (40) is lost directly to space through the so-called atmospheric window. The atmosphere is transparent to radiation in this wavelength interval of the spectrum. Of the radiation entering the atmosphere (356), a fraction (326) is absorbed by the gases in the atmosphere and the clouds while the rest (30) is reflected to space. The atmosphere, in turn emits long wave radiation due to its temperature. A portion of this radiation (169) is lost to space, while a significant portion (333) is reradiated back to earth. The overall energy balance gives the net energy absorbed by the earth as 1 Wm^{-2}. It is interesting to perform the following energy balances for the system.

Applying the energy balance to the atmosphere as a closed-system we have:

$$Q_{net,atm.} = 78 + 17 + 80 + 356 - 169 - 30 - 333 = -1 \text{ Wm}^{-2}$$

Applying the energy balance to the earth's surface we obtain:

$$Q_{net,earth} = 161 - 17 - 80 - 396 + 333 = 1 \text{ Wm}^{-2}$$

The net energy gained by the earth, which is called the planetary energy imbalance [3], is about 1 W per m^2. This quantity is a critical metric, in that it predicts future global warming already taking place.

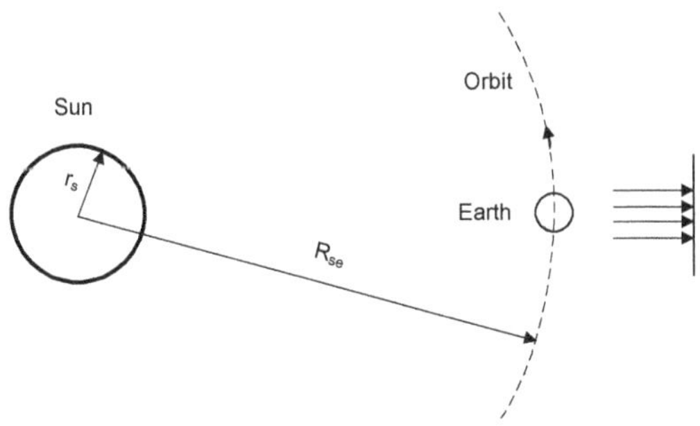

Fig. E1.16 Solar radiation incident on earth

Example 1.16 Develop a simple model to estimate the temperature of the earth, ignoring the effect of the atmosphere. The following data are given: (i) the power output of the sun is 3.84×10^{26} W, (ii) the distance of the sun from the earth = 14.9×10^{10} m, (iii) the diameter of the earth = 12.74×10^{6} m.

Solution Solar radiation incident on unit area of a sphere with the sun at the center, and the radius equal to the sun-earth distance is (see Fig. E1.16)

$$q_s = \frac{Q}{4\pi R_{se}{}^2} = \frac{3.84\times10^{26}}{4\pi\times14.9^2\times10^{20}} = 1376\ \mathrm{Wm}^{-2}$$

The above value is called the solar constant. However, the *standard value of the solar constant* is 1367 Wm^2.

The above radiation intensity is based on unit diametral area of the earth as indicated in Fig. E1.16. Since the earth is a sphere and is rotating about its axis, the energy incident on the diametral area is distributed over the outer surface area of the sphere. Therefore the intensity based on the surface area of the earth is

$$I_s = \frac{q_s\times\pi\times r_e{}^2}{4\pi r_e{}^2} = \frac{1376}{4} = 344\ \mathrm{Wm}^2$$

In order to estimate the earth's temperature we consider it as a sphere with reflectivity (albedo) α for incident solar radiation. The earth's surface is assumed to radiate as a black body. Hence for equilibrium

$$4\pi r_e{}^2(1-\alpha)I_s = 4\pi r_e{}^2\sigma T_e{}^4$$

where, $\sigma = 5.67 \times 10^{-8}$ $\mathrm{Wm^{-2}K^{-4}}$, is the Stefan-Boltzmann constant.

Hence the temperature of the earth is

$$T_e = \left[\frac{(1-\alpha)I_s}{\sigma}\right]^{1/4}$$

The earth's albedo is taken as 0.3. Substituting numerical data in the above equation we obtain the earth's temperature as 255 K or -18°C. Since the atmosphere is not included in this analysis, the temperature obtained can be thought as the temperature at the edge of the atmosphere. It is usually called the emission temperature.

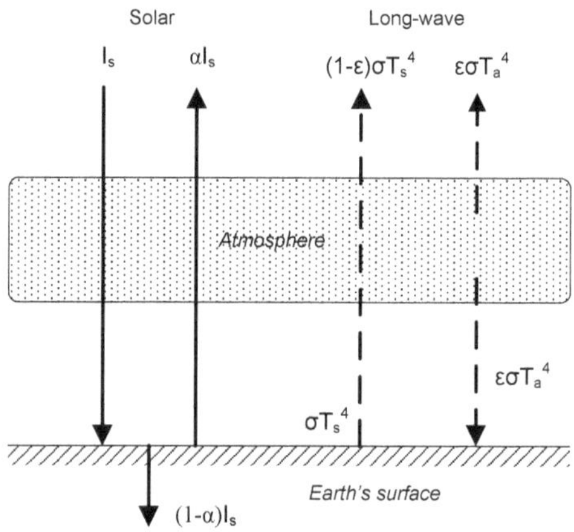

Fig. E1.17 Simplified model for global energy exchange [9]

Example 1.17 Develop a simple model to estimate the mean temperature of the earth's surface and the temperature of the atmosphere.

Solution Figure E1.17 shows a simple model of the earth and the atmosphere [9]. The atmosphere is fully transparent to solar radiation, I_S while the earth's reflectivity or albedo for solar radiation is α. The earth's surface has a temperature T_S, and it emits and absorbs long-wave heat radiation like a black body. The atmosphere has an absorptivity and emissivity ε for long-wave radiation. We now write the following radiation balance equations.

At the outer edge of the atmosphere

$$(1-\alpha)I_s = (1-\varepsilon)\sigma {T_s}^4 + \varepsilon\sigma {T_a}^4 \tag{E17.1}$$

The term on the LHS is the net solar energy inflow. The first term on the RHS is the fraction of long-wave radiation emitted by the earth that is transmitted through the atmosphere to space. The second term on the RHS is the long-wave radiation emitted by the atmosphere that leaves to space.

At the earth's surface

$$(1-\alpha)I_s + \epsilon\sigma {T_a}^4 = \sigma {T_s}^4 \tag{E17.2}$$

The first term on the LHS is the solar radiation absorbed by the earth's surface. The second term is the long-wave radiation emitted by the atmosphere that is absorbed by the earth. The term on the RHS is the long-wave radiation emitted by the earth.

Solving Eqs. (E17.1) and (E17.2) simultaneously we obtain the following expressions for the temperatures of the earth and the atmosphere.

$$T_s = \left[\frac{2(1-\alpha)I_s}{\sigma(2-\epsilon)}\right]^{1/4} \quad \text{and} \quad T_a = \left[\frac{(1-\alpha)I_s}{\sigma(2-\epsilon)}\right]^{1/4}$$

The following numerical data are used to compute the above temperatures:

$$\alpha = 0.3, \epsilon = 0.8, I_s = 341.7 \text{ Wm}^{-2}, \sigma = 5.67 \times 10^{-8} \text{ Wm}^{-2}\text{K}^{-4}$$

Hence we obtain the earth's temperature as 290 K or 17°C, and the atmospheric temperature as 244 K or -29°C. These results are relatively close to the average temperatures obtained by detailed simulations.

Comparing the more realistic processes indicated in Fig. E1.15 with those used in the above simplified model, we note that this model could be further improved by including more details. However, such a model would involve more computation, and require additional data, that is not readily available.

Cost analysis

Example 1.18 Calculate the present worth of 25 yearly payments, the first of which is \$750. The inflation rate is 5 percent per year and the market discount rate is 12 percent. What is the present value of the fifth payment?

Solution The effective discount rate including inflation is given is

$$r_e = \frac{(1+r)}{(1+s)} - 1 = \frac{1.12}{1.05} - 1 = 0.067$$

The present worth of the annual payments over 25 years is

$$S = P_a\left[\frac{1-(1+r)^{-n}}{r}\right]$$

$$S = 750\left[\frac{1-(1+0.067)^{-25}}{0.067}\right] = \$8981.6$$

The present value of the fifth payment is

$$A = \frac{F}{(1+r)^n} = \frac{750}{1.067^5} = \$542.3$$

Example 1.19 The installation cost of a proposed project is \$3.2 million, and it is expected to generate a net annuity of \$150,000 per year for 20 years. The salvage value of the equipment \$400,000. (i) Calculate the net present worth of the project using simple payback. (ii) If the minimum attractive rate of return (MARR) is 5 percent, is the project viable?

Solution (i) When simple payback is used the annual sums are not discounted. The present value of the annuities is $\$150000 \times 20 = 3 \times 10^6$. The salvage value is $\$0.4 \times 10^6$. Therefore *net present worth* of the project is

$$NPW = Income + Salvage\ value - cost$$

$$NPW = 3 \times 10^6 + 0.4 \times 10^6 - 3.2 \times 10^6 = \$200{,}000$$

(ii) Assume a discount rate of 5%. Then the present worth of the salvage value of the equipment is

$$S = \left[\frac{0.4\times10^6}{(1+r)^{20}}\right] = \$0.15075 \times 10^6$$

The present worth of the sum of annuities over 20 years is

$$A = P_a\left[\frac{1-(1+r)^{-n}}{r}\right] = 0.15 \times 10^6\left[\frac{1-(1+0.05)^{-20}}{0.05}\right] = \$1.8693 \times 10^6$$

The net present worth of the project is

$$NPW = Income + Salvage\ value - cost$$

$$NPW = 1.8693 \times 10^6 + 0.15075 \times 10^6 - 3.2 \times 10^6$$
$$= \$ - 1.18 \times 10^6$$

The net present worth is negative and therefore the project is not financially viable at an MARR of 5%.

Example 1.20 A wind turbine system, rated at 1 MW takes one year to build, and costs \$1.8 million. The operating and maintenance costs are \$100,000 per year, and is expected to increase at the rate of 2 percent. There are no fuel costs associated with the project. The discount rate is assumed as 8 percent. Assume a capacity factor of 0.4. The lifetime of the project is 10 years. Calculate the levelized cost of energy.

Solution The capacity factor of an energy system is the fraction of time that it produces the rated power output. Therefore, the annual energy output of the wind turbine system is

$$W = 1.0 \times 24 \times 365 \times 0.4 = 3500\ \text{MWh}$$

We shall first compute the net present worth of the yearly operating and maintenance costs, M. This includes 2% escalation, and discounting to the present. The construction of the plant takes one year during which it does not generate any power. All cash flows are reduced to the beginning of construction, and are expressed in \$1000 units,

$$M = \frac{100}{1.08^2} + \frac{100\times1.02}{1.08^3} + \frac{100\times1.02^2}{1.08^4} + \frac{100\times1.02^3}{1.08^5} + \frac{100\times1.02^4}{1.08^6} + \frac{100\times1.02^5}{1.08^7} + \frac{100\times1.02^6}{1.08^8} + \frac{100\times1.02^7}{1.08^9} + \frac{100\times1.02^8}{1.08^{10}}$$

$$M = \left(\frac{100}{1.08^2}\right)\left(\frac{1-0.944^9}{1-0.944}\right) = 619$$

Hence the total present cost is

$$S = 1800 \times 10^3 + 619 \times 10^3 = \$2419 \times 10^3$$

The yearly electrical energy output (MWh) is discounted to the present at 8 percent per year. Therefore

$$F = \frac{3500}{1.08^2} + \frac{3500}{1.08^3} + \frac{3500}{1.08^4} + \frac{3500}{1.08^5} + \frac{3500}{1.08^6} + \frac{3500}{1.08^7} + \frac{3500}{1.08^8} + \frac{3500}{1.08^9} + \frac{3500}{1.08^{10}}$$

$$F = \left(\frac{3500}{1.08^2}\right)\left(\frac{1-0.9259^9}{1-0.9259}\right) = 20245$$

The levelized cost of energy is given by

$$LCOE = \frac{S}{F} = \frac{2419\times10^3}{20245\times10^3} = \$0.119 \text{ per kWh}$$

Problems

P1.1 Use published data from sources, such as the BP Statistical Review of World Energy [14], to illustrate graphically, (i) how the world reserves to production ratio (R/P) has varied over the recent past for the following fossil fuels: oil, natural gas, and coal, and (ii) how primary energy consumption in different countries has changed recently.

P1.2 An engine in 10 seconds raised a load of 2000 kg through a height of 1.5 m and communicated to it a speed of 3 ms^{-1}. At what average power has the engine been working? [*Answer*: 3.84 kW]

P1.3 An airplane propeller is performing 1800 rpm when the engine of capacity 150 kW is exerting full power. What is the magnitude of the torque exerted by the shaft of the propeller? [*Answer*: 796 Nm]

P1.4 A car of mass 2550 kg is accelerating at 0.8 ms^{-2} up a slope of 1 in 50. The resistance is 0.07 Newtons per kg of mass. The speed is 35 km per hour. Calculate the power exerted by the engine. [*Answer*: 26.4 kW]

P1.5 A mass of 5.5 kg is attached to one end of an elastic string of natural length 1.3 m whose other end is attached to a fixed point A. The elastic modulus of string is such that the 5.5 kg mass hanging vertically would stretch it by 0.15 m. The mass is held at A and allowed to fall vertically. (i) How far below A will it first come to rest?. (ii) Calculate the strain energy stored in the string? [*Answers*: 2.09m, 112.2 J]

P1.6 A jet of water issues from a pipeof cross-sectional area 30 cm^2 at the rate of 1.8 m^3 per min. Calculate the maximum power available in the jet of water. [*Answer*: 1.5kW]

P1.7 A Carnot heat engine has a work output of 500 kJ per cycle, and rejects heat to a reservoir at 15°C. The thermal efficiency is 40%. (a) Calculate the heat source temperature, the heat absorption and rejection per cycle. (b) If the engine is reversed, and operated as a heat pump between heat reservoirs at 15°C and 35°C, calculate the COP. [*Answers*: (a) 207°C, 750 kJ, 1250 kJ, (b) 15.4]

P1.8 A heat pump is used to heat a building maintained at 30°C. It absorbs 20 kW of heat from the ambient at 5°C. The COP of the heat pump is 60% of the COP of a Carnot heat pump operating between the same two temperatures. Calculate the work input to the heat pump, and the heat supplied to the building. [*Answers*: 3.19 kW, 23.19 kW]

P1.9 Methane (CH_4) is burned with 150 percent theoretical air in the combustion system of a gas engine producing an output of 150 kW. The thermal efficiency of the engine is 22%. The heating value of methane is 50,014 kJ kg^{-1} and the density is 0.656 kgm^{-3}. Calculate (i) fuel consumption in m^3s^{-1}. (ii) the air-fuel ratio, (iii) the specific carbon content of the fuel, and (iv) the mass of carbon dioxide produced per hour. [*Answers*: (i) 0.0207 m^3s^{-1}, (ii) 19.4, (iii) 0.75, (iv) 134.6 kgh^{-1}]

P1.10 A solar PV power plant rated at 850 kW, costs $1.4 million to install. The yearly operating and maintenance cost is $13,000. Assume a

capacity factor of 0.18 for the location, and a discount rate of 8%. The life-time of the plant is 25 years. (a) Calculate the levelized cost of energy for this system. (b) Perform a sensitivity evaluation by changing (i) the discount rate by ±25%, and (ii) the capacity factor by ±10%.
[*Answer*: (a) $107.6 per MWh]

References

1. Campbell, Colin J and Laherrere, Jean H, The End of Cheap Oil, *Oil and the Future of Energy*, Scientific American Magazine, The Lyons Press, Connecticut, (2007), 1-7.
2. Duffie, John A, Beckman, William A, and Blair Nathan, *Solar Engineering of Thermal Processes, Photovoltaics and Wind*, John Wiley & Sons, Inc., New Jersey, 2020.
3. Hansen, James, Defusing the Global Warming Time Bomb, *Oil and the Future of Energy*, Scientific American Magazine, The Lyons Press,Connecticut, (2007), 24-33.
4. Jacobson, Mark, Z, *100% Clean, Renewable Energy and Storage for Everything*, Cambridge University Press, Cambridge, 2021.
5. Jenkins, Nick, and Ekanayake, Janaka, *Renewable Energy Engineering*, Cambridge University Press, Cambridge, 2019.
6. Kreith, Frank and Krumdieck, Susan, *Principles of Sustainable Energy Systems*, CRC Press, Taylor & Francis Group, New York, 2015.
7. Lake, James A, Bennett, Ralph, G and Kotek, John F, Next Generation Nuclear Power, *Oil and the Future of Energy*, Scientific American Magazine, The Lyons Press,Connecticut, (2007), 24-33.
8. Shepherd, W and Shepherd, DW, *Energy Studies*, Imperial College Press, London, 2008.
9. Vanek, Francis M, Albright, Louis D, and Angenent, Largus T, *Energy Systems Engineering: Evaluation and Implementation*, McGraw-Hill Book Company, New York, 2012.
10. BP Statistical Review of World Energy 2014, British Petroleum, plc, London, UK, June 2014.
11. BP Statistical Review of World Energy 2015, British Petroleum, plc, London, UK, June 2015.
12. BP Statistical Review of World Energy 2016, British Petroleum, plc, London, UK, June 2016.
13. BP Statistical Review of World Energy 2017, British Petroleum, plc, London, UK, June 2017.

14. BP Statistical Review of World Energy 2018, British Petroleum, plc, London, UK, June 2018.
15. 'Timeline of steam power Wikipedia', [https://en.wikipedia.org/wiki/Timeline_of_steam power]
16. 'Internal combustion engine, Wikipedia' [https://en.wikipedia.org/wiki/Internal_combustion_engine]
17. 'World Energy Consumption, Wikipedia', [https://en.wikipedia.org/wiki/World_energy_consumption]
18. 'Photovoltaic Systems, Wikipedia', [https://en.wikipedia.org/wiki/Photovoltaic_systems]
19. 'Wind power, Wikipedia', [https://en.wikipedia.org/wiki/Wind_power]
20. 'Wave power, Wikipedia', [https://en.wikipedia.org/wiki/Wave_power]
21. 'Ocean thermal energy conversion,Wikipedia' [https://en.wikipedia.org/wiki/Ocean_thermal_energy_conversion]
22. 'Geothermal power,Wikipedia' [https://en.wikipedia.org/wiki/Geothermal_power]
23. 'Tidal power,Wikipedia' [https://en.wikipedia.org/wiki/Tidal_power]
24. Thomas, Alex, *A Simple Model of Global Climate*, [https://youtube/zyd-ynODjgc]

Chapter 2

The Solar Energy Resource

2.1 Introduction

Solar energy received on earth drives the other renewable energy sources such as wind power, hydropower, and wave power. These processes are shown schematically in the energy flow diagram in Fig. 1.1. Solar energy has been used as a heat source for applications like crop drying from ancient times, while space and water heating using solar energy are more recent developments. Large-scale conversion of solar energy to electricity using solar photovoltaic cells is even more recent.

Solar energy is virtually inexhaustible, and it is nonpolluting. However, its use as a primary energy source must deal with its intermittency, and relatively low energy density per unit receiving area.

In this chapter we shall review the various factors influencing the availability of solar energy at a location on earth. The equations derived will be applied in later chapters on solar heating, solar thermal power, and solar photovoltaic systems.

2.2 Fundamentals of Solar Radiation

In this section we will consider several physical aspects of solar radiation including its energy intensity and solar geometry, which are of importance for the design of solar energy conversion systems.

2.2.1 *Solar radiation spectrum*

The radiant energy reaching the earth from the sun is distributed over a range of wavelengths as depicted in Fig. 2.1. The total rate of energy per unit area per unit wavelength interval is called the *spectral irradiance*. The spectral irradiance at the outer edge of the atmosphere, known as extraterrestrial radiation, is well-represented by the spectrum of a black body at 5800 K [1,2]. The energy per unit wavelength interval $E_{\lambda b}$, follows the Planck-distribution as shown Fig. 2.1. The expression for the Planck distribution is

$$E_{\lambda b} = \frac{c_1 \lambda^{-5}}{exp(c_2/\lambda T)-1} \tag{2.1}$$

where the constants are given by:

$$c_1 = 2hc^2 = 1.192 \times 10^8 \text{ W}\mu\text{m}^4/\text{m}^2$$

$$c_2 = \frac{hc}{k} = 1.4389 \times 10^4 \ \mu\text{mK}.$$

The universal constants are: Planck's constant, $h = 6.6238 \times 10^{-34}$ Js, Boltzmann's constant, $k = 1.3807 \times 10^{-23}$ JK^{-1}, and the speed of light, $c = 3 \times 10^8$ ms^{-1}. The wavelength is λ (μm) and the absolute temperature is T (K).

The wavelength, λ_{mas} at which the spectral emissive power, $E_{\lambda b}$ of a blackbody at a given temperature, T is a maximum is given by the *Wien's displacement law*, which may be obtained by differentiating the expression on the RHS of Eq. (2.1). Hence we have

$$\lambda_{max} T = 2897.6 \ \mu m\text{K} \tag{2.2}$$

The *total emissive power* of a surface, defined as the total rate of radiant energy emission in all directions over all wavelengths per unit area of the surface is given by the *Stefan-Boltzmann law*, which may be derived by integrating the expression in Eq. (2.1). Hence we have

$$E_b = \sigma T^4 \tag{2.3}$$

where the constant, $\sigma = 5.67 \times 10^{-8}$ WK^{-4} m^{-2} is the Stefan-Boltzmann constant.

Solar radiation received on the surface of the earth at normal incidence is subject to variations as seen in Fig. 2.1. These variations are caused by: (i) changes in extraterrestrial radiation, (ii) atmospheric scattering by air molecules, water vapor, and dust, and (iii) atmospheric absorption by ozone, water vapor and carbon dioxide. These effects are discussed in detail in Ref.[1,2].

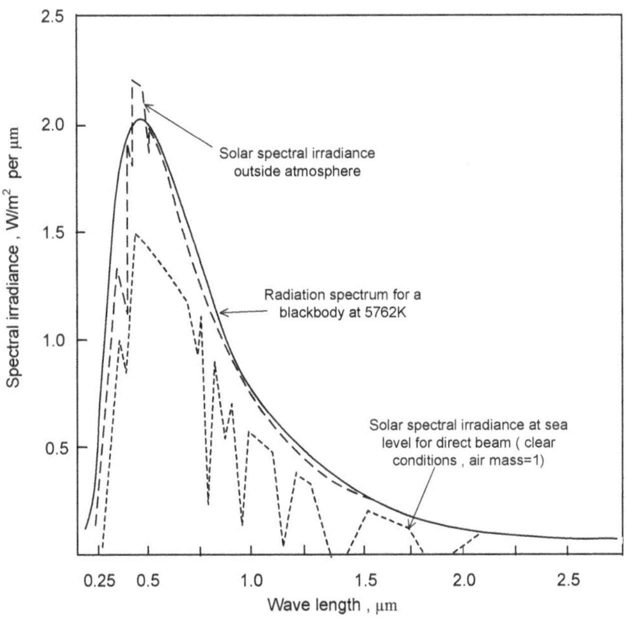

Fig. 2.1 Solar radiation spectrum

2.2.2 *Beam and diffuse radiation*

The intensity of direct normal extraterrestrial solar radiation on the earth's atmosphere is called the *solar constant*. It is the solar radiation striking a unit area normal to the direction of the beam from the sun. Although the solar constant varies from about 1323 Wm^{-2} to 1414 Wm^{-2} due to the slightly elliptical orbit of the earth around the sun, its reference value is treated as constant, and equal to 1367 Wm^{-2}. The *ASHRAE Handbook,* 2013 - *Fundamentals* [6] gives the following approximate expression for the variation of extraterrestrial solar radiation incident on a surface normal to the sun's rays:

$$E_o = E_{sc}\left\{1 + 0.033cos\left[360^0 \frac{(n-3)}{365}\right]\right\} \tag{2.4}$$

where $E_{sc} = 1367$ Wm^{-2}, and n is the day of the year numbered from January 1.

Solar radiation transmitted directly through the atmosphere without change in direction and striking a surface is called *beam radiation*. Beam radiation is also referred to as *direct radiation*. Solar radiation received from the sun after its direction has been changed due to scattering by the atmosphere is called *diffuse radiation*. It does not have a unique direction. The sum of the beam and diffuse radiation on a surface is the *total radiation*.

On a very clear day the fraction of extraterrestrial radiation absorbed or reflected by the upper atmosphere is about 20%. The beam and diffuse radiation fractions striking the surface of the earth are about 70% and 10% respectively. On cloudy days the solar radiation reaching the earth's surface is almost entirely diffuse.

2.3 Direction of Beam Radiation

In this section we shall derive the equations needed to determine the direction of the beam component of solar radiation at any location on the surface of the earth, on any day and time. For this purpose, a brief review of solar geometry is presented below.

The earth revolves about the sun in an orbit that is slightly elliptical. It takes 24 hours to make one rotation about its axis and completes one revolution about the sun in approximately 365.25 days. The axis of rotation of the earth is inclined at an angle of 23.5° to the plane of its orbit around the sun as shown in Fig. 2.2. This tilt of the earth's axis causes the seasons.

The position of the earth at the start of each of the seasons is depicted in Fig. 2.2. At the vernal equinox and the autumnal equinox the plane of the sun's rays is parallel to the equatorial plane of the earth. At the

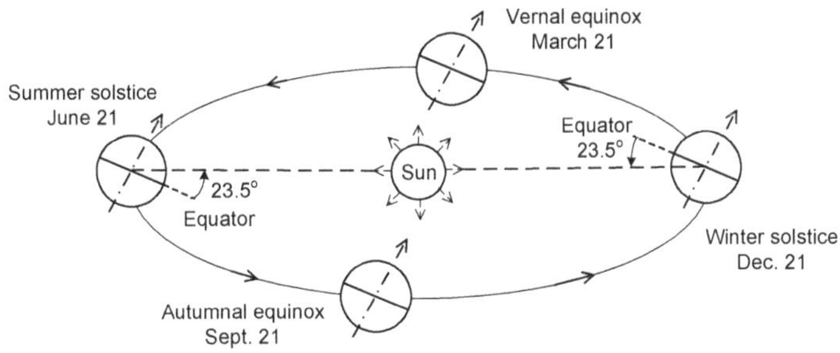

Fig. 2.2 Motion of the earth about the sun

summer solstice and the winter solstice the plane of the sun's rays is inclined at 23.5° and -23.5° respectively to the equatorial plane.

Since all motion is relative, it is convenient to take the earth as fixed and consider the virtual motion of the sun in relation to any location on earth as depicted in Fig. 2.3. To an observer at P on earth, the plane of the sun's rays appears to swing through an angle of 23.5° about a plane parallel to the equatorial plane of the earth.

The location of the sun at any time during the day can be specified by two angles β and ϕ as shown in Fig. 2.4. The angle β, measured from the local horizontal plane upward to the center of the sun is called the *solar altitude angle.* It is the angle between the sun's rays, SP and the horizontal plane at P. The angle ϕ between the due south line at P and the projection PH of PS on the horizontal plane, is called the *solar azimuth angle*. Azimuth angles east of south are positive and west of south are negative by convention.

Although the two angles β and ϕ are convenient in locating the sun's position at any time, they are not fundamental angles. These angles have to be related to the three fundamental angular quantities called: (i) the latitude, (ii) the declination, and (iii) the hour angle.

The latitude, L is the angular distance of a point on the earth measured north or south of the equator as shown in Fig. 2.3. The declination of the sun, δ is the angle between the sun's rays at any time during the day and the zenith direction (directly overhead) at noon on the earth's equator as indicated in Fig. 2.3. The declination is zero at the autumnal and vernal equinoxes. It is +23.5° and -23.5° respectively at the

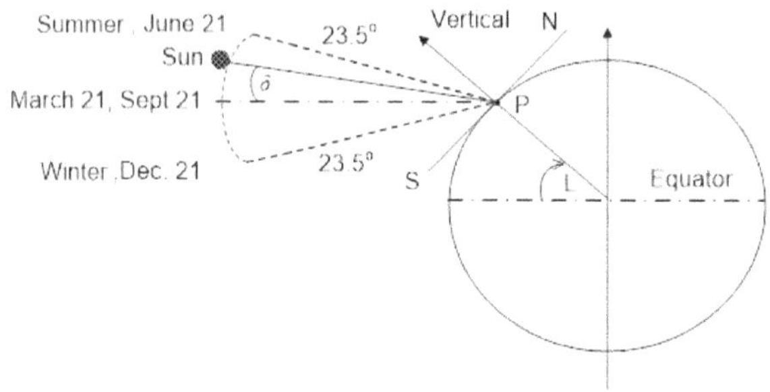

Fig. 2.3 Apparent motion of the sun as observed from a location P on earth

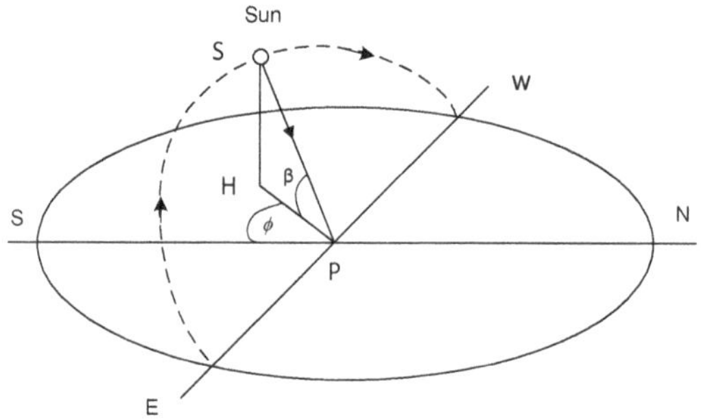

Fig. 2.4 Solar-altitude and solar-azimuth angles

summer solstice and winter solstice in the northern hemisphere. An approximate expression for the solar declination angle on any day of the year is [1]

$$\delta = 23.45 sin\left[\frac{360(284+N_d)}{365}\right] \tag{2.5}$$

where N_d is the day of the year numbered from January 1.

The *hour angle*, *H* at a location is the angular displacement of the sun from solar noon, when the sun is directly overhead at the location. The hour angle is zero at solar noon. For each hour away from solar noon the hour angle is 15° because the earth rotates at the rate of 15° of longitude per hour. In the morning the hour angle is negative, and it is positive in the afternoon. Hence we have

$$H = 15(solar\ time - 12) \text{ degrees} \tag{2.6}$$

The *Solar time* depends on the rotation of the earth and therefore varies continuously with longitude. However, *local standard time*, indicated by a watch, is the same for all locations in a *time zone* which covers a finite longitude interval of about 15°. Each time zone has its assigned *standard longitude* which is used to obtain the *standard* time for the zone. When the sun is directly overhead at a location, it is *solar noon*, whereas when the sun is directly overhead at the *standard* longitude of a time zone it is *local noon*.

Solar time T_{solar} at a location, with a longitude L_{loc} and standard time, T_{std} both measured in minutes, are related by the equation [1].

$$T_{solar} = T_{std} + 4(L_{std} - L_{loc}) + E_{time} - DT \tag{2.7}$$

where L_{std} is the longitude used to obtain the standard time for the time zone of the location.

Longitude is measured positive east of Greenwich where the longitude is zero. The term *DT* is the called the daylight saving time correction, which is the number of hours that the *time is advanced* for daylight saving [1,2].

In Eq. (2.7), E_{time} is a correction factor called the '*equation of time*' which accounts for the perturbations in the earth's rate of rotation. It is given by [1].

$$E_{time} = 9.87sin2B - 7.53cosB - 1.5sinB \tag{2.8}$$

where $$B = \frac{360(N_{day}-81)}{364}$$

and N_{day} (≤ 365) is the day of the year.

The solar altitude β and the solar azimuth ϕ, shown in Fig. 2.4, are related to the latitude L, the solar declination δ, and the hour angle H by the following equations [1]:

$$sin\beta = cosLcosHcos\delta + sinLsin\delta \tag{2.9}$$

$$sin\phi = \frac{cos\delta sinH}{cos\beta} \tag{2.10}$$

$$cos\phi = \frac{(cos\delta sinLcosH - sin\delta cosL)}{cos\beta} \tag{2.11}$$

2.4 Angle of Incidence of Beam Radiation

We shall now obtain an expression for the angle of incidence of beam radiation on a plane surface inclined at angle Σ to the horizontal as shown in Fig. 2.5. The direction of the direct beam of solar radiation is PS and its projection on the horizontal plane is PA. The solar altitude and the solar azimuth at the location are β and ϕ respectively. The normal to the surface is PN and its projection on the horizontal plane, PB makes an angle of ψ with the south.

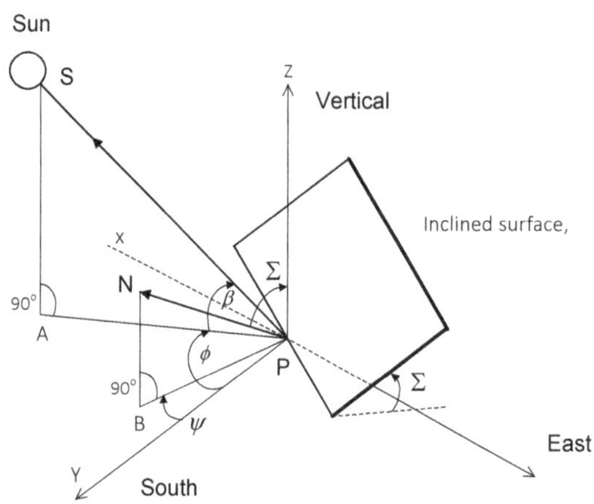

Fig. 2.5 Angle of incidence on an inclined surface

Consider the coordinate system, shown in Fig. 2.5 with the x and y axis directed toward the west and south on the horizontal plane (also see Fig. 2.3) at P. The z-axis is in the vertical direction. The unit vector in the direction PS is given by:

$$\bar{I}_s = \begin{pmatrix} cos\beta sin\phi \\ cos\beta cos\phi \\ sin\beta \end{pmatrix} \tag{2.12}$$

The unit vector in the direction of the surface normal PN is:

$$\bar{n} = \begin{pmatrix} sin\Sigma sin\psi \\ sin\Sigma cos\psi \\ cos\Sigma \end{pmatrix} \tag{2.13}$$

The angle of incidence, θ on the inclined surface is the angle between the direct beam PS and the surface normal PN. This is given by the dot product of the unit vectors $\bar{I}_s$ and $\bar{n}$. Hence we have

$$cos\theta = cos\beta sin\phi sin\Sigma sin\psi + cos\beta cos\phi sin\Sigma cos\psi + sin\beta cos\Sigma$$

$$cos\theta = cos\beta sin\Sigma cos(\phi - \psi) + sin\beta cos\Sigma$$

$$cos\theta = cos\beta sin\Sigma cos\gamma + sin\beta cos\Sigma \quad (2.14)$$

The angle, $\gamma = (\phi - \psi)$ is called the *surface-solar azimuth.*

We note from Eq. (2.14) that for a *horizontal surface* like a roof, $\Sigma = 0$ and the angle of incidence is given by:

$$\theta = (90 - \beta).$$

For a *vertical surface* like a wall, $\Sigma = 90^0$ and the angle of incidence is given by:

$$cos\theta = cos\beta cos\gamma \quad (2.15)$$

The angles β and ϕ in Eq. (2.14) are related to the latitude L, the solar declination δ, and the hour angle H by Eqs. (2.9) to (2.11) respectively.

2.5 Total Radiation on an Inclined Surface

The total solar radiation incident on a solar collection device consists of three components. These are: (i) the direct beam solar radiation G_{db},

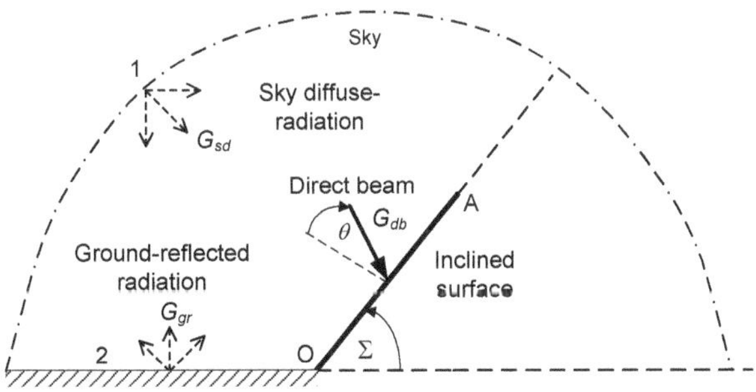

Fig. 2.6 Radiation incident on an inclined surface

(ii) the diffuse radiation from the sky G_{sd}, and (iii) the radiation reflected from the ground and the surrounding buildings G_{gr}. The sky-diffuse radiation and the ground-reflected radiation may be treated as *isotropic*.

Shown schematically in Fig. 2.6 is a three-surface enclosure consisting of a rectangular surface OA, the sky-dome 1, and the large horizontal ground surface 2. The fraction of diffuse radiation emitted by OA that lands on the ground surface 2, considered an infinite plane, is the view factor F_{OA-2}, which is given by the expression [5]

$$F_{OA-2} = \frac{(1-cos\Sigma)}{2} \tag{2.16}$$

where Σ is the inclination of the surface OA to the horizontal.

The fraction of diffuse radiation emitted by OA that lands on the sky dome 1 is given by

$$F_{OA-1} = 1 - F_{OA-2} = \frac{(1+cos\Sigma)}{2} \tag{2.17}$$

Using the reciprocity relation for diffuse view factors [5] we obtain the following expression for the sky radiation striking the surface OA per unit area

$$G_{s-OA} = \frac{(1+cos\Sigma)G_{ds}}{2} \tag{2.18}$$

where G_{ds} is the sky-diffuse radiation incident on a horizontal surface per unit area. Note that the view factor from the horizontal surface to the sky is unity.

Similarly, the ground-reflected diffuse radiation striking the surface OA per unit area is given by

$$G_{g-OA} = \frac{(1-cos\Sigma)G_{gr}}{2} = \frac{(1-cos\Sigma)\rho_g G_{tg}}{2} \tag{2.19}$$

where G_{tg} is the total solar radiation incident on the ground, which includes the direct and diffuse components. The reflectivity of the ground is ρ_g.

The intensity of total solar radiation incident on the ground G_{tg} is given by

$$G_{tg} = (G_{ds} + G_{db} sin\beta) \tag{2.20}$$

where β is the solar altitude angle, and G_{db} is the intensity of direct-beam radiation.

The direct-beam radiation striking the surface OA per unit area is

$$G_{d-OA} = G_{db} cos\theta \tag{2.21}$$

where θ is the angle of incidence on the surface.

The total radiation incident on the surface OA per unit area G_{g-OA} is the sum of the direct radiation, the sky-diffuse radiation, and the ground-reflected radiation. Hence we have

$$G_{t-OA} = G_{d-OA} + G_{s-OA} + G_{g-OA} \tag{2.22}$$

2.6 Clear-sky Model

Measured values of hourly averaged direct and diffuse solar radiation intensities are commonly used in the simulation of solar energy conversion systems. However, when such measured data are not readily available, an alternative design approach is to use mathematical correlations that have been developed using 'clear-sky radiation intensities'. The first such solar radiation model was introduced in the *ASHRAE Handbook,* 1967 - *Fundamentals* [6].

The above model has a simple mathematical form and incorporates three coefficients that depend on the month of the year, and are applicable to any location [1,2]. The model includes an important parameter called the *relative air mass* (AM) m. It is defined as the ratio of the distance travelled by the direct beam when the solar altitude angle is β, to the distance travelled when the sun is overhead at any location. The air mass is given by the approximate expression

$$m = \frac{1}{sin\beta}$$

The shortcomings and limitations of the above model are discussed in the study by Gueymard and Thevenard [3]. They developed a new 'clear-sky model', which is recommended in the *ASHRAE Handbook,*

2013 - *Fundamentals* [6], for computing the intensities of the direct beam component E_b, and diffuse component E_d, of clear-sky solar radiation. The main equations of the new model are [3,6]

$$E_b = E_o exp[-\tau_b m^{ab}] \tag{2.23}$$

$$E_d = E_o exp[-\tau_d m^{ad}] \tag{2.24}$$

where E_o is the extraterrestrial normal radiation intensity, given by Eq. (2.4), and m is the *relative air mass*.

The dependence of the air mass on solar altitude angle may be expressed in the form

$$m = [sin\beta + 0.50572(6.07995 + \beta)^{-1.6364}]^{-1} \tag{2.25}$$

The optical depths τ_b and τ_d included in Eqs. (2.23) and (2.24) are *location-specific* and vary during the year. The air mass exponents ab and ad may be expressed in terms of the optical depths as:

$$ab = 1.454 - 0.406\tau_b - 0.268\tau_d + 0.021\tau_b\tau_d \tag{2.26}$$

$$ad = 0.507 + 0.205\tau_b - 0.080\tau_d - 0.190\tau_b\tau_d \tag{2.27}$$

The monthly values of τ_b and τ_d for many locations around the world are tabulated in the compact disc version of the *ASHRAE Handbook*, 2013 - *Fundamentals* [6]. A few representative values are listed in Table 2.1 for purposes of illustration.

Table 2.1 Monthly values of optical depths for selected locations (values extracted from the data CD of the *ASHRAE Handbook*, 2013 - *Fundamentals* [6])

Optical depths		τ_b			τ_d	
Location/Month	Jan.	May	Sept.	Jan.	May	Sept.
New York, USA	0.318	0.417	0.402	2.514	2.179	2.326
Dallas, USA	0.332	0.376	0.373	2.588	2.350	2.465
Toronto, Canada	0.299	0.391	0.390	2.432	2.250	2.338
Beijing, China	0.382	0.700	0.511	2.222	1.485	1.883
Bangalore, India	0.370	0.448	0.419	2.526	2.161	2.373
Sydney, Australia	0.414	0.293	0.330	2.561	2.660	2.556

The *direct beam solar radiation* incident on a surface of area *A* is

$$H_{db} = AE_b cos\theta \tag{2.28}$$

where θ is the angle of incidence.

The *diffuse radiation* falling on a surface is more difficult to determine because of the anisotropy of the radiation from the sky. For a vertical surface the following expression, based on the empirical relations recommended in Ref.[6], may be used to estimate the diffuse radiation incident on the surface:

$$H_{dif} = AE_d Y \tag{2.29}$$

where Y is a function of the angle of incidence θ of the direct beam. It is given by the expression

$$Y = max.\,[0.45, (0.55 + 0.437cos\theta + 0.313cos^2\theta)] \tag{2.30}$$

To estimate the *diffuse radiation* falling on a surface inclined at an angle Σ to the horizontal the following expression may be used

$$H_{dif} = AE_d(Ysin\Sigma + cos\Sigma), \quad \Sigma \leq 90^o \tag{2.31}$$

$$H_{dif} = AE_d Ysin\Sigma, \Sigma > 90^o \tag{2.32}$$

The *ground-reflected radiation* falling on a surface is given by

$$H_{gr} = \frac{A(E_d + E_b sin\beta)\rho_{gr}(1 - cos\Sigma)}{2} \tag{2.33}$$

where Σ is the inclination and ρ_{gr} is the ground reflectance, taken as 0.2 for typical ground surfaces.

The equations obtained above for the 'clear-sky model' may be used to calculate the maximum solar radiation that could be expected on a clear day. This data is useful for estimating the cooling load of buildings, which is an important aspect of energy-efficient building design. However, the clear sky model is of limited usefulness in the design of solar energy systems, but it has been employed in developing correlations for actual radiation data.

2.7 Estimation of Incident Radiation Under Real Conditions

2.7.1 *Clearness index*

The diurnal and yearly patterns of insolation are stochastic in nature. The frequency of occurrence of different radiation levels depends on local atmospheric conditions, such as the cloudiness of the sky, and the presence of various gases, which are difficult to predict. However, the performance of solar energy devices are strongly effected by these variations in radiation intensity.

Liu and Jordan [1,2] developed a simplified procedure to estimate the solar radiation intensity under different sky conditions by introducing the concept of '*clearness index*'. Their method allows the decomposition of the measured total radiation intensity into a diffuse component and a beam component.

The clearness index K_T, in general, is defined as the average radiation on a horizontal surface to the extraterrestrial radiation on a horizontal surface. Its monthly average value, $\overline{K}_T$, the daily average value, K_T and the hourly average value, k_T are obtained by using measurements of total solar radiation on a horizontal surface. Hence we have

$$\overline{K}_T = \frac{\overline{H}}{\overline{H}_0},\ K_T = \frac{H}{H_0},\ \text{and } k = \frac{I}{I_0} \tag{2.34}$$

where I, H and $\overline{H}$ are hourly, daily and monthly average values of solar radiation on a horizontal surface. The corresponding quantities of extraterrestrial radiation at the same location are I_o, H_o, and $\overline{H}_o$ respectively.

Shown in Fig. 2.7 is the cumulative distribution of the number of days during which the daily clearness index, K_T is above a certain value for different values of the monthly clearness index $\overline{K}_T$. Liu and Jordan [1,2] found these distributions to be nearly identical for locations with same value of $\overline{K}_T$, although the locations had widely different latitudes and elevations.

This information enabled them to develop generalized correlations to represent the curves in Fig. 2.7. In order to interpret the curves, consider

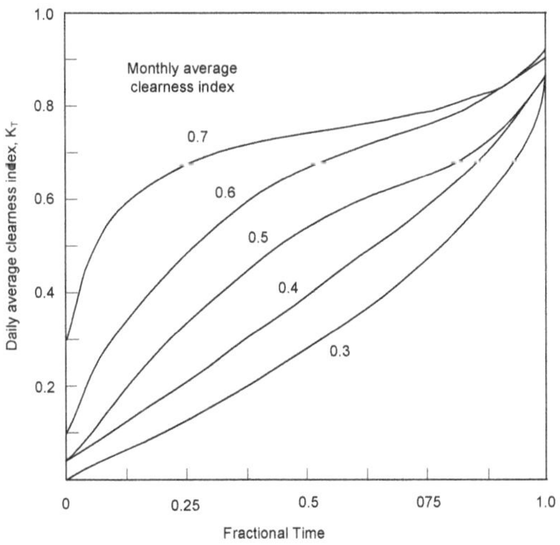

Fig. 2.7 Generalized cumulative distribution of days with various values of K_T

a horizontal line drawn through the K_T value of 0.5. It intersects the $\overline{K}_T = 0.7$ curve at a time fraction of 0.12, and the $\overline{K}_T = 0.5$ curve at a time fraction of 0.44. This shows that when the monthly value, $\overline{K}_T$ is larger, the number of days when K_T exceeds 0.5 is correspondingly larger.

The design and performance evaluation of solar energy systems require knowledge of the solar radiation levels that the system would experience during its life time. Such data can only be obtained by some form of predictive model. We have thus far encountered two such models. The extraterrestrial solar radiation at the edge of atmosphere can be computed using Eq. (2.4). A more realistic model, including the atmosphere is the clear-sky model. Using several location-specific parameters, this model could be used to predict both beam and diffuse radiation at a location. However, this model does not include the effect of clouds, and would therefore overestimate the performance of the system.

More realistic models, based on actual radiation data, including the effects of clouds, have been developed using the clearness index. Such models are available in the form of empirical correlations in most standard textbooks on solar energy systems [1,2]. The data used and the accuracy of the predictions of the correlations are discussed in detail in these sources.

2.7.2 *Correlations for predicting solar radiation under real atmospheric conditions*

In this section, we shall list a few representative correlations that have been developed by various researchers to estimate solar radiation intensity at a location. These could be used to split the measured total radiation into the diffuse and beam components. The computed quantities used in the correlations are the extraterrestrial radiation, and the total radiation predicted by the clear-sky model.

1. *Estimation of hourly diffuse and beam radiation on a horizontal surface using measured hourly total radiation and hourly extraterrestrial radiation*

When the ratio of the hourly diffuse radiation to the hourly total radiation, I_d/I_t is plotted against the ratio of the hourly total radiation to the hourly extraterrestrial radiation, I_t/I_o i.e. (k_T) the resulting curve has a negative slope because the diffuse component of radiation decreases as the clearness index increases. Also, it displays three distinct sections, which have been correlated using the following equations [1]:

$$\frac{I_d}{I_t} = 1.0 - 0.249k_T \quad \text{for} \quad k_T < 0.35 \tag{2.35}$$

$$\frac{I_d}{I_t} = 1.557 - 1.84k_T \quad \text{for} \quad 0.35 < k_T < 0.75$$

$$\frac{I_d}{I_t} = 0.117 \quad \text{for} \quad k_T > 0.75$$

2. *Estimation of the hourly diffuse and beam radiation on a horizontal surface using measured hourly total radiation and hourly total 'clear-sky' radiation*

When the ratio of the hourly diffuse radiation to the hourly total radiation, I_d/I_t is plotted against the ratio of the hourly total radiation to the hourly 'clear-sky' radiation, I_t/I_c the resulting curve displays three distinct sections. These have been correlated using the following equations [1]:

$$\text{For } 0 \le \frac{I_t}{I_c} < 0.48 \qquad \frac{I_d}{I_t} = 1.0 - 0.1\left(\frac{I_t}{I_c}\right) \tag{2.36}$$

$$\text{For } 0.48 \le \frac{I_t}{I_c} < 1.1 \qquad \frac{I_d}{I_t} = 1.11 + 0.0396\left(\frac{I_t}{I_c}\right) - 0.789\left(\frac{I_t}{I_c}\right)^2$$

$$\text{For } 1.1 \le \frac{I_t}{I_c} \qquad \frac{I_d}{I_t} = 0.2$$

3. *Estimation of the hourly total radiation on a horizontal surface using measured daily total radiation.*

When measured hourly total radiation on a horizontal surface at a location is not available, it could be estimated using measured daily total radiation. The following equations may be used for this purpose. The ratio of the hourly total radiation to the total daily radiation is a function of the day length and the hour in question. The ratio of the hourly total radiation, I_t to the daily total radiation, H on a horizontal surface has been correlated using the following equation [1]:

$$\frac{I_t}{H} = \frac{\pi}{24}(a + b\cos\omega)\frac{(\cos\omega - \cos\omega_s)}{\sin\omega_s - r\cos\omega_s} \tag{2.37}$$

where

$$a = 0.409 + 0.5016\sin(\omega_s - 60)$$

$$b = 0.6609 - 4767\sin(\omega_s - 60)$$

$$r = \frac{2\pi\omega_s}{360}$$

ω = hour angle in degrees, and ω_s = sunset angle in degrees

4. *Estimation of the daily diffuse and beam components of radiation on a horizontal surface using measured daily total radiation and daily extraterrestrial radiation.*

When the ratio of the daily diffuse radiation to the daily total radiation, H_d/H is plotted against the ratio of the daily total radiation to the daily extraterrestrial radiation, H/H_0 i.e (K_T), the resulting curve displays four distinct sections. These have been correlated using the following equations [1]:

$$\frac{H_d}{H} = 0.99 \quad \text{for} \quad K_T \leq 0.17 \tag{2.38}$$

$$\frac{H_d}{H} = 1.188 - 2.272K_T + 9.473{K_T}^2 - 21.865{K_T}^3 + 14.648{K_T}^4$$
$$\text{for} \quad 0.17 < K_T < 0.75$$

$$\frac{H_d}{H} = 0.632 - 0.54K_T$$
$$\text{for} \quad 0.75 < K_T < 0.80$$

$$\frac{H_d}{H} = 0.2 \quad \text{for} \quad K_T \geq 0.8$$

5. *Estimation of the diffuse and beam components of monthly average total radiation on a horizontal surface.*

Monthly average diffuse radiation and total radiation can be obtained summing the corresponding hourly or daily data. The monthly total extraterrestrial radiation may be obtained in a similar manner. The day of the month for which the daily extraterrestrial radiation is closest to the monthly average is defined as the 'average day' of the month.

The average days for different months of the year are tabulated in Ref.[1], which may also be used to estimate monthly total extraterrestrial radiation. The ratio of the monthly average diffuse radiation, $\overline{H}_d$ to the monthly average total radiation $\overline{H}$ has been correlated with the monthly average clearness index $\overline{K}_T$, and the sunset hour angle of the mean day of the month, ω_s. The latter is a function of the latitude of the location and the declination and hence accounts for seasonal dependence. The correlation is given by [1]:

$$\frac{\overline{H}_d}{\overline{H}} = 0.775 + 0.00653(\omega_s - 90) -$$
$$[0.505 + 0.00455(\omega_s - 90)]cos[115\overline{K}_T - 103] \tag{2.39}$$

2.8 Estimation of Incident Solar Radiation on an Inclined Surface

The correlations listed above estimate the diffuse and beam radiation on a horizontal surface. However, to evaluate the performance of solar energy systems we require the radiation on inclined surfaces. For instantaneous values of beam radiation the quantities on a horizontal surface could be converted to the corresponding quantities on an inclined surface by simply using the relevant angles of incidence. Hence from Eq. (2.14) we have

$$R_{b,\theta} = \frac{I_{b,\theta}}{I_{b,h}} = \frac{cos\theta}{sin\beta} = \frac{cos\beta sin\Sigma cos\gamma + sin\beta cos\Sigma}{sin\beta} \tag{2.40}$$

The ratio, $R_{b,\theta}$ is the called beam radiation *tilt factor*, which is the ratio of the beam radiation on an inclined surface to that on a horizontal surface.

When hourly beam radiation is used, the angles β and γ in Eq. (2.40) can be calculated at the middle of the hour. However, Eq. (2.40) cannot be used directly to calculate long-term beam radiation on an inclined surface because the radiation intensity and angle of incidence change continuously during the day.

Therefore to be strictly correct, the long-term tilt factor, $\bar{R}_{b,\theta}$ has to be obtained by integrating the instantaneous tilt factors $R_{b,\theta}$, using instantaneous beam radiation as a weighting factor. The computational procedure to obtain the monthly tilt factors is discussed in detail in Ref.[1,2]. To calculate diffuse radiation and ground-reflected radiation incident on an inclined surface, Eqs. (2.16) to (2.22), obtained earlier may be used.

Application of the various correlations listed above to the design of solar energy systems will be illustrated in the worked examples to follow.

2.9 Worked Examples

Example 2.1 The sun has a mean diameter of 1.392×10^6 km, and it may be assumed to be a black body with a mean temperature of 5777 K.

Calculate (i) the daily energy output of the sun, and (ii) the wavelength of radiation for which the energy per unit wavelength is maximum.

Solution The total power output of the sun per unit surface area is obtained from the *Stefan-Boltzmann law,* assuming the sun to be a black body. Hence we have

$$E_b = \sigma T^4$$

The total power output is $P = \pi D_s^{\,2} \sigma T_s^{\,4}$

Substituting numerical values we have

$$P = \pi \times (1.392 \times 10^9)^2 \times 5.67 \times 10^{-8} \times 5777^4$$

$$P = 3.844 \times 10^{26}\ \mathrm{Js^{-1}}$$

The daily energy output is

$$E = 3.844 \times 10^{26} \times 3600 \times 24$$

$$E = 3.32 \times 10^{31}\ \text{J per day.}$$

The wavelength at which the energy is a maximum is given by Eq. (2.2).

$$\lambda_{max} T = 2897.6\ \mu mK$$

$$\lambda_{max} = \frac{2897.6}{5777} = 0.5015\ \mu m$$

Example 2.2 For the solar spectrum, calculate the (i) the power per unit wavelength at 0.6 μm, and (ii) the total power of the spectrum form wavelengths of 0.38 μm to 0.78 μm, which constitutes the visible region. Assume that the temperature of the sun is 5762 K.

Solution (i) The power per unit wavelength at 0.6 μm is obtained by substituting in the Planck-distribution, shown in Fig. 2.1. The expression for the Planck distribution is

$$E_{\lambda b} = \frac{c_1 \lambda^{-5}}{exp(c_2/\lambda T) - 1}$$

where the constants are given by:

$$c_1 = 1.19 \times 10^8 \; W\mu m^4/m^2 \text{ and } c_2 = 1.4389 \times 10^4 \; \mu mK.$$

The wavelength is λ (μm) and the absolute temperature is T (K).

$$E_{\lambda b} = \frac{1.192 \times 10^8 \times 0.6^{-5}}{exp(1.4389 \times 10^4 / 0.6 \times 5762) - 1} = 24.25 \text{ MWm}^{-2} \text{ per } \mu m$$

(ii) To calculate the total energy of the black body spectrum from 0.38 μm to 0.78 μm we need to integrate Eq. (2.1) between these limits.

A polynomial expression to evaluate this integral has been developed by Pivovonsky and Nagel [1]. These are given by

For $\gamma \geq 2$

$$f_{0\to\lambda T} = \frac{E_{0\to\lambda T}}{\sigma T^4} = \frac{15}{\pi^4} \sum_{m=1}^{N} \frac{e^{-m\gamma}}{m^4} [(m\gamma)^3 + 3(m\gamma)^2 + 6(m\gamma) + 6] \tag{E2.2.1}$$

For $\gamma < 2$

$$f_{0\to\lambda T} = \frac{E_{0\to\lambda T}}{\sigma T^4} = 1 - \frac{15\,\gamma^3}{\pi^4} \left[\frac{1}{3} - \frac{\gamma}{8} + \frac{\gamma^2}{60} - \frac{\gamma^4}{5040} + \frac{\gamma^6}{272{,}160} - \frac{\gamma^8}{13{,}305{,}600}\right] \tag{E2.2.2}$$

where $\gamma = c_2/\lambda T$

For $\lambda = 0.38 \; \mu m \quad \gamma = 6.571$

For $\lambda = 0.78 \; \mu m \quad \gamma = 3.201$

A simple MATLAB code was used to evaluate the expression in Eq. (E2.2.1). Hence we obtain

$$f_{0\to 0.38\mu m} = 0.0990 \quad \text{and} \quad f_{0\to 0.78\mu m} = 0.5634.$$

The fraction of power in the wavelength interval from 0.38 μm to 0.78 μm is

$$f_p = 0.5634 - 0.0990 = 0.4644$$

The total power per unit area between the above wavelengths is

$$\Delta E_b = \sigma T^4 f_p$$

$$\Delta E_b = 5.67 \times 10^{-8} \times 5762^4 \times 0.4644 = 29.025 \text{ MWm}^{-2}$$

Example 2.3 Estimate the solar constant using the following data: the mean distance from the sun to the earth is 149×10^9 km, and the mean diameter of the earth is 12.742×10^3 km. What is the mean solar power over the surface of the earth?

Solution Consider a sphere with the sun at the center, and the earth located at its outer surface. The radius of the sphere is equal to the sun-earth distance. The energy incident on unit area of the sphere per unit time is

$$q_s = \frac{Q}{4\pi R_{se}{}^2} = \frac{3.84\times10^{26}}{4\pi\times14.9^2\times10^{20}} = 1376\ \mathrm{Wm^{-2}}$$

The above value is called the solar constant. However, the *standard value of the solar constant* is 1367 $\mathrm{Wm^2}$.

The above radiation intensity is based on unit diametral area of the earth. Since the earth is a sphere and is rotating about its axis, the energy incident on the diametral area is distributed over the outer surface area of the sphere. Therefore the solar power based on the surface area of the earth is

$$I_s = \frac{q_s\times\pi\times r_e{}^2}{4\pi r_e{}^2} = \frac{1376}{4} = 344\ \mathrm{Wm^2}$$

Example 2.4 For a location with a northern latitude of 40°, calculate the extraterrestrial radiation on a surface normal to the sun's rays, and the declination, on August 10.

Solution For August 10 the day number is, $N_d = 222$

The extraterrestrial radiation intensity is given by Eq. (2.4) as:

$$E_o = E_{sc}\left\{1 + 0.033cos\left[360^o\,\frac{(n-3)}{365}\right]\right\}$$

where E_{sc} = 1367 $\mathrm{Wm^{-2}}$ and n is the day of the year.

Substituting in the above equation we get

$$E_o = 1367\left\{1 + 0.033cos\left[360^0\,\frac{(222-3)}{365}\right]\right\} = 1330.5\ \mathrm{Wm^{-2}}$$

The declination is given by Eq. (2.5) as:

$$\delta = 23.45 sin\left[\frac{360(284+N_d)}{365}\right]$$

Substituting in the above equation we have

$$\delta = 23.45 sin\left[\frac{360(284+222)}{365}\right] = 15.36^0$$

Example 2.5 The longitude, and the standard longitude of the time zone, of a particular location are 95°W, and 90°W respectively. The clock time has been advanced by one hour for daylight saving. Calculate the solar time and the hour angle at the location when the standard clock time is 3.30 p.m. on July 14.

Solution The solar time in minutes is given by Eq. (2.7) as

$$T_{solar} = T_{std} + 4(L_{std} - L_{loc}) + E_{time} - DT$$

T_{std} = 3.30 pm = 15.5 hrs, DT =1, L_{std} = 90°W and L_{loc} = 95°W.

The equation of time is given by Eq. (2.8) as

$$E_{time} = 9.87 sin2B - 7.53 cosB - 1.5 sinB$$

where $$B = \frac{360(N_{day}-81)}{364}$$

For July 14, N_{day} = 195.

Substituting in the above equations we obtain B as 112.75 and E_{time} = -5.51 min.

Substituting the above data in Eq. (2.7) we have

$$T_{solar} = 15.5 \times 60 + 4(90 - 95) - 5.51 - 60 = 844.5 \text{ min}$$

Therefore the local solar time is 14.08 hr.

The hour angle H is given by Eq. (2.6) as

$$H = 15(solar\ time - 12) = 31.1^{o}$$

Example 2.6 (a) For a location with a northern latitude of 40°, calculate the solar altitude angle, and the solar azimuth angle at 9 a.m. solar time on August 10. (b) Calculate the solar time at sunrise and sunset, and the day-length on August 10 at the same location.

Solution (a) The hour angle is given by Eq. (2.6) as

$$H = 15(T_{sol} - 12) = -15 \times 3 = -45^o$$

For August 10, $N_{day} = 222$. The declination is given by Eq. (2.5) as

$$\delta = 23.45 sin\left[\frac{360(284+N_d)}{365}\right] = 15.36^o$$

The solar altitude angle is given by Eq. (2.9) as

$$sin\beta = cosLcosHcos\delta + sinLsin\delta$$

$$sin\beta = cos40\cos(-45)\,cos15.36 + sin40sin15.36 = 0.6926$$

Therefore the solar altitude is 43.8 degrees. The solar azimuth angle is given by Eq. (2.10) as

$$sin\phi = \frac{cos\delta sinH}{cos\beta} = \frac{cos15.36\times\sin(-45)}{cos43.8} = -0.9447$$

Therefore the solar azimuth is 70.85 degrees east of south.
(b) At sunrise the solar altitude, β is zero. Hence we have from Eq. (2.9)

$$cosH_{sr} = -tanLtan\delta = -0.2306$$

Therefore $H_{sr} = -103.3$ degrees. The solar time at sunrise T_{sr} is given by Eq. (2.6) as

$$H_{sr} = 15(T_{sr} - 12) = -103.3 \text{ degrees}$$

Hence $T_{sr} = 5.07$ a.m. solar time. Now the hour angles at sunrise and sunset are symmetrical about solar noon. Hence the solar time at sunset is 6.53 p.m. The day-length is 13.46 hours.

Example 2.7 For a location with a northern latitude of 40°, calculate the total extraterrestrial solar radiation incident on a horizontal surface, from sunrise to sunset on August 10.

Solution The extraterrestrial radiation intensity is given by Eq. (2.4) as:

$$E_o = E_{sc}\left\{1 + 0.033cos\left[360^0\frac{(n-3)}{365}\right]\right\}$$

For a horizontal surface the angle of incidence is

$$\theta = \frac{\pi}{2} - \beta$$

The extraterrestrial radiation intensity on a horizontal surface is

$$G = E_o cos\theta = E_o\, sin\beta$$

Substituting from Eq. (2.9) we obtain

$$G = E_o(cosLcosHcos\delta + sinLsin\delta)$$

The total daily extraterrestrial incident on the horizontal surface is obtained by integrating above equation from sunrise to sunset. Hence we have

$$\bar{G} = \int_{t_{sr}}^{t_{ss}} E_o(cosLcosHcos\delta + sinLsin\delta)dt$$

$$\bar{G} = E_o cosLcos\delta \int_{t_{sr}}^{t_{ss}} cosHdt + sinLsin\delta \int_{t_{sr}}^{t_{ss}} dt$$

Now 1 hour of clock time corresponds to 15° or $(\pi/12)$ radians. Therefore

$$dt = \frac{3600}{(\pi/12)}dH$$

Substituting for dt and integrating the above equation for $\bar{G}$ we obtain the daily total extraterrestrial radiation on a horizontal surface as:

$$\bar{G} = \left(\frac{24\times3600}{\pi}\right)E_o\left[cosLcos\delta sinH_{ss} + \frac{2\pi H_{ss}}{360}sinLsin\delta\right] \quad \text{(E2.7.1)}$$

For August 10, $N_{day} = 222$. The declination is given by Eq. (2.5) as

$$\delta = 23.45sin\left[\frac{360(222+284)}{365}\right] = 15.36 \text{ degrees}$$

The extraterrestrial solar radiation incident on a surface normal to the sun's rays on August 10 is given by Eq. (2.4) as

$$E_o = E_{sc}\left\{1 + 0.033cos\left[360^0 \frac{(222-3)}{365}\right]\right\}$$

where E_{sc} = 1367 Wm^{-2} and n is the day of the year.

Hence the extraterrestrial solar radiation intensity, E_o = 1330.5 Wm^{-2}. At sunrise the solar altitude, β is zero. Hence we have from Eq. (2.9)

$$cosH_{sr} = -tanLtan\delta = -0.2306$$

Therefore $H_{sr} = -103.3$ degrees. The sunset hour angle = 103.3 degrees. Substituting in Eq. (E2.6.1) we obtain the total extraterrestrial radiation as:

$$\bar{G} = \left(\frac{24\times3600}{\pi}\right) \times 1330.5\left[cos40cos15.36sin103.3 + \frac{2\pi\times103.3}{360}sin40sin15.36\right]$$

Therefore the total daily extraterrestrial radiation is, $\bar{G} = 37.53$ MJm^{-2}.

Example 2.8 A south-facing vertical surface is placed at the outer edge of the atmosphere where the latitude is 30°. Calculate the total solar energy falling on the surface per unit area from 8 a.m. to 4 p.m. solar time, on 28 May.

Solution For May 28, N_{day} = 148. The declination is given by Eq. (2.5) as

$$\delta = 23.45sin\left[\frac{360(148+284)}{365}\right] = 21.44 \text{ degrees}$$

The extraterrestrial solar radiation incident on a surface normal to the sun's rays on May 28 is given by Eq. (2.4) as

$$E_o = E_{sc}\left\{1 + 0.033cos\left[360^0 \frac{(148-3)}{365}\right]\right\}$$

where E_{sc} = 1367 Wm^{-2}, and $N_{day} = 148$.

Hence the extraterrestrial solar radiation intensity, E_o = 1330.9 Wm^{-2}.

The hour angles at 8 a.m. and 4 p.m. solar time are -60° and 60° respectively. Now the rate of incidence of extraterrestrial radiation on the surface per unit area is given by

$$q = E_o cos\theta$$

where θ is the angle of incidence.

The total energy incident on the surface from 8 a.m. to 4 p.m. is given by

$$Q = \int_{8am}^{4pm} qdT = \int_{8am}^{4pm} E_o \, cos\theta dT \quad \text{(E2.8.1)}$$

where T is solar time in seconds.

For a surface facing south, the surface-solar azimuth, γ is equal to ϕ, the solar azimuth. Substituting for the incidence θ from Eq. (2.15) in Eq. (E2.1.1) we have

$$Q = \int_{8am}^{4pm} E_o \, cos\beta cos\phi dT \quad \text{(E2.8.2)}$$

Substituting in Eq. (E2.8.2) from Eq. (2.11) we obtain

$$Q = \int_{8am}^{4pm} E_o \, (cos\delta sinLcosH - sin\delta cosL)dT \quad \text{(E2.8.3)}$$

Substituting the pertinent numerical values in Eq. (E2.8.3) we have

$$Q = \int_{8am}^{4pm} 1330.9 \, (cos21.44sin40cosH - sin21.44cos40)dT$$

$$Q = \int_{8am}^{4pm} 1330.9 \, (0.598cosH - 0.28)dT \quad \text{(E2.8.4)}$$

The change in hour angle, dH in radians is related to the change in solar time dT, in seconds by the equation

$$dH = \frac{(\pi/12)dT}{3600}$$

Substituting for dT in Eq. (E2.8.4) we obtain the total energy as

$$Q = \left(\frac{180\times3600\times1330.9}{15\times\pi}\right) \int_{-\pi/3}^{\pi/3} (0.598cosH - 0.28) \, dH$$

$$Q = \left(\frac{180\times3600\times1330.9}{15\times\pi}\right) \times 0.449 = 8.22 \times 10^3 \text{ kJm}^{-2}$$

The total extraterrestrial solar energy incident is 8.22 MJm^{-2}.

Example 2.9 For a location with a northern latitude of 38°, calculate the angle of incidence of the direct solar beam at 2.00 p.m. solar time on August 5, on the following surfaces: (i) a flat-surface facing 30° east of south, and tilted 50° from the horizontal, (ii) a vertical wall facing 30° east of south, and (iii) a horizontal surface.

Solution The hour angle is given by Eq. (2.6) as

$$H = 15(T_{sol} - 12) = 15 \times 2 = 30 \text{ degrees}$$

For August 5, n_{day} = 217. The declination is given by Eq. (2.5) as

$$\delta = 23.45 sin\left[\frac{360(217+284)}{365}\right] = 16.83 \text{ degrees}$$

The solar altitude angle is given by Eq. (2.9) as

$$sin\beta = cosLcosHcos\delta + sinLsin\delta$$

$$sin\beta = cos38 \cos 30\, cos16.83 + sin38sin16.83 = 0.831$$

Therefore the solar altitude is 56.25 degrees.

The solar azimuth angle is given by Eq. (2.10) as

$$sin\phi = \frac{cos\delta sinH}{cos\beta} = \frac{cos16.83 \times \sin 30}{cos56.25} = 0.861$$

Therefore solar azimuth is 59.48 degrees west of south.

The surface-solar azimuth is

$$\gamma = 30 + 59.48 = 89.48 \text{ degrees.}$$

The tilt angle of the surface is, Σ = 50°. The angle of incidence, θ is given by Eq. (2.14) as

$$cos\theta = cos\beta sin\Sigma cos\gamma + sin\beta cos\Sigma$$

$$cos\theta = cos56.25sin50cos89.48 + sin56.25cos50 = 0.538$$

Hence the angle of incidence is 57.4°.

(ii) For a vertical wall, Σ = 90°. Therefore the angle of incidence is

$$cos\theta = cos56.25sin90cos89.48 + sin56.25cos90 = 0.00504$$

Hence the angle of incidence is 89.7°.
(iii) For a horizontal surface, Σ = 0°. Therefore the angle of incidence is

$$cos\theta = cos56.25sin0cos89.48 + sin56.25cos0 = 0.831$$

Hence the angle of incidence is 33.75°.

Example 2.10 The latitude of Toronto, Canada is 43.63° N. Calculate the clear-sky beam and diffuse radiation intensities in Toronto, at 3.00 p.m. on July 31.

Solution We calculate the solar altitude angle at 3 p.m. on July 31 using the procedure outlined in Example 2.8. Hence we have

latitude, L = 43.63°, declination, δ = 18.17°, hour angle, H = 45°.

The solar altitude is given by Eq. (2.9) as

$$sin\beta = cos43.63 \cos 45\, cos18.17 + sin43.63sin18.17 = 0.7015$$

Hence the solar altitude angle is 44.55°.

The solar azimuth angle is given by Eq. (2.10) as

$$sin\phi = \frac{cos18.17\times\sin 45}{cos44.55} = 0.943$$

Hence the azimuth angle is 70.56°.

The air mass is given by Eq. (2.25) as

$$m = [sin\beta + 0.50572(6.07995 + \beta)^{-1.6364}]^{-1}$$

$$m = [sin44.55 + 0.50572(6.07995 + 44.55)^{-1.6364}]^{-1} = 1.4238$$

The extraterrestrial solar radiation incident on a surface normal to the sun's rays is given by Eq. (2.4) as

$$E_o = \left\{1 + 0.033cos\left[360^0\frac{(n-3)}{365}\right]\right\}$$

where E_{SC}=1367 Wm^{-2}, and n is the day of the year. Substituting numerical values in the above equation we have

$$E_o = 1367\left\{1 + 0.033cos\left[360^0\frac{(212-3)}{365}\right]\right\} = 1326.5 \text{ Wm}^{-2}$$

The beam and diffuse radiation optical depths for July 31 have been obtained from the data available in Ref.[6]. This gives

$$\tau_b = 0.386, \qquad \tau_d = 2.282$$

Substituting these values in Eqs. (2.26) and (2.27) we have

$$ab = 1.454 - 0.406\tau_b - 0.268\tau_d + 0.021\tau_b\tau_d = 0.704$$

$$ad = 0.507 + 0.205\tau_b - 0.080\tau_d - 0.190\tau_b\tau_d = 0.236$$

The intensities of beam and diffuse radiation are given by Eqs. (2.23) and (2.24) respectively. Substituting the relevant numerical values in the above equations we obtain

$$E_b = E_o exp[-\tau_b m^{ab}] = 1326.5exp[-0.386 \times 1.4238^{0.703}]$$

$$E_d = E_o exp[-\tau_d m^{ad}] = 1326.5exp[-2.282 \times 1.4238^{0.2353}]$$

Hence the intensity of beam radiation is 808.6 Wm^{-2} and the intensity of diffuse radiation is 111.0 Wm^{-2}.

Example 2.11 A vertical surface of area 4 m^2 in Toronto, Canada, faces 50° west of south. The latitude of Toronto is 43.63° N. The ground reflectivity at the location is 0.2. At 3.00 p.m. solar time on July 31, calculate (i) direct beam solar radiation, (ii) sky-diffuse radiation, and (iii) ground-reflected radiation ($\rho = 0.2$) incident on the vertical surface. Use the clear-sky radiation model.

Solution The conditions for this example are the same as those for example 2.10, where we obtained the solar altitude angle, $\beta = 44.55°$ and the solar azimuth angle $\phi = 70.56°$. Hence the surface-solar azimuth angle is

$$\gamma = 70.56 - 50 = 20.56 \text{ degrees.}$$

The angle of incidence of the direct beam on the vertical surface is given by

$$cos\theta = cos44.55sin90cos20.56 + sin44.556cos90 = 0.667$$

Hence the angle of incidence is 48.15°.

For the conditions in example 2.10, we obtained the beam and diffuse radiation intensities using the clear-sky model as, $E_b = 808.6$ Wm^{-2} and $E_d = 111.0$ Wm^{-2} respectively.

(i) The direct radiation incident on the window surface is given by Eq. (2.28) as

$$H_{db} = AE_b cos\theta = 4 \times 808.6cos48.15 = 2158.0 \text{ W}$$

(ii) The diffuse radiation incident on the surface is given by Eq. (2.29) as

$$H_{dif} = AE_d Y$$

where Y is a function of the angle of incidence θ of the direct beam. It is given by the expression in Eq. (2.30) as

$$Y = max.[0.45, (0.55 + 0.437cos\theta + 0.313cos^2\theta)]$$

Substituting the pertinent numerical data in the above expressions we have

$$Y = max[0.45, (0.55 + 0.437cos48.15 + 0.313cos^2 48.15)] = 0.981$$

$$H_{dif} = AE_d Y = 4 \times 111.0 \times 0.981 = 435.6 \text{ W}$$

(iii) The ground-reflected radiation falling on a surface is given by Eq. (2.33) as

$$H_{gr} = \frac{A(E_d + E_b sin\beta)\rho_{gr}(1 - cos\Sigma)}{2}$$

$$H_{gr} = 4(111.0 + 808.6sin44.55) \times \frac{0.2(1 - cos90)}{2}$$

Hence $H_{gr} = 271.3$ W.

Example 2.12 A south-facing flat-plate solar collector at a location with a northern latitude of 30° is inclined at an angle of 75° to the horizontal. (i) Calculate the angle of incidence of the direct solar beam at 2.00 p.m. solar time on 18 June. (ii) Determine the number of hours on 18 June during which the collector is sunlit.

Solution For June 18, N_{day} = 169. The declination is given by Eq. (2.5) as

$$\delta = 23.45sin\left[\frac{360(169+284)}{365}\right] = 23.41 \text{ degrees}$$

The latitude is 30° and the hour angle at 2 p.m. solar time is 30°.

The solar altitude angle is given by Eq. (2.9) as

$$sin\beta = cosLcosHcos\delta + sinLsin\delta$$

$$sin\beta = cos30\cos 30\, cos23.41 + sin30sin23.41 = 0.8887$$

Therefore the solar altitude is 62.49 degrees.

The solar azimuth angle is given by Eq. (2.10) as

$$sin\phi = \frac{cos\delta sinH}{cos\beta} = \frac{cos23.41\times \sin 30}{cos62.49} = 0.993$$

Therefore solar azimuth is 83.4 degrees west of south.

The surface-solar azimuth angle is

$$\gamma = 0 + 83.4 = 83.4 \text{ degrees.}$$

The angle of incidence, θ is given by Eq. (2.14) as

$$cos\theta = cos\beta sin\Sigma\, cos\gamma + sin\beta cos\Sigma$$

where Σ is the tilt angle of the surface.

$$cos\theta = cos62.49sin75cos83.4 + sin62.49cos75 = 0.281$$

Hence the angle of incidence is 73.69°.

(ii) Now the surface-azimuth angle for a south-facing wall is ϕ. When the direct beam *just strikes* the wall, $\theta = 90^o$. Therefore we have

$$cos\theta = cos\beta sin\Sigma\, cos\phi + sin\beta cos\Sigma = 0$$

Substituting from Eqs. (2.11) and (2.9) in the above equation we have

$$\sin\Sigma\,(cos\delta sinLcosH - sin\delta cosL) = -cos\Sigma\,(cosLcosHcos\delta + sinLsin\delta)$$

The above equation may be rearranged to the form

$$cosH = tan\delta tan\left(\Sigma - L\right)$$

Substituting the relevant numerical values in the above equation we have

$$cosH = tan23.41tan(75 - 30) = 0.4329$$

The hour angle at which the beam just strikes the wall is $H = 64.34°$, which corresponds to 4.29 hours before solar noon. Hence the total number of hours during which the wall is sunlit is 8.58 hours. The sunlit period is therefore from 7.43 a.m. to 4.17 p.m. solar time.

Example 2.13 For a location with northern latitude 24°, the measured total hourly radiation on a horizontal surface at 2.30 p.m. on May 15 is 2.7 MJm^{-2}. Calculate (i) the hourly clearness index, and (ii) the hourly diffuse and direct beam components of the radiation.

Solution For May15, $n_{day} = 135$. The extraterrestrial solar radiation incident on a surface normal to the sun's rays is given by Eq. (2.4) as

$$E_o = 1367\left\{1 + 0.033cos\left[360^0 \frac{(135-3)}{365}\right]\right\} = 1337.9 \text{ Wm}^{-2}$$

The declination is given by Eq. (2.5) as

$$\delta = 23.45sin\left[\frac{360(135+284)}{365}\right] = 18.83 \text{ degrees}$$

The solar altitude angle is given by Eq. (2.9) as

$$sin\beta = cosLcosHcos\delta + sinLsin\delta$$

$$sin\beta = cos24\cos 37.5\, cos18.83 + sin24sin18.83 = 0.817$$

Therefore the solar altitude is 54.81 degrees.

The rate of incidence of extraterrestrial radiation on a horizontal surface per unit area is given by

$$q = E_o cos\theta = E_o sin\beta$$

$$q = 1337.9 \times 0.817 = 1093.1 \text{ Wm}^{-2}$$

The hourly extraterrestrial radiation is $1093.1 \times 3600 = 3.935$ MJm^{-2}. The hourly clearness index is given by

$$k_T = \frac{2.7}{3.935} = 0.686$$

We use the correlation given by Eq. (2.35)

$$\frac{I_d}{I_t} = 1.0 - 0.249k_T \quad \text{for} \quad k_T < 0.35$$

$$\frac{I_d}{I_t} = 1.557 - 1.84k_T \quad \text{for} \quad 0.35 < k_T < 0.75$$

$$\frac{I_d}{I_t} = 0.117 \quad \text{for} \quad k_T > 0.75$$

Hence diffuse radiation is given by the equation

$$\frac{I_d}{I_t} = 1.557 - 1.84k_T \quad \text{for} \quad 0.35 < k_T < 0.75$$

$$\frac{I_d}{I_t} = 1.557 - 1.84 \times 0.686 = 0.295$$

Therefore the hourly diffuse and beam radiation are 0.797 MJm^{-2} and 1.9 MJm^{-2} respectively.

Example 2.14 For a location with northern latitude of 40°, the measured total daily radiation on a horizontal surface on August 10 is 24.7 MJm^{-2}. Calculate (i) the daily average clearness index for the day, and (ii) the total diffuse and beam radiation for the day.

Solution For the same location and day we computed the total extraterrestrial radiation on a horizontal surface in example 2.7, as 37.53 MJm^{-2}. The total measured radiation for the day is 24.7 MJm^{-2}.

Therefore the daily average clearness index is

$$K_T = \frac{24.7}{37.53} = 0.658$$

To split the total daily radiation into its components we use the correlation in Eq. (2.38).

$$\frac{H_d}{H} = 1.188 - 2.272K_T + 9.473{K_T}^2 - 21.865{K_T}^3 + 14.648{K_T}^4$$

for $0.17 < K_T < 0.75$

Substituting $K_T = 0.658$ in the above equation we have

$$\frac{H_d}{H} = 1.188 - 1.4949 + 4.1015 - 6.2291 + 2.7458 = 0.3113$$

Hence the daily diffuse radiation is 7.69 MJm^{-2} and the beam radiation is 17.01 MJm^{-2}.

Example 2.15 At a location with a latitude of 30° N, the hourly total solar radiation on a horizontal surface on May 28, at 2 p.m. is 2.6 MJm^{-2}. Calculate (i) the clearness index for the hour, and (ii) the diffuse and beam components of radiation on a horizontal surface.

Solution The extraterrestrial solar radiation incident on a surface normal to the sun's rays on May 28 is given by Eq. (2.4) as

$$E_o = E_{sc}\left\{1 + 0.033cos\left[360^0\frac{(148-3)}{365}\right]\right\}$$

where E_{sc} = 1367 Wm^{-2} and n_d=148.

Hence extraterrestrial solar radiation intensity, E_o = 1330.9 Wm^{-2}. Using the middle of the hour, the hour angle is obtained from Eq. (2.6) as

$$H = 15(T_{sol} - 12) = 15 \times 2.5 = 37.5^o$$

For May 28, N_{day} = 148. The declination is given by Eq. (2.4) as

$$\delta = 23.45sin\left[\frac{360(284+N_d)}{365}\right] = 21.44^o$$

The solar altitude angle is given by Eq. (2.9) as

$$sin\beta = cosLcosHcos\delta + sinLsin\delta$$

$$sin\beta = cos30\cos(37.5)\,cos21.44 + sin30sin21.44 = 0.8223$$

Therefore the solar altitude is 55.3°.

The rate of incidence of extraterrestrial radiation on a horizontal surface per unit area is given by

$$q = E_o cos\theta = E_o sin\beta$$

$$q = 1330.9 \times 0.8223 = 1094.4 \text{ Wm}^{-2}$$

The hourly extraterrestrial radiation is $1094.4 \times 3600 = 3.94$ MJm^{-2}. The hourly clearness index is given by

$$k_T = \frac{2.6}{3.94} = 0.6598$$

To obtain diffuse and beam components we use the relevant correlation given by Eq. (2.35)

$$\frac{I_d}{I_t} = 1.557 - 1.84k_T \text{ for } 0.35 < k_T < 0.75$$

$$\frac{I_d}{I_t} = 1.557 - 1.84 \times 0.6598 = 0.343$$

Therefore the hourly diffuse and beam components of radiation are 0.892 MJm^{-2} and 1.708 MJm^{-2}.

Example 2.16 The total radiation measured on a horizontal surface in Toronto, Canada, on July 31 at 3.00 p.m. solar time is 1.915 MJm^{-2}. The latitude of Toronto is 43.63 degrees north. Estimate the diffuse and beam components of radiation using the total horizontal radiation obtained from the clear-sky model.

Solution The conditions for this example are the same as those in example 2.10 above. Using the clear-sky model we obtained the beam and diffuse radiation at 3 p.m. as 808 Wm^{-2} and 111 Wm^{-2}. The solar altitude angle is 44.55°. Hence the total clear-sky radiation intensity on a horizontal surface is given by Eq. (2.20) as:

$$I_{cs} = 808 \sin\beta + 111 = 677.8 \text{ Wm}^{-2}$$

The hourly clear-sky radiation is 2.44 MJm^{-2}.

Therefore

$$I_t/I_c = 1.915/2.44 = 0.784$$

To estimate the beam and diffuse components of the total measured radiation we use the relevant correlation in Eq. (2.36)

$$\frac{I_d}{I_t} = 1.11 + 0.0396\left(\frac{I_t}{I_c}\right) - 0.789\left(\frac{I_t}{I_c}\right)^2$$

$$\text{for} \qquad 0.48 \leq \frac{I_t}{I_c} < 1.1$$

Substituting numerical values in the above equation we have

$$\frac{I_d}{I_t} = 1.11 + 0.0396 \times 0.784 - 0.789 \times 0.784^2 = 0.656$$

Therefore the hourly diffuse and beam components of the measured radiation are 1.256 MJm^{-2} and 0.658 MJm^{-2}.

Example 2.17 The total radiation measured on a horizontal surface in Toronto, Canada, on July 31 at 3.00 p.m. solar time is 1.87 MJm^{-2}. The latitude of Toronto is 43.63 degrees north. Estimate the diffuse and beam components of radiation using the hourly clearness index.

Solution The conditions for this example are the same as those in worked example 2.10 above. The extraterrestrial radiation was computed as 1326.5 Wm^{-2}. The solar altitude angle was 44.55°. Hence the extraterrestrial radiation on a horizontal surface is

$$q = E_o sin\beta = 1326.5 sin 44.55 = 930.58 \text{ Wm}^{-2}$$

The hourly extraterrestrial radiation is 3.35 MJm^{-2}. The hourly clearness index is

$$k_T = \frac{1.87}{3.35} = 0.558$$

We use the relevant correlation in Eq. (2.35) to estimate the diffuse component of the hourly radiation. Hence we have

$$\frac{I_d}{I_t} = 1.557 - 1.84k_T \text{ for } 0.35 < k_T < 0.75$$

$$\frac{I_d}{I_t} = 1.557 - 1.84 \times 0.558 = 0.53$$

Therefore the hourly diffuse and beam components of the measured radiation are 1.0 MJm^{-2} and 0.87 MJm^{-2}.

Example 2.18 The daily average total radiation on a horizontal surface for the month of June at a location with a latitude of 30° N is 24 MJm^{-2}. (i) Calculate the monthly average clearness index for June. (ii) Estimate the monthly diffuse radiation fraction for June.

Solution The day of the month for which the daily extraterrestrial radiation is closest to the monthly average is defined as the average day of the month. The average days for different months are tabulated in Ref.[1]. For June the average day of the month, n_{ad} = 162.

The declination for the average day is:

$$\delta = 23.45 sin\left[\frac{360(284+n_{ad})}{365}\right] = 23.08$$

The sunrise angle is given by Eq. (2.9) as:

$$cosH_{sr} = -tanLtan\delta = -0.246$$

The sunset angle is 104.2°.

The extraterrestrial radiation for the average day is given by Eq. (2.4) as:

$$E_o = E_{sc}\left\{1 + 0.033cos\left[360^0\frac{(n-3)}{365}\right]\right\} = 1325.5 \text{ Wm}^{-2}$$

The total daily extraterrestrial radiation for the average day is obtained using Eq. (E2.7.1). Hence we have

$$\bar{G} = \left(\frac{24\times3600}{\pi}\right) \times 1325.5\left[cos30cos23.08sin104.2 + \frac{2\pi\times104.2}{360} sin30sin23.08\right] = 41.1 \text{ MJm}^{-2}$$

The monthly average clearness index is given by

$$\bar{K}_T = \frac{24}{41.1} = 0.584$$

We use the correlation in Eq. (2.39) to estimate the monthly diffuse fraction. Hence we get

$$\frac{\bar{H}_d}{\bar{H}} = 0.775 + 0.00653(\omega_s - 90) - [0.505 + 0.00455(\omega_s - 90)]cos[115\bar{K}_T - 103]$$

$$\frac{\bar{H}_d}{\bar{H}} = 0.775 + 0.00653(104.2 - 90) - [0.505 + 0.00455(104.2 - 90)]cos[115 \times 0.584 - 103]$$

Therefore the monthly average diffuse radiation fraction is 0.406.

Example 2.19 The daily total radiation received on a horizontal surface on 28 May, at a location with a latitude of 36° N is 29.8 MJm^{-2}. Estimate the total radiation received from 2 to 3 p.m. on day.

Solution The hour angle at the middle of the hour is 37.5°. The declination on May 28 is 21.44°.

The sunrise angle is given by Eq. (2.9) as

$$cosH_{sr} = -tanLtan\delta = -0.2853$$

Hence the sunrise angle is 106.58°.

To estimate the hourly total radiation we use the correlation in Eq. (2.37). Substituting we get

$$a = 0.409 + 0.5016sin(\omega_s - 60) = 0.7733$$

$$b = 0.6609 - 0.4767sin(\omega_s - 60) = 0.3147$$

$$r = \frac{2\pi\omega_s}{360} = \frac{2\pi \times 106.58}{360} = 1.86$$

$$\frac{I_t}{H} = \frac{\pi}{24}(0.7733 + 0.3147 \times cos37.5)\frac{(cos37.5 - cos106.58)}{sin106.58 - 1.86 \times cos106.58} = 0.097$$

Hence the total hourly radiation from 2 to 3 p.m. is 2.89 MJm^{-2}.

Example 2.20 A flat-plate collector facing 30° east of south, is inclined at 50° to the horizontal. The latitude of the location is 38° north. At 2.30 p.m. solar time on August 5, calculate (i) the angle of incidence of the direct beam on the collector, and (ii) the tilt factor of the collector.

Solution For August 5, n_d = 217. The declination is given by Eq. (2.5) as

$$\delta = 23.45 sin\left[\frac{360(217+284)}{365}\right] = 16.83 \text{ degrees}$$

Taking the middle of the hour, the hour angle is 37.5°.

The solar altitude angle is given by Eq. (2.10) as

$$sin\beta = cosLcosHcos\delta + sinLsin\delta$$

$$sin\beta = cos38 \cos 37.5\, cos16.83 + sin38sin16.83 = 0.772$$

Therefore the solar altitude is 50.53 degrees.

The solar azimuth angle is given by Eq. (2.8) as

$$sin\phi = \frac{cos\delta sinH}{cos\beta} = \frac{cos16.83 \times \sin 37.5}{cos50.53} = 0.9166$$

Therefore solar azimuth is 66.43 degrees west of south.

The surface-solar azimuth is

$$\gamma = 30 + 66.43 = 96.43 \text{ degrees.}$$

The tilt angle of the collector is, Σ = 50°. The angle of incidence, θ is given by Eq. (2.14) as

$$cos\theta = cos\beta sin\Sigma cos\gamma + sin\beta cos\Sigma$$

$$cos\theta = cos50.53sin50cos96.43 + sin50.53cos50 = 0.4417$$

Hence the angle of incidence on the collector surface is 63.78°.

The hourly tilt factor is given by Eq. (2.40) as

$$R_{b,\theta} = \frac{I_{b,\theta}}{I_{b,h}} = \frac{cos\theta}{sin\beta} = \frac{0.4417}{0.772} = 0.572$$

Problems

P2.1 A solar collector emits long wave radiation at 70°C. Assume it to be a black body. (i) Find the wavelength at which the power per unit wavelength is a maximum. (ii) Calculate the total energy radiated. (iii) What fraction of emitted radiation is between the 5 μm and 10 μm? [*Answers*: (i) 8.45 μm, (ii) 784.8 Wm^{-2}, (iii) 265.3 Wm^{-2}]

P2.2 The longitude and the latitude of a location are 93°W and 35°N respectively. The standard longitude of the time zone is 90°W. When the standard clock time is 11.30 a.m. on August 10, calculate (i) the solar time, (ii) the hour angle, (iii) the solar altitude angle, and (iv) the solar azimuth angle at the location. The clock time is advanced by one hour for daylight saving.
[*Answers*: (i) 10.21 a.m., (ii) -26.85°, (iii) 58.9°, (iv) -57.48°]

P2.3 A particular location has a northern latitude of 35°. (a) Calculate the solar altitude angle and the solar azimuth angle at 2 p.m. solar time on January 15. (b) Calculate the solar time at sunrise and sunset on January 15 at the same location.
[*Answers*: (a) 26.9°, 31.5°, (b) 7.03 a.m., 4.57 p.m.]

P2.4 A solar collector facing south-east is tilted at 60° from the horizontal. The latitude of the location is 42°N. Calculate (i) the angle of incidence of the direct beam on the collector at 1.00 p.m. solar time on September 8, and (ii) the tilt factor. [*Answers*: (i) 54.3°, (ii) 0.75]

P2.5 Calculate the clear-sky beam and diffuse radiation at 1.00 p.m. on September 8 at a location with a latitude of 42°N. For the location, τ_b = 0.402 and τ_d = 2.326. [*Answers*: 835 Wm^{-2}, 115 Wm^{-2}]

P2.6 Calculate the total daily extraterrestrial radiation on a horizontal surface on the March 8, at a location with a latitude of 35°N.
[*Answer*: 34.2 MJm^{-2}]

P2.7 The average hourly radiation on a horizontal surface at 10 a.m. on March 20 at a location with latitude of 30°N is 2.1 MJm^{-2}. Estimate the diffuse and beam components of radiation.
[*Answers*: 1.07 MJm^{-2}, 1.03 MJm^{-2}]

P2.8 The measured daily total radiation on a horizontal surface at a location with a latitude of 35°N on March 8 is 22.93 MJm^{-2} (see P2.6). Calculate (i) the daily average clearness index for the day, and (ii) the daily total diffuse and beam components of the radiation.
[*Answers*: (i) 0.67, (ii) 6.72 MJm^{-2}, 16.21 MJm^{-2}]

P2.9 The daily total radiation received on a horizontal surface on 3 March, at a location with a latitude of 28°N, is 25.8 MJm^{-2}. Estimate the total radiation received from 1 to 2 p.m. on the same day.
[*Answer*: 3.38 MJm^{-2}]

P2.10 A solar collector, inclined at 35° to the horizontal, faces 20° east of south. The latitude of the location is 40°N. At 11.00 a.m. solar time on June 5, calculate (i) the angle of incidence of the direct beam on the collector, and (ii) the tilt factor of the collector.
[*Answers*: (i) 16.4°, (ii) 1.03]

References

1. Duffie, John A, Beckman, William A, and Blair Nathan, *Solar Engineering of Thermal Processes, Photovoltaics and Wind,* John Wiley & Sons, Inc., New Jersey, 2020.
2. Goswami, Y., Kreith Frank and Kreider Jan F. K., *Principles of Solar Energy Engineering*, 2nd Edition, Taylor & Francis Group, Philadelphia, 2000.
3. Gueymard, Christian A and Thevenard Didier, 'Monthly average clear-sky broadband irradiance database for worldwide solar heat gain and building cooling load calculations', *Solar Energy*, 83(2009), 1998-2018.
4. Kreith Frank and Krumdieck Susan, *Principles of Sustainable Energy Systems*, CRC Press, Taylor & Francis Group, New York, 2015.
5. Siegel, Robert and Howell, John R, *Thermal Radiation Heat Transfer*, Hemisphere Publishing Corporation, Washington, 1992.

Chapter 3

Solar Process Heat Production

3.1 Introduction

One of the most direct methods of using solar energy for practical purposes is the conversion of solar radiation incident on a surface to heat. Once converted heat energy has numerous applications, such as, domestic hot water heating, space heating, crop drying, industrial process heating, and others. Depending on the solar collection device and the end-use application, the heat delivered can be at different temperatures, the limiting value being the temperature of the sun.

For domestic water heating and crop drying, the required temperature is about 50°C to 60°C, while industrial process heat applications may require temperatures in excess of 100°C. Solar air conditioners working on the absorption cycle can be operated with heat supplied at about 80°C. Solar thermal energy is also being used for electrical energy production using Rankine cycle power plants. In this case the required heat delivery temperatures are about 300°C to 700°C.

The different applications of solar thermal energy mentioned above have been deployed on a commercial scale with varying degrees of success. The most well-established use of solar heat is for domestic hot water production. The solar collectors used for this purpose are relatively simple devices that are usually mounted on roof tops of homes and buildings. They can make use of both the direct beam and the diffuse component of solar radiation. On the other hand, solar thermal power systems, that require concentrators to achieve the desired high temperatures can only convert beam radiation to useful heat.

The main component of a low-temperature solar heating system is the flat-plate collector. When higher temperatures are required, as for absorption-cycle solar air conditioning, a more advanced device called the evacuated-tube collector may be used. These collectors are stationary and therefore their output is effected by the motion of the sun during the day. Collectors used in solar thermal power systems usually employ concentrators located on flexible mountings that enable the device to be moved to follow the sun's motion during the day.

In this chapter we shall consider the basic principles underlying the design and operation of solar process heat systems. Solar thermal power generation will be introduced in the next chapter.

3.1.1 *Generic solar thermal system*

In the typical arrangement of a solar thermal system, shown in Fig. 3.1, the solar collector transfers heat to the storage unit. Under favorable conditions, the end-use application that is either a process requiring heat, or a power plant operating on the Rankine Cycle, will receive heat from the storage unit.

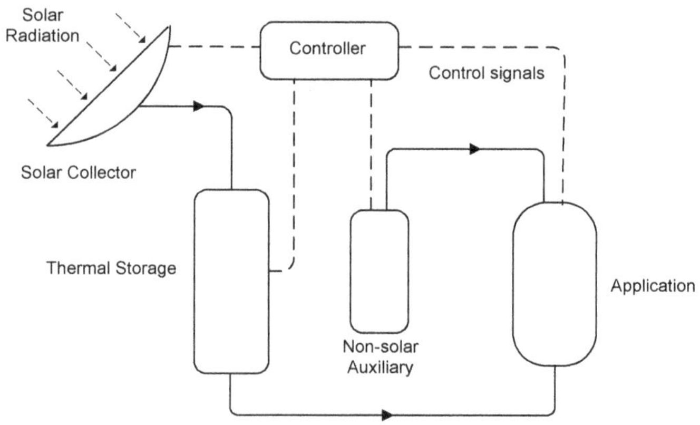

Fig. 3.1 Generic arrangement of a solar thermal system

When heat is not available from the storage unit at the desired temperature, an auxiliary energy source such as a fossil fuel heating unit will supply the required heat to the application. The energy flow paths will be decided by the controller which receives inputs from the collector, the storage unit, and the application. When the solar radiation intensity is not adequate for energy collection to be possible, the collector will be isolated from the system to prevent energy loss to the ambient from the circulating fluid. The auxiliary heating unit will be turned on at this stage.

In this chapter we shall discuss the function of each of the components of the system depicted in Fig. 3.1 and derive the equations necessary to design such components. A few typical solar process heat systems will also be presented.

3.2 Solar Collectors

In this section two basic types of solar collectors are considered. These are flat-plate collectors and evacuated-tube collectors.

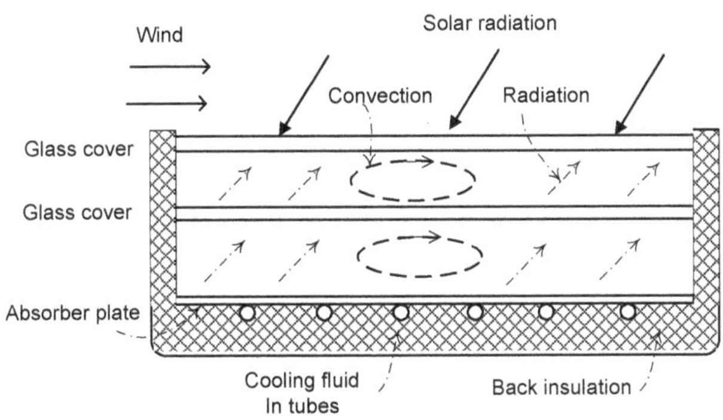

Fig. 3.2 Schematic of a basic flat-plate collector

3.2.1 *Flat-plate collectors*

The solar collector is a key component of the solar thermal system depicted in Fig. 3.1. A typical solar collector is shown schematically in Fig. 3.2. It consists of a metal plate on which solar radiation is incident, and as a result the plate temperature rises gradually. A cooling fluid flowing under the plate removes heat from the plate either by direct contact or through tubes attached to it. In the latter case fluid is distributed to the tubes from the inlet header, and the hot fluid leaving the tubes is collected in the outlet header. A few alternative designs of absorbers are depicted schematically in Fig. 3.5. As the temperature of the absorber rises it tends to loose heat to the ambient by convection and thermal radiation. To decrease this heat loss a cover system is installed above the absorber plate.

The cover system must be transparent to let solar radiation to pass through it to reach the absorber. A typical cover system consists of one or two sheets of glass or plastic. Glass has the desirable radiative property that it allows short-wave solar radiation to pass through with minimal absorption, but absorbs long-wave heat radiation emitted by the absorber plate. A familiar example of this type of radiative behavior is the roof of a green house used for growing plants. The glass roof lets in solar radiation but prevents long wave radiation emitted by the plants and other items in the green house from escaping. Thus the glass roof helps maintain a high temperature inside even when the ambient outside is relatively cold.

In the multi-cover system shown in Fig. 3.2, the air spaces between the glass plates help to decrease the convective heat loss from the absorber plate to the ambient. Many novel design improvements have been incorporated in cover systems and absorbers to minimize the loss of heat to the ambient, and simultaneously increase the solar radiation transmitted and absorbed by the plate. Some of these are: (i) selective paints applied on the absorber to increase the absorption of solar radiation and at the same time decrease the emission of long-wave heat radiation, (ii) coatings applied on transparent covers to decrease the reflection of incident solar radiation, and (iii) improved fin designs that enhance the heat transfer from the absorber to the cooling fluid (see Fig. 3.5).

3.2.2 *Efficiency of a solar collector*

The overall energy balance for a solar collector may be obtained by applying the first law of thermodynamics. This may be written in the general form:

$$Q_u = A_c I_s \bar{\tau} \alpha_s - Q_l - \frac{dE_c}{dt} \tag{3.1}$$

where

Q_u = rate of useful energy collection, which is the rate of heat flow from the absorber plate to the cooling fluid

A_c = area of the collector receiving solar radiation

I_s = solar radiation incident on the collector, per unit area

$\bar{\tau}$ = effective solar transmittance of the cover system

α_s = solar radiation absorptance of the absorber plate

Q_l = rate of heat loss from the collector to the surroundings

$\frac{dE_c}{dt}$ = rate of internal energy storage of the collector

The instantaneous efficiency of a solar collector is defined as the rate of useful energy delivered to the total incident solar radiation, which can be expressed in the form

$$\eta_c = \frac{Q_u}{A_c I_s} \tag{3.2}$$

For practical purposes, the efficiency of the collector must be computed over a finite period. In standard test methods a time interval of 15 to 20 minutes is used [1]. However, for design applications, the efficiency is measured over a longer period of about a day. Then the average efficiency is expressed in the form

$$\eta_c = \frac{\int_0^t Q_u dt}{\int_0^t A_c I_s dt} \tag{3.3}$$

where t is the time interval over which the efficiency is averaged.

3.3 Thermal Analysis of Flat-plate Collectors

In this section we shall develop the basic design equation for a flat-plate collector. Although there are a wide variety of flat-plate solar collectors, the governing equation that is most useful for design purposes has a general form that can be applied to most collectors by suitably modifying some of the design parameters. The thermal analysis of a flat-plate collector consist of three aspects. These are: (i) solar radiation transmission through the covers and absorption by the plate, (ii) heat loss through the cover system, and (iii) useful heat removal from the absorber plate. These energy flows can be represented by the thermal network shown in Fig. 3.3. We shall consider each of these aspects in the following sections.

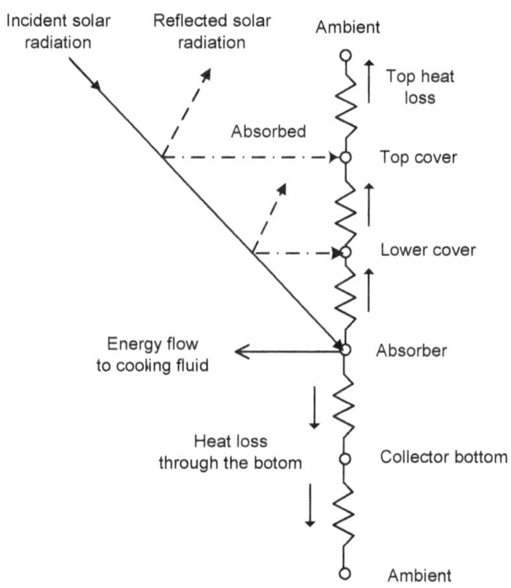

Fig. 3.3 Thermal network for a two-cover collector. The component temperatures and thermal resistances between them are shown

3.3.1 *Absorption of solar radiation*

The determination of the absorption of solar radiation by the absorber plate requires the detailed analysis of solar radiation transfer through semi-transparent layers. In view of the many algebraic equations involved we have included the analysis in Appendix A1. The final expression that needs to be used to calculate the transmissivity of the cover system is relatively compact, which we shall use below. However, we need to include the effects of the absorber plate in the analysis.

In the schematic diagram shown in Fig. 3.4 the covers are considered as a single unit whose overall transmittance T and reflectivity R are known. The expressions needed to determine these quantities for different number of covers are available in Appendix A1. The net-radiation fluxes at the different surfaces are indicated in Fig. 3.4. We write the following net-radiation balance equations at the different surfaces

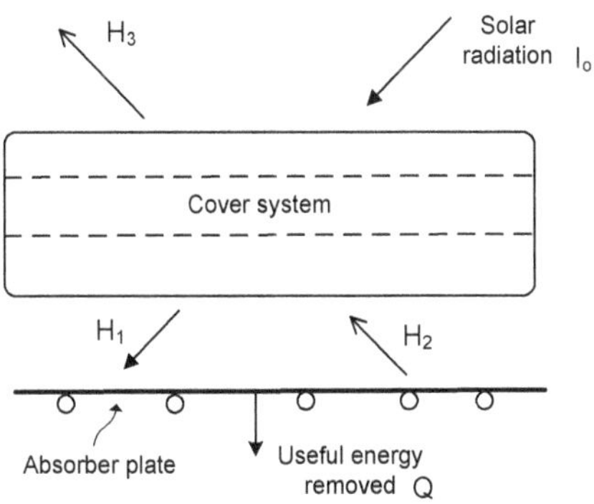

Fig. 3.4 Net radiation fluxes at the surfaces

At the bottom surface of the cover system:

$$H_1 = I_o T + H_2 R \tag{3.4}$$

At the surface of the absorber plate:

$$H_2 = H_1 r_a = H_1(1 - \alpha_a) \tag{3.5}$$

where r_a and α_a are the reflectivity and absorptivity of the absorber plate.

Energy balance for the absorber gives

$$Q = \alpha_a H_1 \tag{3.6}$$

Manipulating Eqs. (3.4)–(3.6) we obtain the absorbed solar energy as

$$Q = \frac{T \alpha_a I_o}{[1 - R(1 - \alpha_a)]} \tag{3.7}$$

In Ref.[1] it is stated that for practical flat-plate collectors the above expression for solar radiation absorption is well approximated by the expression

$$Q_a = 1.01T\alpha_a I_o \tag{3.8}$$

3.3.2 *Heat loss through the cover system*

For the two-cover system shown in Fig. 3.2, the heat loss to the ambient consists of a thermal radiation component and a convective heat transfer component. The former is given by a relation that involves the fourth power of the absolute temperature, while the latter is directly proportional to the temperature difference. For purposes of illustration, consider the radiative exchange between the top surface of the lower cover and the bottom surface of the top cover. Assuming that the surfaces are gray we can write the following expression for the net radiation heat transfer between the surfaces:

$$Q_{r21} = A_c\sigma\epsilon_{e12}\left(T_{g2}{}^4 - T_{g1}{}^4\right) \tag{3.9}$$

where the effective emissivity is given by the expression [1,3]

$$\frac{1}{\epsilon_{e12}} = \frac{1}{\epsilon_{g1}} + \frac{1}{\epsilon_{g2}} - 1 \tag{3.10}$$

The absolute temperatures of the two glass covers are respectively T_{g1} and T_{g2}, and their emissivities are ϵ_{g1} and ϵ_{g2}.

The difference between the absolute temperatures of the covers is generally small. Therefore we could linearized the RHS of Eq. (3.9) to obtain the approximate form

$$Q_{r21} = A_c h_{r12}\left(T_{g2} - T_{g1}\right) \tag{3.11}$$

where the radiation heat transfer coefficient is defined as [1]

$$h_{r12} = \sigma\epsilon_{e12}\left(T_{g1}{}^2 + T_{g2}{}^2\right)\left(T_{g1} + T_{g2}\right) \tag{3.12}$$

The convective heat transfer between the lower and top covers is given by

$$Q_{c21} = A_c h_{c12} \left(T_{g2} - T_{g1} \right) \tag{3.13}$$

where h_{c12} is the convective heat transfer coefficient for the air gap.

The total heat transfer between the surfaces is

$$Q_{t21} = Q_{r21} + Q_{c21} \tag{3.14}$$

Substituting from Eqs. (3.11) and (3.13) in Eq. (3.14) we have

$$Q_{t21} = A_c h_{t21} \left(T_{g2} - T_{g1} \right) \tag{3.15}$$

where the total heat transfer coefficient is given by

$$h_{t21} = h_{r12} + h_{c12} \tag{3.16}$$

The expressions for heat transfer from the absorber plate to the bottom cover are summarized below.

$$Q_{rc2} = A_c \sigma \epsilon_{ec2} \left(T_c{}^4 - T_{g2}{}^4 \right) \tag{3.17}$$

where the effective emissivity is given by the expression

$$\frac{1}{\epsilon_{ec2}} = \frac{1}{\epsilon_c} + \frac{1}{\epsilon_{g2}} - 1 \tag{3.18}$$

The absolute temperature and emissivity of the absorber plate are T_c and ϵ_c respectively.

The linearized form of the RHS of Eq. (3.17) is

$$Q_{rc2} = A_c h_{rc2} \left(T_c - T_{g2} \right) \tag{3.19}$$

where the radiation heat transfer coefficient is defined as

$$h_{rc2} = \sigma \epsilon_{ec2} \left(T_c{}^2 + T_{g2}{}^2 \right) \left(T_c + T_{g2} \right) \tag{3.20}$$

The convective heat transfer between the absorber and the bottom cover is given by

$$Q_{cc2} = A_c h_{cc2} \left(T_c - T_{g2} \right) \tag{3.21}$$

The total heat transfer between the absorber and the bottom cover is

$$Q_{tc2} = Q_{rc2} + Q_{cc2} \tag{3.22}$$

Substituting from Eqs. (3.19) and (3.21) in Eq. (3.22) we have

$$Q_{tc2} = A_c h_{tc2}(T_c - T_{g2}) \tag{3.23}$$

where the total heat transfer coefficient is given by

$$h_{tc2} = h_{rc2} + h_{cc2} \tag{3.24}$$

The total heat loss to the surroundings from the top cover surface can be obtained in a similar manner. The cover looses heat by radiation to the sky that is taken as a large black surface, at a temperature T_{sky}. Therefore

$$Q_{r1s} = A_c \sigma \epsilon_{g1}(T_1{}^4 - T_{sky}{}^4) \tag{3.25}$$

where the effective emissivity is that of the cover surface because the sky is assumed black.

The linearized form of the RHS of Eq. (3.25) is

$$Q_{r1s} = A_c h_{r1s}(T_1 - T_{sky}) \tag{3.26}$$

where the radiation heat transfer coefficient is defined as

$$h_{r1s} = \sigma \epsilon_{g1}(T_{sky}{}^2 + T_{g1}{}^2)(T_{sky} + T_{g1}) \tag{3.27}$$

The convective heat transfer from the top cover to the ambient is given by

$$Q_{c1a} = A_c h_{c1a}(T_{g1} - T_a) \tag{3.28}$$

where T_a is the ambient temperature.

The convective heat transfer coefficient from the top surface to the ambient is a function of the wind speed. Correlations to compute this quantity are available in most standard text books on solar energy [1,3]. A few of these correlations are included in Appendix A2.

The total heat loss from the top cover to the ambient is

$$Q_{t1a} = Q_{r1s} + Q_{c1a} \tag{3.29}$$

For practical purposes we could assume that the sky temperature and the ambient air temperature are equal. Then the total heat transfer from the top cover to the ambient may be expressed as

$$Q_{t1a} = A_c h_{t1a}\left(T_{g1} - T_a\right) \tag{3.30}$$

where the total heat transfer coefficient is given by

$$h_{t1a} = h_{r1a} + h_{c1a} \tag{3.31}$$

Under steady conditions of operation, the above individual heat transfer rates are equal as seen in the thermal network in Fig. 3.3. Therefore

$$Q_{tc2} = Q_{t21} = Q_{t1a} = Q_l \tag{3.32}$$

Substituting from Eqs. (3.15), (3.23) and (3.29) in Eq. (3.32) we obtain

$$Q_l = A_c U_l (T_c - T_a) \tag{3.33}$$

where the overall heat transfer coefficient is given by

$$\frac{1}{U_l} = \frac{1}{h_{tc2}} + \frac{1}{h_{t21}} + \frac{1}{h_{t1a}} \tag{3.34}$$

Equation (3.34) gives the overall thermal resistance between the absorber and the ambient, which is the sum of the individual thermal resistances shown in Fig. 3.3.

Computation of the overall heat loss coefficient using the above procedure is illustrated in worked examples (3.8) to (3.11). It is somewhat tedious because of the number of variables involved.

Several empirical equations have been developed [1] to estimate U_l and thereby simplify the computation. Here we include the relation introduced by Klein [1] for U_l ($Wm^{-2}C^{-1}$).

$$U_l = \left[\frac{N}{\frac{C}{T_p}\left(\frac{T_p - T_a}{N+f}\right)^e} + \frac{1}{h_w}\right]^{-1} + \sigma\left(T_p^{\ 2} + T_a^{\ 2}\right)\left(T_p + T_a\right)\left[\frac{1}{\varepsilon_p + 0.00591 N h_w} + \frac{2N + f - 1 + 0.133\varepsilon_p}{\varepsilon_g} - N\right]^{-1} \qquad (3.35)$$

$$f = \left(1 + 0.089h_w - 0.1166h_w\varepsilon_p\right)(1 + 0.07866N)$$

$$C = 520(1 - 0.000051s^2), \qquad e = 0.43\left(1 - \frac{100}{T_p}\right)$$

s = collector tilt (degrees), ε_g = emittance of glass (0.88),
ε_p = emittance of plate, N = number of glass covers
T_a = ambient temperature (K), T_p = mean plate temperature (K),
h_w = wind heat transfer coefficient.

It is convenient to calculate the loss coefficient using Eq. (3.35), but it is necessary to assume the mean plate temperature. The application of Eq. (3.35) is illustrated in worked examples 3.12 and 3.13.

3.3.3 *Useful energy collection*

We now analyze the heat transfer process from the absorber plate to the cooling fluid, which gives the useful energy collection rate. There are a wide variety of absorber plate designs, a few of which are shown schematically in Fig. 3.5.

Figure 3.5(a) is a simple design where a corrugated sheet is welded to a flat sheet to form a series of parallel flow passages. Solar radiation is incident on the top sheet. In the designs in Fig. 3.5(b) and (c) tubes are bonded to flat sheets. A roll bond design with oval shaped flow channels is shown in Fig. 3.5(d). A plate-fin design of an absorber is depicted in Fig. 3.5(e).

Most of these designs have been analyzed and expressions for important design parameters are available in the form of charts and tables in standard textbooks on solar energy [1,3]. The analysis is straightforward in principle but is often tedious because of the complex heat flow paths involved. In view of this situation, we shall consider a simple absorber design often used in air-cooled collectors.

The analysis in this case is straightforward, and it brings out all the important design parameters, which usually involve complex algebraic expressions for most of the designs shown in Fig. 3.5.

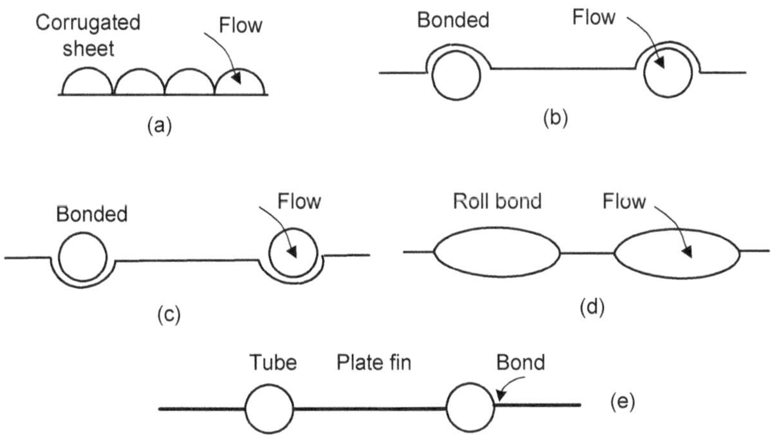

Fig. 3.5 Absorber designs

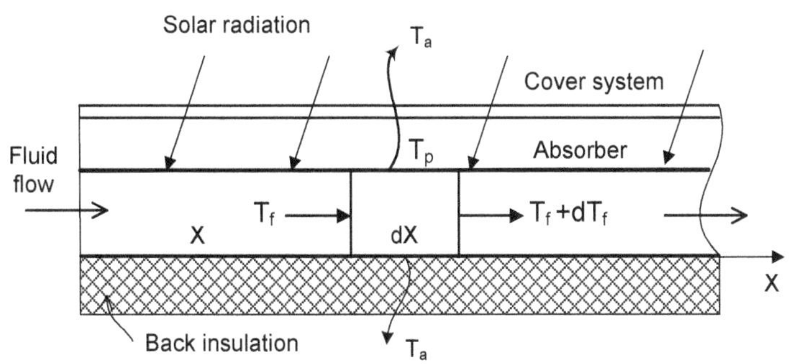

Fig. 3.6 Analysis of a solar air heater

The simple solar air heater shown in Fig. 3.6 has a rectangular flow channel for air flow, and the absorber plate forms the top surface of the channel. The length and width of the channel are L and W respectively, and the mass flow rate of air is m_f.

Consider an elemental control volume of length dx at a distance x from the entrance of the channel. The air receives heat by forced convection from absorber plate, and its temperature increases from T_f to $(T_f + dT_f)$ as it passes through the control volume. The heat loss from the air to the surroundings across the back of the flow channel is neglected. We assume steady conditions of operation, and write the following energy balance equations.

$$m_f c_p(T_f + dT_f) - m_f c_p T_f = W h_i(T_p - T_f)dx$$

The term on the RHS of the above equation is the rate of heat transfer by forced convection from the plate to the air, where h_i is the convective heat transfer coefficient.

Hence
$$m_f c_p \frac{dT_f}{dx} = W h_i(T_p - T_f) \tag{3.36}$$

where c_p is the specific heat capacity of air.

Assuming that the absorber plate is thin and that heat conduction along the plate is negligible, we can write the following energy balance equation for a section of length *dx*

$$I_s(\tau\alpha)W dx = U_l(T_p - T_a)W dx + h_i(T_p - T_f)W dx \tag{3.37}$$

The term on the LHS of Eq. (3.37) is the rate of absorption of solar radiation. The first term on the RHS is the rate of heat loss to the ambient from the plate and the second term is the convective heat transfer rate to the air from the plate.

From Eqs. (3.36) and (3.37) we obtain

$$mc_p \frac{dT_f}{dx} = W\left(\frac{h_i U_l}{h_i + U_l}\right)\left[\frac{(\tau\alpha)I_s}{U_l} + T_a - T_f\right] \tag{3.38}$$

The solution of the first order differential equation (3.38) gives the following expression involving the outlet temperature of air T_{fo}:

$$\left[\frac{\frac{\tau\alpha I_s}{U_l} + T_a - T_{fo}}{\frac{\tau\alpha I_s}{U_l} + T_a - T_{fi}}\right] = Exp\left[\frac{-WLh_i U_l}{mc_p(h_i + U_l)}\right] \tag{3.39}$$

where T_{fi} is the inlet air temperature and L is the length of the flow channel.

A generalized design parameter that is applicable to different absorbers is the *collector efficiency factor*, F' [1,3]. It is defined as the ratio of the thermal resistance from the plate to the ambient, $1/U_l$ to the thermal resistance from the fluid to the ambient $1/U_o$.

For the simple design considered here, we have

$$F' = \frac{1/U_l}{1/U_o} = \frac{1/U_l}{1/U_l+1/h_i} = \frac{h_i}{U_l+h_i} \tag{3.40}$$

Substituting from Eq. (3.40) in Eq. (3.39) we obtain the equation

$$\left[\frac{\frac{\tau\alpha I_s}{U_l}+T_a-T_{fo}}{\frac{\tau\alpha I_s}{U_l}+T_a-T_{fi}}\right] = \mathrm{Exp}\left[\frac{-A_c U_l F'}{mc_p}\right] \tag{3.41}$$

where $A_c = WL$, is the area of the collector.

Another generalized design parameter is the *collector heat removal factor*, F_R. It is the ratio of the actual heat removal rate to the maximum possible heat removal rate. This parameter is like the effectiveness used in heat exchanger design. The maximum useful heat removal is defined as the heat removal rate if the entire collector plate is at the temperature of the inlet fluid. For this idealized condition, the heat loss to the ambient is a minimum. Hence we have

$$\frac{Q_u}{Q_{max}} = F_R = \frac{mc_p\left(T_{fo}-T_{fi}\right)}{A_c\left[\tau\alpha I_s-U_l\left(T_{fi}-T_a\right)\right]} \tag{3.42}$$

The useful heat removal rate may be expressed in the form

$$Q_u = F_R A_c\left[\tau\alpha I_s - U_l\left(T_{fi} - T_a\right)\right] \tag{3.43}$$

Manipulating Eqs. (3.41), (3.42) and (3.43) we obtain the following expression for F_R:

$$F_R = \left(\frac{mc_p}{A_c U_l}\right)\left[1 - Exp\left(\frac{-A_c U_l F'}{mc_p}\right)\right] \tag{3.44}$$

Equations (3.43) and (3.44) are two of the most important equations used to design flat-plate solar collectors. Expressions for F' have been obtained for different absorber plate designs [1,3], some of which are depicted in Fig. (3.5).

Substituting from Eq. (3.43) in Eq. (3.2) we get the instantaneous efficiency of the collector as

$$\eta_c = F_R(\tau\alpha) - \frac{F_R U_l (T_{fi} - T_a)}{I_s} \tag{3.45}$$

The variation of collector efficiency with the input parameter, $(T_{fi} - T_a)/I_s$, for a few typical flat-plate air heating collectors is shown in Fig. 3.7. These collectors have different numbers of covers as indicated.

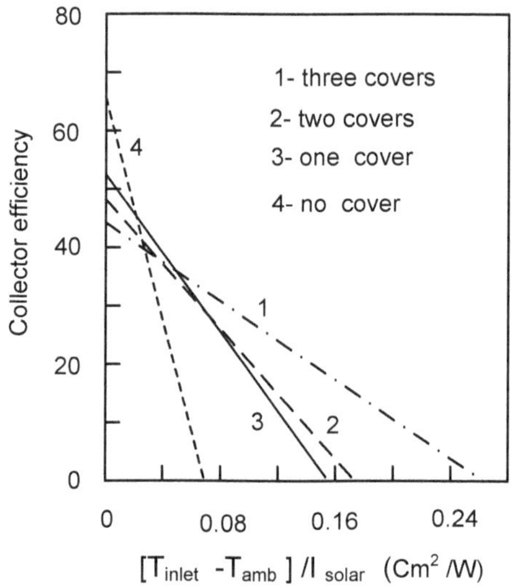

Fig. 3.7 Efficiency of solar collectors

In general U_l is a function of the inlet fluid temperature and F_R is a function of U_l. Hence Eq. (3.45), which is plotted in Fig. 3.7 for different designs of air heaters, does not represent a straight line. However, if we assume that U_l and F_R are nearly constant, then the slope of the linear function represented by Eq. (3.45) is $F_R U_l$ and the intercept on the vertical axis is $F_R(\tau\alpha)$. When the collector has no covers (4), $F_R(\tau\alpha)$ and $F_R U_l$ are both large as expected. As more covers are added, both the solar radiation transmitted and the heat loss from the absorber are decreased. This is reflected in the decrease of the intercept and the slope of the graphs in Fig. 3.7.

3.4 Evacuated Tube Collectors

There are two ways to improve the performance of flat-plate collectors significantly. The first method is to increase the solar radiation incident on the absorber, which will now have a smaller area. This can be achieved by using concentrators.

The second method is to decrease the heat loss from the absorber to the ambient. This can be realized by replacing air in the space between covers with an high molecular weight gas such as Argon or Krypton. This technique has been used widely to reduce the heat transfer through double-glazed windows in buildings, but its application in solar collectors has been limited.

The third and more common way to reduce the heat loss to the ambient is to decrease the pressure in the air space surrounding the absorber. This is difficult to achieve with flat glass covers, described in the previous section, because the flat glass surface may not withstand the large pressure difference across it. However, tubular glass covers offer an excellent opportunity to implement this method because of the inherently high compressive strength of the glass tube. In such collectors, the absorber is placed centrally inside the cover and the space inside the transparent tube is evacuated. These collectors are called *evacuated tube collectors* or *concentric-tube collectors* (CTC).

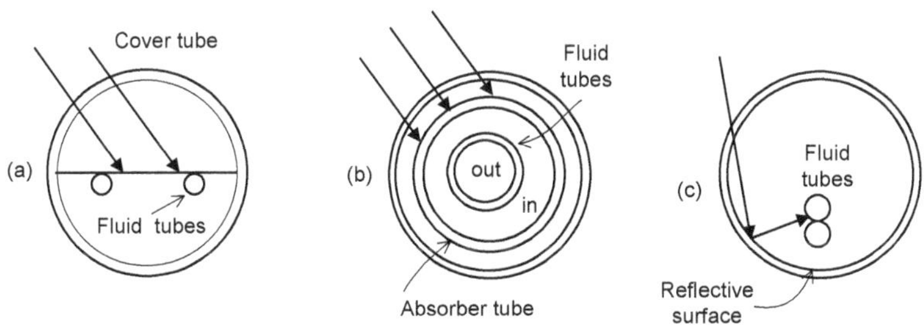

Fig. 3.8 Concentric-tube-collectors (CTC)

Three designs of evacuated tube collectors that have been built and tested are shown schematically in Figs. 3.8(a), (b) and (c). The collector shown in Fig. 3.8(a) has a flat absorber plate, located centrally inside the glass tube, with the heat removing fluid flowing in a tube attached to the plate. The space inside the tube is evacuated. This arrangement requires a metal to glass seal at one end, with may provide a thermal bridge for heat leaks.

The second design shown in Fig. 3.8(b), is an all-glass tubular design, which avoids the need for a metal-to-glass seal. In the commercially-available collector [3] of this type, the heat removing fluid enters through the annulus between the absorber tube and the inner fluid delivery tube. The flow direction reverses at the end of the tube and hot fluid leaves through the inner tube. The operating pressures are limited by the all-glass construction of the collector.

The third configuration depicted in Fig. 3.8(c) is similar to the design in Fig. 3.8(a). It has a vacuum-deposited reflecting surface on the bottom half of the cover tube. The absorber located inside the cover receives solar radiation directly as well as through reflection by the circular mirror as indicated in Fig. 3.8(c).

The main aim of the tubular collector design is the suppression of convection and conduction in the enclosed space between the absorber and the outer glass cover. This is achieved by reducing the air pressure in the enclosed space. The effect of pressure on the heat transfer in enclosed spaces has been investigated extensively [3]. It occurs in two identifiable stages depending on the Rayleigh number.

Initially, the effect of convection decreases at a rate proportional to the square root of the air density. Beyond the lower threshold for convection, thermal conduction becomes the only mode of heat loss. At this stage, the thermal conductivity of air becomes independent of pressure if the mean free path, λ is larger than the conduction path length, d. When the pressure in the narrow air gap is very low, the heat transfer rate is given by

$$q = \frac{k\Delta T}{d+2\lambda} \tag{3.46}$$

At ambient pressure, λ is about 70 nm and at a low pressure of 10^{-3} torr its value is 7 cm, which is much larger than d. At this low pressure conduction, which occurs by molecular collisions and diffusion, is suppressed. From Eq. (3.46) the ratio of the heat transfer rate to the conduction heat transfer rate is obtained as

$$\frac{q}{(k\Delta T/d)} = \frac{1}{1+2\lambda/d} \tag{3.47}$$

With modern vacuum technology low pressures on the order of 10^{-3} to 10^{-4} torr can be achieved.

The analysis of the tubular collectors shown in Fig. 3.8(a) and (c) may be performed using the methods developed for flat-plate collectors in section 3.3. The analysis of the all-glass collector, shown in Fig. 3.8(b) is outlined below.

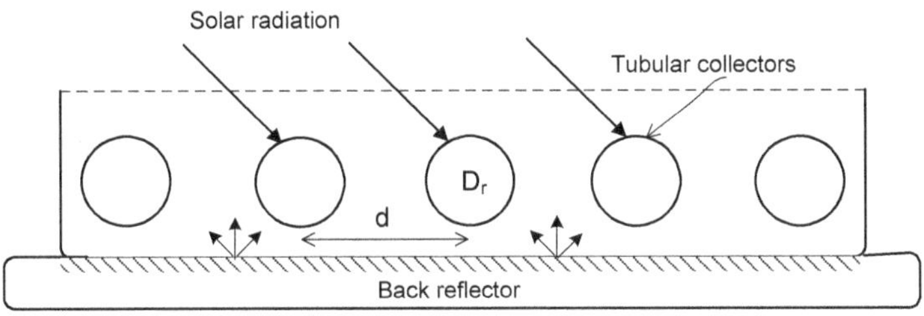

Fig. 3.9 Collector with multiple glass evacuated - tube units

A collector consisting of an array of glass-tube (CTC) units of the type shown in Fig. 3.8(c) is depicted in Fig. 3.9. Its overall arrangement is similar to that of a flat-plate collector where cold fluid is distributed by an inlet header and hot fluid leaving through an outlet header. The array can collect both direct and diffuse radiation because of the back reflector that diffusely reflects any beam radiation missing the tubes. A comprehensive analysis of the collector is available in Ref. [3]. We shall list the main aspects of the analysis below.

The optical efficiency of the CTC can be written in the form:

$$\eta_o = \frac{\tau_e \alpha_r I_{eff}}{I_{bc}+I_{dc}} \tag{3.48}$$

where τ_e is the effective envelope transmittance, α_r is the receiver absorptance, I_{eff} is the effective solar radiation reaching the absorber, including directly incident radiation and radiation reflected from the back. I_{bc} and I_{dc} are the beam and diffuse radiation components intercepted by the aperture area of the collector. Computation of the transmittance τ_e and absorptivity α_r in Eq. (3.48) is beyond the scope of the present chapter. However, the derivation of the relevant equations is available in Ref.[3].

The thermal analysis of the collector to determine the heat loss coefficient can be done using the method described in section 3.3 for flat-plate collectors. The useful energy removal rate from the collector may be written as

$$q_u = \tau_e \alpha_r I_{eff} \left(\frac{A_t}{A_c}\right) - U_c (T_r - T_a) \left(\frac{A_r}{A_c}\right) \tag{3.49}$$

where A_t is the projected area of the tubes, A_r is the receiver or absorber area, and A_c is the aperture area of the collector. The first term on the RHS is the total solar radiation absorption rate and the second term is the rate of heat loss to the ambient.

3.5 Applications of Solar Thermal Systems

In this section we shall present the design and analysis of solar thermal systems for low and medium temperature applications. Solar thermal power generation, which is a high temperature application, will be considered in Chapter 4.

The sub-components common to most practical solar thermal systems are shown schematically in the generic arrangement depicted in Fig. 3.1. The main applications include [1,3]: solar water heating, solar space heating, solar space cooling, solar drying, solar desalination, swimming pool heating.

Solar thermal systems used in these applications include several common sub-components. Therefore we shall consider a few representative designs of solar water heating systems to illustrate the design procedure and operation.

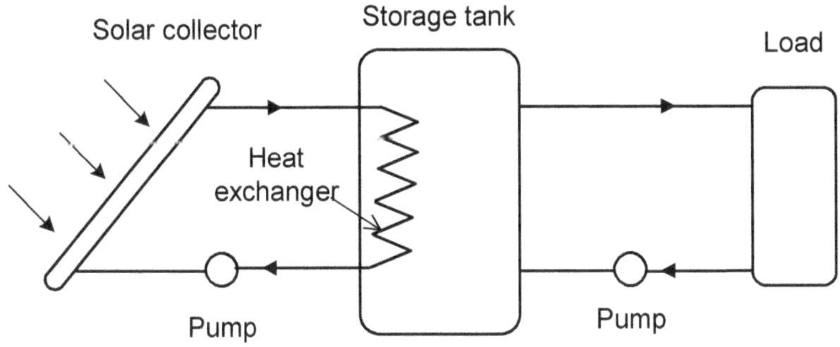

Fig. 3.10 Basic solar heating system with storage

3.5.1 *Solar water heating*

A basic solar water heating system is shown schematically in Fig. 3.10. The collector-loop consists of the solar collector which is connected to a heat exchanger immersed in the water storage tank. A pump circulates the fluid between the collector and the heat exchanger. Usually, the fluid in the collector-loop is a mixture of water and an anti-freeze liquid such as glycol. This is to deal with freezing ambient temperatures that could occur during winter in temperate regions. In the simple design shown, the storage tank contains water that is circulated between the tank and load by the load-loop pump.

We shall now develop the energy balance equations for the different components of the system that enable us to simulate the daily performance of the system. Several simplifying assumptions are made in the analysis. (i) Heat losses to the ambient from the storage tank, the piping and the pumps are neglected. These losses could be easily included in the energy equations, but it would make the algebra more tedious. (ii) All components of the system, except the storage tank, operate under quasi-steady conditions. (iii) The fluid in the tank is well-mixed so that its temperature is uniform. (iv) The load is quasi-steady. (v) The work input to the pump is transferred to the fluid.

Subject to the above assumptions we can write the following energy balance equations.

For the solar collector efficiency, we use Eq. (3.45).

$$\eta_c = F_R(\tau\alpha) - \frac{F_R U_l (T_{fi} - T_a)}{I_s} \tag{3.50}$$

Hence the useful energy collection rate is

$$Q_u = A_c I_s \eta_c = A_c I_s F_R(\tau\alpha) - A_c F_R U_l (T_{fi} - T_a) \tag{3.51}$$

where T_{fi} is the temperature of the fluid entering the collector.

The energy balance for the heat exchanger coil immersed in the tank gives

$$Q_u = mc(T_{fo} - T_{fi}) = mc\epsilon(T_{fo} - T_s) \tag{3.52}$$

where T_{fo} is the temperature of fluid entering the tank from the collector and T_s is the storage tank temperature. The mass flow rate and specific heat capacity of the fluid are m and c respectively. The heat transfer effectiveness of the coil, which is defined as the ratio of the actual heat transfer to the maximum possible heat transfer, is ϵ.

Eliminating T_{fi} and T_{fo} between Eqs. (3.51) and (3.52) we obtain the following expression for the useful energy collection rate in terms of the storage tank fluid temperature T_s.

$$Q_u = A_c I_s \eta_c = A_c \bar{F}_R(\tau\alpha) I_s - A_c \bar{F}_R U_l (T_s - T_a) \tag{3.53}$$

The modified collector heat removal factor $\bar{F}_R$ in Eq. (3.53), that accounts for the inclusion of a heat exchanger in the tank, is given by the expression

$$\bar{F}_R = F_R \left[1 + \frac{A_c F_R U_l (1 - \varepsilon)}{mc\varepsilon}\right]^{-1} \tag{3.54}$$

The above expression can be easily derived by manipulating Eqs. (3.51) and (3.52). (see worked example 3.18)

The transient energy equation for the storage tank is

$$(MC)_s \left[\frac{dT_s}{dt}\right] = f_c Q_u - Q_l \tag{3.55}$$

where $(MC)_s$ is heat capacity of the storage tank and Q_l is the rate of energy flow from the storage tank to the load.

The collector control factor f_c is defined by

$$f_c = 1 \quad \text{if} \quad I_s > U_l(T_s - T_a),$$
$$f_c = 0 \quad \text{otherwise.}$$

This implies that useful energy collection is possible only if the solar radiation intensity exceeds the heat loss through the collector. The control factor is used to switch-off the collector-loop pump, if the first condition is not satisfied. Then the load will be supplied by the stored energy in the tank.

If the load is a heat load, then the water from tank must be at a temperature that exceeds the minimum temperature required by the load. If this is not the case, the load-loop pump is switched-off. Under such conditions, the load must be supplied by an auxiliary heat source as shown in Fig. 3.1.

Several variations of the solar heating system considered above have been designed and operated depending on the application [2,5,7]. Three different solar process heat systems are presented below.

3.5.2 *Closed systems without storage*

The closed system without storage shown in Fig. 3.11(a) is a relatively simple heating system. For most industrial processes the hot water loads are uniform during the period of operation of the plant. In such a situation, the cost of the system could be reduced by operating without a storage tank [2].

When useful solar energy collection is possible, water leaving the process plant is pumped directly through the collector. The three-way valve blocks any water flow through the by-pass path. However, in such a design, depending on the collector field area, there may be periods when the temperature of the fluid leaving the collector may exceed the required inlet temperature of the load. Then some energy may have to be dumped to the ambient. This is particularly true for loads that require water at a fixed temperature.

If the temperature of water leaving the collector is below the temperature required by the load, then the auxiliary heat source supplies the necessary heat to raise the temperature to the desired value. When the

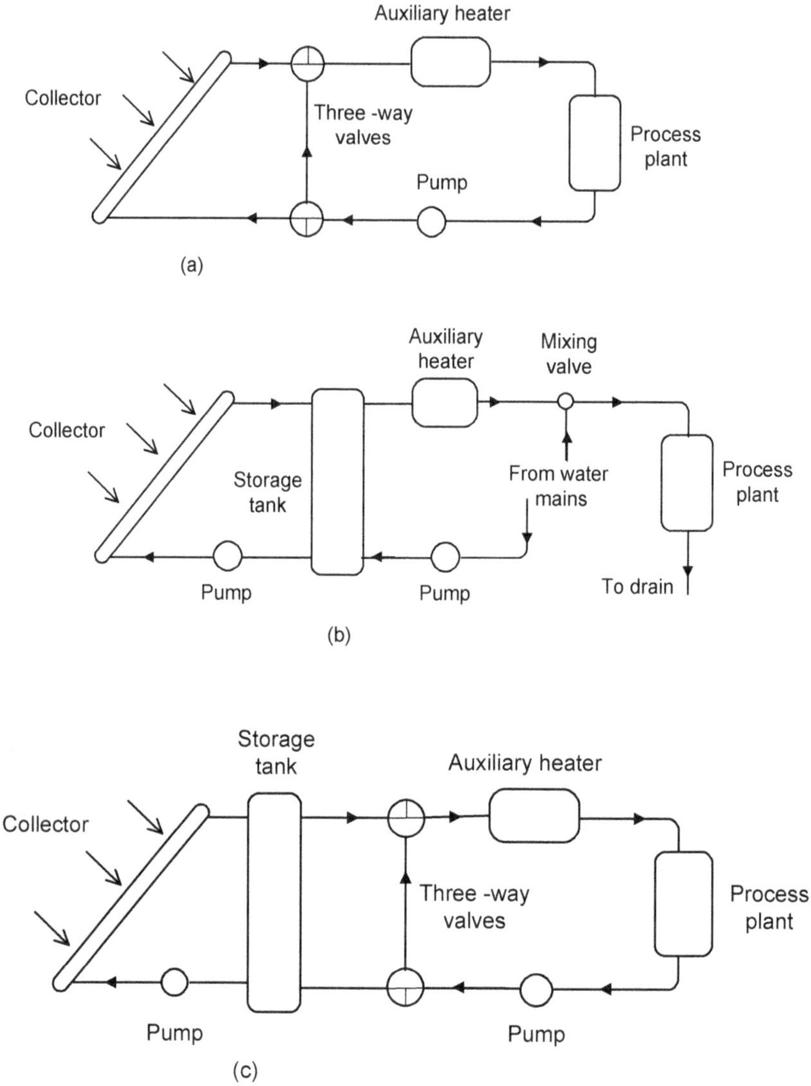

Fig. 3.11 Solar process heat systems (a) closed system without storage, (b) open system with storage, (c) closed system with storage

solar radiation level is insufficient for useful energy collection, water leaving the process is sent through the by-pass path to be heated by auxiliary heat source. The collector loop is inactive during this period.

3.5.3 *Open systems with storage*

The open solar heating system, depicted in Fig. 3.11(b), is a common design, widely used in domestic hot water systems and other applications like hot water systems in hotels. When solar radiation intensity is sufficient for useful energy collection, water is circulated between the storage tank and the collector by the collector- loop pump. Such systems operating in temperate climates, where freezing ambient conditions could occur, a heat exchanger is usually included in the collector-loop as shown in Fig. 3.10. The fluid in the collector-loop is then a mixture of water and an anti-freeze liquid.

Hot water required to supply the load is obtained from the storage tank, and cold water flows from the mains supply to replenish the water in the tank. If the temperature of the water flowing from the tank is lower than the temperature required by the process, then an auxiliary heater is used to boost the temperature. If the solar radiation intensity is insufficient for useful energy collection, the collector-loop pump is switched off.

If the temperature in the storage tank is too high for the load, then the desired inlet water temperature is achieved by mixing cold water from the mains supply as shown in Fig. 3.11(b).

3.5.4 *Closed systems with storage*

A closed heating system including a storage tank is depicted in Fig. 3.11(c). These systems usually operate at a fixed load temperature [6]. Water leaving the plant is recirculated for energy transport from the storage tank to the process plant. The outlet fluid from the plant leaves at a preset temperature determined by the process. For instance, the pasteurizer unit of a food processing plant described in Ref.[5] operates at about 74°C.

When the storage tank has not reached the process inlet temperature, solar energy cannot be utilized. In such a situation, if an auxiliary heat source is placed in series, it would heat the water in the storage tank. Therefore the collector pump is switched-off, and the three-way valve is used to send the water leaving the process directly to the auxiliary heater.

A comparison of the three heating systems shows that open systems taking in water at ambient temperature from the mains supply, are better able to utilize the solar energy input. Worked examples 3.18 to 3.20 illustrate the analysis of typical solar process heat systems.

3.6 Worked Examples

Example 3.1 A flat plate solar collector with an aperture area of 3.2 m^2 receives solar radiation at an average rate of 780 Wm^{-2} for a period of one hour. The average transmittance-absorptance product is 0.85. The average rate of heat loss from the collector to the ambient has been estimated to be 442 Wm^{-2}. Calculate (i) the average useful energy collection rate, and (ii) the average efficiency of the collector.

Solution The average rate of absorption of solar radiation is

$$Q_S = A_c I_S (\tau\alpha)_{av} = 3.2 \times 780 \times 0.85 = 2121.8 \text{ W}$$

The average rate of heat loss

$$Q_l = A_c q_l = 3.2 \times 442 = 1414.4 \text{ W}$$

From the overall energy balance of the collector, the average rate of useful energy collection is

$$Q_u = Q_s - Q_i = 2121.2 - 1414.4 = 706.8 \text{ W}$$

The average hourly efficiency is

$$\eta_{av} = \frac{Q_u}{Q_s} = \frac{706.8}{2496} = 28.3\%$$

Example 3.2 Water flows through a flat plate collector of area 2.2 m^2 at the rate of 0.036 kgs^{-1}. The temperature of water increases from 45°C to 51°C. Solar radiation incident on the collector is 871 Wm^{-2}, and the transmittance-absorptance product of the collector is 0.85. Calculate (i) the rate of heat loss from the collector, and (ii) the efficiency of the collector.

Solution The rate of useful heat removal from the collector is

$$Q_u = \dot{m}c_p(T_{wo} - T_{wi})$$

$$Q_u = 0.036 \times 4190 \times (51 - 45) = 905\ \text{W}$$

The total rate of absorption of solar energy is

$$Q_s = A_c \tau \alpha I_s = 2.2 \times 0.85 \times 871 = 1628\ \text{W}$$

The rate of heat loss from the collector is

$$Q_l = Q_s - Q_u = 1628 - 905 = 723\ \text{W}$$

The efficiency of the collector is

$$\eta_c = \frac{905}{2.2 \times 871} = 47\%$$

Example 3.3 Water enters a flat-plate collector of area 3.2 m^2 at 35°C and leaves at 42°C. The mass flow rate of water is 0.04 kgs^{-1}. The absorber plate temperature is 79.6°C and the ambient temperature is 21°C. The solar radiation incident on the collector is 934 Wm^{-2}. The transmittance-absorptance product of the collector is 0.87. The top loss coefficient of the collector is estimated as 7.88 Wm^{-2}K^{-1}. Calculate (i) the rate of useful energy removal, (ii) the rate of heat loss, (iii) the rate of absorption of solar energy, and (iv) the efficiency of the collector.

Solution The rate of useful heat removal from the collector is

$$Q_u = \dot{m}c_p\left(T_{wo} - T_{wi)}\right)$$

$$Q_u = 0.04 \times 4190 \times (42 - 35) = 1173.2\ \text{W}$$

The rate of heat loss from the collector is

$$Q_l = A_c U_l\left(T_p - T_a\right)$$

$$Q_l = 3.2 \times 7.88 \times (79.6 - 21) = 1477.65\ \text{W}$$

The rate of absorption of solar energy is

$$Q_{abs} = Q_l + Q_u = 1173.2 + 1477.65 = 2650.85\ \text{W}$$

Total solar radiation incident on the collector is

$$Q_s = A_c I_s = 3.2 \times 934 = 2988.8 \text{ W}$$

The efficiency of the collector is

$$\eta_c = \frac{Q_u}{Q_s} = \frac{1173.2}{2988.8} = 39.2\%$$

Example 3.4 Calculate the surface reflectance of glass at an angle of 50°. The refractive index of glass for solar radiation is 1.53.

Solution From Appendix A1 we obtain Fresnel's expressions for the reflectivity of unpolarized radiation passing through medium 1, with a refractive index n_1, to medium 2 with a refractive index n_2.

For the perpendicular component of unpolarized light, the reflectance is

$$r_p = \frac{sin^2(\theta_2 - \theta_1)}{sin^2(\theta_2 + \theta_1)}$$

For the parallel component of unpolarized light, the reflectance is

$$r_a = \frac{tan^2(\theta_2 - \theta_1)}{tan^2(\theta_2 + \theta_1)}$$

The overall reflectance is

$$r = \frac{(r_p + r_a)}{2}$$

The laws of refraction gives:

$$n_1 sin\theta_1 = n_2 sin\theta_2$$

Substituting numerical data in the above equations we obtain:

$$sin\theta_2 = \frac{sin50}{1.53} = 0.50068$$

Therefore $\theta_2 = 30.04$. The perpendicular and parallel components are:

$$r_p = \frac{sin^2(30.04 - 50)}{sin^2(30.04 + 50)} = 0.12$$

$$r_a = \frac{tan^2(30.04 - 50)}{tan^2(30.04 + 50)} = 0.004$$

The overall reflectance is

$$r = \frac{(0.12+0.004)}{2} = 0.062$$

Hence the *surface reflectivity* is 0.062.

Example 3.5 A sheet of glass used as a cover of a solar collector has a thickness of 3.2 mm and a refractive index of 1.53. The extinction coefficient is 20 m^{-1}. Calculate the effective transmittance and the effective reflectivity of the glass sheet for solar radiation incident at 50° to the surface.

Solution The absorptivity of a glass layer is a function of the extinction coefficient *K*, which is a property of the material, the thickness *L*, and the angle of incidence of the radiation beam to the normal of the surface. The intrinsic transmittance of the layer is given in Appendix A1 as:

$$\tau = \exp(-KL/cos\theta_2)$$

The angle of incidence and the refractive index are the same as in Example 3.4. Therefore $\theta_2 = 30.04$.

Substituting numerical data in the above equation we obtain

$$\tau = \exp(-20 \times 10^{-3} \times 3.2/cos30.04) = 0.9287$$

The following expressions for the *effective transmittance,* T and the *effective reflectivity* R_o, for a single layer were derived in Appendix A1.

$$T = \frac{Q_t}{Q_i} = \frac{\tau(1-r_o)(1-r_i)}{1-\tau^2 r_o r_i}$$

$$R_o = \frac{Q_r}{Q_i} = r_o + \frac{\tau r_i(1-r_o)^2}{1-\tau^2 r_o r_i}$$

The two surfaces of the glass sheet have the same reflectivity of 0.062, which was obtained in Example 3.4. Substituting numerical values in the above expressions we have

$$T = \frac{0.9287(1-0.062)(1-0.062)}{1-0.9287^2 \times 0.062 \times 0.062}$$

$$T = \frac{0.8171}{0.9966} = 0.8198$$

$$R_o = 0.062 + \frac{0.9287 \times 0.062(1-0.062)^2}{1-0.9287^2 \times 0.062 \times 0.062}$$

$$R_o = 0.062 + \frac{0.0507}{0.9966} = 0.113$$

Example 3.6 A flat plate collector has a single glass cover of thickness 4 mm and a collector plate with an absorptivity of 0.85 for solar radiation. A beam of solar radiation of intensity 625 Wm^{-2} is incident on the collector at an angle of 50°. The extinction coefficient of glass 20 m^{-1}. Calculate the rate of energy absorption by the absorber plate.

Solution The glass cover of the collector is the same as the glass sheet considered in worked example 3.5. The rate of absorption of solar radiation by the collector plate given by Eq. (3.7) as:

$$Q_a = \frac{T\alpha_a I_o}{[1-R(1-\alpha_a)]} = \frac{0.8198 \times 0.85 \times 625}{[1-0.113 \times (1-0.85)]} = 440 \text{ W}$$

In Ref.[1] it is stated that for practical flat plate collectors the above expression for solar radiation absorption is well approximated by the expression

$$Q_a = 1.01 T\alpha_a I_o$$

$$Q_a = 1.01 \times 0.8198 \times 0.85 \times 625 = 449 \text{ W}$$

The difference is about 0.7 percent.

Example 3.7 A flat plate collector has two glass covers of thickness 4 mm each, and a collector plate with an absorptivity 0.88 for solar radiation. A beam of solar radiation of intensity 735 Wm^{-2} is incident on the collector at an angle of 50°. The extinction coefficient of glass 20 m^{-1}. Calculate the rate of energy absorption by the absorber plate.

Solution Each of the two glass covers of the collector has the same properties as the single glass sheet in example 3.5. Therefore we use the overall transmittance and overall reflectivity obtained earlier for a single layer. The expressions for overall properties of multiple layers are available in Appendix A1. These are given below.

For 1 layer

$$\frac{1}{T_{o1}} = \frac{1}{T_1}$$

In Example 3.5, we obtained,

$$T_{o1} = T_1 = 0.8198$$

$$R_{2i} = R_{1o} = 0.113$$

$$T_2 = T_1 = 0.8198$$

For 2 layers

$$\frac{1}{T_{o2}} = \frac{(1-R_{2i}R_{1o})}{T_2}\left(\frac{1}{T_{o1}}\right)$$

Substituting numerical data we have

$$\frac{1}{T_{o2}} = \frac{(1-0.113\times0.113)}{0.8198}\left(\frac{1}{0.8198}\right) = 1.4689$$

Hence the overall transmittance of two layers is

$$T_{o2} = 0.681$$

From Appendix A1, the overall reflectivity of layer 1 is

$$\bar{R}_1 = 0.113, \qquad R_{2i} = 0.113$$

The overall reflectivity of 2 layers is

$$\bar{R}_2 = R_{2o} + \frac{\bar{R}_1 T_2{}^2}{(1-R_{2i}\bar{R}_1)}$$

Substituting numerical data we obtain

$$\bar{R}_2 = 0.113 + \frac{0.113\times 0.8198^2}{(1-0.113\times0.113)} = 0.19$$

The rate of absorption of solar radiation by the collector plate is given by Eq. (3.7) as:

$$Q_a = \frac{T\alpha_a I_o}{[1-R(1-\alpha_a)]} = \frac{0.618\times0.88\times735}{[1-0.19\times(1-0.88)]} = 409 \text{ W}$$

In Ref.[1] it is stated that for practical flat-plate collectors the above expression for solar radiation absorption is well approximated by the expression

$$Q_a = 1.01T\alpha_a I_o = 1.01 \times 0.618 \times 0.88 \times 735 = 404 \text{ W}$$

The difference is about 1.2 percent.

Example 3.8 A single-cover solar collector is inclined at 35° to the horizontal. The air gap between the absorber plate and the glass cover is 24 mm. The temperatures of the absorber and the cover are 90°C and 50°C respectively. The wind speed over the collector surface is 7.2 ms^{-1}. Calculate (i) the convective heat transfer coefficient between the plate and the cover, and (ii) the wind heat transfer coefficient.

Solution For free convection in an inclined rectangular cavity we use the following correlation given in Appendix A2.

$$Nu = 1 + 1.44\left[1 - \frac{1708}{Ra\cos\beta}\right]^{+}\left(1 - \frac{1708(\sin 1.8\beta)^{1.6}}{Ra\cos\beta}\right) + \left[\left(\frac{Ra\cos\beta}{5830}\right)^{1/3} - 1\right]^{+} \quad \text{(E3.8.1)}$$

where the meaning of the + exponent is that only *positive* values of the term in the square brackets are to be used. (if the term is negative then its value is taken as zero).

The meanings of the symbols are:

$$\text{Nusselt number: } Nu = \frac{hL}{k},$$

$$\text{Rayleigh number: } Ra = \frac{g\beta'\Delta T L^3}{\alpha\nu}$$

$$\text{Prandtl number: } Pr = \frac{\nu}{\alpha}$$

ν = kinematic viscosity,
α = thermal diffusivity,
h = heat transfer coefficient,
L = plate spacing,
k = thermal conductivity,
g = acceleration due to gravity,
$\beta' = \frac{1}{T}$ = volumetric coefficient of expansion,
ΔT = temperature difference between plates.

The *average temperature* of the air in the cavity is 70°C. We obtain the following properties of air at the mean temperature from tabulated data in Ref.[6]. Hence we have

$$k = 2.95 \times 10^{-5}\ \mathrm{kWm^{-1}K^{-1}}, \quad \nu = 1.986 \times 10^{-5}\ \mathrm{m^2s^{-1}},$$

$$Pr = 0.698, \quad \alpha = \frac{\nu}{Pr},$$

$$\beta' = \frac{1}{T_{av}} = \frac{1}{(273+70)} = 2.915 \times 10^{-3}$$

$$\Delta T = 40^{\circ}\mathrm{C}, \quad g = 9.81\ \mathrm{ms^{-2}}$$

$$Ra = \frac{g\beta'\Delta T L^3}{\alpha\nu} = \frac{g\beta'\Delta T L^3 Pr}{\nu^2}$$

$$Ra = \frac{9.81\times2.915\times10^{-3}\times40\times0.024^3\times0.698}{1.986^2\times10^{-10}} = 2.789 \times 10^4$$

$$Ra\cos(\beta) = 27890\cos(35) = 22846$$

Substituting numerical data in Eq. (E3.8.1) we have

$$Nu = 1 + 1.44\left[1 - \frac{1708}{22846}\right]^{+}\left(1 - \frac{1708(\sin 63)^{1.6}}{22846}\right) + \left[\left(\frac{22846}{5830}\right)^{1/3} - 1\right]^{+}$$

$$Nu = \frac{h_c L}{k} = \frac{h_c\times0.024}{2.95\times10^{-5}} = 2.828$$

Therefore the convective heat transfer coefficient between the plate cover is 3.48 $\mathrm{Wm^{-2}K^{-1}}$

The wind heat transfer coefficient is given by Eq. (A2.4) in Appendix A2 as:

$$h = 2.8 + 3.0V$$

The wind speed is $V = 7.2$ ms^{-1}. Substituting in the above equation we obtain the wind heat transfer coefficient as 24.4 Wm^{-2}K^{-1}.

Example 3.9 A flat-plate solar collector with a single glass cover is inclined at 35° to the horizontal. The emissivity of the absorber plate and the glass cover for long wave radiation are 0.94 and 0.88 respectively. The plate temperature, the glass temperature, and the ambient temperature are 70°C, 40°C and 20°C respectively. Calculate (i) the radiation heat transfer coefficient between the plate and the glass, and (ii) the radiation heat transfer coefficient between the glass cover and the ambient.

Solution The absolute temperatures of the plate, cover, and the ambient are:

$$T_p = 273 + 70 = 343 \text{ K}, \qquad T_c = 273 + 40 = 313 \text{ K},$$

$$T_a = 273 + 20 = 293 \text{ K}.$$

The effective emissivity between the plate and the cover is given by Eq. (3.18) as

$$\frac{1}{\epsilon_{epc}} = \frac{1}{\epsilon_p} + \frac{1}{\epsilon_c} - 1 = \frac{1}{0.94} + \frac{1}{0.88} - 1$$

Hence $\epsilon_{epc} = 0.833$

The Stefan-Boltzmann constant is, $\sigma = 56.7 \times 10^{-12}$ kWm^{-2}K^{-4}

The radiation heat transfer coefficient between the plate and the cover is given by Eq. (3.20) as

$$h_{rpc} = \sigma\epsilon_{epc}\left(T_p{}^2 + T_c{}^2\right)\left(T_p + T_c\right)$$

Substituting numerical data in the above equation we obtain

$$h_{rpc} = 56.7 \times 10^{-12} \times 0.833(343^2 + 313^2)(343 + 313)$$

Hence we have

$$h_{rpc} = 6.68 \text{ Wm}^{-2}\text{K}^{-1}$$

The radiation heat transfer coefficient between the cover and the ambient is given by Eq. (3.27) as

$$h_{rca} = \sigma\epsilon_{g1}\left({T_a}^2 + {T_c}^2\right)(T_a + T_c)$$

Substituting numerical data in the above equation we obtain

$$h_{rca} = 56.7 \times 10^{-12} \times 0.88(293^2 + 313^2)(293 + 313)$$

Hence we have

$$h_{rca} = 5.56 \text{ Wm}^{-2}\text{K}^{-1}$$

Example 3.10 A flat plate solar collector has a single glass cover, and it is inclined at 35° to the horizontal. The spacing between the plate and the cover is 24 mm. The plate is at a temperature of 70°C, and the ambient temperature is 24°C. The wind speed above the cover is 9 ms^{-1}. The emissivities of the plate and the cover for long wave radiation are 0.94 and 0.88 respectively. Calculate the top heat loss coefficient for the collector.

Solution The absolute temperatures of the plate and the ambient are:

$$T_p = 273 + 70 = 343 \text{ K}, \quad \text{and} \quad T_a = 273 + 24 = 297 \text{ K}.$$

Unlike the previous two examples, the glass cover temperature is not known in this example. Therefore to compute the heat transfer coefficients we assume a cover temperature of 40°C as a first guess. Subsequently this value will be updated in an iterative manner to improve the estimate of the heat transfer coefficients.

We now compute the relevant heat transfer coefficients. For free convection in an inclined rectangular cavity formed by the cover and the plate we use the correlation given in Eq. (E3.8.1).

The *average temperature* of the air in the cavity is: $(70 + 40)/2 =$ 55°C. We obtain the properties of air at this temperature from tabulated data in Ref.[6]. Hence we have

$$k = 2.818 \times 10^{-5}\ \text{kWm}^{-1}\text{K}^{-1}, \qquad \nu = 1.808 \times 10^{-5}\ \text{m}^2\text{s}^{-1},$$

$$Pr = 0.7, \quad g = 9.81\ \text{ms}^{-2}$$

$$\alpha = \frac{\nu}{Pr}, \quad \beta' = \frac{1}{T_{av}} = \frac{1}{(273+55)} = 3.048 \times 10^{-3}, \quad \Delta T = 30^\circ\text{C}$$

$$Ra = \frac{g\beta'\Delta T L^3 Pr}{\nu^2} = \frac{9.81 \times 3.048 \times 10^{-3} \times 30 \times 0.024^3 \times 0.7}{1.808^2 \times 10^{-10}} = 2.656 \times 10^4$$

$$Racos(\beta) = 26560cos(35) = 21757$$

Substituting numerical data in Eq. (E3.8.1) we get

$$Nu = 1 + 1.44\left[1 - \frac{1708}{21757}\right]^+ \left(1 - \frac{1708(sin63)^{1.6}}{22757}\right) + \left[\left(\frac{22846}{5830}\right)^{1/3} - 1\right]^+$$

$$Nu = \frac{h_c L}{k} = \frac{h_c \times 0.024}{2.818 \times 10^{-5}} = 2.817$$

Therefore $\quad h_{cpc} = 3.31\ \text{Wm}^{-2}\text{K}^{-1}$

The effective emissivity between the plate and the cover is given by Eq. (3.18) as

$$\frac{1}{\epsilon_{epc}} = \frac{1}{\epsilon_p} + \frac{1}{\epsilon_c} - 1 = \frac{1}{0.94} + \frac{1}{0.88} - 1$$

Hence $\epsilon_{epc} = 0.833$

The Stefan-Boltzmann constant is, $\quad \sigma = 56.7 \times 10^{-12}\ \text{kWm}^{-2}\text{K}^{-4}$

The radiation heat transfer coefficient between the plate and the cover is given by Eq. (3.20) as

$$h_{rpc} = \sigma\epsilon_{epc}\left(T_p{}^2 + T_c{}^2\right)\left(T_p + T_c\right)$$

Substituting numerical data in the above equation we obtain

$$h_{rpc} = 56.7 \times 10^{-12} \times 0.833(343^2 + 313^2)(343 + 313)$$

Hence we have $\quad h_{rpc} = 6.68\ \text{Wm}^{-2}\text{K}^{-1}$

The radiation heat transfer coefficient between the cover and the ambient is given by Eq. (3.27) as

$$h_{rca} = \sigma \epsilon_{g1}\left({T_a}^2 + {T_c}^2\right)(T_a + T_c)$$

Substituting numerical data in the above equation we obtain

$$h_{rca} = 56.7 \times 10^{-12} \times 0.88(297^2 + 313^2)(297 + 313)$$

Hence we have

$$h_{rca} = 5.67 \text{ Wm}^{-2}\text{K}^{-1}$$

The wind heat transfer coefficient is given by Eq. (A2.4) in Appendix A2 as

$$h_w = 2.8 + 3.0V$$

The wind speed is $V = 9$ ms^{-1}. Substituting in the above equation we obtain the wind heat transfer coefficient as 29.8 Wm^{-2}K^{-1}.

The total heat transfer coefficients are:

$$h_{tpc} = 3.31 + 6.68 = 9.99 \text{ Wm}^{-2}\text{K}^{-1}$$

$$h_{tca} = 29.8 + 5.67 = 35.47 \text{ Wm}^{-2}\text{K}^{-1}$$

For the overall heat transfer from the plate to the ambient we have

$$Q = 9.99(70 - t_c) = 35.47(t_c - 24)$$

Hence the cover temperature is 34.1°C. Note that our first guess of the cover temperature was 40°C.

Now repeat the above calculation with the new cover temperature of 34.1°C. For brevity we will list below the main results of the computation.

$h_{cpc} = 3.36$ Wm^{-2}K, $h_{rpc} = 6.5$ Wm^{-2}K^{-1}, $h_{rca} = 5.5$ Wm^{-2}K^{-1},

$h_w = 29.8$ Wm^{-2}K^{-1}, $h_{tpc} = 9.86$ Wm^{-2}K,

$h_{tca} = 35.3$ Wm^{-2}K, $t_c = 34.04$°C.

The new value obtained for the cover temperature very close to previous value. We substitute the values obtained above for the various heat transfer coefficients in the following expression for the loss coefficient.

$$\frac{1}{U_l} = \frac{1}{h_{tpc}} + \frac{1}{h_{tca}} = \frac{1}{9.86} + \frac{1}{35.3} = 0.1296$$

Hence the top loss coefficient is 7.71 Wm^{-2}K.

Example 3.11 A flat plate solar collector has two glass covers, and it is inclined at 40° to the horizontal. The spacing between the plate and the inner cover is 22 mm, and the spacing between the two covers is 24 mm. The plate is at a temperature of 70°C, and the ambient temperature is 24°C. The wind speed above the cover is 9 ms^{-1}. The emissivities of the plate and the covers for long wave radiation are 0.94 and 0.88 respectively. Calculate the top heat loss coefficient for the collector.

Solution A two-cover collector is shown schematically in Fig. 3.2. The absolute temperatures of the plate and the ambient are:

$$T_p = 273 + 70 = 343 \text{ K, and } T_a = 273 + 24 = 297 \text{ K}$$

The temperatures of the two glass covers are not known. To compute the heat transfer coefficients we assume, as a first guess, the inner and outer cover temperatures as 57°C and 42°C respectively. Subsequently these values will be updated in an iterative manner to improve the estimate of the heat transfer coefficients.

We now compute the relevant heat transfer coefficients. For free convection in the inclined rectangular cavity formed by the cover and the plate we use the following correlation given in Eq. (E3.8.1).

The *average temperature* of the air in the two cavities are: (70+57)/2 = 63.5°C, and (42+57)/2 = 49.5°C. We first consider the cavity between the plate and the inner cover. The properties of air at 63.5°C are obtained from tabulated data in Ref.[6]. Hence we have

$$k = 2.902 \times 10^{-5}\ \mathrm{kWm^{-1}K^{-1}}, \quad \nu = 1.922 \times 10^{-5}\ \mathrm{m^2s^{-1}},$$

$$Pr = 0.699, \quad g = 9.81\ \mathrm{ms^{-2}}$$

$$\alpha = \frac{\nu}{Pr}, \quad \beta' = \frac{1}{T_{av}} = \frac{1}{(273+63.5)} = 2.972 \times 10^{-3}, \quad \Delta T = 13^{\circ}\mathrm{C}$$

$$Ra = \frac{g\beta'\Delta T L^3 Pr}{\nu^2}$$

$$Ra = \frac{9.81 \times 2.972 \times 10^{-3} \times 13 \times 0.022^3 \times 0.699}{1.922^2 \times 10^{-10}} = 7.636 \times 10^3$$

$$Racos(\beta) = 7636cos(40) = 5849.5$$

Substituting numerical data in Eq. (E3.8.1) we have

$$Nu = \frac{h_c L}{k} = \frac{h_c \times 0.022}{2.902 \times 10^{-5}} = 1.745$$

Therefore $\qquad h_{cpc1} = 2.3\ \mathrm{Wm^{-2}K^{-1}}$

Next we consider the cavity between the outer and the inner covers. The following properties of air at 49.5°C are obtained from tabulated data in Ref.[6]. Hence we have

$$k = 2.797 \times 10^{-5}\ \mathrm{kWm^{-1}K^{-1}}, \quad \nu = 1.783 \times 10^{-5}\ \mathrm{m^2s^{-1}},$$

$$Pr = 0.7016, \quad g = 9.81\ \mathrm{ms^{-2}}$$

$$\alpha = \frac{\nu}{Pr}, \quad \beta' = \frac{1}{T_{av}} = \frac{1}{(273+49.5)} = 3.104 \times 10^{-3}, \quad \Delta T = 15^{\circ}\mathrm{C}$$

$$Ra = \frac{g\beta'\Delta T L^3}{\alpha\nu} = \frac{g\beta'\Delta T L^3 Pr}{\nu^2}$$

$$Ra = \frac{9.81 \times 3.104 \times 10^{-3} \times 15 \times 0.024^3 \times 0.7016}{1.783^2 \times 10^{-10}} = 1.392 \times 10^4$$

$$Racos(\beta) = 13920cos(40) = 10663$$

Substituting numerical data in Eq. (E3.8.1) we have

$$Nu = \frac{h_c L}{k} = \frac{h_c \times 0.024}{2.797 \times 10^{-5}} = 2.253$$

Therefore $\qquad h_{c12} = 2.63\ \mathrm{Wm^{-2}K^{-1}}$

We now compute the radiation heat transfer coefficients. The effective emissivity between the plate and the inner cover is given by Eq. (3.18) as

$$\frac{1}{\epsilon_{ep1}} = \frac{1}{\epsilon_p} + \frac{1}{\epsilon_{c1}} - 1 = \frac{1}{0.94} + \frac{1}{0.88} - 1$$

Hence $\epsilon_{epc} = 0.833$

The Stefan-Boltzmann constant is, $\sigma = 56.7 \times 10^{-12}$ kWm^{-2}K^{-4}

The radiation heat transfer coefficient between the plate and the cover is given by Eq. (3.20)

$$h_{rp1} = \sigma\epsilon_{ep1}\left(T_p{}^2 + T_{c1}{}^2\right)\left(T_p + T_{c1}\right)$$

Substituting numerical data in the above equation we obtain

$$h_{rpc} = 56.7 \times 10^{-12} \times 0.833(343^2 + 330^2)(343 + 330)$$

Hence we have $h_{rp1} = 7.201$ Wm^{-2}K^{-1}

The effective emissivity between the inner cover and the outer cover is given by Eq. (3.10) as

$$\frac{1}{\epsilon_{e12}} = \frac{1}{\epsilon_p} + \frac{1}{\epsilon_c} - 1 = \frac{1}{0.88} + \frac{1}{0.88} - 1$$

Hence $\epsilon_{e12} = 0.786$

The radiation heat transfer coefficient between the inner cover and the outer cover is given by Eq. (3.12)

$$h_{r12} = \sigma\epsilon_{e12}\left(T_{c1}{}^2 + T_{c2}{}^2\right)\left(T_{c1} + T_{c2}\right)$$

Substituting numerical data in the above equation we obtain

$$h_{r12} = 56.7 \times 10^{-12} \times 0.786(330^2 + 315^2)(330 + 315)$$

Hence we have $h_{r12} = 5.98$ Wm^{-2}K^{-1}

The radiation heat transfer coefficient between the outer cover and the ambient is given by Eq. (3.27)

$$h_{r2a} = \sigma\epsilon_{g2}\left(T_a{}^2 + T_{c2}{}^2\right)\left(T_a + T_{c2}\right)$$

Substituting numerical data in the above equation we obtain

$$h_{rca} = 56.7 \times 10^{-12} \times 0.88(297^2 + 315^2)(297 + 315)$$

Hence we have $h_{rca} = 5.723$ Wm^{-2}K^{-1}

The wind heat transfer coefficient is given by Eq. (A2.4) in Appendix A2.

$$h_w = 2.8 + 3.0V$$

The wind speed is $V = 9$ ms^{-1}. Substituting in the above equation we obtain the wind heat transfer coefficient as 29.8 Wm^{-2}K^{-1}. The total heat transfer coefficients are:

$$h_{tp1} = 2.3 + 7.20 = 9.5$$

$$h_{t12} = 2.63 + 5.98 = 8.61$$

$$h_{t2a} = 29.8 + 5.723 = 35.5$$

The overall heat transfer coefficient is given by

$$\frac{1}{U_o} = \frac{1}{9.5} + \frac{1}{8.61} + \frac{1}{35.5} = 0.2495$$

Therefore the overall top loss coefficient is

$$U_o = 4.0 \text{ Wm}^{-2}\text{K}^{-1}$$

For the overall heat transfer from the plate to the ambient we have

$$Q = U_o(70 - 24) = 4 \times 46 = 184 \text{ Wm}^{-2}$$

From the thermal network in Fig. 3.3 we note that

$$Q = 9.5(70 - t_{c1}) = 8.617(t_{c1} - t_{c2}) = 35.5(t_{c2} - 24) = 184$$

Hence the temperature of the covers are 50.6°C and 29.18°C.

We now repeat the above computation using the new cover temperatures. For brevity the main results obtained are listed below:

$$h_{cp1} = 2.77, \quad h_{rp1} = 7.0, \quad h_{c12} = 3.0$$

$$h_{r12} = 5.47, \quad h_{r2a} = 5.37, \quad h_w = 29.8$$

$$h_{tp1} = 9.77, \quad h_{t12} = 8.47, \quad h_{t2a} = 35.17$$

The overall top loss coefficient is given by:

$$\frac{1}{U_o} = \frac{1}{9.77} + \frac{1}{8.47} + \frac{1}{35.17} = 0.2488$$

$$U_o = 4.02 \text{ Wm}^{-2}\text{K}^{-1}$$

The temperatures of the covers are 51:08°C and 29.25°C. These temperatures are close to the temperatures obtained after the first iteration. We could therefore use these as the final values to obtain the loss coefficient and the rate of heat loss as

$$U_o = 4.02 \text{ Wm}^{-2}\text{K}^{-1} \text{ and } Q = 184.8 \text{ Wm}^{-2}$$

Example 3.12 Calculate the top heat loss coefficient using the empirical equation, for the single cover collector described in example 3.10.

Solution The empirical relation for the loss coefficient [1] is given in Eq. [3.35] as:

$$U_l = \left[\frac{N}{\frac{C}{T_p}\left(\frac{T_p - T_a}{N+f}\right)^e} + \frac{1}{h_w}\right]^{-1} + \sigma\left(T_p{}^2 + T_a{}^2\right)\left(T_p + T_a\right)\left[\frac{1}{\varepsilon_p + 0.00591 N h_w} + \frac{2N + f - 1 + 0.133\varepsilon_p}{\varepsilon_g} - N\right]^{-1} \quad \text{(E3.12.1)}$$

where

$$f = \left(1 + 0.089 h_w - 0.1166 h_w \varepsilon_p\right)(1 + 0.07866N),$$

$$C = 520(1 - 0.000051 s^2),$$

$$e = 0.43\left(1 - \frac{100}{T_p}\right)$$

s = collector tilt (degrees), ε_g = emittance of glass (0.88),
ε_p = emittance of plate,
T_a = ambient temperature (K), T_p = mean plate temperature (K),
h_w = wind heat transfer coefficient.

For the present problem the numerical values are:

$N = 1, \quad s = 40^{\circ}, \quad \varepsilon_g = 0.88, \quad \varepsilon_p = 0.94, \quad T_a = 297\text{K},$

$T_p = 343\ \text{K}, \quad h_w - 29.8\ \text{Wm}^{2}\ \text{K}^{-1}.$

$e = 0.43\left(1 - \frac{100}{343}\right) = 0.304,$

$C = 520(1 - 0.000051 \times 35^2) = 487.5$

$f = (1 + 0.089 \times 29.8 - 0.1166 \times 29.8 \times 0.94)(1 + \ 0.07866 \times 1)$
$= 0.416$

Substituting numerical data in Eq. (3.35) we get

$$U_l = \left[\frac{1}{\frac{477.57}{343}\left(\frac{46}{1+0.416}\right)^{0.304}} + \frac{1}{29.8}\right]^{-1} +$$

$$56.7 \times 10^{-12}(343^2 + \ 297^2)(343 + 297) \times$$

$$\left[\frac{1}{0.94+0.00591\times 29.8} + \frac{2+0.416-1+0.133\times 0.94}{0.88} - 1\right]^{-1}$$

Hence we have

$$U_l = 3.539 + \frac{7.47}{1.647} = 8.07\ \text{Wm}^{-2}\ \text{K}^{-1}$$

The difference between the above value and that obtained by the detailed analysis in example 3.10 is about 0.36 W, which is 4.6%.

Example 3.13 Calculate the heat loss coefficient using the empirical equation, for the two- cover collector described in Example 3.11.

Solution The empirical relation for the loss coefficient [1] is given in Eq. (3.35) and Eq. (E3.12.1).

For the present problem the numerical values are:

$N = 2$, $s = 40^\circ$, $\varepsilon_g = 0.88$, $\varepsilon_p = 0.94$, $T_a = 297\text{K}$,

$T_p = 343$ K, $h_w = 29.8$ Wm^{-2} K^{-1}.

$$e = 0.43\left(1 - \frac{100}{343}\right) = 0.304,$$

$$C = 520(1 - 0.000051 \times 40^2) = 477.57$$

$$f = (1 + 0.089 \times 29.8 - 0.1166 \times 29.8 \times 0.94)(1 + 0.07866 \times 2) = 0.4467$$

Substituting numerical data in Eq. (3.12.1) we have

$$U_l = \left[\frac{2}{\frac{477.57}{343}\left(\frac{46}{2+0.4467}\right)^{0.304}} + \frac{1}{29.8}\right]^{-1} + 56.7 \times 10^{-12}(343^2 + 297^2)(343 + 297) \times \left[\frac{1}{0.94+0.00591\times 2\times 29.8} + \frac{4+0.4467-1+0.133\times 0.94}{0.88} - 2\right]^{-1}$$

Hence we have

$$U_l = 1.606 + \frac{7.47}{2.83} = 4.24 \text{ Wm}^{-2}\text{ K}^{-1}$$

The difference between the above value and that obtained by detailed analysis in example 3.11 is about 0.2 W, which is 4.9%.

Example 3.14 A flat plate collector has a plate-fin absorber as shown in Fig. E3.14.1. The copper tubes are 10 mm ID and 0.5 mm thick, and are connected by a 0.5 mm thick plate with a center-to-center distance of 15 cm. The heat transfer coefficient for water flow in the tubes is 1500 Wm^{-2}K^{-1}. The water flow rate in the tubes is 0.02 kgs^{-1}. The transmittance of the cover is 0.9 and the absorptance of the copper surface is 0.9. The width and length of the collector are 1 m and 2 m respectively. The total solar radiation incident on the collector is 730 Wm^{-2}, and the ambient temperature is 17°C. The heat loss coefficient of the collector is 8 Wm^{2}K^{-1}. Calculate (i) the collector efficiency factor, and (ii) the heat removal factor.

Solution

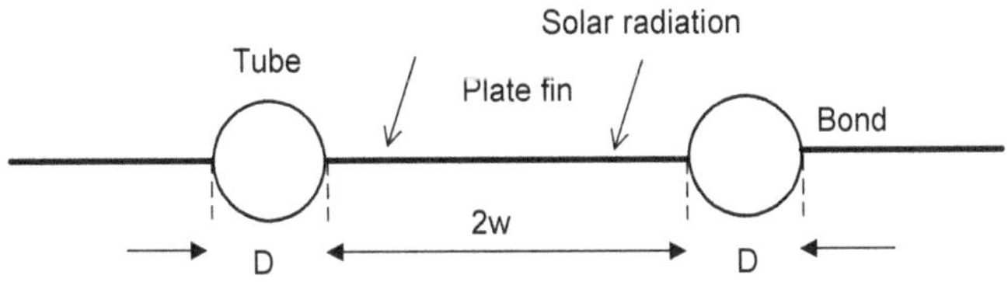

Fig. E3.14.1 Tube-fin absorber

The expressions for the collector efficiency and the heat removal factor for a tube-fin absorber are available in most standard text books on solar energy [1,3]. The various symbols used are indicated in Fig. E3.14.1. The collector efficiency factor is given by the expression [1,3]

$$F' = \left(\frac{1}{WU_L}\right)\left[\frac{1}{U_L\{D+(W-D)F\}} + \frac{1}{C_B} + \frac{1}{\pi D_i h_f}\right]^{-1}$$

The fin efficiency is given by

$$F = \frac{\tanh m(W-D)/2}{m(W-D)/2} \quad \text{where } m^2 = \frac{U_L}{k\delta}$$

The heat removal factor is given by

$$F_R = \left(\frac{\dot{m}c_p}{A_c U_L}\right)\left[1 - exp\left(-A_c U_L F'/\dot{m}c_p\right)\right]$$

Substituting the given numerical data in the above equations we obtain the following:

$$m^2 = \frac{8}{390\times 0.5\times 10^{-3}} = 41.02$$

Therefore $m = 6.4$ m^{-1}.

$$\frac{(W-D)}{2} = \frac{0.15-0.01}{2} = 0.07$$

The fin efficiency is given by

$$F = \frac{\tanh m(W-D)/2}{m(W-D)/2} = \frac{tanh(0.07\times 6.4)}{0.07\times 6.4} = 0.938$$

The collector efficiency factor is

$$F' = \frac{1/WU_L}{1/U_o} = \left(\frac{1}{WU_L}\right)\left[\frac{1}{U_L\{D+(W-D)F\}} + \frac{1}{\pi D_i h_f}\right]^{-1}$$

$$F' = \left(\frac{1}{8\times 0.15}\right)\left[\frac{1}{8\{0.01+(0.15-0.01)0.938\}} + \frac{1}{\pi\times 0.01\times 1500}\right]^{-1} = 0.92$$

The heat removal factor is

$$F_R = \left(\frac{0.02\times 4190}{2\times 8}\right)[1 - exp(-2\times 8\times 0.92/0.02\times 4190)] = 0.84$$

Example 3.15 An air heating collector has a rectangular duct of height 15 mm, width 1m, and length 5 m. The flow rate of air is 3.5 $m^3\ min^{-1}$. The following properties of air at the mean temperature have been obtained: density, 1.1 kgm^{-3}, specific heat, 1.0 $kJkg^{-1}K^{-1}$, viscosity, 1.8×10^{-5} $kgm^{-1}s^{-1}$. The heat loss coefficient of the collector is 5.2 $Wm^{-2}\ K^{-1}$. Calculate (i) collector efficiency factor, and (ii) the heat removal factor.

Solution We first calculate the mean convective heat transfer coefficient form the plate to the air in the flow channel. The hydraulic mean diameter is given by

$$D_h = \frac{4A}{p} = \frac{4\times 1\times 0.015}{(2+0.03)} = 0.0296\ \text{m}$$

The mean air velocity is

$$V = \frac{3.5}{60\times 1\times 0.015} = 3.89\ \text{ms}^{-1}$$

The Reynolds number is

$$Re = \frac{\rho V D_h}{\mu} = \frac{1.1\times 3.89\times 0.0296}{1.8\times 10^{-5}} = 7037$$

To calculate the mean Nusselt number we use the correlation recommended in Ref.[3] for flow in short rectangular ducts.

$$Nu_m = \frac{0.0192 Re^{3/4} Pr}{1+1.22 Re^{-1/8}(Pr-2)}$$

where $Pr = \frac{c_p \mu}{k}$, is the Prandtl number for air, taken as 0.72.

$$Nu_m = \frac{0.0192 \times 7037^{3/4} \times 0.72}{1 + 1.22 \times 7037^{-1/8} \times (0.72 - 2)} = 21.94$$

The heat transfer coefficient is given by

$$h_c = \frac{k Nu_m}{D_h} = \left(\frac{c_p \mu}{Pr}\right)\left(\frac{Nu_m}{D_h}\right)$$

$$h_c = \frac{21.94 \times 1.8 \times 10^{-5}}{0.72 \times 0.0294} = 0.0187 \text{ kWm}^{-2}\text{K}^{-1}$$

The collector efficiency factor is given by Eq. (3.40) as

$$F' = \frac{h_i}{h_i + U_l} = \frac{18.7}{18.7 + 5.2} = 0.782$$

The heat removal factor is given by Eq. (3.44) as

$$F_R = \left(\frac{m c_p}{A_c U_l}\right)\left[1 - Exp\left(\frac{-A_c U_l F'}{m c_p}\right)\right]$$

$$\frac{m c_p}{A_c U_l} = \frac{3.5 \times 1.1 \times 1000}{60 \times 5 \times 5.2} = 2.47$$

Hence

$$F_R = (2.47)[1 - Exp(-0.782/2.24)] = 0.67$$

Example 3.16 The solar radiation incident on an air heater, inclined at 40° to the horizontal, is 865 Wm^{-2}. The effective transmittance-absorptance product is 0.84, and the top heat loss coefficient is 4.1 Wm^{-2}K^{-1}. The inlet air temperature is 58°C, and the air mass flow rate is 0.054 kgs^{-1}. The ambient temperature is 20°C. The width and length of the collector are 1m and 4 m respectively. The air flow channel depth is 15 mm. Neglect heat losses from the bottom of the collector. Calculate (i) the outlet temperature of the air, (ii) the useful rate of energy collection, and (iv) the efficiency of the collector.

Solution We first calculate the convective heat transfer coefficient from the plate to the air in the flow channel. Air properties are evaluated at a mean temperature of 62°C, assuming a temperature rise of 8°C of the air in the duct. Hence we have [6]:

$$\rho = 1.048 \text{ kgm}^{-3}, \quad \mu = 2.0185 \times 10^{-5} \text{ kgm}^{-1}\text{s}^{-1},$$

$$Pr = 0.699, \quad k = 2.901 \times 10^{-5} \text{ kJm}^{-1}\text{s}^{-1}, \quad c_p = 1.007 \text{ kJkg}^{-1}\text{K}^{-1}.$$

The hydraulic mean diameter is given by

$$D_h = \frac{4A}{p} = \frac{4\times1\times0.015}{(2+0.03)} = 0.0296 \text{ m}$$

The mean air velocity is

$$V = \frac{0.054}{1.048\times0.015} = 3.44 \text{ ms}^{-1}$$

The Reynolds number is

$$Re = \frac{\rho V D_h}{\mu} = \frac{1.048\times3.44\times0.0296}{2.0185\times10^{-5}} = 5286$$

To calculate the Nusselt number in a rectangular channel we use the correlation recommended in Ref.[3]

$$Nu_m = \frac{0.0192 Re^{3/4} Pr}{1+1.22 Re^{-1/8}(Pr-2)}$$

$$Nu_m = \frac{0.0192\times5286^{3/4}\times0.699}{1+1.22\times5286^{-1/8}\times(0.699-2)} = 18.23$$

The heat transfer coefficient is given by

$$h_i = \left(\frac{Nu_m}{D_h}\right) k = \frac{18.23\times2.901\times10^{-5}}{0.029} = 0.018 \text{ kWm}^{-2}\text{K}^{-1}$$

To calculate the outlet air temperature we use Eq. (3.38).

$$mc_p \frac{dT_f}{dx} = W\left(\frac{h_i U_l}{h_i+U_l}\right)\left[\frac{(\tau\alpha)I_s}{U_l} + T_a - T_f\right] \quad \text{(E3.16.1)}$$

The following substitution is made to make the solution simpler. Let

$$\theta = \frac{(\tau\alpha)I_s}{U_l} + T_a - T_f \quad \text{(E3.16.2)}$$

Then Eq. (E3.16.1) may be written in the form

$$mc_p \frac{d\theta}{dx} = -W\left(\frac{h_i U_l}{h_i + U_l}\right)\theta \tag{E3.16.3}$$

Substituting numerical data in Eq. (E3.16.3) we obtain

$$\frac{d\theta}{dx} = -0.0614\theta \tag{E3.16.4}$$

The solution of Eq. (E3.16.4) gives the outlet air temperature as

$$\theta_o = 0.7822\theta_i$$

Substituting the given data in Eq. (E3.16.2) we obtain,

$$\theta_i = 139.2, \text{ and therefore } \theta_o = 108.9$$

Substituting in Eq. (E3.16.2), we have $T_o = 88.3°C$.
The rate of useful energy collection is

$$Q_u = mc_p(T_o - T_i) = 0.054 \times 1007(88.3 - 58) = 1647 \text{ J}$$

Solar radiation incident on the collector surface is

$$Q_s = A_c I_t = 4 \times 865 = 3460 \text{ J}$$

Hence the efficiency of the collector is

$$\frac{Q_u}{Q_s} = \frac{1647}{3460} = 47.6\%$$

Example 3.17 An evacuated tube absorber of diameter 60mm has a glass cover of diameter 90 mm. The space between them is evacuated. The absorber surface has an emittance of 0.92 and its temperature 180°C. The wind speed and air temperature are 6 ms^{-1} and 15°C respectively. Calculate overall heat loss coefficient.

Solution The heat transfer processes in an evacuated tube collector are depicted in Fig. E13.17.1. The only mode of heat transfer between the metal flow tube and the glass cover is thermal radiation. The cover looses heat by radiation to the atmosphere, and by convection due to wind blowing across the glass tube.

Let the absolute temperatures of the tube and the cover be T_1 and T_2 respectively. The ambient temperature is 288K. To evaluate the radiation heat loss from the cover we need to know its temperature. Since this is not known, we shall assume an initial value of 40°C. Using the first estimate of the heat loss we could obtain a more accurate value of the cover temperature.

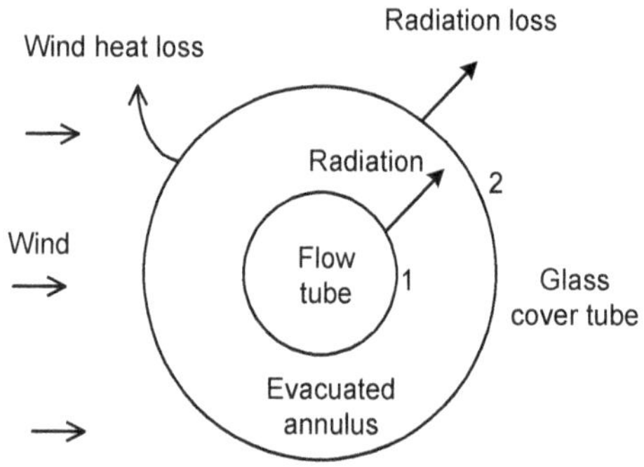

Fig. E3.17.1 Evacuated tube collector heat loss

Radiation heat transfer from the flow tube to glass cover

The following data are used to calculate the radiation heat transfer coefficient h_{r12} from the flow tube to the glass cover (see Fig. E3.17.1).

$$T_1 = 453K, \quad T_2 = 313K, \quad D_1 = 60 \text{ mm}, D_2 = 90 \text{ mm},$$

$$\epsilon_1 = 0.92, \quad \epsilon_2 = 0.88, \quad \sigma = 56.7 \times 10^{-12} \text{ kWm}^{-2}\text{K}^{-4}$$

$$h_{r12} = \sigma \epsilon_{e12}(T_1{}^2 + T_2{}^2)(T_1 + T_2)$$

where the effective emissivity is given by the expression [1,3]

$$\frac{1}{\epsilon_{e12}} = \frac{1}{\epsilon_1} + \left(\frac{1}{\epsilon_2} - 1\right)\left(\frac{D_1}{D_2}\right)$$

Substituting the relevant numerical data in the above equations we get

$$\epsilon_{e12} = 0.849 \quad \text{and} \quad h_{r12} = 11.18 \text{ Wm}^{-2}\text{K}^{-1}$$

$$Q_{r12} = A_1 h_{r12}(T_1 - T_2)$$

$$Q = Q_{r12} = \pi \times 0.06 \times 11.18(453 - T_2) \qquad \text{(E3.17.1)}$$

Heat transfer from the glass cover to the ambient

We first calculate the wind heat transfer coefficient from the glass cover to the ambient. The properties of air like the viscosity, the thermal conductivity and the density are obtained from tabulated data [6] at the mean temperature of, $0.5(15 + 40) = 27.5$°C.

$$\mu = 1.846 \times 10^{-5}\ \text{kgm}^{-1}\text{s}^{-1} \quad k = 2.624 \times 10^{-5}\ \text{kWm}^{-1}\text{K}^{-1},$$

$$\rho = 1.177\ \text{kgm}^{-3}.$$

The Reynolds number is given by

$$Re = \frac{\rho VD}{\mu} = \frac{1.177 \times 6 \times 0.09}{1.846 \times 10^{-5}} = 34430$$

The applicable wind heat transfer coefficient is given by Eq. (A2.6) as

$$Nu = 0.24 Re^{0.6}$$

Hence the Nusselt number is

$$Nu = \frac{hD}{k} = 0.24 \times 34430^{0.6} = 126.6$$

The convective heat transfer coefficient is

$$h = \frac{126.6 \times 2.624 \times 10^{-5}}{0.09} = 37\ \text{Wm}^{-2}\text{K}^{-1}$$

In Ref.[1] it is recommended that for wind induced forced convection the above heat transfer coefficient should be increased by 25%. Therefore $h = 46$ Wm^{-2}K^{-1}.

The radiation heat loss from the outer glass cover to the ambient is given by

$$Q_{r12} = A_2 h_{r2a}(T_2 - T_a)$$

where the radiation heat transfer coefficient is defined as

$$h_{r2a} = \sigma\epsilon_g\left({T_a}^2 + {T_2}^2\right)(T_a + T_2)$$

We substitute the following data in the above equation to obtain the radiation heat transfer coefficient.

$$T_a = 288K, \qquad T_2 = 313K,$$
$$\epsilon_g = 0.88, \qquad D_2 = 90 \text{ mm}$$

Hence we have $$h_{r2a} = 5.43 \text{ Wm}^{-2}\text{K}^{-1}$$

The total heat transfer per meter from the absorber tube to the ambient can be expressed in the form

$$Q = \pi \times 0.09 \times (46 + 5.43)(T_2 - 288) \qquad \text{(E3.17.2)}$$

Note that the heat loss from the cover to the ambient is by wind convection and thermal radiation. These two heat transfer rates are added to obtain the total heat transfer coefficient in Eq. (E3.17.2). From Eqs. (E3.17.1) and (E3.17.2) it follows that

$$T_1 - T_a = Q\left(\frac{1}{\pi\times0.06\times1\times11.18} + \frac{1}{\pi\times0.09\times1\times51.43}\right) = 453 - 288$$

Therefore the rate of total heat loss is

$$Q = 304 \text{ Wm}^{-1}$$

The overall heat loss coefficient based on the inner tube area is given by

$$U = \frac{304}{\pi\times0.06\times1\times(180-15)} = 9.8 \text{ Wm}^{-2}\text{K}^{-1}$$

The temperature T_2 of the glass cover is obtained from Eq. (E3.17.1) as 35.7°C, which is close to the initially assumed temperature of 40°C. We could now repeat the above calculation with the new cover temperature of 35.7°C. However this does not change U by much.

Example 3.18 The solar process heat system shown in Fig. E3.18.1 incorporates a heat exchanger, with an effectiveness of 0.6 between the collector and the storage tank. The fluid capacitance rates on the collector side and the storage tank side of the heat exchanger are 1015 WK^{-1} and 1120 WK^{-1} respectively. The area of the collector is 20 m^2, $F_R U_L$ is 5.3 Wm^{-2}K^{-1}, and $F_R(\tau\alpha)$ is 0.74. The heat removal factor is 0.87. The flow rate through the collector is 0.21 kgs^{-1}. Solar radiation intensity on the collector surface is 680 Wm^{-2}, and the ambient temperature is 15°C. The storage tank temperature is 40°C. Calculate (i) useful energy collection rate, and (ii) the collector efficiency.

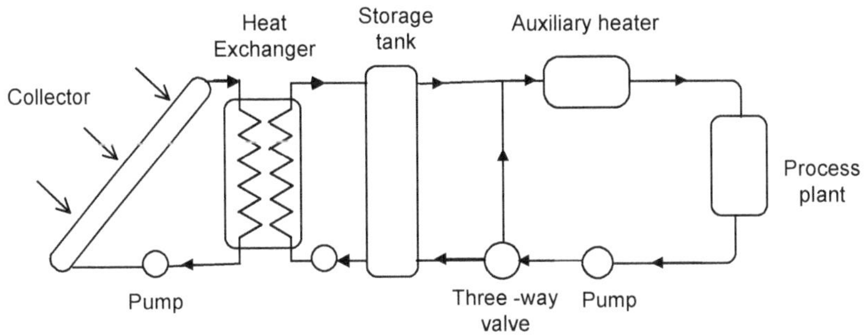

Fig. E3.18.1 Process plant with collector-storage heat exchanger

Solution Let the inlet and outlet temperatures of the collector be T_{fi} and T_{fo} respectively, and the temperatures of the fluid entering and leaving the heat exchanger from the storage tank be T_s and T_1 respectively. The mass flow rates and heat capacities of the collector fluid and the storage tank fluid be m_1, c_1 and m_2, c_2 respectively. Assume that the heat losses to the ambient from the piping and the heat exchanger are negligible.

Let the rate of useful energy collected by the solar collector and transferred to the storage tank via the heat exchanger be Q. We now write the energy balance equations for the heat exchanger and the collector.

For the heat exchanger:

$$Q = m_2 c_2 (T_1 - T_s) = \epsilon (mc)_{min} (T_{fo} - T_s) \qquad \text{(E3.18.1)}$$

$$Q = m_1 c_1 (T_{fo} - T_{fi}) = m_2 c_2 (T_1 - T_s) \qquad \text{(E3.18.2)}$$

From Eqs. (E3.18.1) and (E3.18.2) we get

$$\epsilon = \frac{m_1 c_1 (T_{fo} - T_{fi})}{(mc)_{min} (T_{fo} - T_s)} = \frac{m_2 c_2 (T_1 - T_s)}{(mc)_{min} (T_{fo} - T_s)} \qquad \text{(E3.18.3)}$$

where $(mc)_{min}$ is the smaller of the two capacitance rates $(m_1 c_1)$ and $(m_2 c_2)$.

For the collector:

$$Q = A_c F_R [I_s (\tau\alpha) - U_L (T_{fi} - T_a)] \qquad \text{(E3.18.4)}$$

Manipulating Eqs. (E3.18.1) to (E3.18.3) we obtain the following equations for the collector:

$$(T_{fo} - T_s) = \frac{Q}{\epsilon(mc)_{min}}$$

$$(T_{fo} - T_{fi}) = \frac{Q}{m_1 c_1}$$

From the two equations above we obtain

$$(T_{fi} - T_s) = \frac{Q}{\epsilon(mc)_{min}} - \frac{Q}{m_1 c_1} \qquad \text{(E318.5)}$$

Eliminating T_{fi} between Eqs. (E3.18.4) and (E3.18.5) we have

$$Q = A_c F'_R [I_s(\tau\alpha) - U_L(T_s - T_a)] \qquad \text{(E3.18.6)}$$

where the modified flow factor, F'_R that accounts for the heat exchanger is obtained as:

$$\frac{F'_R}{F_R} = \left[1 + A_c F_R U_L \left(\frac{1}{\epsilon(mc)_{min}} - \frac{1}{m_1 c_1}\right)\right]^{-1} \qquad \text{(E3.18.7)}$$

We substitute the following numerical data in the above equations:

$\epsilon = 0.6$, $m_1 c_1 = 1015$ WK^{-1}, $m_2 c_2 = 1120$ WK^{-1},

$A_c = 20$ m^2, $F_R U_L = 5.3$ Wm^{-2}K^{-1}, $F_R = 0.87$

$F_R(\tau\alpha) = 0.74$, $m_1 = 0.2$ kgs^{-1}, $I_s = 680$ Wm^{-2},

$(mc)_{min} = 1015$ WK^{-1}.

From Eq. (E3.18.7) we obtain

$$\frac{F'_R}{F_R} = \left[1 + 20 \times 5.3 \left(\frac{1}{0.6 \times 1015} - \frac{1}{1015}\right)\right]^{-1} = 0.93 \qquad \text{(E3.18.8)}$$

Therefore $$F'_R = 0.813$$

Substituting in Eq. (E3.18.6) we have

$$Q = 20 \times 0.813 \times [680(0.74/0.87) - (5.3/0.87)(40 - 15)]$$

$$Q = 6928.25 \text{ W}$$

Total radiation incident on the collector is

$$Q_s = A_c I_s = 20 \times 680 = 13600 \text{ W}$$

Hence the efficiency of the collector is

$$\eta_s = \frac{Q}{Q_s} = \frac{6928.25}{13600} = 50.9\%$$

Example 3.19 A roof-top water heating flat plate collector of area 2 m^2 has a heat removal factor, F_R of 0.82 and a loss coefficient of 7.5 $Wm^{-2}K^{-1}$. The collector faces 30° east of south, and is tilted at 35° from the horizontal. The latitude of the location is 38°N.

At 2.00 p.m., on August 5 the measured direct beam solar radiation on a horizontal surface at the location is 640 Wm^{-2}, and the diffuse radiation is 210 Wm^{-2}.

The transmittance-absorptance for the direct beam and diffuse radiation are 0.74 and 0.65 respectively. The ambient temperature is 18°C, and the inlet water temperature to the collector is 50°C. Calculate (i) the rate of useful energy collection, and (ii) the efficiency of the collector.

Solution The hour angle at 2 p.m. is given by Eq. (2.6) as

$$H = 15(T_{sol} - 12) = 15 \times 2 = 30 \text{ degrees}$$

For August 5, $N_{day} = 217$. The declination is given by Eq. (2.5) as

$$\delta = 23.45 sin\left[\frac{360(217+284)}{365}\right] = 16.83 \text{ degrees}$$

The solar altitude angle is given by Eq. (2. 9) as

$$sin\beta = cosLcosHcos\delta + sinLsin\delta$$

$$sin\beta = cos38 \cos 30\, cos16.83 + sin38 sin16.83 = 0.831$$

Therefore the solar altitude is 56.25 degrees.

The solar azimuth angle is given by Eq. (2.10) as

$$sin\phi = \frac{cos\delta sinH}{cos\beta} = \frac{cos16.83 \times \sin 30}{cos56.25} = 0.861$$

Therefore solar azimuth is 59.48 degrees west of south.

The surface-solar azimuth of the collector is

$$\gamma = 30 + 59.48 = 89.48 \text{ degrees.}$$

The tilt angle of the collector is, $\Sigma = 35^{\circ}$. The angle of incidence, θ_c is given by Eq. (2.14) as

$$cos\theta_c = \ cos\beta sin\,\Sigma\, cos\gamma + sin\beta cos\,\Sigma$$

$$cos\theta_c = \ cos56.25 sin35 cos89.48 + sin56.25 cos35 = 0.684$$

Hence the angle of incidence on the collector surface is 46.84°

For a horizontal surface $\Sigma = 0^{\circ}$. Therefore the angle of incidence is

$$cos\theta_h = \sin\beta = \ 0.831$$

Hence the angle of incidence on a horizontal surface is 33.75°.

The direct beam radiation intensity on the collector surface is

$$I_b = H_d\left(\frac{cos\theta_c}{cos\theta_h}\right) = 640\left(\frac{0.684}{0.831}\right) = 526.8 \text{ Wm}^{-2}$$

The diffuse radiation intensity on the collector surface is given by Eq. (2.18) as

$$I_{df} = \frac{(1+cos\Sigma\,)H_{df}}{2} = \frac{(1+cos35)\times 210}{2} = 191 \text{ Wm}^{-2}$$

Total radiation absorbed by the collector is

$$I_t = (\tau\alpha)_b I_b + (\tau\alpha)_d I_{df}$$

$$I_t = 0.74 \times 526.8 + 0.65 \times 191 = 514 \text{ Wm}^{-2}$$

Rate of useful energy collection is

$$Q = A_c F_R\left[I_t(\tau\alpha) - U_L\left(T_{fi} - T_a\right)\right]$$

$$Q = 2 \times 0.82[514 - 7.5(50 - 18)] = 449.4 \text{ W}$$

The efficiency of the collector is

$$\eta = \frac{Q/A}{I_b + I_{df}} = \frac{449.4/2}{526.8+191} = 31.3\%$$

Example 3.20 The solar process heat system shown in Fig. E3.20.1 incorporates a heat exchanger between the collector and the storage tank. The collector area is 4 m^2, the loss coefficient is 8 Wm^{-2}, and the heat removal factor is 0.8. The modified heat removal factor accounting for the heat exchanger is 0.76. The storage tank has 150 kg of water, and the heat loss coefficient - outside tank area product, $(UA)_{tank}$ is 1.5 WK^{-1}. The tank is in a room at 22°C.

The constant water flow rate to the process plant is 8 kg per hour. The tank is replenished with water at 20°C from the mains. The system begins operating at 7 a.m. when the tank temperature is 42°C. Based on the solar radiation at the location and the transmittance-absorptance of the collector, the radiation absorbed by the collector, S (MJm^{-2}) was calculated using the procedure outlined in worked example 3.19 above.

These values and the hourly ambient temperatures are listed in Table E3.20.1, for each hour until 5 p.m. calculate the following

Table E3.20.1 Hourly energy absorbed by the collector and the hourly ambient temperature. S -MJm^{-2}, T_{amb} – °C

Time	7-8	8-9	9-10	10-11	11-12
S	-	0.35	0.81	3.26	3.15
T_{amb}	-10	-8	-4	3	4

Time	12-1	1-2	2-3	3-4	4-5
S	3.32	3.12	1.62	0.97	-
T_{amb}	7	7.5	8	9	7

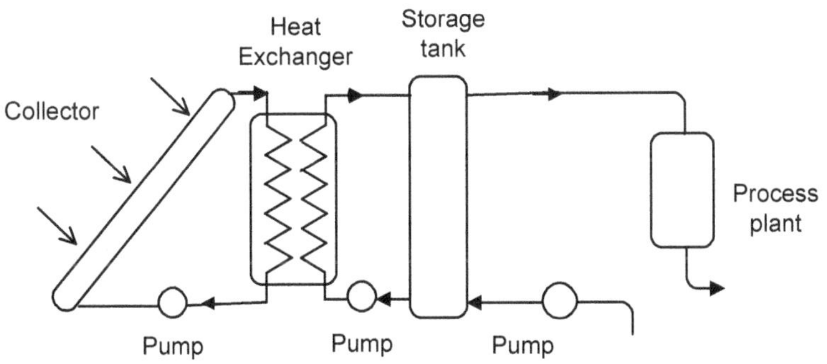

Fig. E3.20.1 Process plant with collector-storage heat exchanger

quantities for each hour: (i) the heat supplied to the process from the storage tank, (ii) the hourly heat loss from the storage tank, (iii) the hourly useful energy collected and transferred to the tank, and (iv) the variation of the storage tank temperature.

Solution We assume that all sub-components of the water heating system except the storage tank operate in a quasi-steady manner. In other words only the time variation of the storage tank temperature is included in the analysis of the system. All other variables that characterize the system are hourly averaged quantities. Heat losses from the piping, pumps, and the heat exchanger are neglected.

Let the mean temperature of the storage tank at the beginning of the hour be T_{si} and at the end of the hour T_{sf}. The heat supplied to the process and the heat loss from the tank during an hour is based on the temperature at the beginning of the hour.

Heat supplied to the process during an hour is

$$Q_{pr} = m_p c_{pw}(T_{si} - T_{lr}) \tag{E3.20.1}$$

$$Q_{pr} = 8 \times 4190(T_{si} - 20)/10^6 \text{ MJ} \tag{E3.20.1a}$$

Heat loss from the storage tank during an hour is

$$Q_{tl} = (AU)_t(T_{si} - T_{room}) \tag{E3.20.2}$$

$$Q_{tl} = 1.5(T_{si} - 22) \times 3600/10^6 \tag{E3.20.2a}$$

Useful energy supplied from the collector to the storage tank is

$$Q_u = A_c F'_R[S_h - U_L(T_{si} - T_a)]^+ \tag{E3.20.3}$$

$$Q_u = 4 \times 0.76[S_h - 8(T_{si} - T_a)]^+ \tag{E3.20.3a}$$

In the above equations the + sign indicates that only positive values of the quantity within square brackets are acceptable. If the value is negative then no energy collection from the collector is possible and therefore collector pump and the heat exchanger pump are switched off.

The change in storage tank temperature during an hour is given by

$$T_{sf} - T_{si} = \frac{Q_u - Q_{pr} - Q_{tl}}{(Mc)_{tank}} = \frac{Q_u - Q_{pr} - Q_{tl}}{150 \times 4190} \tag{E3.20.4}$$

To illustrate the numerical computation we shall consider two typical hours of operation. For the hour from 7 to 8 a.m. the following quantities apply.

Heat supplied to the process during an hour is given by Eq. (E3.20.1a)

$$Q_{pr(7-8)} = 8 \times \frac{4190(42-20)}{10^6} = 0.737 \text{ MJ}$$

Heat loss from the storage tank during an hour is given by Eq. (E3.20.2a)

$$Q_{tl(7-8)} = 1.5(T_{si} - 22) \times \frac{3600}{10^6} = 0.108 \text{ MJ}$$

The useful energy collected is zero because there no radiation input.

The change in temperature of the storage tank is given by Eq. (E3.20.4)

$$T_{sf} - T_{si} = \frac{Q_u - Q_{pr} - Q_{tl}}{(Mc)_{tank}}$$

$$T_{sf} - T_{si} = \frac{(0-0.737-0.108)\times 10^6}{150\times 4190} = -1.35°\text{C}$$

Hence the temperature of the storage tank at 8 a.m. is:

$$T_{s-8} = 42 - 1.35 = 40.65°\text{C}$$

We apply the same set of equations to the period from 8 to 9 a.m. and 9 to 10 a.m. The temperature at 9 a.m. is 39.39°C and the temperature at 10 a.m. is 38.21°C. No solar energy is collected during these two hours. Consider the period from 10 a.m. to 11 a.m.

Heat supplied to the process from 10 to 11 a.m. is given by Eq. (E3.20.1a)

$$Q_{pr(10-11)} = 8 \times \frac{4190(38.21-20)}{10^6} = 0.61 \text{ MJ}$$

Heat loss from the storage tank from 10 to 11 a.m. given by Eq. (E3.20.2a)

$$Q_{tl(10-11)} = 1.5(38.21 - 22) \times \frac{3600}{10^6} = 0.088 \text{ MJ}$$

The useful energy collected is given by Eq. (E3.20.3a) as

$$Q_{u(10-11)} = 4 \times 0.76 \times \frac{3600[905-8(38.21-3)]^+}{10^6} = 6.82 \text{ MJ}$$

The change in temperature of the storage tank is given by Eq. (E3.20.4)

$$T_{sf} - T_{si} = \frac{Q_u - Q_{pr} - Q_{tl}}{(Mc)_{tank}} = \frac{6.82 - 0.61 - 0.088}{0.15 \times 4.19} = 9.74°\text{C}$$

Hence the temperature of the storage tank at 11 a.m. is:

$$T_{s-11} = 38.21 + 9.74 = 47.95°\text{C}$$

The above computational procedure is continued to cover all time periods up to 5 p.m. The results obtained are tabulated below.

Table E3.20.2 Computed hourly values of variables

Time	7-8 am	8-9 am	9-10 am	10-11am	11-12 am
S (MJm^{-2})	-	0.35	0.81	3.26	3.15
$T_{amb.}$ (°C)	-10	-8	-4	3	4
T_{si} (°C)	42	40.65	39.4	38.2	47.95
T_{sf} (°C)	40.65	39.4	38.2	47.95	55.3
Q_{useful} (MJ)	0	0	0	6.82	5.72
Q_{loss} (MJ)	0.108	0.101	0.094	0.088	0.14
Q_{load} (MJ)	0.737	0.692	0.65	0.61	0.94

Time	12-1 pm	1-2 pm	2-3 pm	3-4 pm	4-5 pm
S (MJm^{-2})	3.32	3.12	1.62	0.97	-
$T_{amb.}$ (°C)	7	7.5	8	9	7
T_{si} (°C)	55.3	62.5	67.3	64.4	61.65
T_{sf} (°C)	62.5	67.3	64.4	61.65	59.1
Q_{useful} (MJ)	5.86	4.67	0	0	0
Q_{loss} (MJ)	0.18	0.22	0.24	0.23	0.21
Q_{load} (MJ)	1.18	1.42	1.59	1.49	1.40

The total energy quantities for the 10 hours are as follows: (i) the total useful energy collected is 23.07 MJ, (ii) the total energy supplied to the load is 10.71 MJ, (iii) the total energy lost from the tank is 1.16 MJ, (iv) the net energy added to the storage tank during the period is 10.75 MJ. The total energy consumed and stored in the tank is the sum of items (ii) to (iv), which is equal to 23.07MJ. This agrees with total useful energy collected during the day.

Problems

P3.1 A water heating flat plate collector of area 2 m^2 has a water flow rate of 0.038 kgs^{-1}. The water temperature increases from 40°C to 47°C. Solar radiation absorbed by the collector is 780 Wm^{-2}. The average plate temperature is 55°C, and the ambient temperature is 18°C. Calculate (i) the rate of heat loss from the collector, and (ii) the top heat loss coefficient. [*Answers:* (i) 445.5 W, (ii) 6 $Wm^{-2}K^{-1}$]

P3.2 A flat plate collector has two glass covers of thickness 4 mm each, and a collector plate with an absorptivity of 0.86 for solar radiation. A beam of solar radiation of intensity 815 Wm^{-2} is incident on the collector at an angle of 35°. The extinction coefficient of glass is 20 m^{-1}, and the refractive index of glass for solar radiation is 1.53. Calculate the rate of energy absorption by the absorber plate. [*Answer*: 591.7 Wm^{-2}]

P3.3 A flat plate solar collector with a single cover is inclined at 40° to the horizontal. The air gap between the absorber plate and the glass cover is 24 mm. The temperatures of the absorber plate and the cover are 80°C and 40°C respectively. The wind speed over the collector surface is 6.2 ms^{-1}. Calculate (i) the convective heat transfer coefficient between the plate and the cover, and (ii) the wind heat transfer coefficient. [*Answers*: (i) 3.44 $Wm^{-2}K^{-1}$, (ii) 21 $Wm^{-2}K^{-1}$]

P3.4 A flat plate solar collector has a single glass cover, and it is inclined at 30° to the horizontal. The spacing between the plate and the cover is 24 mm. The plate is at a temperature of 70°C, and the ambient temperature is 20°C. The wind speed above the cover is 7.2 ms^{-1}. The emissivities of the plate and cover for long wave radiation are 0.92 and 0.88 respectively. Calculate the top heat loss coefficient.
[*Answer*: 7.4 $Wm^{-2}K^{-1}$]

P3.5 A flat plate solar collector has two glass covers, and it is inclined at 40° to the horizontal. The spacing between the plate and the inner cover is 24mm, and the spacing between the two covers is 20 mm. The plate is at a temperature of 75°C, and the ambient temperature is 15°C.

The wind speed above the cover is 6 ms^{-1}. The emissivities of the plate and cover for long wave radiation are 0.92 and 0.88 respectively. Calculate the heat loss coefficient using the empirical equation.
[*Answer*: 3.91 $Wm^{-2}K^{-1}$]

P3.6 A single-cover collector has a plate-fin absorber as shown in Fig. 3.5(e). Copper plates of thickness 0.5 mm connect the copper flow tubes of internal diameter 12 mm and thickness 0.5 mm. The center-to-center distance of the tubes is 15 cm. The heat transfer coefficient for water flow in the tubes is 620 $Wm^{-2}K^{-1}$. The water flow rate in the tubes is 0.02 kgs^{-1}.

The transmittance of the cover is 0.9 and the absorptance of the copper surface is 0.86. The width and length of the collector are 1.5 m and 2.5 m respectively. The total solar radiation incident on the collector is 730 Wm^{-2} and the ambient temperature is 17°C. The heat loss coefficient is 6 $Wm^{-2}K^{-1}$, and the thermal conductivity of copper is 390 $Wm^{-1}K^{-1}$. The inlet water temperature to the collector is 48°C. Calculate (i) the collector efficiency factor, (ii) the heat removal factor, (iii) the rate of useful energy collection, and (iv) the efficiency of the collector. [*Answers*: (i) 0.92, (ii) 0.815, (iii) 1158 W, (iv) 42.3%]

P3.7 An evacuated tube collector has an absorber tube of diameter 50 mm and a glass cover of diameter 80 mm. The space between them is evacuated. The absorber surface has an emittance of 0.92 and its temperature 160°C. The emittance of the cover is 0.88. The wind speed and ambient temperature are 6 ms^{-1} and 15° C respectively. Calculate (i) the overall heat loss coefficient based on the inner tube area, and (ii) the rate of heat loss. [*Answers*: (i) 8.27 $Wm^{-2}K^{-1}$, (ii) 188.3 Wm^{-1}]

P3.8 A flat plate solar air heater is inclined at 30° to the horizontal. Solar radiation incident on the collector surface is 865 Wm^{-2}. The top heat loss coefficient of the collector is 4.5 $Wm^{-2}K^{-1}$, and the effective transmittance-absorptance product is 0.82. The inlet air temperature is 50°C and the air mass flow rate is 0.056 kgs^{-1}. The ambient temperature is 25°C.

The width and length of the collector are 1 m and 3 m respectively. The depth of the air flow channel is 15 mm. Neglect heat losses from the back of the collector. Calculate (i) the outlet temperature of the air, (ii) the useful rate of energy collection, and (iii) the efficiency of the collector. [*Answers*: (i) 73°C, (ii) 1303 W, (iii) 50.2%]

P3.9 The flat plate collector of a water heating system has an area 2.4 m^2, a heat removal factor F_R of 0.84 and a loss coefficient of 6.5 $Wm^{-2}K^{-1}$. The transmittance-absorptance for direct beam and diffuse radiation are 0.76 and 0.68 respectively. The collector faces 15° east of south, and is tilted at 35° from the horizontal.

The latitude of the location is 30°N. At 11.00 a.m. on April 12, the measured direct beam solar radiation on a horizontal surface at the location is 720 Wm^{-2}, and the diffuse radiation is 185 Wm^{-2}. The ambient temperature is 22°C and the inlet water temperature to the collector is 45°C. Calculate (i) the rate of useful energy collection, and (ii) the efficiency of the collector. [*Answers*: (i) 1118.7 W, (ii) 49.3%]

P3.10 A solar heating system has a heat exchanger that transfers heat from the collector to the storage tank as shown in Fig. E3.20.1. The total collector area is 6 m^2, and the heat loss coefficient is 7.5 Wm^2. The modified heat removal factor accounting for the heat exchanger is 0.78. The storage tank has 200 kg of water, and the heat loss coefficient - outside area product is 2.2 WK^{-1}. The tank is in a room at 18°C.

Hot water is supplied from the storage tank to a heating process at the rate of 10 kg per hour. The tank is replenished with water at 18°C from the mains. The system begins operating at 7 a.m. when the tank temperature is 46°C. Based on the solar radiation at the location and the transmittance-absorptance of the collector, the radiation absorbed by the collector, S (MJm^{-2}) was calculated using the procedure outlined in worked example 3.19 above. These values and the hourly ambient temperatures are listed in Table P3.20.1, for each hour until 5 p.m.

(a) Calculate the following quantities for each hour: (i) the heat supplied to the process from the storage tank, (ii) the hourly heat loss from the storage tank, (iii) the hourly useful energy collected and transferred to the tank, and (iv) the variation of the storage tank temperature.

(b) For the 10-hour period of operation, calculate the total useful energy supplied to the tank from the collector, the total heat supplied to the load, and the total heat loss from the storage tank.

Table P3.20.1 Hourly energy absorbed by the collector and the hourly ambient temperature. Units: S - MJm^{-2}, T_{amb} - °C

Time	7-8	8-9	9-10	10-11	11-12
S	-	0.42	0.92	3.8	3.6
$T_{amb.}$	-4	-2	2	5	6

Time	12-1	1-2	2-3	3-4	4-5
S	4.2	3.9	2.2	1.2	-
$T_{amb.}$	8	8.5	8	9	7

References

1. Duffie John A., Beckman William A., with Blair Nathan, *Solar Engineering of Thermal Processes, Photovoltaics and Wind*, John Wiley & Sons, New Jersey, USA (2020).
2. Gordon J.M. and Rabl A., Design, Analysis and Optimization of Solar Industrial Process Heat Plants without Storage, *Solar Energy* 28(6), (1982), pp 519-530.
3. Goswami, Y., Kreith Frank, and Kreider Jan F., *Principles of Solar Energy Engineering*, 2nd Edition, Taylor & Francis, Philadelphia, PA, USA (2000).
4. Kreith Frank and Krumdieck Susan, *Principles of Sustainable Energy Systems*, CRC Press, Taylor & Francis Group, New York, USA (2015).
5. Proctor D and Morse R. N, Solar Energy for the Australian Food Processing Industry, *Solar Energy*, 19, (1977), pp 63-77.
6. Rogers G.F.C. and Mayhew Y.R., *Thermodynamic and Transport Properties of Fluids.* 5th ed. Blackwell, Oxford, U.K. 1998.
7. Wijeysundera, N.E. and Ho, J.C., Long Term Thermodynamic Performance of Solar Heat and Power Systems, *Applied Energy*, 24 (1986), pp 43-63.

Chapter 4

Solar Thermal Power Generation

4.1 Introduction

An important application of solar energy is the production electricity, which can be done using the photoelectric effect or conventional heat engine cycles. The practical device that uses the former is the photovoltaic (PV) cell that converts solar radiation directly to electricity. On the other hand, solar thermal power systems use the heat absorbed by a solar collector to operate a Rankine-cycle power plant [1,2]. In essence, the solar collector system becomes the heat source of the power cycle.

Solar thermal power plants may be classified as low, medium and high temperature cycles, based on the operating temperature of the solar collector. Low temperature cycles, typically use flat plate collectors, and the operating temperatures are limited to about 100°C. The temperature range of operation for medium temperature cycles is from about 100°C to 300°C. High temperature cycles work at temperatures above 300°C. Early attempts to harness solar energy to produce mechanical power using various designs in different temperature ranges are reviewed in Ref.[2].

4.1.1 *Low-temperature solar power systems*

A low temperature solar thermal power plant using a solar pond as the heat source is depicted in Fig. 4.1. Such systems have been built and tested in Israel and Australia. The solar pond is a natural or artificial water body, where salt has been mixed to create a concentration gradient along its depth. The salt concentration increases with depth up to a certain level and then remains constant.

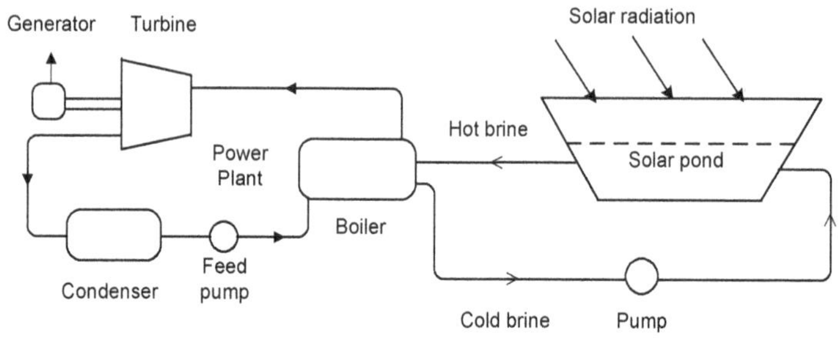

Fig. 4.1 Low temperature solar pond power plant

Usually, as solar radiation is absorbed in a body of water, the resulting increase in temperature promotes convection through which heat is lost to the ambient. Thus the water temperature remains relatively low. However, in a solar pond the concentration gradient in the upper layer, compensates for the density change caused by the increase in temperature, and thereby convection is suppressed. The upper layer behaves like an insulator and thus heat is stored in the lower convection layer. Consequently, its temperature rises. In practical solar ponds temperatures of up to 100°C have been reached under favorable conditions.

In the solar power system, brine from the bottom layer of the solar pond is circulated through the boiler of the Rankine cycle power plant to generate vapor (see Fig. 4.1). The vapor expands through the turbine to operate the electric generator. Cold brine exiting from the boiler is returned to the bottom layer of the solar pond. The working fluid of the power plant is usually a low boiling liquid, such as a hydrocarbon or an organic fluid. An advantage of the solar pond is that it provides heat storage that helps deal with the intermittency of solar radiation.

A low temperature solar thermal power system using flat plate collectors is shown in Fig. 4.2. It is like the plant described above, except that the hot water produced by a collector is used to generate vapor in the boiler. Such a system was built in France, using butane as the working fluid, to pump water [1]. Although not shown in the figure, this type of system is likely to incorporate a storage tank, and an auxiliary energy source such as a natural gas heater to deal with periods when useful energy collection is not possible due to low radiation levels. Such a system is depicted in Fig. 3.11(c).

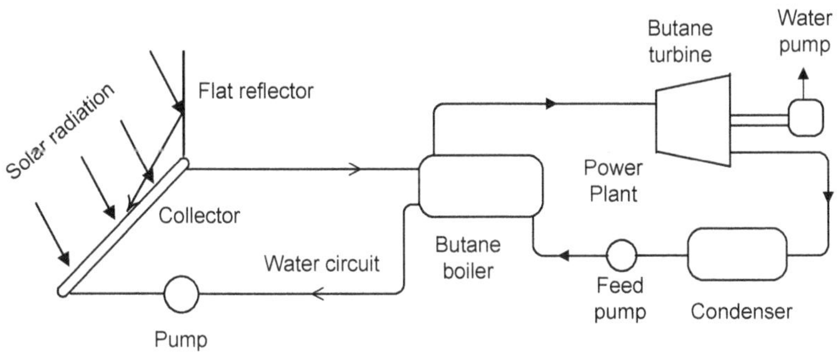

Fig. 4.2 Low temperature solar power plant for water pumping

4.1.2 *Medium and high temperature power systems*

Trough-concentrator systems

The medium temperature solar power plant shown schematically in Fig. 4.3, has a long parabolic reflector, commonly called a trough, that focuses the solar beam onto a cylindrical tube located at the focal-line of the parabola. A field of collectors are arranged in rows, with their axis typically aligned in the north-south orientation to maximize solar energy collection. This arrangement enables the concentrator to track the sun from east to west during the day and concentrate sunlight continuously onto the receiver tube. The solar receiver, which usually includes various design innovations like selective absorbers, is a key component of this system. It is located concentrically within a glass tube to minimize convective heat losses from the absorber tube. Under favorable conditions, the heat-transfer fluid flowing through the collector-tubes is able to reach temperatures of about 350°C [9].

This type of solar-thermal power plant includes a heat storage system to deal with the intermittent nature of solar radiation, and for night time operation. Several innovative storage designs have been tested in practice. In the two-tank direct system, depicted in Fig. 4.3, the cooling fluid is pumped between two tanks through the solar collector in a closed loop. Fluid from the hot-fluid tank is pumped through a heat exchanger to generate steam for use in a Rankine-cycle thermal power plant.

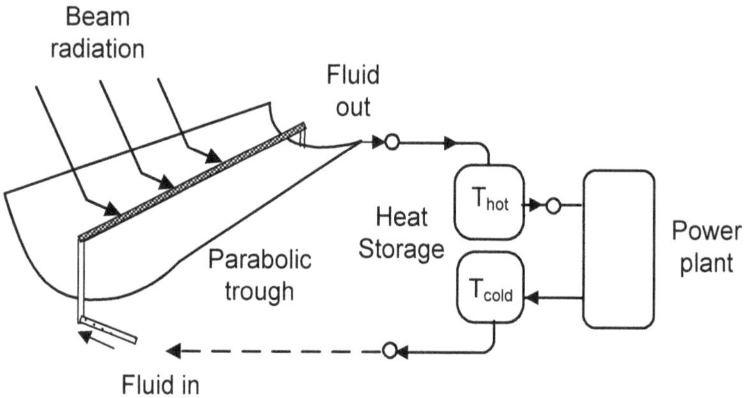

Fig. 4.3 Parabolic trough solar thermal power system

The low temperature fluid exiting from the heat exchanger is stored in the cold-fluid storage tank, to be heated by pumping it through the collectors. Commonly, mineral oil is used as the heat-transfer and storage fluid. Several modifications have been made to this basic heat storage tank design to improve its performance and reduce cost. The power ratings of a number operating parabolic trough power plants in different countries are listed in Table 4.1.

Fresnel mirror systems

A variation of a linear concentrating system, called a *Fresnel* mirror design is depicted schematically in Fig. 4.4. In this case a series of mirrors, either flat or slightly curved is located in a horizontal row at the bottom of a structure which supports a receiver tube at the top. The tube and the mirrors are oriented in the north-south direction, and the mirrors track the sun as it moves from east to west. In some designs a secondary reflector is placed above the absorber and parallel to it to reflect any radiation that misses the absorber as illustrated in Fig. 4.4. The rest of the design is essentially the same as the parabolic trough system shown in Fig. 4.3.

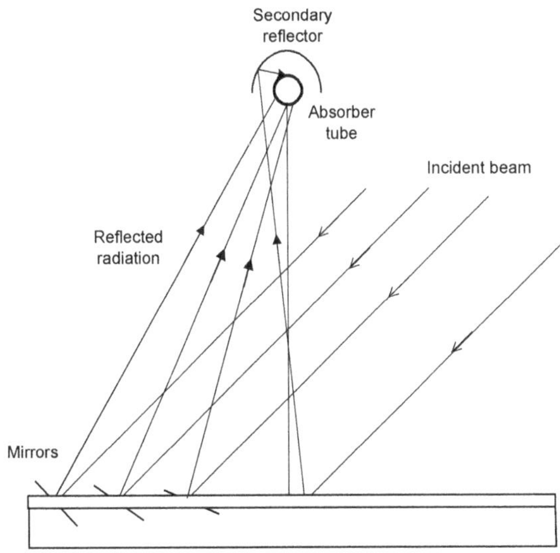

Fig. 4.4 A linear Fresnel mirror system

Solar power towers

The solar power tower [SPT] is a thermal power system which uses a large field of flat or slightly curved mirrors, called heliostats, to reflect and concentrate sunlight onto a receiver located at the top of a tall tower, as shown schematically in Fig. 4.5. Optically the solar power tower functions like a circular Fresnel reflector. The mirrors track the sun continuously with the aid of an automated control system. Molten salt, which is a heat transfer and storage medium, is pumped up the tower, where concentrated sunlight heats it to about 1000°C. Hot molten salt is stored in tanks, from which it is pumped through a heat exchanger to generate steam necessary to operate a conventional Rankine-cycle power plant.

In current designs the areas of heliostats are between 40 to 70 m^2 [1]. Ideally, the surface should be slightly parabolic with focal length equal to the distance between the mirror and the receiver. Because the latter distance is large compared to the dimensions of the mirror, a slightly spherical or a flat surface offers a more convenient alternative. At present

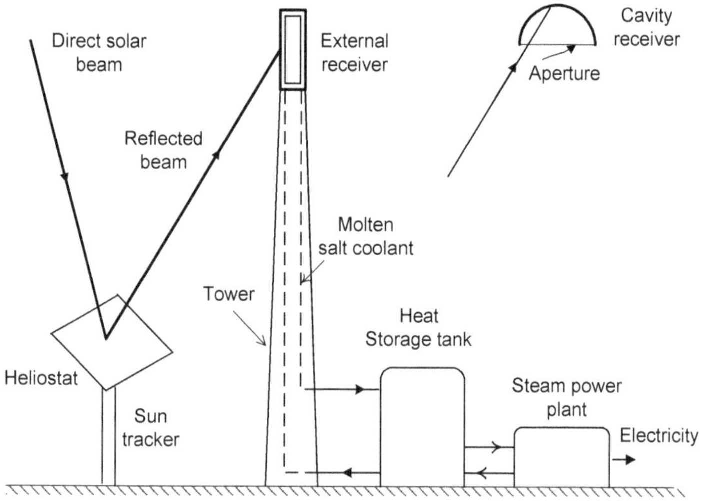

Fig. 4.5 Solar power tower - CSP system

glass and plastic are the two materials used as reflective surfaces. Plastic surfaces are light but less reflective than glass surfaces, that are much heavier. The use of plastic makes the support structure and the drive mechanism lighter and cheaper.

Tracking the sun about two axis can be done using commands from a computer that calculates the exact position of the sun at any moment using an algorithm. Alternative, a sensor on the mirror may be used to determine the position of the sun. However, the latter method is effected by passing clouds that may obscure the sun, and the sensor would become ineffective for a period.

The central receiver at the top of the tower has a radiation absorbing surface as in all solar collectors. Two possible designs are shown schematically in Fig. 4.5. In the external receiver, the absorber surface is on the outside of the cylindrical structure which is lined with cooling tube panels. These are flat for small plants and slightly curved for larger receivers.

In the cavity-type receiver, solar radiation enters the receiver through an aperture, and the cooling pipes line the interior of the cavity. Focusing solar radiation may be more critical because it has to enter through the aperture to be absorbed. However, the cavity receiver has lower reflection and convection losses because the absorbing surface is not exposed directly to the ambient. The latter losses are larger for the external receiver, but it has the advantage of being a simpler design that has a larger absorber area and allows for easier access for maintenance [4].

SPT systems have the advantage that the smaller receiver looses less heat compared to parabolic trough systems that have long lengths of piping through which the cooling fluid is pumped across the collector field to the heat storage tank. On the other hand, the transfer of reflected solar radiation from the heliostat field to the receiver of the SPT system is direct, and involves fewer losses. Moreover, in a large distributed system, significant amount of energy is used to pump the heat transfer fluid through the parabolic collector field. These pumping costs, however, will be offset by the somewhat high cost of heliostats and the central receiver tower of the SPT system.

Dish concentrators

A third solar thermal power system, called the solar dish-engine system [2,9] is depicted in Fig. 4.6. It uses a mirrored dish with a parabolic shape to concentrate sunlight onto a receiver at the focus of the parabola. The dish tracks the sun continuously about two axis with the aid of a control system. Heat collected at the receiver is used as the heat source of a Stirling engine that operates an electric generator. Temperatures of about 700°C have been achieved with these systems [9].

The operating temperature of all concentrator-type solar thermal power plants has to satisfy two opposing requirements. As the operating temperature increases, the efficiency of the solar collector decreases because of increased heat loss to the ambient. However, the efficiency of the Rankine cycle power plant increases with temperature [2,9]. The overall system efficiency, which is the fraction of the total incident solar radiation converted to electrical energy, is the product of the collector efficiency and the Rankine cycle efficiency. Hence, a compromise between the above two opposing effects has to be found when choosing the operating temperature of the collector system. Currently operating

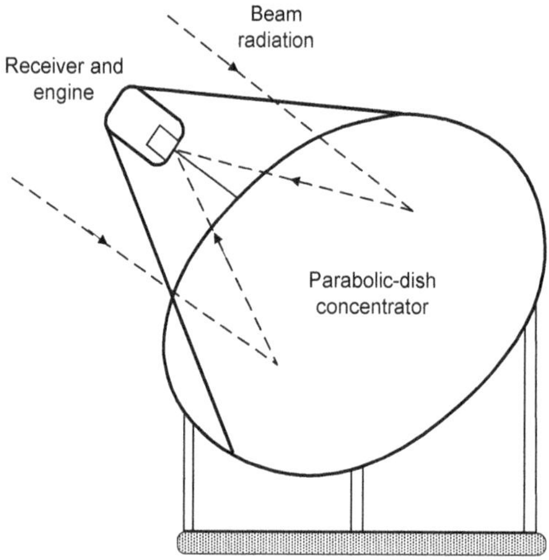

Fig. 4.6 Parabolic dish - CSP system

Table 4.1 Some of the larger operating solar power plants (Data from Wikipedia, Ref.[8])

Country	Location	Coordinates (approx.)	Electrical Capacity	Technology Type
Morocco	Ghassate	31° N, 7° W	510 MW	Parabolic trough, Power tower
USA	San Bernadino, CA	35° N, 115° W	392 MW	Power tower
USA	Mojave Desert, CA	35° N, 117° W	310 MW	Parabolic trough
USA	Barstow, CA	35° N, 117° W	280 MW	Parabolic trough
USA	Nye county	38° N, 117° W	125 MW	Power tower
Spain	Logrosan	39° N, 5° W	200 MW	Parabolic trough
Spain	Sanlucar la Mayor	37° N, 6° W	150 MW	Parabolic trough
Spain	Guadix	37° N, 3° W	150 MW	Parabolic trough
India	Dhursar	27° N, 72° E	125 MW	Fresnel reflector
Isreal	Ashalim	31° N, 35° E	121 MW	Parabolic trough
Isreal	Ashalim	31° N, 35° E	121 MW	Power tower
South Africa	Northern Cape	28° N, 23° W	100 MW	Parabolic trough

solar thermal power plants have overall efficiencies up to about 30%. In worked examples 4.2 and 4.3 the above optimum condition is computed.

A list including some of the larger concentrating solar power (CSP) systems that are in operation in different countries is given in Table 4.1[8]. It is seen that more CSP systems have been constructed based on the parabolic trough (PT) design compared to the solar power tower (SPT) design.

In this section we presented a brief introduction to solar thermal power generation systems. The design concepts used in large-scale plants, that have been built and are in operation, are the parabolic trough, the Fresnel mirror, and the power tower. These systems include several common sub-components like, solar concentrators, thermal power plants, and heat removal and storage systems. In the following sections we shall discuss the important design aspects of these sub-components.

4.2 Solar Concentrators

Some applications of solar energy, in particular, solar thermal power generation and high-temperature industrial processes require heat delivery at elevated temperatures. A solar receiver maybe raised to a high temperature by decreasing its area and focusing on to it solar radiation incident on a much larger area. This process is called *solar concentration*. Reflecting and refracting optical devices of various designs are commonly used for this purpose. Such devices are called *solar concentrators*. Solar power generation systems introduced in the previous section employ different types of concentrators to raise the temperature of the working fluid. These include two-dimensional parabolic troughs, Fresnel reflectors, paraboloidal surfaces of revolution, and arrays of heliostats. A few low-concentration devices like flat mirrors, conical reflectors and non-imaging reflectors are depicted in Fig. 4.7.

The ratio of the *aperture area* of the reflector to the *area of the receiver* is called the *concentration ratio*, which can range from low values of 1.5 to 2 for flat mirrors, to high values of the order of 10,000 for parabolic concentrators. Increasing the concentration ratio increases

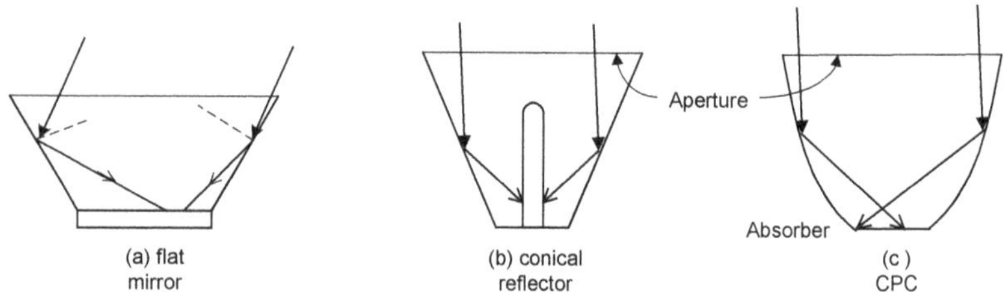

Fig. 4.7 (a) Flat mirror (b) Conical reflector (c) Compound-parabolic (CPC)

the temperature of the receiver. However, in practice this requires very precise orientation of the concentrator and tracking the sun continuously, which increases the complexity and cost of the optical system. The effect of angular misorientation on the performance of conical, spherical and parabolic concentrators is discussed in Ref.[7]. Moreover, the reflecting surfaces must be kept clean and free of dust in an outdoor setting, that is not always conducive. The reflector surface and the support structure need to be sufficiently rigid to avoid vibrations due to wind forces, which could make accurate focusing difficult.

4.2.1 *Upper limit of concentration ratio*

The area-based concentration ratio (CR) of a concentrator is defined as the ratio of the aperture area to receiver area. It is possible to obtain the upper limit of CR based on the second law of thermodynamics [1,2].

Consider the ideal three-dimensional concentrator shown in Fig. 4.8, a practical version of which is a paraboloidal concentrator depicted in Fig. 4.6. The aperture and receiver areas of the energy collecting device are A_a and A_r respectively, the radius of the sun is r, and the distance of the concentrator from the center of the sun is R. Let the temperatures of the sun and the receiver be T_s and T_r respectively. Assuming that the sun and receiver are black bodies we can write the following expressions for radiation transfer between them.

Solar radiation incident on the aperture of the concentrator is given by

$$Q_{s-a} = \frac{\sigma T_s{}^4 r^2 A_a}{R^2} \tag{4.1}$$

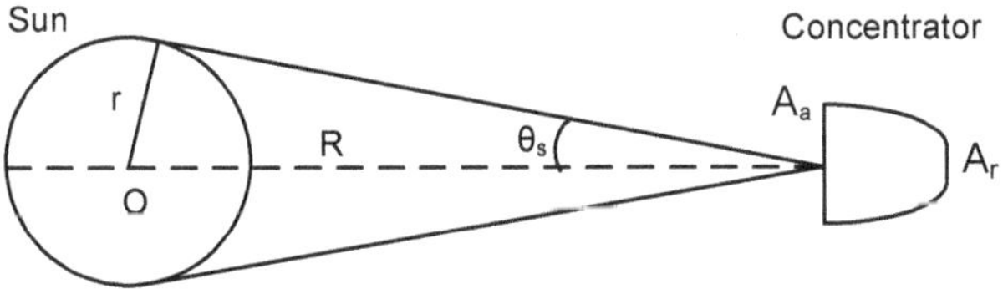

Fig. 4.8 Maximum concentration ratio of an ideal concentrator

Radiation emitted by the receiver that is incident on the sun is given by

$$Q_{r-s} = \sigma T_r{}^4 A_r F_{r-s} \tag{4.2}$$

The view factor between the receiver and the sun is F_{r-s}. For thermodynamic equilibrium

$$Q_{s-a} = Q_{r-s} \tag{4.3}$$

Now the view factor between the receiver and the sun, $F_{r-s} = 1$. From the three equations above it follow that

$$\frac{\sigma T_s{}^4 r^2 A_a}{R^2} = \sigma T_r{}^4 A_r \tag{4.4}$$

Equation (4.4) is valid under all conditions. Therefore at thermal equilibrium, when $T_r = T_s$, we obtain the upper limit of the concentration ratio as

$$(CR)_{3D} = \frac{A_a}{A_r} = \frac{R^2}{r^2} = \frac{1}{sin\theta_s{}^2} \tag{4.5}$$

where θ_s is the half-angle subtended by the sun on the earth, which is equal to 0.27°. Hence we obtain the upper limit of CR for a three-dimensional concentrator as 45,000. For a two-dimensional concentrator like the parabolic trough shown in Fig. 4.3, the upper limit of the concentration ratio is given by [2]

$$(CR)_{2D} = \frac{R}{r} = \frac{1}{sin\theta_s} \tag{4.6}$$

The numerical value of the upper limit in this case is 212.

4.2.2 *Nonimaging concentrators*

There are two categories of concentrators used in solar energy applications, commonly called imaging concentrators and nonimaging concentrators. The parabolic concentrators shown in Figs. 4.3 and 4.4 belong to the former category, where solar radiation falling on the receiver surface forms an image of the sun. On the other hand, non-imaging concentrators, as the name implies, do not form a distinct image, but instead funnel the radiation from the aperture to the receiver.

For a three dimensional nonimaging concentrator, all radiation within a cone of a particular angle reaches the receiver, either directly or by multiple reflections. In the case of a two-dimensional nonimaging concentrator, radiation within a wedge with a specified angle will be directed to the receiver. The above angles of the cone and the wedge are known as the *acceptance angles*. The actual design of a nonimaging concentrator depends on the shape of the receiver. We shall illustrate the important design aspects of a nonimaging concentrator by referring to the flat receiver depicted in Fig. 4.9. This two-dimensional concentrator consists of two parabolic reflector-sections AB and CD, and it is called a *compound parabolic concentrator* (CPC).

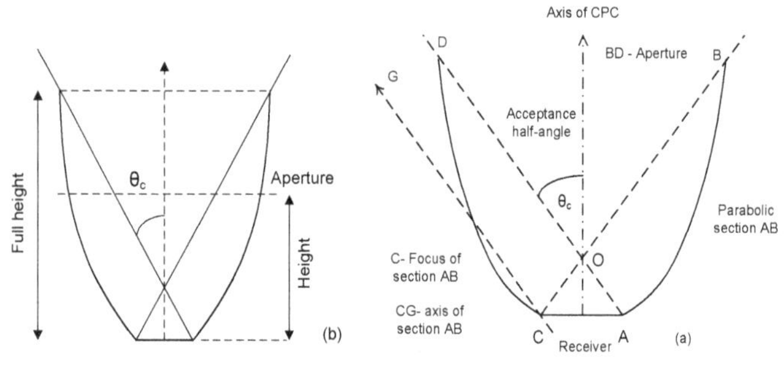

Fig. 4.9 (a) A compound parabolic concentrator (CPC) for a flat absorber (b) Truncated CPC

The focus of the parabolic segment AB is at C and the focus of segment CD is at A, and the axis of segment AB is along CG. The acceptance half-angle of the CPC is θ_c. All radiation incident on the aperture with directions within the angle DOB will reach the receiver surface AC under ideal conditions. The concentration ratio of the CPC can be obtained by using a procedure similar to that used earlier to derive Eq. (4.6) [1]. It is given by the expression

$$(CR)_{CPC} = \frac{1}{sin\theta_c} \tag{4.7}$$

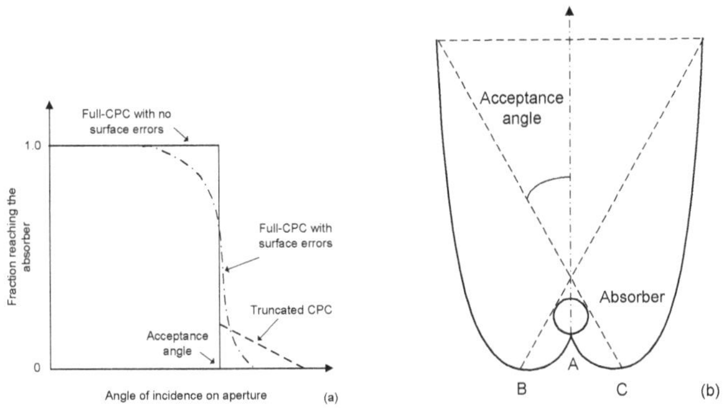

Fig. 4.10 (a) Effect of truncation and surface errors (b) CPC for a round absorber

The concentration ratio of a full-CPC for different acceptance angles is listed in Table 4.2. For the full-CPC the upper ends of the parabolic reflector sections are parallel to the axis of the CPC as seen in Fig. 4.9(a). Different parts of the reflector segments contribute differently to the total radiation reaching the receiver surface, with the upper part contributing the least amount. Therefore the reflector sectors could be truncated as indicated in Fig. 4.9(b) without diminishing the performance significantly. The advantage of truncation is that it reduces both the size and the cost of the reflector. Truncation does not affect the acceptance angle but changes the average number of reflections undergone by the radiation beam before reaching the receiver. Several optical and geometrical properties, like the height to aperture ratio, the reflector length to aperture ratio, and the average number of reflections for a full-CPC and a truncated-CPC are listed in Table 4.2[2].

The variation of the fraction of radiation incident on the aperture that reaches the receiver with the angle of incidence is shown graphically in

Fig. 4.10(a). For the ideal CPC with no surface errors a sharp cut-off occurs when the incidence angle exceeds the acceptance half-angle, θ_c. With the truncated CPC some radiation incident at angles larger θ_c also reaches the receiver. When surface errrors are present, as happens in practice, the fraction of radiation begins to decrease before the incident angle reaches θ_c, but some radiation is accepted for angles exceeding θ_c.

Table 4.2 Optical and geometrical properties of full and truncated CPC concentrators (Data from Ref.[2])

Acceptance half-angle, θ_c	Concentration Ratio, CR	Height/ Aperture ratio	Reflector/ Aperture ratio	Average number of reflections
36°	1.7 (full-height)	1.09	2.24	0.61
36°	1.6 (truncated)	0.65	1.36	0.42
36°	1.4 (truncated)	0.38	0.80	0.25
11.5°	5.0 (full-height)	2.94	6.00	0.99
11.5°	4.9 (truncated)	2.25	4.26	0.91
11.5°	3.65 (truncated)	1.04	2.22	0.60
5.7°	10 (full-height)	5.47	11.05	1.25
5.7°	9.8 (truncated)	4.17	8.44	1.17
5.7°	7.28 (truncated)	1.86	3.86	0.87

The CPC concentrators discussed above were based on the design for a flat receiver. However, for other receiver shapes such as a round tube, special geometrical constructions are needed to determine the ideal reflector design that admits all radiation incident within an acceptance angle. The reflector shape for a round-tube receiver is shown in Fig. 4.10(b), and the details of the geometrical construction used are available in Ref.[1,2].

4.2.3 *Parabolic trough concentrators*

Parabolic trough concentrators are used widely in commercial-scale solar thermal power systems as seen from the data in Table 4.1. These are two-dimensional imaging concentrators, commonly constructed as modules and spread over a wide area as shown in Fig. 4.3.

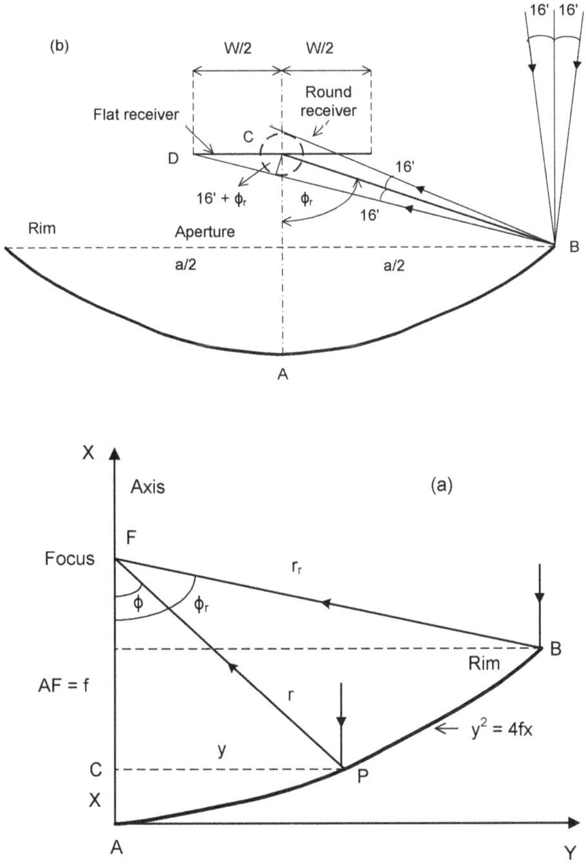

Fig. 4.11 (a) Dimensions of parabolic concentrator, (b) Images on a flat receiver and a round-tube receiver

A cross sectional view of a parabolic trough concentrator is shown in Fig. 4.11. Some of the more relevant geometrical properties of the parabola, including its axis, the focal point, the rim, and the equation describing the curve are indicated in Fig. 4.11. The equation of the parabola in the X-Y coordinate system is given by

$$y^2 = 4fx \tag{4.8}$$

where the focal distance, $AF = f$, and x and y are the coordinates of any point P.

The polar coordinates of the point P are r and ϕ. The polar equation of the parabola maybe obtained by considering the right-angle triangle FCP, where

$$AF = f - r\cos\phi = x \qquad CP = r\sin\phi = y$$

Substituting in Eq. (4.8) we get

$$(r\sin\phi)^2 = 4f(f - r\cos\phi)$$

Solution of the above quadratic equation gives the polar equation of the parabola as

$$r = \frac{2f}{1+\cos\phi} \tag{4.8a}$$

A beam of solar radiation of total angular width 0.53° (32'), which is the small angle subtended by the sun at any point on earth, is incident on the rim of a parabolic trough concentrator as shown in Fig. 4.11(b). The beam is parallel to the axis of the parabola and normal to the aperture of the concentrator.

Consider the beam of solar radiation reflected at point P on the rim of the concentrator under ideal conditions. The image formed on a flat receiver surface of width W, placed at the focus, parallel to the aperture is shown in Fig. 4.11(b). Using the properties of triangle BCD it can be shown that

$$\frac{W}{2} = \frac{r_r \sin 16'}{\cos(\phi_r + 16')} \tag{4.9}$$

From Eq. (4.9) we see that even under ideal conditions, the image formed on the receiver has a finite size. Moreover, Eq. (4.9) shows that as the point of incidence of the radiation beam moves from the center towards the rim of the reflector the image increases in size.

For the round cylindrical receiver shown in Fig. 4.11(b), the image is formed on the circumference of a circle whose diameter D is given by

$$\frac{D}{2} = r_r \sin 16' \tag{4.10}$$

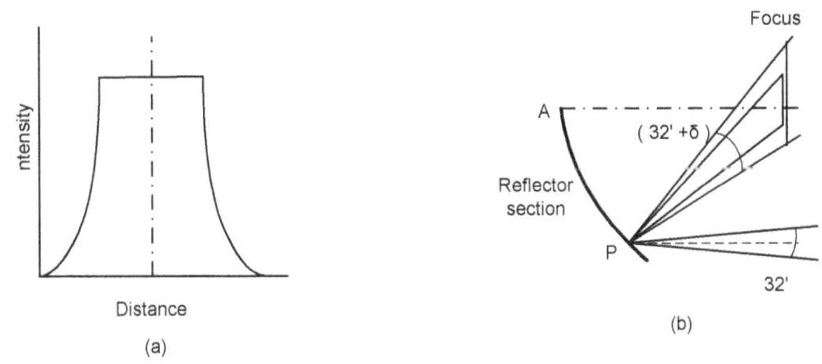

Fig. 4.12 (a) Intensity at focus, (b) Dispersion due reflector irregularities

Even an ideal optical system produces a finite-size image on the receiver due to the angle subtended by the sun at any point on earth. The radiation intensity at the image, which is based on the assumption that solar intensity over the surface of the sun is uniform, is shown graphically in Fig. 4.12(a). With real optical systems, especially those used as solar concentrators, the ideal image is distorted due to a number reasons, one of which is small-scale irregularities in the reflector shape. This effect can be accounted for by including a *dispersion angle* δ for the reflected beam of radiation, which further spreads the image over the receiver as shown in Fig. 4.12(b). When the dispersion angle is included, Eqs. (4.9) and (4.10) may be written as

$$\frac{W}{2} = \frac{r_r \sin(16' + \delta/2)}{cos(\phi_r + 16' + \delta/2)} \tag{4.11}$$

For the round cylindrical receiver

$$\frac{D}{2} = r_r \sin(16' + \delta/2) \tag{4.12}$$

Application of the above equations is illustrated in the worked examples (4.9) to (4.11).

Parabolic trough concentrators are oriented with the axis either in the east-west direction or the north-south direction. The concentrator is rotated about the respective axis to maximize the solar radiation incident on the aperture as the sun moves across the sky. This process is called *sun tracking*, and different modes of rotation are employed based on the concentration ratio. In general, the beam radiation incident on the aperture will be parallel to the central plane of the reflector, but not necessarily normal to the aperture.

The angle of incidence of the beam on the aperture depends on the mode of tracking. The image from any point on the reflector will be enlarged by a factor $1/cos\theta$, where θ is the angle of incidence. Errors in the orientation of the concentrator, and the position of the receiver with respect to the concentrator will further distort the image formed.

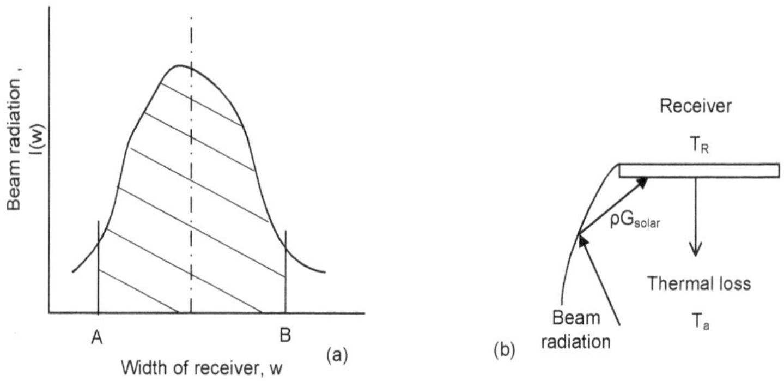

Fig. 4.13 (a) Intensity distribution on receiver, (b) Optical and thermal losses

Due the various reasons mentioned above the image formed at the receiver will not have well-defined boundaries. There are two opposing factors that influence the size of the receiver as shown in Fig. 4.13(b). Increasing the area of the receiver helps intercept much of the reflected solar radiation but this would increase the heat loss to the ambient. The optimum design of receivers has been studied by Lof and Duffie [1] who concluded that it is desirable to intercept about 90 to 95 percent of the radiation incident on the receiver. Hence the optical loss incurred in this type of collector is about 5 to 10 percent.

The *intercept factor*, γ is the design parameter used to quantify optical loss. Figure 4.13(a) shows an arbitrary radiation intensity distribution on a flat receiver whose width is AB. The total energy incident on the receiver is the shaded-area between A and B. We define the intercept factor as

$$\gamma = \frac{\int_A^B I(w)dw}{\int_{-\infty}^{\infty} I(w)dw} \tag{4.13}$$

The denominator gives the total energy that would be intercepted if the receiver was infinitely long. Similar expressions may be obtained for other types of concentrators and receivers. The radiation absorbed by the receiver may be written as

$$S = A_a G_b \rho \gamma \tau \alpha \tag{4.14}$$

where A_a is the unshaded area of the aperture, G_b is the beam radiation intensity on the plane of the aperture, γ is the intercept factor, ρ is the specular reflectivity of the reflector, τ is the transmittance of the cover system and α is the absorptivity of the receiver. The useful energy collection rate Q_u is given by

$$Q_u = S - Q_{rl} \tag{4.15}$$

where Q_{rl} is the rate of heat loss from the receiver. The computation of Q_{rl} is like flat plate collectors and evacuated tube collectors, presented in chapter 3, except that the radiation heat loss at high temperatures may have to include the exact radiation transfer equation. We shall illustrate the calculation in worked example 4.3.

Three different types of receivers used in practical parabolic trough systems are given in Ref.[1]. These include: (i) A cylindrical receiver consisting of black chrome coated steel pipe with the top half insulated and covered by a steel jacket. The radiation absorbing bottom half is surrounded by a glass semicylinder with an air space in between. (ii) A black chrome plated steel tube surrounded by a borosilicate glass tube with an air gap in between. (iii) A flat-plate receiver with two black chrome plated fluid-flow pipes running down and back. The back of the receiver is insulated and the front is covered with a glass plate, similar to a flat-plate collector.

The maximum concentration ratio C_m for a parabolic trough concentrator may be obtained from the ratio of the aperture area, a to the total area W of a receiver that is large enough to intercept all the reflected solar radiation. From Fig. 4.11(b) we obtain the ratio a/W for a round tube receiver as

$$\frac{a}{W} = \frac{2r_r \sin\phi_r}{2\pi r_r \sin(16' + \delta/2)} = \frac{\sin\phi_r}{\pi \sin(16' + \delta/2)} \tag{4.16}$$

For a flat receiver we have

$$\frac{a}{W} = \frac{2r_r \sin\phi_r \cos(\phi_r + 16' + \delta/2)}{2r_r \sin(16' + \delta/2)} = \frac{\sin\phi_r \cos(\phi_r + 16' + \delta/2)}{\sin(16' + \delta/2)} \quad (4.17)$$

The maximum CR for a flat receiver when the reflector is shaded by the receiver is given by

$$C_m = \frac{a - W}{W} = \frac{a}{W} - 1 = \frac{\sin\phi_r \cos\left(\phi_r + 16' + \frac{\delta}{2}\right)}{\sin\left(16' + \frac{\delta}{2}\right)} - 1 \quad (4.18)$$

4.2.4 *Paraboloidal concentrators*

Paraboloidal concentrators, commonly called dish concentrators, are surfaces of revolution as seen in the solar power system depicted in Fig. 4.6. The cross section of the dish is like the view shown in Fig. 4.11. The aperture is a circular disc of diameter *a*. The receivers for this type of concentrator are either spheres or flat discs. The maximum concentration ratio can be obtained from the ratio of aperture area to the total receiver area. Both these area are circular in shape. The expression for a spherical receiver is [1]

$$C_m = \frac{A_{ap} - A_{rc}}{A_{rc}} = \left[\frac{\sin\phi_r}{2\sin(16' + \delta/2)}\right]^2 - 1 \quad (4.19)$$

For a flat disc receiver [1]

$$C_m = \frac{A_{ap} - A_{rc}}{A_{rc}} = \left[\frac{\sin\phi_r \cos(\phi_r + 16' + \delta/2)}{\sin(16' + \delta/2)}\right]^2 - 1 \quad (4.20)$$

The radiation absorbed by the receiver may be written as

$$S = A_a G_b \rho \gamma \alpha \quad (4.21)$$

where A_a is the unshaded area of the aperture.

4.2.5 *Central receiver solar collectors*

A schematic of a central receiver based solar power system is shown in Fig. 4.5. It is commonly called a power tower. In concept the reflector system is a large paraboloidal surface which is broken up into small segments, commonly called heliostats. These are usually located close to the ground as shown in Fig. 4.14. Such an arrangement is called a Fresnel-reflector, a two dimensional version of which is depicted in Fig. 4.4. The spacing of the heliostats is of critical importance to avoid mutual shading and blocking of reflected radiation.

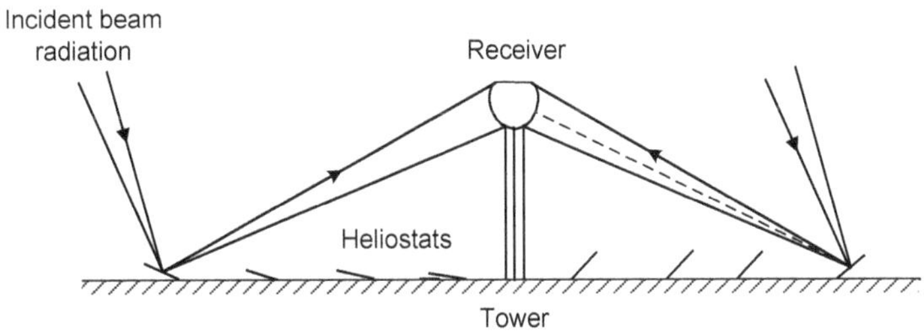

Fig. 4.14 Cross sectional view of solar power tower with heliostat field

In practical systems the fraction ψ of the total ground area occupied by the mirrors is about 0.3 to 0.5 [1]. The maximum concentration ratio C_m may be expressed in terms of ψ by considering the heliostat array as a large paraboloidal reflector. For a spherical receiver that intercepts all the reflected radiation, the expression for C_m is [1]

$$C_m = \frac{A_{ap} - A_{rc}}{A_{rc}} = \psi \left[\frac{\sin\phi_r}{2\sin\,(16' + \delta/2)}\right]^2 - 1 \tag{4.22}$$

For a flat disc receiver the maximum concentration ratio is given by [1]

$$C_m = \frac{A_{ap} - A_{rc}}{A_{rc}} = \psi \left[\frac{\sin\phi_r cos(\phi_r + 16' + \delta/2)}{\sin\,(16' + \delta/2)}\right]^2 - 1 \tag{4.23}$$

The heliostat array need not be symmetrical, and the ground cover does not have to be uniform or in one plane.

4.3 Thermal Storage and Heat Transport

All solar thermal power plants include heat storage systems to deal with periods when the sun is blocked by clouds and when the sun sets. This is also an advantage of thermal power systems compared to PV systems because it is easy to store heat with minimum losses, compared to storing electrical energy on a large scale. Several thermal energy storage systems have been built and tested since 1985. The more common heat storage systems are the two-tank direct system, the two-tank indirect system, and the single-tank thermocline system [3]. The arrangement of storage tanks in parabolic trough and power tower based power plants are shown schematically in Figs. 4.3 and 4.5 respectively.

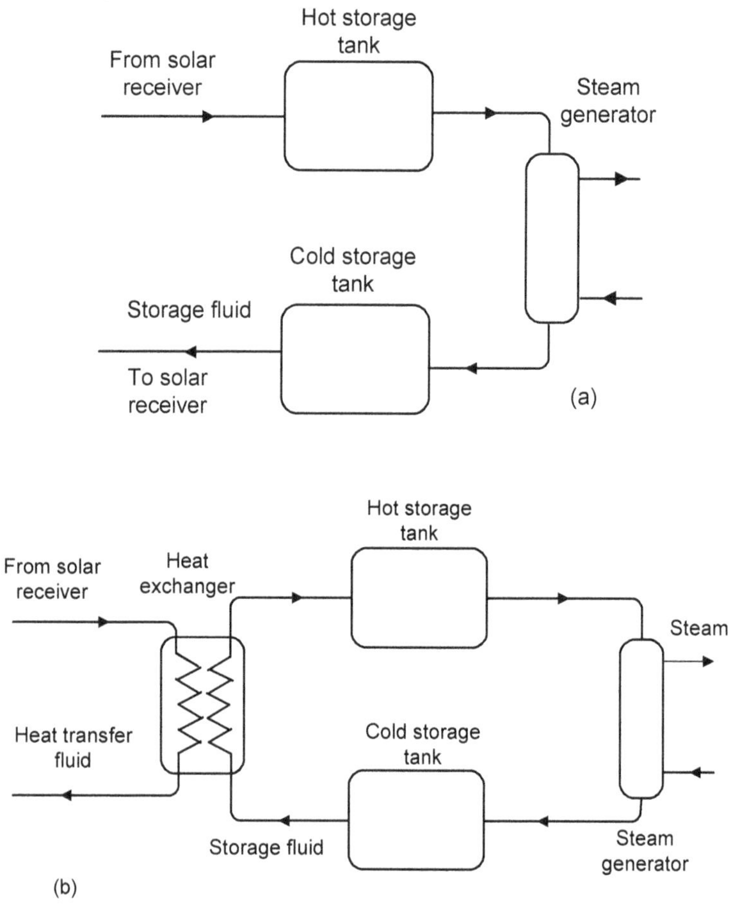

Fig. 4.15 (a) Two-tank direct storage system, (b) Two-tank indirect storage system

The two-tank direct system shown in Fig. 4.15(a) consists of two storage tanks. The hot tank stores the hot heat transfer fluid returning from the solar receiver after absorbing energy from the solar collectors. Fluid from the hot tank is pumped through the steam generator which supplies steam to the Rankine cycle power plant. Fluid from the steam generator is returned to the cold storage tank to be pumped back to the solar receiver. This arrangement has been used in the early parabolic trough and power tower based power plants, where the same fluid is used to transfer heat from the receiver to the storage tank and to store the collected heat.

The two-tank indirect system shown in Fig. 4.15(b) has a similar arrangement of hot and cold tanks but different fluids are used to transport heat from the receiver to the hot tank and to store the collected heat. The advantage of this system is that a cheaper fluid could be used to store heat compared to the heat transfer fluid.

However, the system needs an additional heat exchanger, where storage fluid from the cold tank is heated by the hot heat transfer fluid returning from the solar receiver. The heated fluid is stored in the hot tank from which it is pumped to the steam generator and then returned the cold tank. The two-tank indirect system is been used in solar thermal power systems constructed more recently.

In the single-tank thermocline system heat is stored in a solid medium like silica sand contained in a single tank. High temperature fluid from the receiver is returned to the top of the tank and fluid from the top is withdrawn to generate steam. Cold fluid from the bottom of the tank is circulated to the solar receiver and cold fluid from the steam generator is returned to the bottom of the tank.

In this arrangement a temperature gradient or a thermocline is maintained in the vertical direction in the tank. The resulting buoyancy effects create thermal stratification of the fluid within the tank which helps to stabilize and maintain the thermocline. The single tank system costs less than the two tank systems. The single tank arrangement has been tested in early power tower systems where steam was used as the heat transfer fluid and mineral oil was used as the storage fluid.

The efficiency and cost of a solar power plant depends on the choice of the heat transfer fluid and the storage fluid. The efficiency of the power plant increases with the temperature of the heat source but this could be limited by the availability of satisfactory heat transfer fluids that can operate at high temperatures.

In advanced solar power plants the heat transfer fluid might be liquid sodium or a molten mixture of salts at about atmospheric pressure. Heat transfer oils used in solar central receiver systems are Therminol-66, Therminol VP- 1 (eutectic mixture of biphenyloxide and diphenyl oxide) and Carolin HT-43 which have an operating temperature range of -7 to 315°C [3]. These oils are also used to transport heat from the heat storage tank to the steam generator. Oils have the advantage of low corrosion characteristics with most materials. However, they are flammable, can cause flow problems at low temperatures, and they can decompose at high temperatures.

The heat transfer fluid used most widely in parabolic trough solar plants is a eutectic mixture of organic compounds biphenyl oxide and diphenyl oxide. The behavior of these fluids limits the operating temperature of solar power plants to about 400°C because they tend to degrade above this temperature. The efficiency of solar thermal plants could be increased significantly if more stable heat transfer fluids such as binary or ternary mixtures of salts that operate satisfactorily at higher temperatures are used.

These eutectic mixtures of salts, which include Solar salt, HiTec, and HitecXL function well at temperatures of 500°C or higher. The compositions, freezing points (FP), and upper temperatures (UT) of these salt mixtures are [3]:

(i) Solar salt – 60% $NaNO_3$, 40% KNO_3, FP-220°C, UT-600°C,
(ii) Hitec-40% $NaNO_2$, 7% $NaNO_3$, 53% KNO_3, FP-142°C, UT-535°C,
(iii) Hitec XL-, 7% $NaNO_2$, 45% KNO_3, 48% $Ca(NO_3)_2$, FP-120°C, UT-500°C,
(iv) $LiNO_3$ – mixture, FP-120°C, UT-550°C.

The main limitation of these salts is their high freezing points which cause them to freeze inside the flow tubes and therefore freeze protection has to be considered when these salts are used.

4.4 Rankine Cycle

Many designs of solar thermal power systems use traditional vapor power cycles to produce the mechanical energy required to drive electric generators, as seen from the typical plants described in section 4.1. The ideal cycle on which the vapor power plant operates is called the *Rankine cycle*. In the flow diagram of the Rankine cycle shown in Fig. 4.16(a), vapor in a superheated state enters the turbine at 1. During the expansion process 1-2 through the turbine a part of the initial vapor enthalpy is converted to work. At 2 the wet vapor enters the condenser, in which heat is rejected to a cooling medium. The fully condensed liquid exiting from the condenser at 3 is pumped into the boiler by the feed pump to be heated in a constant pressure process from 4 to 1.

In the more common version of the Rankine cycle, the vapor is superheated to a temperature much higher than the saturation temperature corresponding to the boiler pressure. This process (1a-1) is carried out in a separate heater called the *superheater* as shown in Fig. 4.16(a). Superheating results in a higher average temperature of heat supply to the cycle which in turn increases its thermal efficiency.

Liquid entering the boiler is usually in a sub-cooled liquid state and the section of the boiler in which it is heated to the saturation temperature (4-4a) is sometimes termed the *economizer*. The complete unit consisting of the economizer, the boiler and the superheater is commonly called the *steam generator*.

In a typical turbine, the vapor expands over a series of blades attached to a wheel mounted on a shaft. During this process (1-2) the vapor transfers a part of its enthalpy to the shaft as mechanical work. In the modern version of the Rankine cycle, the exit temperature of the steam from the turbine is lowered by introducing a *condenser* in which the steam rejects heat to a cooling medium whose temperature is much closer to the ambient temperature. The working fluid entering the feed pump is in a liquid state and it could therefore be pumped to the boiler pressure conveniently with a relatively small work input.

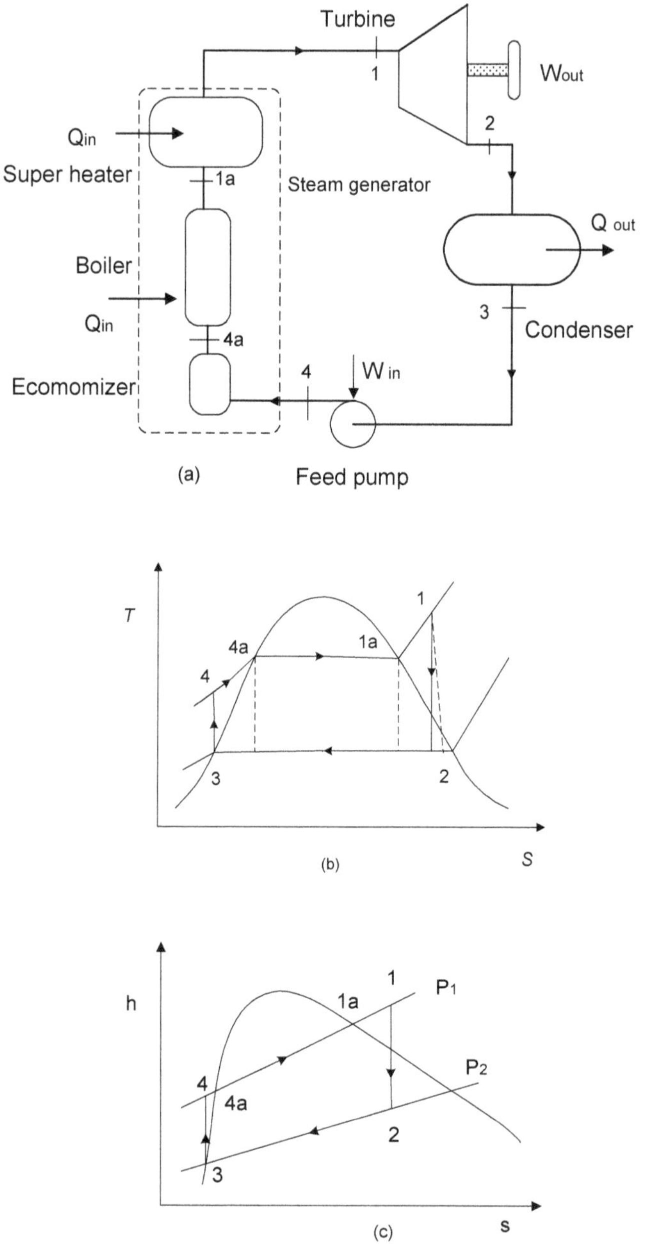

Fig. 4.16 (a) Rankine cycle power plant, (b) temperature-entropy diagram, (c) enthalpy-entropy diagram

The *temperature-entropy* (*T*-*s*) diagram and the *enthalpy-entropy* (*h*-*s*) diagram of the ideal Rankine cycle are shown in Figs. 4.16(b) and (c). The process 4-1 where heat is supplied and the process 2-3 during which heat is rejected are both constant pressure processes. The expansion 1-2 in the turbine and pumping process 3-4 are reversible adiabatic processes. With these ideal processes the ideal Rankine cycle is *internally* reversible.

4.4.1 *Analysis of the Rankine cycle*

We apply the energy equation to each component of the Rankine cycle heat engine depicted in Fig. 4.16(a). The important assumptions are: (i) the kinetic energy and potential energy of the working fluid are negligible compared to the enthalpy, (ii) the heat interaction between the working fluid and the surroundings is negligible, and (iii) the pressure drop due to fluid friction is negligible.

The work output for the reversible adiabatic expansion ($s_1 = s_2$) in the turbine 1-2 is

$$W_{1-2} = m(h_1 - h_2) \tag{4.24}$$

where m is the mass flow rate of the working fluid. The specific enthalpy and entropy of the fluid are h and s respectively.

The constant-pressure heat input in the boiler is

$$Q_{1a-4} = m(h_{1a} - h_4) \tag{4.25}$$

The constant-pressure heat input in the superheater is

$$Q_{1a-1} = m(h_1 - h_{1a}) \tag{4.26}$$

The constant-pressure heat rejection in the condenser is

$$Q_{2-3} = m(h_2 - h_3) \tag{4.27}$$

The work input to the *adiabatic* feed pump is

$$W_{3-4} = m(h_4 - h_3) \tag{4.28}$$

We assume the compression process in the pump to be *isentropic*. Since the liquid has a very low compressibility, its specific volume may be assumed constant. Hence the pump work is

$$W_{3-4} = mv(P_4 - P_3) \tag{4.29}$$

where v is the mean specific volume of the liquid and P is the pressure.

The thermal efficiency of the cycle is

$$\eta = \frac{W_{net}}{Q_{in}} = \frac{(W_{1-2} - W_{3-4})}{(Q_{4-1a} + Q_{1a-1})} \tag{4.30}$$

The above analysis of the ideal Rankine cycle can be easily extended to cover deviations from ideal conditions such as pressure drops in the piping, heat losses to the surroundings, and the inefficiencies in the turbine and the pump. Worked examples 4.14 and 4.15 illustrate the analysis of ideal and nonideal Rankine cycle power plants.

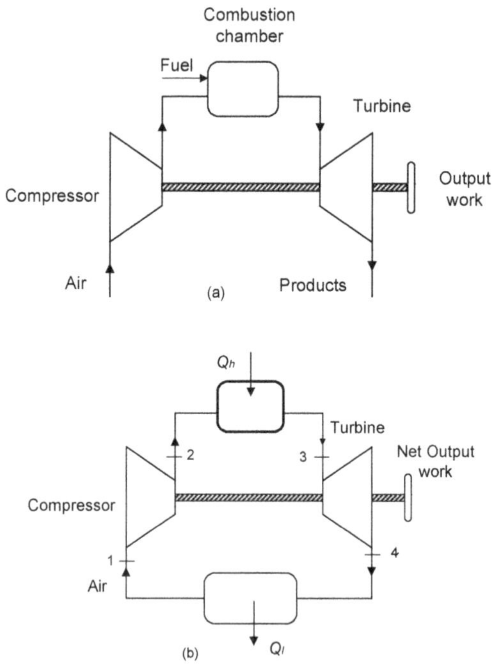

Fig. 4.17 (a) Simple gas turbine power plant, (b) Air standard Brayton cycle

4.5 Gas Turbine Engine Cycles

Gas turbine power plants are also being considered for use in solar thermal power systems. In this case concentrated solar radiation is uscd to heat the gas that operates the turbine. A conventional gas turbine plant, shown schematically in Fig. 4.17(a), consists of a compressor, a turbine and a combustion chamber. Atmospheric air flows in steadily through the compressor where it is pressurized and then enters the combustion chamber. Fuel flows steadily to the combustion chamber where continuous combustion occurs.

As the products of combustion, with a high pressure and temperature, flow through the turbine a part of its enthalpy is converted to work. Of the total work produced by the turbine a significant fraction is used to drive the compressor. The ratio of the compressor work input to the total turbine work output is called the *back-work-ratio*.

In the case of a steam power plant, the work required by the feed-pump is a only a very small fraction of the work output of the turbine. In contrast, the compressor work input in a gas turbine plant could be more than half the work produced by the turbine. This is because the compression of a gas requires much more work than pressurizing a liquid as it happens in the steam plant. The efficiency of the compressor is therefore of prime importance in the efficient operation of gas turbine plants.

In many applications gas turbine plants do not operate in closed-cycles. Moreover, the working fluid passing through the compressor and the turbine is not the same gas due to the combustion process. These factors make the analysis of real gas turbines complex. We shall analyze the gas turbine cycles by resorting to the somewhat simplified approach based on *air-standard* cycles.

4.5.1 *Analysis of the Brayton cycle*

The ideal air-standard cycle on which the gas turbine operates is known as the *Brayton cycle*. Figure 4.17(b) shows a schematic diagram of the air standard cycle which is a closed system with air as the working fluid where heat is supplied from a high temperature reservoir and heat is

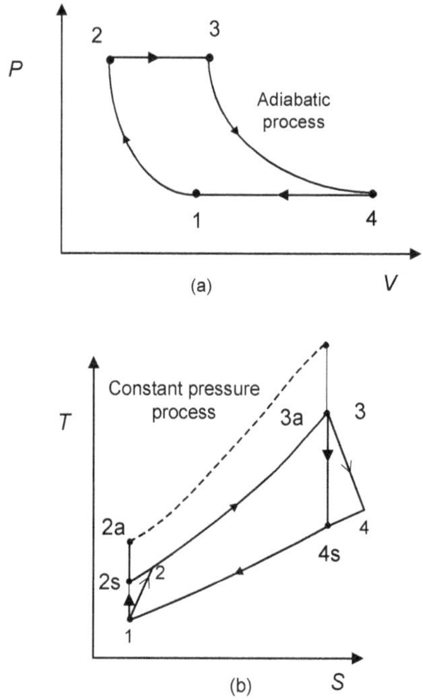

Fig. 4.18 (a) Brayton cycle P-V diagram, (b) Brayton cycle T-S diagram

rejected to a heat sink reservoir. These heat interactions between the air and the two reservoirs occur in heat exchangers under constant pressure. The expansion in the turbine and the compression in the compressor are isentropic processes and the working fluid is assumed to be an ideal gas. All processes undergone by the air are therefore *internally* reversible.

The *P-V* and *T-s* diagrams of the Brayton cycle are shown in Figs. 4.18(a) and (b) respectively. The ideal cycle is 1-2s-3-4s where the expansion process in the turbine and the compression process in the compressor are isentropic processes.

The *T-s* diagram also shows the cycle where the turbine and the compressor are non-ideal. The processes in these two components are adiabatic but they involve internal frictional losses which result in entropy increases as indicated. These inefficiencies are accounted for by using the concept of an *isentropic efficiency*. It is defined as the actual work output to the ideal work output for the turbine, and for the compressor it is the ratio of the ideal work input to the actual work input.

Consider the ideal cycle 1-2s-3-4s and apply the energy equation to the high temperature and low temperature heat exchangers, neglecting the kinetic and potential energy. The rate of heat input is

$$Q_h = mc_p(T_3 - T_{2s}) \tag{4.31}$$

The heat rejection rate is

$$Q_c = mc_p(T_{4s} - T_1) \tag{4.32}$$

where m is the steady air flow rate through the system. The temperature, pressure and specific heat capacity of air are T, P and c_p respectively.

The work output of the ideal turbine is

$$W_t = mc_p(T_3 - T_{4s}) \tag{4.33}$$

The work input to the ideal compressor is

$$W_c = mc_p(T_{2s} - T_1) \tag{4.34}$$

For the *isentropic* compression process 1-2s we have

$$\frac{P_2}{P_1} = \left(\frac{T_{2s}}{T_1}\right)^{\frac{\gamma}{\gamma-1}} \tag{4.35}$$

where $\gamma = \frac{c_p}{c_v}$, is the ratio of the specific heat capacities.

Similarly, for the *isentropic* expansion 3-4s in the turbine

$$\frac{P_3}{P_4} = \left(\frac{T_3}{T_{4s}}\right)^{\frac{\gamma}{\gamma-1}} \tag{4.36}$$

Neglecting pressure losses in the heat exchangers,

$$P_3 = P_2 \text{ and } P_4 = P_1 \tag{4.37}$$

From Eqs. (4.35), (4.36) and (4.37) it follows that

$$\frac{T_{4s}}{T_1} = \frac{T_3}{T_{2s}} \tag{4.38}$$

The efficiency of the cycle is given by

$$\eta = 1 - \frac{Q_c}{Q_h} = 1 - \frac{mc_p(T_{4s} - T_1)}{mc_p(T_3 - T_{2s})} \tag{4.39}$$

Substituting in Eq. (4.39) from Eqs. (4.35) and (4.36) we obtain the efficiency as:

$$\eta = 1 - \frac{T_1}{T_{2s}} = 1 - \left(\frac{P_1}{P_2}\right)^{\frac{\gamma}{\gamma - 1}} \tag{4.40}$$

$$\eta = 1 - \frac{1}{r^{\frac{\gamma}{\gamma - 1}}} \tag{4.41}$$

where the pressure ratio r of the cycle is given by

$$r = \frac{P_2}{P_1} \tag{4.42}$$

Substituting from Eqs. (4.35) to (4.37) in Eqs. (4.33) and (4.34) we obtain the turbine work output and the compressor work input as:

$$W_t = mc_p T_3 \left(1 - \frac{1}{r^{\frac{\gamma}{\gamma - 1}}}\right) \tag{4.43}$$

$$W_c = mc_p T_1 \left(r^{\frac{\gamma}{\gamma - 1}} - 1\right) \tag{4.44}$$

The net work output of the Brayton cycle is given by

$$W_{net} = W_t - W_c \tag{4.45}$$

We note from Eq. (4.41) that the efficiency of the ideal Brayton cycle increases with pressure ratio. This is also clear from the *T-s* diagrams in Fig. 4.18(b) where the cycle 1-2a-3a-4 is for a higher pressure than the cycle 1-2-3-4. Since these cycles are internally reversible, the area under the constant pressure lines gives the heat transfer. The heat rejection rates of the two cycles are equal while for the cycle 1-2a-3a-4 the heat input rate is larger. Therefore the efficiency, given by Eq. (4.39), is larger for the latter cycle.

Form Eq. (4.43) we see that the work output of the turbine increases with the pressure ratio, and the turbine inlet temperature T_3. The work input to the compressor also increases with the pressure ratio and decreases with decreasing air inlet temperature T_1 according to Eq. (4.44). However, the net work output of the cycle, given by Eq. (4.45), increases with the pressure ratio up to a maximum value and then decreases due to the increase of the compressor work input. The analysis of ideal and non-ideal gas turbine cycles is illustrated in worked examples 4.17 and 4.18.

4.6 Stirling Cycle

The Rankine cycle and the Brayton cycle that we discussed above are ideal cycles but they are called *externally irreversible* cycles. This is because the heat flow from the hot source to the working fluid and the heat rejection from the working fluid to the cold sink occur across finite temperature differences. This is unlike the Carnot cycle introduced in chapter 1, where the above temperature differences are ideally zero, and therefore called an externally reversible cycle. Another *ideal* cycle that is externally reversible is the *Stirling cycle*. Actual engines based on this ideal cycle have been included in solar power generation systems that use parabolic dish reflectors as shown in Fig. 4.6.

The operation of the ideal Stirling cycle may be demonstrated using the two-piston arrangement shown schematically in Fig. 4.19(a). During the process 1-2 heat is added to the gas, ideally at constant temperature, and the left piston moves out as the gas expands doing work. During process 2-3 both pistons move to the right maintaining the gas volume constant. Heat is added to the regenerator while the gas passes through it. The regenerator can be a plug of wire gauze or a porous plug made by placing small metal spheres between two wire screens.

The regenerator as a whole is a poor conductor of heat and it maintains a temperature gradient in the direction of gas flow. When the gas enters the space on the right in state 3, its temperature is equal to the temperature of the cold source to which it rejects heat, ideally at a constant temperature.

During the process 3-4, the right piston moves to the left compressing the gas and thus doing work on the gas. During the process 4-1 both

(a)

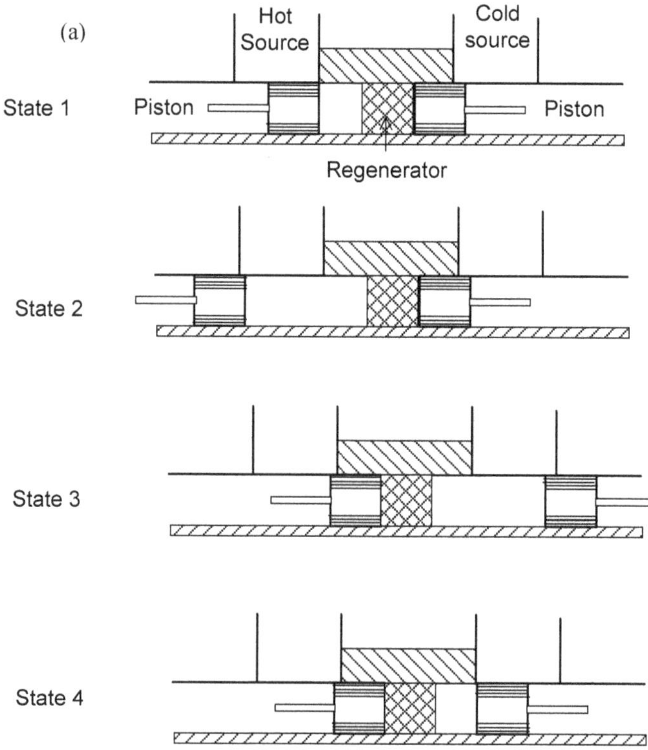

(b)

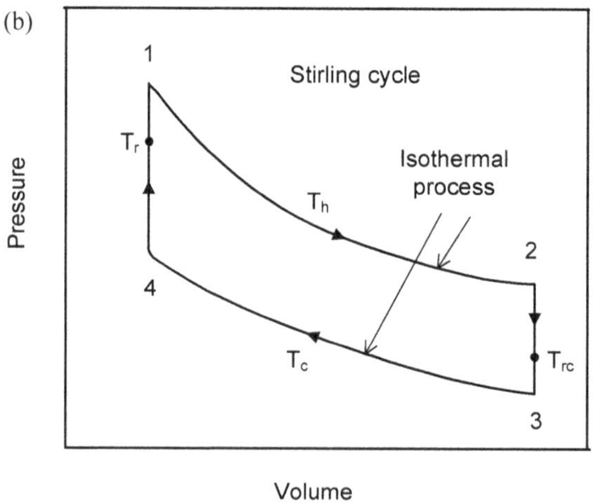

Fig. 4.19 (a) The operation of the ideal Stirling cycle, (b) P-V diagram of Stirling cycle

pistons move to the left maintaining the gas volume constant, and the gas absorbing heat from the regenerator to reach a temperature equal to the temperature of the hot source. Thus, ideally, the heat transfer between the gas and the two heat reservoirs occurs at constant temperature at every stage, making the cycle externally reversible. The regenerator does not retain any net heat during the cycle.

4.6.1 *Analysis of the ideal Stirling cycle*

The work done during the isothermal processes 1-2 and 3-4 are given by [4]

$$W_{12} = mRT_h ln\left(\frac{V_2}{V_1}\right) \tag{4.46}$$

$$W_{34} = mRT_c ln\left(\frac{V_3}{V_4}\right) \tag{4.47}$$

where m is the mass of gas, R is the gas constant, T is the absolute temperature and V is the volume. The subscripts represent the different states indicated in Fig. 4.19(b).

For an isothermal process with an ideal gas the heat interactions are equal to the work interactions. Therefore

$$Q_{12} = -W_{12} \tag{4.48}$$

$$Q_{34} = -W_{34} \tag{4.48a}$$

Heat flow from the *ideal regenerator* to the working fluid during process, 4-1 is equal to the heat transferred to the regenerator during the process 2-3. Therefore

$$Q_{41} = Q_{23} = mc_v(T_h - T_c) \tag{4.49}$$

The specific heat capacity is c_v.

The efficiency of the Stirling cycle is given by

$$\eta = \frac{W_{net}}{Q_{in}} = \frac{W_{12} - W_{34}}{Q_{12}} \tag{4.50}$$

Substituting from Eqs. (4.46) to (4.48) in Eq. (4.50), and noting that, $V_2 = V_3$ and $V_4 = V_1$ we obtain the ideal efficiency as

$$\eta = \frac{T_h - T_c}{T_h} \tag{4.51}$$

As seen from Eq. (4.51) the efficiency of the ideal Stirling cycle is equal to the efficiency of the Carnot cycle operating between same hot and cold reservoirs. However, this is only true for an *ideal regenerator*.

4.6.2 *Effect of non-ideal regenerator*

In practice the regenerator is not ideal and therefore the cold fluid is not heated to the temperature of the hot reservoir when it passes through it. This is shown in Fig. 4.19(b) where the regenerator temperature is indicated as T_r at the hot end and T_{rc} at the cold end. Therefore the working fluid has to be heated from T_r to T_h at constant volume by absorbing heat from the hot reservoir, which an irreversible process. The heat input to the cycle with the *non-ideal* regenerator is

$$Q_{12} = mRT_h ln\left(\frac{V_2}{V_1}\right) + mc_v(T_h - T_r) \tag{4.52}$$

Energy balance for the cycle gives

$$mc_v(T_h - T_r) = mc_v(T_{rc} - T_c) \tag{4.53}$$

The efficiency of the Stirling cycle with the *non-ideal regenerator* is given by

$$\eta = \frac{W_{net}}{Q_{in}} = \frac{W_{12} - W_{34}}{Q_{12}}$$

$$\eta = \frac{mRT_h ln\left(\frac{V_2}{V_1}\right) - mRT_c ln\left(\frac{V_3}{V_4}\right)}{mRT_h ln\left(\frac{V_2}{V_1}\right) + mc_v(T_h - T_r)} \tag{4.54}$$

The above equation can be simplified to the form

$$\eta = \frac{T_h ln\left(\frac{V_2}{V_1}\right) - T_c ln\left(\frac{V_3}{V_4}\right)}{T_h ln\left(\frac{V_2}{V_1}\right) + c_v(T_h - T_r)/R} \tag{4.55}$$

The effectiveness of the regenerator is defined as [4]

$$e = \frac{T_r - T_c}{T_h - T_c} \tag{4.56}$$

Eliminating T_r between Eqs. (4.55) and (4.56) we obtain the efficiency as

$$\eta = \frac{T_h - T_c}{T_h + \left(\frac{C_v}{R}\right)(1-e)(T_h - T_c)\left[ln\left(\frac{V_2}{V_1}\right)\right]^{-1}} \tag{4.57}$$

Note that when $e = 1$ the efficiency is the Carnot efficiency. When $e = 0$, the efficiency has a finite value because the cycle has a positive work output even without the regenerator.

4.6.3 *Real Stirling engines*

From the conceptual arrangement shown in Fig. 4.19(a) we see that the two pistons in the Stirling cycle must move in a predetermined manner to achieve the necessary energy transfers. In real Stirling engines this is achieved by using various mechanical linkages. Three designs commonly used in actual engines are designated as alfa, beta and gamma engines. The three designs follow the Stirling cycle and share some basic elements, but they differ in construction.

The alfa-type engines have the simplest configuration, where the hot and cold chambers are clearly separated, usually in separate cylinders as shown schematically in Fig. 4.20(a). One cylinder is in contact with the heat source while the other rejects heat to the heat sink. A common crank shaft, connected to the two pistons by separate connecting rods 1-A and 1-B, delivers power. The regenerator is in a duct connecting the two cylinders and the gas passes across it periodically releasing and absorbing heat alternatively.

We can follow the clockwise rotation of the crank to identify the various phases of the cycle. From 1-2, both pistons move in and the gas is compressed while heat is rejected to the heat sink. From 2-3, piston A moves out and B moves in maintaining the volume constant and absorbing heat from the regenerator. From 3-4 both pistons move out and the gas expands absorbing heat from the heat source. From 4-1 piston A moves in and B moves out maintaining the volume constant and transferring heat to the regenerator.

The above processes are like those shown in Fig. 4.19(b). When hot gas passes across the regenerator, heat is added to it and cold gas passing

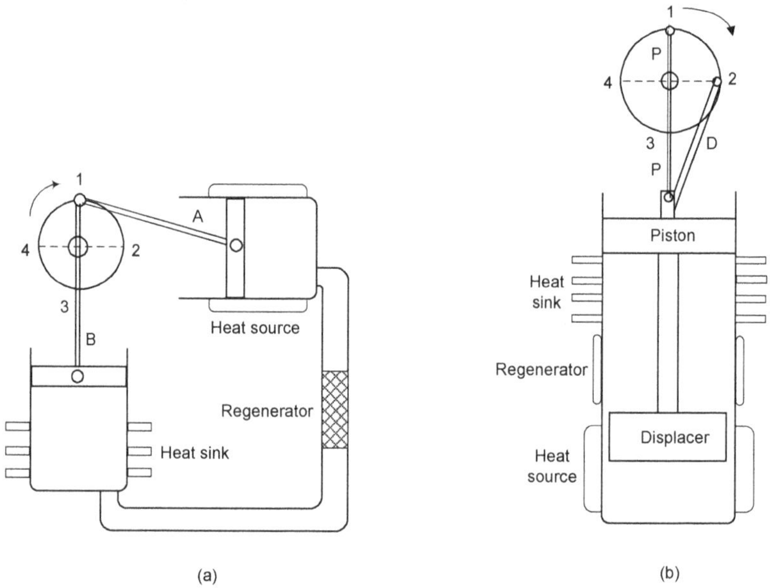

Fig. 4.20 (a) Schematic of alfa-type Stirling engine, (b) Schematic of beta-type Stirling engine

through it absorbs heat from it. A concern about the alfa-design is the possible heat leak from the hot region to the cold region through the various linkages.

A schematic of the beta-type engine is depicted in Fig. 4.20(b), which has a piston (P) that delivers power, and a displacer (D) that helps transfer gas between the hot and cold regions. These have separate connecting rods and crank shafts P and D, which are out of phase by 90°.

The operation of the engine may be understood by following the rotation of the crank shafts in the clockwise direction. When P is at 4 (P4) and D at1 (D1), the power piston has compressed the gas, the displacer has moved so that most of the gas is adjacent to the heat source. At P1, D2 positions the heated gas increases in pressure and pushes the power piston to the topmost position of the power stroke.

At P2, D3 positions, the displacer moves the gas to the upper end towards the heat sink. At P3, D4 positions the cooled gas is compressed while rejecting heat to the sink. Usually, the engine may also have a regenerator, which can be incorporated either in the cylinder wall or the displacer.

The function of the gamma-type engine is similar to that of the beta-type described above. The main difference in construction is that in the gamma-type engine, the power piston is not located in line with the displacer. The latter is in a separate cylinder that is connected to the compression section of the main cylinder which incorporates the heat reservoirs. This diminishes the limitations imposed by the size and orientation of the displacer on the movement of the power piston.

4.7 Worked Examples

Example 4.1 A parabolic trough solar collector system of average efficiency 35% is used as the heat source of the steam power plant with an efficiency of 40%. The average solar radiation intensity on the aperture of the collectors is 780 Wm^{-2}. The efficiency of the electric generator is 85% and its output is 8 MW. Calculate (i) the total aperture area of the collector field, (ii) the heat energy delivered to the steam plant, and (iii) the overall efficiency of the system.

Solution The total solar radiation incident on the aperture is given by

$$H = AI_s = 780A \text{ W}$$

The energy absorbed by the solar system and delivered to the steam plant is

$$Q_{abs} = AI_s\eta_c = 780A \times 0.35 \text{ W}$$

The output of the steam power plant is

$$Q_{out} = Q_{abs}\eta_{th} = AI_s\eta_c\eta_{th} \text{ W}$$

$$Q_{out} = 780A \times 0.35 \times 0.4 \text{ W}$$

The output of the generator is

$$E_{out} = Q_{out}\eta_{gen} = AI_s\eta_c\eta_{th}\eta_{gen}$$

$$E_{out} = 780A \times 0.35 \times 0.4 \times 0.85 = 8 \times 10^6 \text{ W}$$

The total aperture of the collector field is obtained from the above equation as 86,000 m^2.

The energy delivered to the steam plant is 23.5 MW

The overall efficiency is given by

$$\eta_{overall} = \frac{E_{out}}{H} = \eta_c \eta_{th} \eta_{gen}$$

$$\eta_{overall} = 0.35 \times 0.4 \times 0.85 = 12\%.$$

Example 4.2 The average direct beam radiation level at a location is 790 Wm^{-2} and the average ambient temperature is 29°C. A compound-parabolic concentrator (CPC) collector with the following specifications is used to operate a low temperature power plant: concentration ratio = 3, reflectivity = 0.9, heat loss coefficient = 2.4 Wm^{-2}K^{-1}, transmittance-absorptance product = 0.8. Assume that the power plant operates on an ideal Carnot cycle. Calculate (i) the temperature of the receiver that gives the maximum overall efficiency, (ii) the maximum overall efficiency, and (iv) the efficiency of the Carnot engine.

Solution This is an idealized analysis to demonstrate the existence of an optimum collector temperature that maximizes the overall efficiency. As the receiver temperature increases the efficiency of the thermal power cycle increases but so does the receiver heat loss to the ambient. The latter decreases the collector efficiency. Hence, the overall efficiency, which is the product of the collector efficiency and the power cycle efficiency, has a maximum value. We shall write the energy equations for a general solar collector which could be a flat plate collector or a low temperature concentrator-collector such as a CPC.

For solar radiation transfer from aperture to receiver:

$$I_s A_a \rho = A_r q_{sr} \tag{E4.2.1}$$

The subscript r represents the receiver and a the aperture. The effective reflectivity of the concentrator is ρ. The term on the LHS of Eq. (E4.2.1) is the solar radiation incident on the aperture, and the RHS is the radiation reaching the receiver.

For overall energy balance of the receiver:

$$Q_r = (\tau\alpha)_r A_r q_{sr} - U_r A_r (T_r - T_a) \tag{E4.2.2}$$

where $(\tau\alpha)_r$ is the transmittance-absorptance product of the receiver.

Note that Eq. (E4.2.2) is similar to the energy balance equation for a flat plate collector.

The efficiency of the collector is given by

$$\eta_c = \frac{Q_r}{A_a I_s} = \frac{(\tau\alpha)_r A_r q_{sr} - U_r A_r (T_r - T_a)}{A_a I_s} \tag{E4.2.3}$$

The above equation can be rearranged as

$$\eta_c = \frac{A_a \rho (\tau\alpha)_r I_s - A_r U_r (T_r - T_a)}{A_a I_s}$$

$$\eta_c = \rho(\tau\alpha)_r \left[1 - \frac{U_r (T_r - T_a)}{CR\rho(\tau\alpha)_r I_s}\right] \tag{E4.2.4}$$

where $CR = \frac{A_a}{A_r}$, is the concentration ratio.

The efficiency of the Carnot heat engine is given by

$$\eta_{carnot} = 1 - \frac{T_a}{T_r} \tag{E4.2.5}$$

We have taken the receiver temperature of the solar collector as the heat source temperature, and the ambient temperature as the heat sink temperature of the Carnot cycle.

The overall efficiency of the power system is given by

$$\eta_o = \eta_c \eta_{carnot} \tag{E4.2.6}$$

Manipulating Eqs. (E4.2.4), (E4.2.5) and (E4.2.6) we obtain

$$\eta_o = \rho(\tau\alpha)_r \left[1 - \frac{U_r (T_r - T_a)}{CR\rho(\tau\alpha)_r I_s}\right] \left(1 - \frac{T_a}{T_r}\right) \tag{E4.2.7}$$

To obtain the condition for the maximum efficiency we differentiate Eq. (E4.2.7) w.r.t T_r and equate the resulting expression to zero. Hence we obtain the optimum receiver temperature as

$$T_{ro} = \left[\left(T_a + \frac{CR\rho(\tau\alpha)_r I_s}{U_r}\right) T_a\right]^{\frac{1}{2}} \tag{E4.2.8}$$

Substituting the given numerical data in Eq. (E4.2.8) we have

$$T_{ro} = \left[\left(302 + \frac{3\times0.9\times0.8\times790}{2.4}\right)302\right]^{\frac{1}{2}} = 553\ \text{K}$$

Substituting the above value for T_r in Eq. (E4.2.7) we obtain the maximum overall efficiency as 21%. Substituting in Eq. (E4.2.5) we get the Carnot efficiency at the optimum temperature as 45.3%.

Example 4.3 An ideal Carnot heat engine is operated using a high temperature solar collector system as the heat source. The average solar radiation intensity on the aperture is 900 Wm^{-2} and the average ambient temperature is 15°C. The concentration ratio is 500. Assume that the receiver surface is a black body, and that the reflectivity of the concentrator is one. Calculate (i) the optimum temperature of the receiver that maximizes the overall efficiency of the system, (ii) the stagnation temperature of the receiver, (iii) the overall efficiency, and (iv) the efficiency of Carnot heat engine.

Solution Due to the high concentration ratio the temperature of the receiver would be relative high. Therefore the main mode of heat loss from the receiver to the ambient is by thermal radiation. Hence we could ignore the convective heat loss for this simplified analysis.

We obtained an expression for the efficiency of a concentrator-collector in Eq. (E4.2.3). When the only mode of heat loss from the collector is thermal radiation, the efficiency can be written as

$$\eta_c = \frac{Q_u}{Q_S} = \frac{A_r(\tau\alpha)_r q_{sr} - A_r\sigma\left(T_r{}^4 - T_a{}^4\right)}{A_a I_s} \tag{E4.3.1}$$

where subscript r represents the receiver a the aperture. The above equation can be rearranged as

$$\eta_c = \rho(\tau\alpha)_r\left[1 - \frac{\sigma\left(T_r{}^4 - T_a{}^4\right)}{CR\rho(\tau\alpha)_r I_S}\right] \tag{E4.3.2}$$

where CR is the concentration ratio.

Assume that the receiver temperature is much larger than the ambient temperature. This gives the simpler form of Eq. (E4.3.2) as:

$$\eta_c = \left[1 - \frac{\sigma T_r{}^4}{CRI_S}\right] \tag{E4.3.3}$$

The efficiency of the Carnot heat engine is given by

$$\eta_{carnot} = 1 - \frac{T_a}{T_r} \tag{E4.3.4}$$

The overall plant efficiency is given by

$$\eta_o = \eta_c \eta_{carnot} \tag{E4.3.5}$$

$$\eta_o = \left[1 - \frac{\sigma T_r{}^4}{CRI_S}\right]\left(1 - \frac{T_a}{T_r}\right) \tag{E4.3.6}$$

Equation (E4.3.6) is differentiated and equated to zero. Hence we get

$$T_r{}^5 - 0.75 T_a T_r{}^4 - 0.25 T_a \left(\frac{CRI_S}{\sigma}\right) = 0 \tag{E4.3.7}$$

Substituting numerical values in the above equation we obtain

$$T_r{}^5 - 0.75 T_a T_r{}^4 - 0.25 T_a \left(\frac{CRI_S}{\sigma}\right) = 0 \tag{E4.3.8}$$

$$T_r{}^5 - 216 T_r{}^4 - 5.71 \times 10^{14} = 0$$

In order to solve the above equation we make the substitution:

$$\theta = T_r/10^2$$

$$\theta^5 - 2.16\theta^4 - 5.71 \times 10^4 = 0$$

The solution to the above equation is obtained by trial-and-error as 9.4175.

Hence the optimum receiver temperature is 941.75 K. From Eq. (E4.3.4) we obtain the Carnot efficiency as 69.4%. From Eq. (E4.3.6) the overall efficiency is 62%.

The stagnation temperature is the maximum temperature of the receiver when no useful energy is collected. This happens when the solar radiation absorbed is equal to the radiation heat loss from the receiver to the ambient. Then the collector efficiency is zero. Hence from Eq. (E4.3.3) we have

$$T_{stg}{}^{4} = \frac{CRI_S}{\sigma} = \frac{500\times 900}{5.67\times 10^{-8}}$$

The stagnation temperature is 1678.4 K.

Example 4.4 Show that all beams of solar radiation incident on a parabolic concentrator parallel to its axis passes through the focus after specular reflection.

Solution

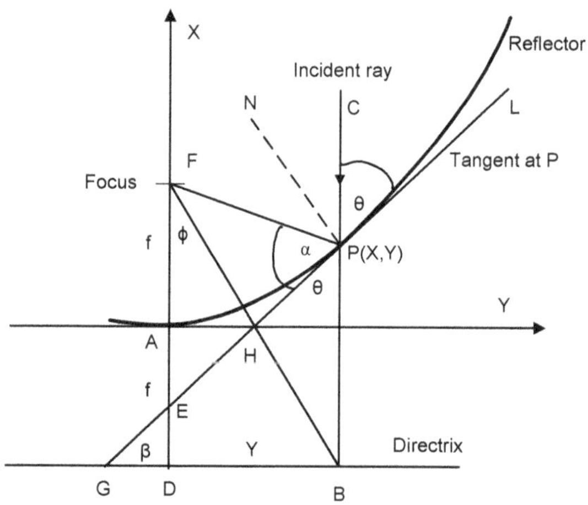

Fig. E4.4.1 Geometrical parameters of parabolic reflector

The geometric parameters of a parabolic concentrator are indicated in Fig. E4.4.1. An important property of the parabolic curve is that any point on it, such as P, is equidistant from the *focus* and the *directrix*. The directrix is a straight line parallel to the y-axis and at a distance equal to the focal length. Hence we have AD = AF = *f*, and PF = PB.

Consider the light ray CP incident on the reflector at P. The tangent at P is PG and the normal is PN. We need to show that angle CPL = angle HPF, that is $\theta = \alpha$. The equation of the parabola is

$$y^2 = 4fx$$

Now the slope of the tangent at P:

$$\tan\beta = \frac{dx}{dy} = \frac{y}{2f}$$

From ΔDFB, $$\tan\phi = \frac{y}{2f}$$

Therefore $$\beta = \phi$$

Consider the two triangles FEH and GED. Two of their angles are clearly equal. Hence

$$angle\ EHF = angle\ GDE = \pi/2$$

Now ΔFPB is an isosceles triangle, with FP = PB, and PH is perpendicular to FB. Therefore angle FPH is equal to angle BPH, that is $\alpha = \theta$.

Hence PF is the direction of the reflected beam, which passes through the focus F. We also observe the important fact that the rim-angle at P, angle DFP is 2β, where $tan\beta$ is the slope of the tangent at P.

Example 4.5 A trough concentrator at a location with a northern latitude 30° faces south and it is rotated about an east-west axis continuously in such a manner that the angle of incidence of beam radiation on the aperture is a minimum. At 11 a.m. solar time on April 12, the intensity of beam radiation on a horizontal plane is 680 Wm^{-2}. Calculate (i) the angle of incidence on the aperture, and (ii) the beam radiation incident on a unit area of the aperture.

Solution We shall make use of a number of equations derived in chapter 2 to solve this problem. The angles involved are shown in Fig. 2.5. The angle of incidence of the direct beam on an inclined surface is given by Eq. (2.14) as:

$$cos\theta = cos\beta sin\Sigma\, cos\gamma + sin\beta cos\Sigma \qquad \text{(E4.5.1)}$$

The angle, $\gamma = (\phi - \psi)$ is called the *surface-solar azimuth.*

For a south facing surface $\psi = 0$, and Eq. (E4.5.1) becomes

$$cos\theta = cos\beta sin\Sigma\, cos\phi + sin\beta cos\Sigma \qquad \text{(E4.5.2)}$$

Substituting in Eq. (E4.5.2) from Eq. (2.11) and Eq. (2.9) we get

$$cos\theta = sin\Sigma\,(cos\delta sinLcosH - sin\delta cosL) + (cosLcosHcos\delta + sinLsin\delta)cos\Sigma$$

The above equation can be simplified to the form

$$cos\theta = sin\delta sin(L - \Sigma) + cos\delta cosHcos(L - \Sigma) \qquad \text{(E4.5.3)}$$

In order to obtain the minimum angle of incidence we differentiate Eq. (E4.5.3) w.r.t to Σ and equate the resulting expression to zero. Hence we have

$$\tan(L - \Sigma) = \frac{tan\delta}{cosH} \qquad \text{(E4.5.4)}$$

We substitute from Eq. (E4.5.4) in Eq. (E4.5.3) and simplify the resulting equation to obtain the minimum angle of incidence as:

$$cos\theta_m = [1 - (cos\delta sinH)^2]^{1/2} \qquad \text{(E4.5.5)}$$

Now for April 12, $N_{day} = 102$. The declination is given by Eq. (2.5) as

$$\delta = 23.45sin\left[\frac{360(284+N_d)}{365}\right] = 8.29 \qquad \text{(E4.5.6)}$$

At 11 a.m. the hour angle, $H = -15$ degrees.

Substituting in Eq. (E4.5.5) we obtain the angle of incidence on the aperture as 14.8 degrees.

The solar altitude is obtained from Eq. (2.9) as:

$$sin\beta = cos30cos(-15)cos8.294 + sin30sin8.294 = 0.9$$

The solar altitude is 64.14°. The angle of incidence on a horizontal surface is 25.86°.

The radiation incident on the aperture per unit area is

$$G = \frac{Hcos\theta_m}{sin\beta} = \frac{680cos14.8}{sin64.14} = 730.6\ \text{Wm}^{-2}$$

Example 4.6 A trough concentrator at a location with a northern latitude 30° faces south, and its axis is along the east-west direction. The collector is rotated about its axis once a day to ensure that at noon the beam radiation is normal to the aperture. At 11 a.m. solar time on April 12, the intensity of beam radiation on a horizontal plane is 780 Wm^{-2}. Calculate (i) the angle of incidence of beam radiation on the aperture, and (ii) the beam radiation incident on a unit area of the aperture.

Solution Since the location, the day and time are the same as in example 4.5 we use some of the data obtained earlier. These are $\delta = 8.29$, $H = -15$.

For a south facing surface the angle of incidence is given by Eq. (E4.5.3) as:

$$cos\theta = sin\delta sin(L - \Sigma) + cos\delta cosH cos(L - \Sigma) \quad \text{(E4.6.1)}$$

The angle incidence is given by

$$cos\theta = sin\delta sin(L - \Sigma) + cos\delta cos0 cos(L - \Sigma)$$

$$cos\theta = sin\delta sin(L - \Sigma) + cos\delta cos(L - \Sigma) = cos[\delta - (L - \Sigma)]$$

We are given that the angle of incidence θ is zero at 12 noon. Hence from the above equation we get

$$(L - \Sigma) = \delta$$

Therefore the tilt angle of the aperture is

$$\Sigma = L - \delta$$

Since δ varies with the day, the tilt must be adjusted daily. This will ensure that at 12 Noon the direct beam is normal to the aperture. Substituting this condition in Eq. (E4.6.1) get the angle of incidence as:

$$cos\theta = sin^2\delta + cosH cos^2\delta \quad \text{(E4.6.2)}$$

Substituting numerical values in Eq. (E4.6.2) we obtain the angle of incidence at 11 a.m. as 14.84°. The beam radiation incident on the aperture per unit area is

$$G = \frac{H\cos\theta}{\sin\beta} = \frac{680\cos 14.84}{\sin 64.14} = 730.6 \text{ Wm}^{-2}$$

Example 4.7 A CPC collector has an acceptance half-angle of 26° and it is oriented along the east-west axis at a location with a northern latitude of 30°. The aperture area of the collector is 15m^2 and it is inclined at 35° to the horizontal. The beam radiation intensity on a horizontal surface at the location at 10 a.m. solar time on April 12, is 680 Wm^{-2}. The average transmittance of the aperture is 0.9. Calculate the total beam radiation transmitted through the aperture.

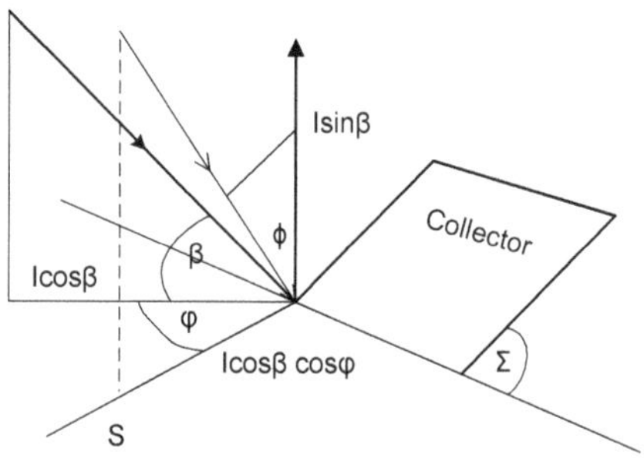

Fig. E4.7.1 Direction of the plane of the sun

Solution Since the aperture is facing south we first find the inclination of the plane of movement of the sun to the plane normal of the aperture. The resolved components of solar beam in this vertical plane are

$I\sin\beta$ (vertical) and $I\cos\beta\cos\varphi$ (horizontal),

their resultant is inclined at an angle ϕ to the vertical as indicated in Fig. E4.7.1. The solar altitude and azimuth are β and φ respectively. From the right angle triangle it follows that

$$\tan\phi = \frac{I\cos\beta\cos\varphi}{I\sin\beta} = \frac{\cos\varphi}{\tan\beta} \qquad \text{(E4.7.1)}$$

The angle between the aperture-surface normal and the vertical is equal to the aperture tilt Σ. The acceptance half-angle measured from the surface normal is θ_c. Therefore if the angle of the resolved solar beam from the vertical lies between $(\Sigma - \theta_c)$ and $(\Sigma + \theta_c)$, the beam will fall within the acceptance angle $2\theta_C$. Then all incident beam radiation will be transmitted to the absorber. Otherwise all radiation will be rejected. This however, is for an ideal untruncated CPC with no surface imperfections. The above condition can be stated as:

$$\Sigma + \theta_c > \phi > \Sigma - \theta_c$$

From the given data we have

$$35 + 26 > \phi > 35 - 26 \quad \text{(E4.7.2)}$$

We need to determine ϕ for the given date and time. Now the declination on April 12 was obtained in example 4.5 as 8.29°. At 10 a.m. the hour angle, $H = -30$. The solar angles are given by

$$sin\beta = cosLcosHcos\delta + sinLsin\delta \quad \text{(E4.7.3)}$$

$$sin\varphi = \frac{cos\delta sinH}{cos\beta} \quad \text{(E4.7.4)}$$

Substituting numerical data we have

$$sin\beta = cos(30)cos(-30)cos(8.294) + sin(30)sin(8.294) = 0.814$$

$$\beta = 54.5$$

$$sin\varphi = \frac{cos(8.294)sin(-30)}{cos(54.5)} = 0.85,\ \varphi = -58.4$$

From Eq. (E4.7.1) we get

$$tan\phi = \frac{cos\varphi}{tan\beta} = 0.3737,\ \phi = 20.5$$

Therefore angle ϕ satisfies the condition (E4.7.2), and the solar beam is within the acceptance angle. Therefore all incident radiation could be transmitted to the absorber. The angle of incidence of beam radiation on the south facing aperture is given by Eq. (E4.6.1) as

$$cos\theta = sin\delta sin(L - \Sigma) + cos\delta cosHcos(L - \Sigma)$$

$$cos\theta = sin(8.29)sin(30 - 35) + cos(8.29)cos(-30)cos(30 - 35) = 0.841$$

The angle of incidence on the aperture is 32.7°. The angle of incidence on a horizontal surface is

$$\theta_h = \frac{\pi}{2} - \beta = 35.5 \text{ degrees}$$

The transmitted radiation is given by

$$G = \frac{680 \times 0.9}{cos(35.5)} \times \cos(32.7) = 632.5 \text{ Wm}^{-2}.$$

Example 4.8 A parabolic trough concentrator is 8 m long and has an aperture width of 1.2 m. The focal length is 1.2 m. The dispersion angle of the reflector is 0.6°. A beam of solar radiation is incident on the aperture at right angles. Calculate (i) the rim angle of the concentrator, (ii) the minimum areas of a flat receiver and a round-tube receiver, placed at the focus on which all the reflected radiation will be incident, (iii) the concentration ratio for each receiver, and (iv) the fraction of the aperture area that would be shaded by the receiver.

Solution The slope of the parabola at the rim is

$$tan\theta = \frac{dx}{dy} = \frac{y}{2f} = \frac{0.6}{2.4}, \qquad \theta = 14.04$$

From the analysis in example 4.4, the rim angle is

$$2\theta = 2 \times 14.04 = 28.08$$

The rim radius is given by Eq. (4.8a) as

$$r = \frac{2f}{1+cos\phi} = \frac{2 \times 1.2}{1+cos(28.08)} = 1.275 \text{ m}$$

For a flat receiver, the width is given Eq. (4.11) as

$$\frac{W}{2} = \frac{r_r \sin(16' + \delta/2)}{cos(\phi_r + 16' + \delta/2)}$$

$$\frac{W}{2} = \frac{1.275 \times sin(0.267+0.3)}{cos(28.08+0.267+0.3)} = 0.0144 \text{ m}$$

Hence the width of the flat receiver is 0.0288 m.

The concentration ratio is

$$CR = \frac{A}{W} = \frac{1.2}{0.0288} = 41.67$$

The fraction of the aperture that is shaded is

$$f_{sh} = \frac{0.0288}{1.2} = 0.024$$

For the round cylindrical receiver, the diameter is given by Eq. (4.12) as

$$\frac{D}{2} = r_r \sin\left(16' + \frac{\delta}{2}\right)$$

$$\frac{D}{2} = 1.275 \times sin(0.3 + 0.267) = 0.0126 \text{ m}$$

Hence the diameter of the round receiver is 0.0252 m.

The concentration ratio is

$$CR = \frac{A}{\pi D} = \frac{1.2}{\pi \times 0.0252} = 15.1$$

The fraction of the aperture that is shaded is

$$f_{sh} = \frac{0.0252}{1.2} = 0.021.$$

Example 4.9 A parabolic dish concentrator has an aperture diameter of 1.4 m and a focal length of 0.6 m. The receiver is a disc located centrally at the focus of the dish. The dispersion angle of the reflector is 0.6°. Calculate (i) the rim angle and the rim radius of the reflector, (ii) the minimum diameter of the receiver disc that would intercept all reflected beam radiation incident on the aperture at right angles, (iii) the concentration ratio, (iv) the fraction of the aperture area that would be shaded by the receiver, and (v) the mass of the reflector if the mass per unit area of the material is 16 kgm^{-2}.

Solution Consider the solar beam reflected at point A on the rim of the paraboloid as shown in Fig. E4.9.1. The reflected beam will be incident at points B and C on the horizontal disc located at the focus. Considering both sections of the parabola, by symmetry, a disc of radius BF will be

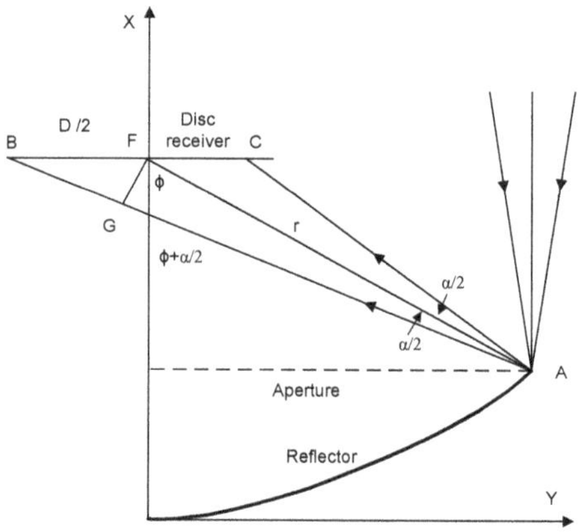

Fig. E4.9.1 Parabolic dish concentrator with disc-type receiver

able to receive all the reflected radiation incident on the aperture. Line FG is drawn perpendicular to AB. Therefore we have

$$FG = rsin(\alpha/2)$$

where r is the rim radius and α is the angle subtended by the sun at any point on earth. From the geometry of Δ BFG, we obtain the relation

$$\frac{D}{2} = BF = \frac{FG}{cos\left(\phi+\frac{\alpha}{2}\right)}$$

From the two equations above we get the diameter of the disc as

$$D = \frac{2rsin(\alpha/2)}{cos\left(\phi+\frac{\alpha}{2}\right)} \tag{E4.9.1}$$

When the dispersion angle is included the above expression becomes

$$D = \frac{2rsin\left(\frac{\alpha}{2}+\frac{\delta}{2}\right)}{cos\left(\phi+\frac{\alpha}{2}+\frac{\delta}{2}\right)} \tag{E4.9.2}$$

The slope of the parabola at the rim is given by

$$tan\theta = \frac{dx}{dy} = \frac{y}{2f} = \frac{0.7}{2\times 0.6} = 0.583$$

$$\theta = 30.26 \text{ degrees}$$

From example 4.4, the rim angle is

$$2\theta = 2 \times 30.26 = 60.52$$

The rim radius is given by Eq. (4.8a) as

$$r = \frac{2f}{1+cos\phi} = \frac{2\times0.6}{1+cos(60.52)} = 0.804 \text{ m}$$

The receiver disc diameter is given by Eq. (E4.9.2) as

$$D_{rc} = \frac{2rsin\left(\frac{\alpha}{2}+\frac{\delta}{2}\right)}{cos\left(\phi+\frac{\alpha}{2}+\frac{\delta}{2}\right)} = \frac{2\times0.804\times sin(0.267+0.3)}{cos(60.5+0.267+0.3)} = 0.0329 \text{ m}$$

The concentration ratio is

$$CR = \frac{A_{ap}}{A_{rc}} = \frac{\pi D_{ap}{}^2}{\pi D_{rc}{}^2} = \frac{1.4^2}{0.0329^2} = 1810.7$$

The fraction of the aperture that is shaded by the receiver is

$$f = \frac{A_{rc}}{A_{ap}} = 0.055\%$$

The weight of the reflector is an important design parameter because it will determine the support structure required. The area of the paraboloid is given by

$$A = \int_0^{r_a} 2\pi y ds = \int_0^{r_a} 2\pi y \left[1 + \left(\frac{dx}{dy}\right)^2\right]^{1/2} dy$$

$$A = \int_0^{r_a} 2\pi y \left[1 + \frac{y^2}{4f^2}\right]^{1/2} dy$$

Integration of the above equation gives the area as

$$A_{ref} = \left(\frac{8\pi f^2}{3}\right)\left[\left(1 + \frac{r_a{}^2}{4f^2}\right)^{3/2} - 1\right]$$

Substituting numerical values we obtain the receiver disc area as 1.664 m^2. Hence the weight is 26.6 kg.

Example 4.10 A parabolic dish concentrator has a focal length of 0.6 m and an aperture diameter of 1.4 m. The receiver is a cylinder of diameter 0.1 m located centrally along the axis of the parabola as depicted in Fig. E4.10.1. The dispersion angle of the reflector is 0.6°. Solar radiation is incident at right angles to the aperture. Calculate the height of the images formed on the receiver cylinder due to beams reflected at (i) the rim, and at points on the reflector distant 0.3 m and 0.1 m from the center, and (ii) the concentration ratio.

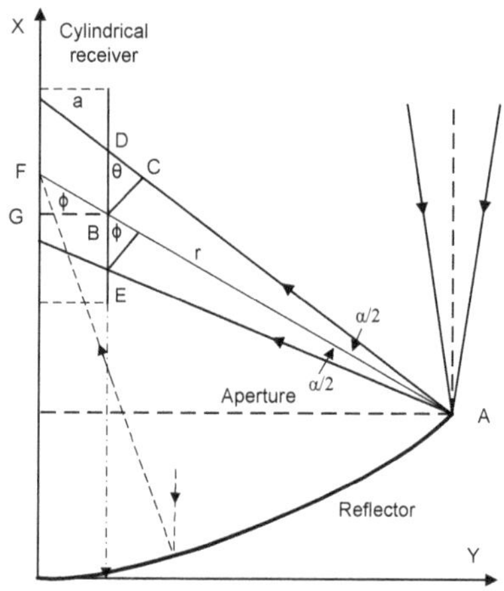

Fig. E4.10.1 Parabolic dish concentrator with cylindrical receiver

Solution Consider the beam reflected at point A on the rim of the parabolic reflector. The image formed on the outer surface of the cylinder is ED as shown in Fig. E4.10.1. From the geometry of triangles, BFG, ABC and BCD we obtain the following:

$$BF = \frac{a}{sin\phi} \qquad AB = r - \frac{a}{sin\phi},$$

$$BC = (r - a/sin\phi)sin(\alpha/2)$$

$$BD = \frac{BC}{sin\theta} = \frac{\left[(r-a/sin\phi)sin\left(\frac{\alpha}{2}\right)\right]}{sin\left(\phi-\frac{\alpha}{2}\right)} \tag{E4.10.1}$$

Similarly, by considering triangle ABE, we can show that

$$BE = \frac{\left[(r-a/sin\phi)sin\left(\frac{\alpha}{2}\right)\right]}{sin\left(\phi+\frac{\alpha}{2}\right)} \tag{E4.10.2}$$

The height of the image is $ED = BE + BD$

The vertical distance of point B from the vertex of the parabola is

$$x_o = f - \frac{a}{tan\phi} \tag{E4.10.3}$$

Using Eqs. (E4.10.1)-(E4.10.3) we can calculate the height of the image of a beam reflected at any point on the reflector. Also we can find its location with respect to the vertex of the parabola. Hence we can estimate the radiation flux distribution on the receiver.

For numerical computations we shall include the dispersion angle of the reflector. We first calculate the rim angle and radius at point A. The slope of the parabola at the rim is

$$tan\theta = \frac{dx}{dy} = \frac{y}{2f} = \frac{0.7}{1.2}, \qquad \theta = 30.25$$

From the analysis of example 4.4, the rim angle is,

$$2\theta = 2 \times 30.25 = 60.5$$

The rim radius is given by Eq. (4.8a) as

$$r = \frac{2f}{1+cos\phi} = \frac{2\times0.6}{1+\cos(60.5)} = 0.804 \text{ m}$$

Substituting in Eqs. (E4.10.1) to (E4.10.3) we obtain

$$BD = \frac{[(0.804-0.05/\sin 60.5)sin(0.267+0.3)]}{sin(60.5-0.267-0.3)} = 8.53 \text{ mm}$$

$$BE = \frac{[(0.804-0.05/sin60.5)sin(0.267+0.3)]}{sin(60.5+0.267+0.3)} = 8.44 \text{ mm}$$

$$x_o = 0.6 - \frac{0.05}{tan60.5} = 0.572 \text{ m}$$

The total height of the image, ED is given by

$$ED = BE + BD = 8.53 + 8.44 = 16.97 \text{ mm}$$

It is spread about a point 0.572 m from the vertex.

We use a similar procedure for the other two cases, and the results are summarized below:

For y = 0.3 m, $\phi_2 = 28.07, r_2 = 0.637$ m, $BD = 11.3$ mm,
$BE = 10.96$ mm, $ED = 22.26$ mm $x_0 = 0.506$ m

For y = 0.1 m, $\phi_3 = 9.53, r_3 = 0.604$ m, $BD = 19$ mm,
$BE = 10.7$ m, $ED = 29.7$ mm, $x_0 = 0.301$ m

Notice that the image is distributed from about 0.3 m all the way up to about 0.6 m. As the center is approached the image formed moves down and it is spread over a larger area of the receiver surface. However, more energy is reflected to the area around the focus because the aperture area gets larger with the radius.

Example 4.11 A parabolic trough concentrator has an aperture width of 1.2 m and a focal length of 1.2 m. The reflectivity of the reflector is 0.85 and dispersion angle is 0.6°. Beam radiation of intensity 900 Wm^{-2} is incident on the aperture at right angles. Estimate the radiation flux distribution on a flat absorber located at the focal plane.

Solution Solar radiation is incident on the aperture at right angles as depicted in Fig. E4.11.1(a). To estimate the flux distribution on the flat receiver, we adopt a numerical procedure. For simplicity we divide the aperture into 6 equal sections which receive equal amounts of radiation. The mid points 1-6 of the 6 sections are indicated on the reflector.

By symmetry sections 1 and 6 reflect the same amount of radiation onto the receiver as shown. The same applies to sections 2 and 5 and 3 and 4. We shall compute the radiation incident on the receiver which is reflected from the different mid points. Consider point 1 as an example. The y-coordinate of 1 is 0.5 m. The slope at 1 is given by

$$tan\theta = \frac{dx}{dy} = \frac{y}{2f} = \frac{0.5}{2.4} = 0.2083, \theta = 11.77 \text{ degrees}$$

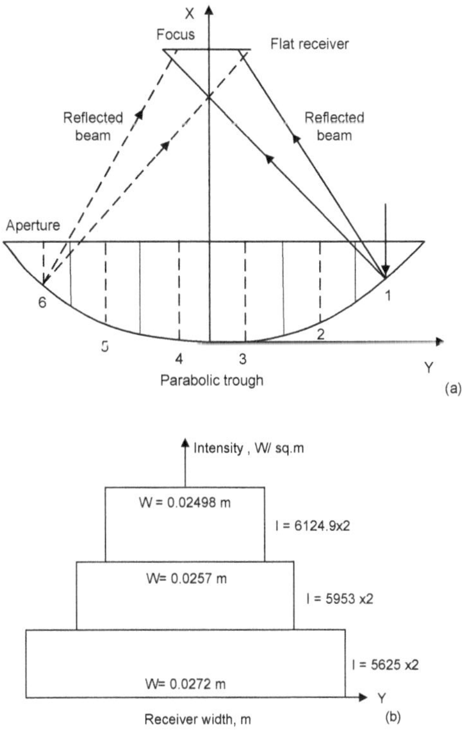

Fig. E4.11.1 Parabolic trough concentrator, (a) schematic diagram of sections, (b) flux distribution

From the analysis in example 4.4, the rim angle is,

$$2\theta = 2 \times 11.77 = 23.54 \text{ degrees}$$

The rim radius is given by Eq. (4.8a) as

$$r = \frac{2f}{1+cos\phi} = \frac{2\times 1.2}{1+cos(23.54)} = 1.25 \text{ m}$$

The width of the flat receiver over which this radiation is distributed is given as

$$W = \frac{2r_r \sin(16'+\delta/2)}{cos(\phi_r+16'+\delta/2)} = \frac{2\times 1.25\times sin(0.267+0.3)}{cos(23.54+0.267+0.3)} = 0.0272 \text{ m}$$

Consider unit length of the reflector. The radiation reflected from the sections from 1 to 6 is given

$$I = (900 \times 1.2 \times 1 \times 0.85)/6 = 153 \text{ W}$$

The radiation intensity on the receiver due to reflection from section 1 is

$$H_1 = \frac{153}{0.0272 \times 1} = 5625 \text{ Wm}^{-2}$$

We have repeated the above calculation for all the sections. The results are summarized in Table E4.11.1 below.

Table E4.11.1 Flux distribution at the receiver

Section	1	2	3	4	5	6
Rim angle (deg)	23.54	14.25	4.77	4.77	14.25	23.54
Radius (m)	1.25	1.219	1.202	1.202	1.219	1.25
Image Width (m)	0.0272	0.0257	0.025	0.025	0.0257	0.0272
Intensity (Wm^{-2})	5625	5953.3	6124.9	6124.9	5953.3	5625

The results are displayed graphically in Fig. E4.11.1(b). Notice that the intensity is highest at the center section of the receiver which has a total intensity of 35406 Wm^{-2}, the next section has a total intensity of 23156 Wm^{-2}.

Example 4.12 A trough concentrator collector has a receiver tube of diameter 60 mm and an evacuated outer glass cover of diameter 90 mm as shown in Fig. E4.12.1. The receiver surface emittance is 0.9 and its temperature is 220°C. The average wind speed is 6 ms^{-1} and the ambient temperature is 20°C. Calculate (i) the rate of heat loss, and (ii) the heat loss coefficient.

Solution To compute the overall heat loss coefficient, we need to assume an initial value for the glass cover temperature. As a first guess assume that the cover temperature is 60°C. The mean temperature between the cover and the ambient is 40°C. We first calculate the wind heat transfer coefficient using the following Eq. (A2.4) [1].

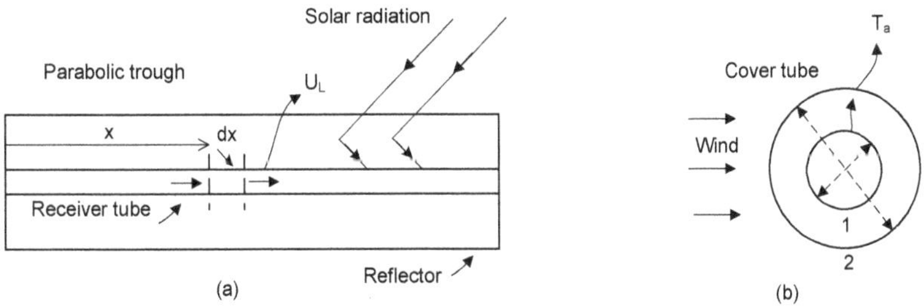

Fig. E4.12.1 (a) Parabolic trough collector, (b) Details of receiver tube and cover

$$Nu = 0.24Re^{06} \tag{E4.12.1}$$

The following properties of air at 40°C are obtained from tabulated data [5].

$$\rho = 1.06 \text{ kgm}^{-3}, \mu = 2 \times 10^{-5} \text{ kgm}^{-1}\text{s}^{-1}, k = 2.87510^{-5} \text{ kWm}^{-1}\text{K}^{-1}.$$

Now the Reynolds number is

$$Re = \frac{\rho VD}{\mu} = \frac{1.06\times6\times90\times10^{-3}}{2\times10^{-5}} = 28620$$

Substituting in the correlation in Eq. (E4.12.1) we get

$$Nu = 0.24Re^{0.6} = 0.24 \times 28620^{0.6} = 113.3$$

This value is multiplied by 1.25 for wind blowing over a cylinder [1]. Hence we have, $Nu = 141.6$.

The heat transfer coefficient is given by

$$h = \frac{Nuk}{D} = \frac{141.6\times2.875\times10^{-5}}{90\times10^{-3}} = 45.2 \text{ Wm}^{-2}\text{K}^{-1}$$

The radiation heat transfer coefficient between the outer cover and the ambient is given by

$$h_{rga} = \sigma\varepsilon\left(T_g^{\ 2} + T_a^{\ 2}\right)\left(T_g + T_a\right)$$

$$h_{rga} = 5.67 \times 10^{-8} \times 0.88(333^2 + 293^2)(333 + 293) = 6.15$$

Total heat transfer coefficient between the outer cover and the ambient is

$$h_{tga} = 45.2 + 6.15 = 51.35 \text{ Wm}^{-2}\text{K}^{-1}$$

Since the space between the tube and the cover is evacuated, the only significant mode of heat transfer is thermal radiation. The radiation heat transfer coefficient between the outer cover and receiver tube is given by

$$h_{rtg} = \sigma\varepsilon_{eff}\left(T_g^{\ 2} + T_t^{\ 2}\right)\left(T_g + T_t\right)$$

$$\frac{1}{\varepsilon_{eff}} = \frac{1}{\varepsilon_t} + \left(\frac{D_t}{D_g}\right)\left(\frac{1}{\varepsilon_g} - 1\right)$$

$$\frac{1}{\varepsilon_{eff}} = \frac{1}{0.9} + \left(\frac{60}{90}\right)\left(\frac{1}{0.88} - 1\right) = 1.202$$

$$\begin{aligned} h_{rtg} &= 5.67 \times 10^{-8} \times 0.832(493^2 + 333^2)(493 + 333) \\ &= 13.8 \text{ Wm}^{-2}\text{K}^{-1} \end{aligned}$$

Consider unit length of the collector and write the following steady state energy balance equations.

Tube to glass: $Q = \pi \times 0.06 \times 13.8 \times \left(493 - T_g\right)$

Glass to ambient: $Q = \pi \times 0.09 \times 51.35 \times \left(T_g - 293\right)$

where Q is the heat transfer rate per unit length.

From above equations we obtain

$$Q = 441 \text{ Wm}^{-1} \text{ and } T_g = 323.4 \text{ K}$$

The overall heat transfer coefficient is given by

$$Q = U_{loss} \times 0.06 \times \pi \times (493 - 293) = 441$$

Hence we get $U_{loss} = 11.7 \text{ Wm}^{-2}\text{K}^{-1}$

The temperature of the glass cover is 50.4°C. The initially assumed value of the cover is 60°C. We could now repeat the computation with the new cover temperature. This however, does not change the loss coefficient by much.

Example 4.13 The length and aperture of a parabolic trough concentrator are 10 m and 2.5 m respectively. It has a stainless steel receiver tube of outer diameter 60 mm and wall thickness 5 mm. The glass tube surrounding the receiver has a diameter of 90 mm. Heat transfer fluid with a specific heat capacity of 1.28 $kJkg^{-1}K^{-1}$ enters the receiver at 200°C and flows through it at the rate of 0.14 kgs^{-1}.

The convective heat transfer coefficient from the tube to the fluid is 320 $Wm^{-2}K^{-1}$. The overall heat loss coefficient from the receiver tube to the ambient is 11.7 $Wm^{-2}K^{-1}$. The ambient temperature is 20°C and the solar radiation absorption rate per unit aperture area is 480 Wm^{-2}. The thermal conductivity of stainless steel is 16 $Wm^{-1}K^{-1}$. Calculate (i) the collector efficiency factor, (ii) the collector heat removal factor, (iii) the useful energy gain, and (iv) the exit temperature of the fluid.

Solution A schematic view of the trough collector is given in Fig. E4.12.1(a). Consider an elemental section of the receiver of length dx. We write the following energy equations for the different components. For the receiver tube element we have

$$S_r \pi D_o dx = U_L(T_o - T_a)\pi D_o dx + h_{fo}(T_o - T_f)\pi D_o dx \qquad \text{(E4.13.1)}$$

where S_r is the rate of absorption of solar radiation *per unit receiver area.*

The LHS of the above equation is the rate of absorption of solar radiation, the first term on the RHS is the rate of heat loss, and the second term is the rate of heat transfer to the fluid. For the flowing fluid the energy equation is

$$mc_p dT_f = h_{fo}(T_o - T_f)\pi D_o dx \qquad \text{(E4.13.2)}$$

The heat transfer coefficient from the outside of the tube to the fluid is given by

$$\frac{1}{h_{fo}} = \frac{D_o}{D_i h_{fi}} + \frac{D_o ln\frac{D_o}{D_i}}{2k} \qquad \text{(E4.13.3)}$$

where k is the thermal conductivity of the tube. Rearranging Eq. (E4.13.1) we obtain

$$(T_o - T_f) = \frac{S_r}{U_L + h_{fo}} - \frac{U_L(T_f - T_a)}{U_L + h_{fo}} \qquad \text{(E4.13.4)}$$

From Eqs. (E4.13.2) and (E4.13.4) we have

$$\frac{dT_f}{dx} = \left(\frac{h_{fo}\pi D_o}{mc_p}\right)\left[\frac{S_r}{U_L+h_{fo}} - \frac{U_L(T_f-T_a)}{U_L+h_{fo}}\right] \tag{E4.13.5}$$

We solve the differential equation (E4.13.5) and substitute the fluid temperatures at the inlet and outlet of the collector to obtain the fluid temperature rise as

$$T_{fo} - T_{fi} = \left[\frac{S_r}{U_L} - (T_{fi} - T_a)\right]\left(1 - e^{\frac{-A_r U_L F'}{mc_p}}\right) \tag{E4.13.6}$$

where the collector efficiency factor F' is defined as,

$$F'U_L = \left(\frac{1}{U_L} + \frac{1}{h_{fo}}\right)^{-1}$$

Now the useful energy collection rate is given by

$$Q_u = mc_p(T_{fo} - T_{fi}) \tag{E4.13.7}$$

Substituting from Eq. (E4.13.6) we get

$$Q_u = mc_p\left[\frac{S_r}{U_L} - (T_{fi} - T_a)\right]\left(1 - e^{\frac{-A_r U_L F'}{mc_p}}\right)$$

$$Q_u = A_r F_R\left[S_r - U_L(T_{fi} - T_a)\right] \tag{E4.13.8}$$

where the collector heat removal factor is defined as

$$F_R = \left(\frac{mc_p}{A_r U_L}\right)\left(1 - e^{\frac{-A_r U_L F'}{mc_p}}\right)$$

Equation (E4.13.8) may be rearranged to the form

$$Q_u = A_a F_R\left[S_a - U_L\left(\frac{A_r}{A_a}\right)(T_{fi} - T_a)\right] \tag{E4.13.9}$$

where S_a is the absorption rate per *unit aperture area*.

We substitute numerical data to obtain the required quantities. Hence we have from Eq. (E4.13.3)

$$h_{fo} = 244.4\ \text{Wm}^{-2}\text{K}^{-1}$$

$$F'U_L = \left(\frac{1}{U_L} + \frac{1}{h_{fo}}\right)^{-1} = 11.16$$

Hence $$F' = \frac{11.16}{11.7} = 0.954$$

$$F_R = \left(\frac{mc_p}{A_r U_L}\right)\left(1 - e^{\frac{-A_r U_L F'}{mc_p}}\right) = 0.9$$

The unshaded area of the aperture is

$$A_a = 10(2.5 - 0.09) = 24.1\ \text{m}^2$$

The useful energy collection rate is given by Eq. (E4.13. 9) as

$$Q_u = 24.1 \times 0.9\left[480 - \left(\frac{\pi \times 0.06 \times 10 \times 11.7}{24.1}\right)(200 - 20)\right]$$

Therefore $$Q_u = 6839\ \text{W}$$

The useful energy collection is related to the temperature rise of the fluid by Eq. (E4.13.7). Substituting numerical values we obtain the exit fluid temperature as 238°C.

Example 4.14 A thermal power plant drives an electric generator with an output of 10 MW and an efficiency of 89%. The plant operates on an ideal Rankine cycle with a boiler pressure of 30 bar and a condenser pressure of 0.04 bar. Steam is superheated to 350°C before it is expanded insentropcally in the turbine.

Steam exiting from the turbine is condensed to a saturated liquid before it is pumped to the boiler isentropically. Calculate (i) the net work output of the cycle per kg of steam, (ii) the heat input in the boiler and superheater, (iii) the heat rejected in the condenser, (iv) the feed pump work, (v) the thermal efficiency of the cycle, and (vi) the steam flow rate through the turbine.

Solution The *T-s* and *h-s* diagrams of the Rankine cycle are shown in Figs. 4.16(b) and (c). The expansion and compression processes are ideal. We obtain the following property data from the steam tables [5].

$$h_1 = 3117 \text{ kJkg}^{-1}, \quad s_1 = 6.744 \text{ kJ K}^{-1}\text{kg}^{-1}, \quad h_3 = 121 \text{ kJkg}^{-1}$$

For the isentropic process $s_1 = s_2$

$$6.744 = 8.473x_{2s} + 0.422(1 - x_{2s})$$

The dryness fraction, $x_{2s} = 0.785$

$$h_2 = 121(1 - x_{2s}) + 2554x_{2s} = 2030.9$$

The work output of the turbine *per kg of steam* is

$$w_{12} = h_1 - h_2 = 3117 - 2030.9 = 1086 \text{ kJkg}^{-1}.$$

The work input to the feed pump is

$$w_{34} = v(P_4 - P_3) = (30 - 0.04) \times 100 \times 10^{-3} = 3 \text{ kJkg}^{-1}.$$

We have assumed that the specific volume of water is 10^{-3} m^{3}kg^{-1}.

Hence $h_4 = 121 + 3 = 124 \text{ kJkg}^{-1}$

The net work output of the cycle is

$$w_{net} = w_{12} - w_{34} = 1086 - 3 = 1083 \text{ kJkg}^{-1}$$

Heat input in the boiler and superheater is

$$q_{41} = h_1 - h_4 = 3117 - 124 = 2993 \text{ kJkg}^{-1}$$

Heat rejected in the condenser is

$$q_{23} = h_2 - h_3 = 2030.9 - 121 = 1909.9 \text{ kJkg}^{-1}$$

Net heat input to the cycle is

$$q_{net} = q_{41} - q_{23} = 2993 - 1909.9 = 1083 \text{ kJkg}^{-1}$$

As expected for the cycle the net work output is equal to the net heat input.

The thermal efficiency of the cycle is

$$\eta = \frac{w_{net}}{q_{in}} = \frac{1083.1}{2993} = 36\%$$

The electrical energy output of the plant is 10 MW and the efficiency of the generator is 89%. Therefore net work output of the turbine that drives the generator is

$$W_{out} = 10 \times \frac{1000}{0.89} = 11236 \text{ kW}.$$

The rate of net work output is

$$m_{st} w_{net} = W_{out}$$

The mass flow rate of steam is given by

$$m_{st} = \frac{W_{out}}{w_{net}} = \frac{11236}{1083} = 10.37 \text{ kgs}^{-1}.$$

Example 4.15 A Rankine cycle power plant with the same conditions as in example 4.14 drives an electric generator whose output is 8 MW and efficiency is 85%. The overall efficiency of the boiler is 82% and the fuel used has a higher heating value of 43 MJkg^{-1}. The temperature rise of the condenser cooling water is 6°C. Calculate (i) the hourly fuel consumption of the boiler, and (ii) the mass flow rate of cooling water.

Solution In example 4.14 we obtained the net work output of the Rankine cycle as 1083 kJkg^{-1} and the thermal efficiency of the cycle as 36%. Since the conditions of the Rankine cycles are the same, the above quantities can be used in the present calculation.

The work input to the generator is given by

$$W_{gi} = \frac{8}{0.85} = 9.41 \text{ MW}$$

The net work output of the Rankine cycle is also equal to 9.41 MW.

The steam flow rate is given by

$$m_s = \frac{9.41 \times 1000}{1083} = 8.69 \text{ kgs}^{-1}.$$

The heat input in the steam generator of the Rankine cycle is given by

$$Q_{bi} = \frac{W_{gi}}{\eta_{Ranine}} = \frac{9.41}{0.36} = 26.14 \text{ MW}$$

The fuel consumption rate in the steam generator is

$$m_f = \frac{Q_{bi}}{\eta_{boiler} CV} = \frac{26.14}{0.82 \times 43} = 0.74 \text{ kgs}^{-1}$$

The hourly fuel consumption is 2664 kg per hour.

The rate of heat rejection in the condenser is given by

$$Q_{rej} = Q_{bi} - W_{no} = 26.14 - 9.41 = 16.73 \text{ MW}$$

Energy balance for the cooling water gives

$$m_{cw} c_w (T_{wo} - T_{wi}) = Q_{rej}$$

$$m_{cw} \times 4.16 \times 6 = 16.73 \times 1000$$

The rate of flow of condenser cooling water is 670.3 kgs^{-1}.

Example 4.16 A solar concentrator-collector system supplies heat to a Rankine cycle power plant producing electricity. The average efficiency of solar collector system, which is the ratio of the heat energy delivered to the total solar radiation incident on the aperture of the collector field, is 38%. The average solar radiation intensity on the aperture is 740 Wm^{-2}.

The boiler and condenser pressures of a steam power plant are 30 bar and 0.04 bar respectively. Steam entering the turbine is heated to 350°C in a superheater. The isentropic efficiencies of the turbine and the feed pump are 80% and 85% respectively. The electrical energy output of the generator is 12 MW and the its efficiency is 85%. Calculate (i) the net work output of the cycle, (ii) the heat input in the steam generator, (iii) the thermal efficiency of the cycle, and (iv) the total aperture area of the solar collector system.

Solution The *T-s* and *h-s* diagrams of the Rankine cycle are shown in Figs. 4.16(b) and (c). The expansion and compression processes are not ideal. We obtained the following property data from the steam tables [5].

$$h_1 = 3117 \text{ kJkg}^{-1}, \quad s_1 = 6.744 \text{ kJ K}^{-1}\text{kg}^{-1}, \quad h_3 = 121 \text{ kJkg}^{-1}$$

For the isentropic process $\quad s_1 = s_2$

$$6.744 = 8.473 x_{2s} + 0.422(1 - x_{2s})$$

The dryness fraction, $x_{2s} = 0.785$

$$h_{2s} = 121(1 - x_{2s}) + 2554x_{2s} = 2030.9$$

For the non-isentropic expansion in the turbine we have

$$\eta_t = \frac{w_{12}}{h_1 - h_{2s}} = \frac{h_1 - h_2}{1086.1} = 0.8$$

Hence we obtain

$$w_{12} = 869 \text{ kJkg}^{-1} \quad \text{and} \quad h_2 = 2248 \text{ kJkg}^{-1}$$

The ideal work input to the feed pump is

$$w_{34} = v(P_4 - P_3) = (30 - 0.04) \times 100 \times 10^{-3} = 3 \text{ kJkg}^{-1}.$$

For the non isentropic compression in the pump we have

$$\eta_P = \frac{h_4 - h_3}{w_{34}} = \frac{3}{w_{34}} = 0.85$$

Hence we obtain

$$w_{34} = 3.53 \text{ kJkg}^{-1} \text{ and } h_4 = 121 + 3.53 = 124.5 \text{ kJkg}^{-1}$$

The net work output of the cycle is

$$w_{net} = w_{12} - w_{34} = 869 - 3.53 = 865.4 \text{ kJkg}^{-1}$$

Heat input in the boiler and superheater is

$$q_{41} = h_1 - h_4 = 3117 - 124.5 = 2992.5 \text{ kJkg}^{-1}$$

Heat rejected in the condenser is

$$q_{23} = h_2 - h_3 = 2248 - 121 = 2127 \text{ kJkg}^{-1}$$

Net heat input to the cycle is

$$q_{net} = q_{41} - q_{23} = 2992.5 - 2127 = 865.5 \text{ kJkg}^{-1}$$

As expected for the cycle the net work output is equal to the net heat input.

The thermal efficiency of the cycle is

$$\eta = \frac{w_{net}}{q_{in}} = \frac{865.4}{2992.5} = 28.9\%$$

Notice that in example 4.14, the Rankine cycle with the same conditions had an ideal turbine and an ideal pump, and the efficiency was 36%. When friction in the turbine and the pump are included the efficiency becomes 29%.

The electrical energy output of the plant is 12 MW and the efficiency of the generator is 85%. Therefore net work output of the turbine that drives the generator is

$$W_{out} = 12 \times \frac{1}{0.85} = 14.1 \text{ MW}.$$

Rate of net work output is

$$m_{st} w_{net} = W_{out}$$

Mass flow rate of steam is given by

$$m_{st} = \frac{W_{out}}{w_{net}} = \frac{14.1 \times 1000}{865.4} = 16.3 \text{ kgs}^{-1}$$

Heat input in the steam generator is

$$Q_{sg} = \frac{W_{out}}{\eta} = \frac{14.1 \times 1000}{0.289} = 48.79 \times 10^3 \text{ kW}$$

We shall assume that the heat input in the steam generator is equal to the useful energy delivered by the solar collector system. Hence we have

$$I_s A_a \eta_{sc} = Q_u = Q_{sg}$$

Substituting numerical values we obtain

$$0.74 \times 0.38 A_a = 48.79 \times 10^3$$

The total aperture area of the collector is 173.5×10^3 m^2.

Example 4.17 In a simple air standard Brayton cycle air enters the compressor at 1 bar and 27°C. The pressure ratios for the compressor and turbine are 5. The compressor and turbine are ideal. At entry to the turbine the air temperature is 1150°C. Calculate (i) the net work input, (ii) the heat input, and (iii) the efficiency of the cycle.

Solution A schematic of a simple gas turbine plant is shown in Fig. 4.17(b). The *P-v* and *T-s* diagrams are shown in Figs. 4.18(a) and (b) respectively. The following data are given

$$T_1 = 300 \text{ K}, \quad T_3 = 1423 \text{ K}, \quad r = 5$$

For the isentropic compression process 1-2s, we have

$$\left(\frac{P_2}{P_1}\right)^{\frac{\gamma-1}{\gamma}} = \frac{T_{2s}}{T_1} = r^{\frac{\gamma-1}{\gamma}}$$

where $\gamma = \frac{c_p}{c_v} = 1.4$, is the ratio of the specific heat capacities.

Substituting numerical values we obtain $\quad T_{2s} = 475.13 \text{ K}$

For the isentropic expansion process 3-4s, we have

$$\left(\frac{P_2}{P_1}\right)^{\frac{\gamma-1}{\gamma}} = \frac{T_3}{T_{4s}} = r^{\frac{\gamma-1}{\gamma}}$$

where $\gamma = \frac{c_p}{c_v} = 1.4$, is the ratio of the specific heat capacities.

Substituting numerical values we obtain $\quad T_{4s} = 898.48 \text{ K}$

Apply the energy equation, to the high temperature and low temperature heat exchangers, neglecting the kinetic and potential energy. The rate of heat input is

$$Q_h = mc_p(T_3 - T_2) = mc_p(1423 - 475.13) = 947.87mc_p$$

$$Q_h = 947.87 \times 1.005 = 952.6 \text{ kJkg}^{-1}$$

The heat rejection rate is

$$Q_c = mc_p(T_4 - T_1) = mc_p(898.48 - 300) = 598.5mc_p$$

where m is the steady air flow rate through the system.

The work output of the turbine is

$$W_t = mc_p(T_3 - T_4) = mc_p(1423 - 898.48) = 524.5mc_p$$

The work input to the compressor is

$$W_c = mc_p(T_2 - T_1) = mc_p(475.13 - 300) = 175.13mc_p$$

The net work output is

$$W_{net} = W_t - W_c = 349.37mc_p$$

$$W_{net} = 349.37 \times 1.005 = 351.1\text{kJkg}^{-1}$$

The efficiency of the cycle is given by

$$\eta = 1 - \frac{Q_c}{Q_h} = 1 - \frac{mc_p(T_4 - T_1)}{mc_p(T_3 - T_2)}$$

$$\eta = 1 - \frac{Q_c}{Q_h} = 1 - \frac{598.5mc_p}{947.9mc_p} = 36.8\%$$

Example 4.18 A gas turbine plant drives an electric generator with an output of 5 MW and efficiency 85%. The gas turbine operates on the air standard Brayton cycle where air enters the compressor at 1 bar and 27°C. The pressure ratios of the compressor and turbine are 5. At entry to the turbine the air is at 1150°C. The isentropic efficiencies of the turbine and compressor are 0.75 and 0.65 respectively. The combustion chamber burns natural gas with a heating value of 48 MJkg^{-1}. Calculate (i) the work output of the turbine, (ii) the heat input in the combustion chamber, (iii) the thermal efficiency, and (iv) the fuel consumption rate.

Solution A schematic of a simple gas turbine plant is shown in Fig. 4.17(b). The *P-v* and *T-s* diagrams are shown in Figs. 4.18(a) and (b) respectively. The following data are given

$$T_1 = 300 \text{ K}, \quad T_3 = 1423 \text{ K}, \quad r = 5$$

For the isentropic compression process 1-2s, we have

$$\left(\frac{P_2}{P_1}\right)^{\frac{\gamma-1}{\gamma}} = \frac{T_{2s}}{T_1} = 5^{\frac{\gamma-1}{\gamma}}, T_{2s} = 475.13 \text{ K}$$

where $\gamma = \frac{c_p}{c_v} = 1.4$, is the ratio of the specific heat capacities.

The isentropic efficiency of the non-ideal compressor is defined as

$$\eta_c = \frac{T_{2s}-T_1}{T_2-T_1} = \frac{475.13-300}{T_2-300} = 0.65,$$

$$T_2 = 569.43 \text{ K}$$

For the isentropic expansion process 3-4s, we have

$$\left(\frac{P_2}{P_1}\right)^{\frac{\gamma-1}{\gamma}} = \frac{T_3}{T_{4s}} = 5^{\frac{\gamma-1}{\gamma}},$$

$$T_{4s} = 898.48 \text{ K}$$

The isentropic efficiency of the non-ideal turbine is

$$\eta_t = \frac{T_3-T_4}{T_3-T_{4s}} = \frac{1423-T_4}{1423-898.48} = 0.75,$$

$$T_4 = 1029.6 \text{ K}$$

Apply the energy equation to the high temperature and low temperature heat exchangers, neglecting the kinetic and potential energy. The rate of heat input is

$$Q_h = mc_p(T_3 - T_2) = mc_p(1423 - 569.43) = 853.57mc_p$$

The heat rejection rate is

$$Q_c = mc_p(T_4 - T_1) = mc_p(1029.1 - 300) = 729.1mc_p$$

where m is the steady air flow rate through the system.

The work output of the turbine is given by

$$W_t = mc_p(T_3 - T_4) = mc_p(1423 - 1029.6) = 393.4mc_p$$

The work input to the compressor is

$$W_c = mc_p(T_2 - T_1) = mc_p(569.43 - 300) = 269.43mc_p$$

The net work output is

$$W_{net} = W_t - W_c = 123.97mc_p$$

The efficiency of the gas turbine cycle is given by

$$\eta = 1 - \frac{Q_c}{Q_h} = 1 - \frac{729.1}{853.57} = 14.6\%$$

In example 4.17 we obtained the efficiency with an ideal compressor and turbine as 36.8%. Note that friction in the turbine and the compressor has reduced the efficiency significantly.

The electrical energy output of the generator is 5 MW and its efficiency is 85%. Therefore the net work output of the gas turbine that drives the generator is

$$W_{out} = 5 \times \frac{1}{0.85} = 5.88 \text{ MW}.$$

The heat input in the combustion chamber is

$$Q_{com} = \frac{W_{out}}{\eta} = \frac{5.88}{0.146} = 40.3 \text{ MW}$$

The fuel consumption rate in the combustion chamber is

$$m_f = \frac{Q_{com}}{CV} = \frac{40.3}{48} = 0.84 \text{ kgs}^{-1}$$

Example 4.19 An ideal Stirling engine operates between a hot reservoir at 300°C and a cold reservoir at 50°C. The compression ratio is 4. Calculate (a) (i) the net work output, (ii) the heat input, (iii) the heat output, and (iv) the thermal efficiency. (b) If the engine has no regenerator calculate the above quantities.

Solution A *P-v* diagram of a Stirling cycle are depicted in Fig. 4.19(b). Consider unit mass of air in the cycle. The work done during the isothermal processes 1-2 and 3-4 are given by

$$W_{12} = mRT_h ln\left(\frac{V_2}{V_1}\right) = 1 \times 0.287(300 + 273)ln4 = 228 \text{ kJkg}^{-1}$$

$$W_{34} = mRT_c ln\left(\frac{V_3}{V_4}\right) = 1 \times 0.287(50 + 273)ln4 = 128.5 \text{ kJkg}^{-1}$$

For an isothermal process with an ideal gas the heat interactions are equal to the work interactions. Therefore the heat input and heat output are given by

$$Q_{12} = W_{12} = 228 \text{ kJkg}^{-1} \quad \text{and} \quad Q_{34} = W_{34} = 128.5 \text{ kJkg}^{-1}$$

Heat flow from the *ideal regenerator* to the working fluid during process, 4-1 is equal to the heat transferred to the regenerator during the process 2-3 because they are given by

$$Q_{41} = Q_{23} = mc_v(T_h - T_c) = 1 \times 0.718(300 - 50) = 179.5 \text{ kJkg}^{-1}$$

The efficiency of the ideal Stirling cycle is given by

$$\eta = \frac{W_{net}}{Q_{in}} = \frac{228-128.5}{228} = 43.6\%$$

(b) If there is no regenerator, additional heat has to be supplied from the heat source during the process 4-1. This is equal to 179.5 kJkg^{-1}. Therefore the total heat input is 407.5 kJkg^{-1} and the heat rejection is 308 kJkg^{-1}. Hence the efficiency is

$$\eta = \frac{W_{net}}{Q_{in}} = \frac{228-128.5}{228+179.5} = 24.4\%$$

Example 4.20 A parabolic dish solar collector supplies heat to a Stirling engine which operates between a heat source at 300°C and a heat sink at 50°C. The output of the engine 10 kW. The compression ratio is 4 and the regenerator effectiveness is 0.7. The average efficiency of the solar collector is 40% and the average solar radiation intensity on the aperture is 800 Wm^{-2}. Calculate (i) the thermal efficiency of the engine, and (ii) the aperture area of the collector.

Solution The efficiency of a Stirling engine with a non-ideal regenerator is given by Eq. (4.55) as

$$\eta = \frac{W_{net}}{Q_{in}} = \frac{RT_h ln\left(\frac{V_2}{V_1}\right) - RT_c ln\left(\frac{V_3}{V_4}\right)}{RT_h ln\left(\frac{V_2}{V_1}\right) + c_v(T_h - T_r)}$$

The effectiveness of the regenerator is given by Eq. (4.56) as

$$e = \frac{T_r - T_c}{T_h - T_c} = \frac{T_r - 323}{573-323} = 0.7$$

Hence $T_r = 498$ K

Substituting in the above expression for efficiency we have

$$\eta = \frac{0.287 \times 573 ln4 - 0.287 \times 323 ln4}{0.287 \times 573 ln4 + 0.718(573 - 498)} = 35.3\%$$

The heat input to the engine from the collector is

$$Q_{in} = \frac{W_o}{\eta_e} = \frac{10}{0.353} = 28.3 \text{ kW}$$

The area of the collector aperture is

$$A_a = \frac{28.3}{0.8 \times 0.4} = 88.43 \text{ m}^2.$$

Problems

P4.1 A trough concentrator at a location with a northern latitude 40° is rotated about a horizontal north-south axis with continuous adjustment to minimize the angle of incidence on the aperture. At 3 p.m. solar time on April 12, the intensity of beam radiation on a horizontal plane at the location is 590 $Wm^{-2.}$ Calculate (i) the tilt angle of the aperture, (ii) the angle of incidence on the aperture, and (iii) the beam radiation incident on a unit area of the aperture.
[*Answers*: (i) 48°, (ii) 19.8°, (iii) 882.8 Wm^{-2}]

P4.2 The acceptance angle of a CPC collector is 23° and it is oriented along the east-west axis at a location with a northern latitude of 35°. The aperture area of the collector is 15 m^2 and it is inclined at 25° to the horizontal. The beam radiation intensity on a horizontal surface at the location at 10 a.m. solar time on August 15, is 560 Wm^{-2}. The average transmittance of the aperture is 0.9. Calculate the total beam radiation transmitted through the aperture. [*Answer*: 531 Wm^{-2}]

P4.3 The length and aperture width of a parabolic trough concentrator are 9 m and 1.4 m respectively. The focal length is 1.2 m. The dispersion angle of the reflector is 0.5°. A beam of solar radiation is incident on the aperture at right angles. Calculate (i) the rim angle of the concentrator, (ii) the minimum areas of a flat receiver and a round tube receiver, placed at the focus on which all the reflected radiation will be incident,

(iii) the concentration ratio for each receiver, and (iv) the fraction of the aperture area that would be shaded by the receiver.
[*Answers*: (i) 32.5°, (ii) 0.25 m^2, 0.66 m^2, (iii) 50, 19, (iv) 2%, 1.67%]

P4.4 The aperture diameter of a parabolic dish concentrator is 1.6 m and the focal length is 0.6 m. The receiver is a disc located centrally at the focus of the reflector dish. The dispersion angle of the reflector is 0.5°. Calculate (i) the rim angle and the rim radius of the reflector, (ii) the minimum diameter of the receiver disc that would intercept all reflected beam radiation incident on the aperture at right angles, and (iii) the concentration ratio. [*Answers*: (i) 67.38°, 0.877 m, (ii) 0.042 m, (iii) 1447]

P4.5 Estimate the radiation flux distribution on the disc receiver in Problem 4.4. The solar radiation incident on the aperture at right angles is 720 Wm^{-2}. Consider the radiation incident on 4 equal concentric areas of the concentrator aperture and use the mid-circle of each area to calculate the incident area on the receiver, and the intensity of radiation in kWm^{-2}. [*Answers*: 3301, 2903, 762, 385]

P4.6 A parabolic trough concentrator is 8 m long and its aperture width is 2 m. It has a stainless steel receiver tube of outer diameter 50 mm and wall thickness 5 mm. The glass tube surrounding the receiver has a diameter of 80 mm. Heat transfer fluid with a specific heat capacity of 1.3 $kJkg^{-1}K^{-1}$ enters the receiver at 108°C and flows through it at the rate of 0.15 kgs^{-1}. The convective heat transfer coefficient from the tube to the fluid is 340 $Wm^{-2}K^{-1}$.

The overall heat loss coefficient from the receiver tube to the ambient is 10.8 $Wm^{-2}K^{-1}$. The ambient temperature is 22°C and the solar radiation absorption rate per unit aperture area is 780 Wm^{-2}. The thermal conductivity of stainless steel is 16 $Wm^{-1}K^{-1}$. Calculate (i) the collector flow efficiency factor, (ii) the collector heat removal factor, (iii) the useful energy gain, and (iv) the exit temperature of the fluid.
[*Answers*: (i) 0.958, (ii) 0.927, (iii) 820.3 W, (iv) 112.2°C]

P4.7 The average efficiency of a parabolic trough solar collector system is 35%. It is used as the heat source of a steam power plant with

an efficiency of 38%. The average solar radiation intensity on the aperture of the collectors is 690 Wm^{-2}. The efficiency of the electric generator is 85% and its output is 6 MW. Calculate (i) the total aperture area of the collector field, (ii) the heat energy delivered to the steam plant, and (iii) the overall efficiency of the system.
[*Answers*: (i) 77,000 m^2, (ii) 18.58 MW, (iii) 11.3%]

P4.8 A concentrator solar collector system with a concentration ratio of 15 and and optical efficiency of 0.78 supplies heat to a Rankine cycle steam plant. The heat loss coefficient from the receiver to the ambient at 20°C is given by the equation: $U_L = 3 + 0.025(T_{receiver} - T_{ambient})$ $Wm^{-2}K^{-1}$. The receiver temperature is 310°C and the average solar radiation intensity on the aperture is 720 Wm^{-2}.

The boiler pressure of the steam cycle is 20 bar and the condenser pressure is 0.04 bar. Steam is superheated to 300°C before it is expanded insentropcally in the turbine. Steam exiting from the turbine is condensed to a saturated liquid before it is pumped to the boiler isentropically. The electric generator of the plant has an output of 6 MW and its efficiency is 84%. Calculate (i) the efficiency of the Rankine cycle, and (ii) the total aperture area of the collectors.
[*Answers*: (i) 34%, (ii) 58,000 m^2]

P4.9 An electric generator with an output of 6 MW and efficiency 88% is driven by a gas turbine. The gas turbine operates on the air standard Brayton cycle where air enters the compressor at 1 bar and 27°C. The pressure ratios of the compressor and turbine are 4. At entry to the turbine the air is at 1050°C. The isentropic efficiencies of the turbine and compressor are 0.75 and 0.65 respectively. The combustion chamber burns natural gas with a heating value of 46 $MJkg^{-1}$. Calculate (i) the thermal efficiency of the cycle, (ii) the work output of the turbine, (iii) the heat input in the combustion chamber, and (iv) the fuel consumption rate.
[*Answers*: (i) 12.5%, (ii) 6.82MW, (iii) 54.5 MW, (iv) 1.186 kgs^{-1}]

P4.10 A Stirling engine receives heat from a dish-concentrator solar collector. The engine operates between a heat source at 200°C and a heat sink at 40°C. The output of the engine 5 kW. The compression ratio is 5 and the regenerator effectiveness is 0.7. The average efficiency of the solar collector is 38% and the average solar radiation intensity on the aperture is 750 Wm^{-2}. Calculate (i) the thermal efficiency of the engine, and (ii) the aperture area of the collector. [*Answers*: (i) 29.2%, (ii) 60 m^2]

References

1. Duffie, John A., Beckman William A., with Blair Nathan, *Solar Engineering of Thermal Processes, Photovoltaics and Wind*, John Wiley & Sons, New Jersey, 2020.
2. Goswami, Y., Kreith Frank, and Kreider Jan F., *Principles of Solar Energy Engineering*, 2nd Edition, Taylor & Francis, Philadelphia, PA, 2000.
3. Kearney D., Kelly B., Cable R., Potrovitza N., Herrmann U., Nava P., Mahoney R., Pacheco J., Blake D., Proce H. Assessment of a molten salt heat transfer fluid in a parabolic trough solar field, *Journal of Solar Energy Engineering*, 125 (2002), 153-159.
4. Kreith Frank and Krumdieck Susan, *Principles of Sustainable Energy Systems*, CRC Press, Taylor & Francis Group, New York, 2015.
5. Rogers G.F.C. and Mayhew Y.R., *Thermodynamic and Transport Properties of Fluids*. 5th ed. Blackwell, Oxford, 1998.
6. Vanek Francis M., Albright Louis D., and Angenent Largus T., *Energy Systems Engineering: Evaluation and Implementation*, McGraw-Hill Book Company, New York, 2012.
7. Wijeysundera N.E, Effect of angular misorientation on the performance conical, spherical and parabolic solar concentrators, *Solar Energy*, 19 (1976), 583-588.
8. 'List of solar thermal power stations-Wikipedia –' http://en.wikipedia.org/wiki/List_of_solar_thermal_power_stations.
9. 'Concentrated solar power-Wikipedia –' http://en.wikipedia.org/wiki/Concentrated_solar_power.

Chapter 5

Solar Photovoltaic Power Generation

5.1 Introduction

At present two methods are used to generate electricity on a commercial scale using solar radiation directly. The first method, presented in chapter 4, is commonly called concentrated solar power or CSP. In a CSP plant solar radiation incident on a collector heats a fluid that is subsequently used as the heat source of a conventional thermal power cycle. The second method, to be presented in this chapter, makes use of the photovoltaic (PV) effect to convert solar radiation directly to electricity.

Historically, the photovoltaic effect was first observed by Edmund Becquerel in 1839. He illuminated one of two identical platinum electrodes immersed in a weak acidic solution containing silver chloride, between which an electrical voltage developed [5]. He observed that the current generated in the cell increased when sunlight was used for irradiation. The same effect was noted in solids by Adams and Day in 1877 when light was incident on selenium.

Schottky, Lang, and Grandhl did pioneering work on the photovoltaic effect using selenium and copper oxide [3]. These cells had an efficiency of about 1% to 2% in converting light to electricity. Selenium cells convert light in the visible part of the sun's spectrum and were therefore adopted as light-measuring devices. The use of *n*-type and *p*-type semiconductors to develop photovoltaic devices was pioneered by researchers at the Bell Laboratory in 1954. These devices achieved an efficiency of about 6%, which was soon bettered to 11%. During this period, the application of photovoltaic devices or solar cell systems as power sources in artificial satellites provided the main impetus for research on such systems.

For converting light or solar radiation partially to electricity, photovoltaic devices have many desirable attributes [3]. They have no moving parts in the usual mechanical sense, which tend to undergo wear during operation. However, the material of the solar cell itself would degrade over time. They contain no gases or fluids that can leak out as in some solar-thermal systems, and they do not require fuels to operate.

These devices have a rapid response time, and therefore achieve full output almost instantly. They operate under moderate temperatures and hence have fewer safety concerns in this regard. While in operation these systems produce no pollution, but some of the materials used in the manufacturing processes may be toxic. If properly manufactured and installed, PV systems require very little maintenance. They can be made from silicon, which is the second most abundant element in the earth's crust.

Solar cell devices are modular, permitting wide range of applications, including residential use and remote applications, intermediate scale systems for use as business and neighborhood power systems, and large scale systems for grid-level application in solar farms. They have a relatively high efficiency for converting solar radiation to electricity, which has been improving steadily during the period from 1960 to 2017, as seen from the data in Table 5.3. They also have a wide power-handling capability from micro-watts to megawatts. Their high power to weight ratio and the amenability to on-site installation makes them suitable for roof-top applications.

5.2 Physical Processes in PV Cells

In this section the important physical phenomena and processes associated with the functioning and design of solar PV cells will be reviewed. For this purpose we shall use the silicon solar cell as the main example because it is widely applied in practice at the present time.

5.2.1 *Atomic structure of materials*

All matter is made from atoms which consist of protons, neutrons and electrons. Positively charged protons and the neutral neutrons reside in the densely packed nucleus at the center of the atom. The negatively charged, much lighter, electrons orbit the nucleus. Since an atom has the same number of protons and electrons, it is electrically neutral. The energy of the electrons depends on the location of their orbitals and the motion about their own axis.

In atoms with multiple electron shells, the innermost electrons require the maximum amount of externally imparted energy to overcome the attraction of the nucleus and become free. The electrons in the outermost band of shells, on the other hand, require much less energy to be freed. These electrons are the only ones that participate in interactions between neighboring atoms. Electrons in the outermost band determine how neighboring atoms react or join, and it is therefore called the *valence band.* If the outermost band of shells of neighboring atoms are not filled with electrons, then these atoms can arrange themselves to share electrons to make the outermost bands full. The bonds formed between neighboring atoms in this manner are called *covalent bonds*.

Some of the more energetic electrons in the valence band may jump to a still higher band that is far away from the nucleus. A small impressed force could cause these electrons to move away from the atom, and move into a band called the *conduction band*. Electrons in the conduction band are responsible for the conduction of electricity and heat. The difference in energy between an electron in the innermost shell of the conduction band and an electron in the valence band is called the *band gap* or the *forbidden gap*.

Materials whose valence bands are full and have large band gaps over about 3 ev are called *insulators*. Materials with relatively empty valence bands and some electrons in the conduction band are good *conductors* of electricity. Metals belong in this category. Materials with partially filled valence bands, and intermediate band gaps less than about 3 ev, are called *semiconductors*. Energy bands of insulators, conductors and semiconductors are shown schematically in Fig. 5.1. Pure semiconductors

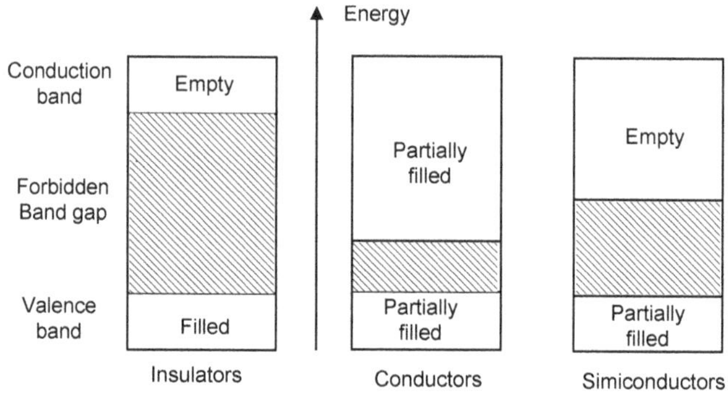

Fig. 5.1 Energy band diagrams of insulators, conductors and semiconductors.

are called *intrinsic* semiconductors and those containing artificially introduced impurities are called *extrinsic* semiconductors.

5.2.2 *Atomic structure of silicon and the effect of light*

The silicon atom has fourteen electrons orbiting the nucleus in three shells. The outermost shell has four electrons and so it is half full because this shell can contain up to 8 electrons. These four electrons can be given to, accepted from, or shared with a neighboring atom. In a solid, each silicon atom shares its four valence electrons with another silicon atom. The atoms are arranged in a tetrahedral unit containing five atoms where one atom shares electrons with four neighboring atoms with covalent bonds, as shown schematically in Fig. 5.2(a). Each silicon atom in the solid is held in place at a fixed distance and angle with each of the atoms with which it shares a bond. The fixed, regular arrangement of atoms in a solid is called a *crystal lattice*.

Light striking a silicon crystal may be reflected, absorbed or transmitted through. When light of relatively low energy is absorbed by a solid, it produces heat without changing the electrical properties of the solid. In the case of a silicon crystal light causes the atoms to vibrate and twist about their fixed locations, but do not break loose. The electrons in the bonds also gain energy and attain a higher energy level. Because these levels are not stable, the electrons soon return to their original energy levels, giving off the excess energy it gained as heat.

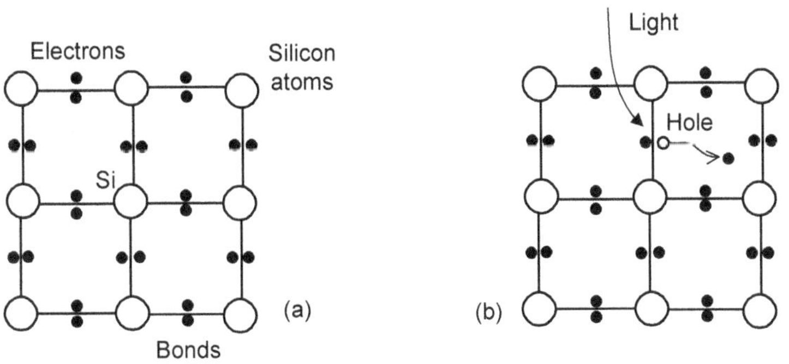

Fig. 5.2 (a) Silicon crystal with covalent bonds (b) Light-generated electron-hole pair

Light of higher energy striking a silicon crystal can alter its electrical properties by dislodging an electron from its position in the crystal as shown in Fig. 5.2(b). A bond missing an electron is called a *hole*. A freed electron that moves about in the crystal is said to be in the conduction band and it is used in the conduction of electricity. Electrons in the conduction band and holes in the bonds play an important role in the operation of PV cells. These are called *light-generated electron-hole pairs*.

Electrons in bonds can jump into a hole in a neighboring bond thus creating a hole in the original bond. In this way the holes can move through the crystal. The movement of electrons and holes increase with the temperature of the material. The generation of electron-holes pairs and their movement in the material is the central process in the production of electric current by the PV effect. However, without a means to separate these free electrons and holes, they will move randomly in the crystal and will eventually return to valence positions and loose their energy as heat. The mechanism used to separate the electron-hole pairs and use them to generate an electric current in an external circuit is called a *built-in potential barrier*.

5. 3 The Potential Barrier and its Function

In the PV cell a potential barrier is created by generating opposite electric charges across a dividing line, that selectively separates the light-generated electrons and holes by transferring most electrons to one side and holes to the other side. The potential barrier sets up an electric field that minimizes the chances of recombination of electrons and holes. The potential difference thus generated across the cell can be used to drive an electric current in an external circuit. Potential barriers in solar cells can be formed by different means. The method commonly used in silicon PV cells is to slightly alter the structure of the crystal on either side of the dividing line. We shall illustrate the formation of the potential barrier by referring to a silicon PV cell.

5.3.1 *Negative-carrier and positive-carrier dopants*

Silicon has four valence electrons that form the bonds in a silicon crystal. Suppose that we introduce an impurity, such as phosphorus with five valence electrons, into the pure silicon crystal. The phosphorus atom would occupy the same position as a silicon atom in the crystal and share four of its valence electrons with four neighboring silicon atoms.

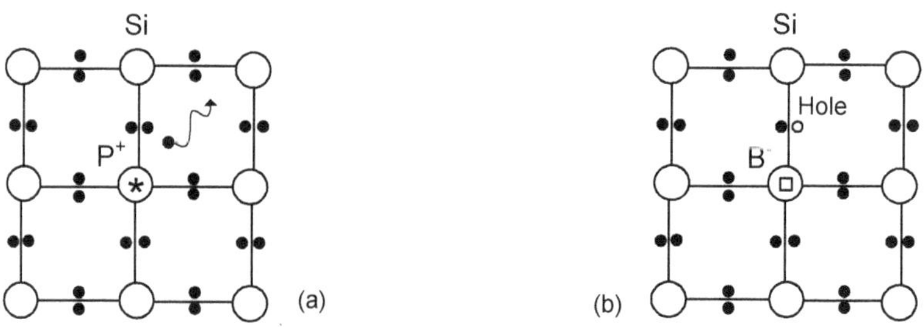

Fig. 5.3 (a) Phosphorous doped silicon (b) Boron doped silicon

The extra electron in phosphorus with no bond to share will be relatively free as illustrated in Fig. 5.3(a). At room temperature this free electron will have sufficient energy to move away from the atom leaving a *positively charged* phosphorus atom in the crystal. This electron will behave like a conduction band electron of the silicon crystal. A crystal with many substituted impurity atoms would have many conduction-band electrons and equal number of positively charged impurity ions in the crystal structure. The whole crystal would be electrically neutral but its electrical properties would be greatly altered. Impurities introduced in this manner are called *dopants*, and dopants with one extra valence electron are called *donors* because they donate an electron to the crystal. A crystal that is *donor-doped* with free negative charges is called *n-type*, and such altered crystals are used in forming the potential barrier.

Another type of altered silicon crystal, used in the potential barrier, is formed by introducing impurity atoms with one valence electron less than silicon as shown in Fig. 5.3(b). The commonly used impurity is boron which has three valence electrons, and it would replace a silicon atom in the crystal. The bonds between the boron atom and its four neighboring silicon atoms would have one bond with a missing electron, which is called a *hole*. The hole would move freely across the crystal when electrons from bonds in near-by atoms jump into the hole.

As the hole moves away the boron atom would be left with a negative ion. A crystal with a large number of substituted impurity atoms would have many holes that move through the crystal as free positive charges. An impurity with three valence electrons is called an *acceptor* because its holes accept electrons from the rest of the crystal. Silicon doped with an acceptor impurity is called *p-type* because it has free positive charges or moving holes. In the *p*-type materials holes are the *majority carriers* and electrons are the *minority carriers*. For *n*-type materials electrons are the majority carriers and holes are the minority carriers.

5.3.2 *Formation of the potential barrier*

In order to establish the potential barrier, essential for the operation of a solar cell, a *n*-type layer of silicon is placed adjacent to a *p*-type layer with two surfaces in contact. Initially, there is an excess of electrons in the *n*-side and an excess of holes in the *p*-side. When the two layers are brought into contact, *free* electrons in the *n*-type material will jump into the many holes in the *p*-type material, filling the holes as indicated in Fig. 5.4(a). Electrons in the *valence band* of the *n*-type material can jump into the holes in the *p*-side thus causing the holes to move to the *n*-side. The driving force for this diffusion process is the concentration difference of electrons and holes in the two materials. This is a rapid process which sends large numbers of electrons to the *p*-side and holes to the *n*-side across the dividing line as depicted in Fig. 5.3(b). This process results in an immediate imbalance of charge, with more positive charges along the *n*-side and more negative charges along *p*-side of the interface.

When electrons move to the *p*-side they jump into the holes in the silicon bonds and the holes that moved to the *n*-side are filled by the extra electrons available on the *n*-side. There are now very few minority carriers, that is, holes on the *n*-side and electrons on the *p*-side to be transferred back by the charge imbalance. Therefore, the charge imbalance remains fixed in place. This diffusion process, however, does not continue indefinitely because the charge carriers that already crossed the junction set up an *electric field* that opposes further flow of free carriers. Eventually, an equilibrium is established where no more electrons or holes switch sides. This sets up a fixed *potential barrier* at the junction with the *n*-side adjacent to the junction *positively* charged and the *p*-side adjacent to the junction *negatively* charged as shown in Fig. 5.4(b). The strength of the electric field at the barrier depends on amount of dopant in the silicon. The potential barrier opposes the movement of *majority carriers* across the junction, that is, electrons to the *p*-side and holes to the *n*-side. However, the movement of *minority carriers* is not hindered by the barrier, that is electrons to the *n*-side and holes to the *p*-side. This *selective barrier* at the junction is the means of separating charges during electron-hole generation under illumination of the solar cell, and it is the key to the production of a PV electric current.

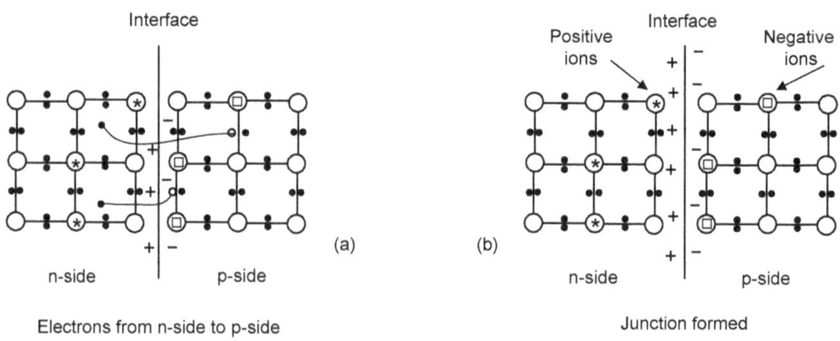

Fig. 5.4 (a) Potential barrier forming, (b) Junction formed

5.3.3 *Action of potential barrier under illumination*

We shall now illustrate the main steps in the production of an electric current in a silicon solar cell due to the incidence of light on it. Assume that the light has sufficient energy to free an electron from the bond in the silicon crystal and create an electron-hole pair. Further assume that this pair is generated on the *p*-side of cell. The electron which will be free for a short time has a high chance of combining with one of the many holes present on the *p*-side. But in a well designed solar cell the electron will move about in the crystal and encounter the junction before it could combine with a hole. However, if it did combine with a hole it would loose its energy as heat and will not contribute to the electric current of the cell.

Once the electron is within the electric field of the junction it will be accelerated across the barrier to the *n*-side. Since there are far fewer holes on the *n*-side, the chances of it combining with a hole is greatly reduced. Also, for the electron to return to the *p*-side it must overcome the repulsion of the potential barrier. The hole partner of the electron-hole pair remains on the *p*-side of the barrier because it is repelled by the barrier at the junction. The chances of it recombining is much less because there are numerous holes on the *p*-side. We observe that because of illumination and charge separation, there is now an excess of uncombined negative charges on the *n*-side and an excess of holes on the *p*-side.

When the electron-hole pair is generated in the *n*-side of the junction, the freed electron is repelled by the barrier and will therefore remain on the *n*-side. The hole-partner will encounter the barrier where it will cross to the *p*-side when normally bound electrons will fill the holes. Once on the *p*-side the chances of recombining is greatly reduced because there are very few *free* electrons on the *p*-side.

If we connect the *n*-side of the cell to the *p*-side by means of an external circuit as shown in Fig. 5.5(a), current flows through the circuit to reduce the charge imbalance in the cell. Negative charges flow through the electrode on the *n*-side and perform useful work on the load. The electrons then enter the *p*-side where they recombine with the holes near the electrode. Thus an equilibrium is established where the incident

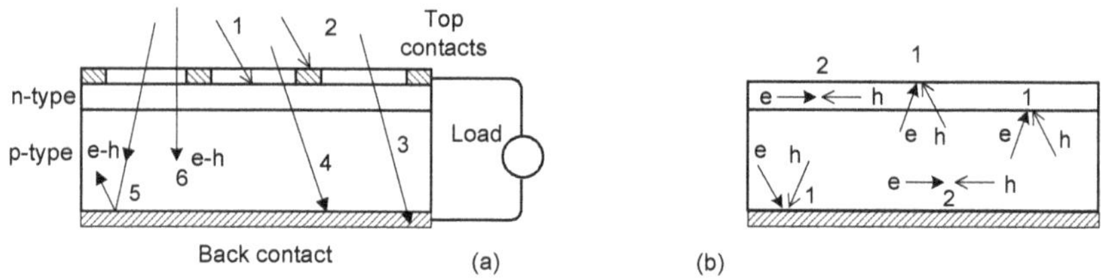

Fig. 5.5 (a) Behavior of incident light (b) Electron-hole recombination processes

light continually generates electron-hole pairs, and causes a charge imbalance which is relieved by the current doing external work.

5.3.4 *Losses limiting solar cell efficiency*

The *conversion efficiency* of a solar cell is the fraction of incident light that is converted to electricity. Some of the major factors that limit the efficiency of a solar cell are [3]:

1. Reflection from the top surface of the cell.
2. Self-shading resulting from the top-surface electric contacts.
3. Light that does not have sufficient energy to generate electron-hole pairs.
4. Light that has more energy than the minimum required to separate an electron from the bond.
5. Random encounter of light-generated electrons and holes that recombine before they can contribute to the output current of the cell.
6. Light-generated electrons and holes that are brought together by material defects in the cell.
7. Resistance to current flow.
8. Poor performance due to nonoptimal operating temperatures.

Light incident on a solar cell can behave in different ways, some of which are depicted in Fig. 5.5(a). It can be partly reflected at the top surface of the cell as in (1). Normal silicon has a reflectivity of about 0.36 for sunlight, which would decrease the efficiency dramatically. However, by treating and texturing the surface the reflectivity can be lowered to 0.05. Self-shading, occurs due to light getting reflected and absorbed by the top surface electric contacts as indicated in (2). A typical shading loss is about 8% but this can be reduced to 3% or less. Light can go right through and be absorbed in the rear electric contact as in (3). Weakly absorbing light could be reflected at the back out of cell as in (4), and can also be absorbed after reflection at the back as in (5). Low energy light will become absorbed, generating heat in the form of atomic vibrations (6).

Band-gap energy

The energy losses associated with factors (3) and (4) listed above are related to the energy of light photons incident on the cell. In the case of sunlight the energy spectrum is very wide as was discussed in chapter 2. In order to be matched the energy of the photons must be exactly equal to the *band-gap energy* of the material, which is 1.12 ev for silicon. Photons with lower energies will not be able generate electron-hole pairs, and such photons would pass through the material. If the energy of the photon is larger than the band-gap, then the excess energy of the electron-hole pair generated will be transferred to the material as heat, and the process is called *thermalization.*

The energy of a light photon is given by the Planck equation [6]

$$E_{ph} = h\upsilon \tag{5.1}$$

where υ [s^{-1}] is the frequency of light, and $h = 6.626 \times 10^{-34}$ [Js], is the Planck constant.

The wavelength of light is related to the frequency by

$$c = \nu\lambda \tag{5.2}$$

where λ [m] is the wavelength of light, and $c = 3 \times 10^{8}$ [ms^{-1}] is the speed of light.

From Eqs. (5.1) and (5.2) we have

$$E_{ph} = \frac{hc}{\lambda} \tag{5.3}$$

For the generation of electron-hole pairs the energy of the photons must be larger than the band gap. Therefore

$$E_{ph} \geq BG \tag{5.4}$$

From Eqs. (5.3) and (5.4) we get

$$\lambda \leq \frac{hc}{BG} \tag{5.5}$$

Important light absorbing properties of a number of semiconductors, including the band gap energy, the frequency and wavelength of photons corresponding to the band gap energy, are listed in Table 5.1.

Table 5.1 Light absorption limits of some semiconductors (data from Ref.[5])

Material	Frequency (THz)	Wavelength μm	Band-gap (eV)	Region
Sn-α	19.3	15.50	0.08	Far infrared
Ge	162	1.850	0.67	Infrared
Si	265	1.130	1.10	Infrared
GaAs	326	0.920	1.35	Near infrared
GaP	540	0.555	2.24	Visible
C	1300	0.230	5.40	Ultraviolet

Recombination losses

The loss of efficiency associated with factors (5) and (6) listed above, are due to inadvertent recombination of electrons and holes before they could contribute to the electric current [6]. Some of these processes are shown schematically in Fig. 5.5(b). In well-designed solar cells these are largely eliminated. Recombination occurs in two main ways called direct and indirect recombination.

Direct recombination, which is rare, occurs when an electron and a hole randomly encounter each other, and the electron falls back into the hole, reforming the bond as in (2) in Fig. 5.5(b). The energies of the hole and electron are lost as heat. This type of recombination is more likely before electrons and holes are separated by the potential barrier. Cells must be designed to minimize the time spent by light-generated electrons on the p-side, and holes generated on the n-side.

Indirect recombination can occur in many ways. In the bulk material, recombination may be caused by empty or dangling bonds from impurities or defects which can capture free electrons or holes [6]. Collisions of free charge carriers reduce their energy and increase the chance of recombination. Surfaces of solar cells can be sites of much recombination as indicated by 1 in Fig. 5.5(b). Minute 'bridges' can exist across the junction, which provides a path for carriers to avoid the junction's field and filter back. This reduces the voltage of the cell. Moreover, surfaces are prone to unconventional atomic structures with numerous sources of collision and with dangling bonds, which provide opportunities for electrons and holes to recombine.

5.4 Analysis of Photovoltaic Cells

In this section we shall develop an electrical analysis of the photovoltaic cell, which is useful in the design and performance evaluation of solar cell power systems. In the preceding section we discussed the formation of the potential barrier at the p-n junction and its role in light-generated charge separation under illumination.

When an external voltage V_a is applied to the p-n junction the potential difference across the barrier changes from V_{bi} to $(V_{bi} - V_a)$. Under equilibrium the potential in the p-region is negative with respect to the n-region (see Fig. 5.4b) When V_a is negative with respect to the p-region potential, the applied voltage increases the potential difference across the barrier and enlarges the barrier region. Under these conditions the p-n junction is under *reverse-bias voltage*. When V_a is positive with respect to the p-region potential, the applied voltage decreases the potential difference across the barrier and the barrier region becomes narrower. Under these conditions the p-n junction is under *forward-bias voltage*.

The balance between the forces responsible for diffusion (concentration gradient) and drift (electric field) is now disturbed. The forward-bias voltage will increase the concentration of minority carriers at the edges of the potential barrier region (electrons at the p-side and holes at the n-side). This concentration gradient causes diffusion of minority carriers from the edge into the bulk region. Since these minority carriers recombine with majority carriers in the bulk, the diffusion current is called the *recombination current* I_r. The recombination current is balanced by the so-called *thermal generation current* I_g, which is caused by the drift of the minority carrier across the junction. Both recombination and generation currents have contributions from electrons and holes [6].

When $V_a = 0$, the above currents are in balance. Hence the net current is zero.

$$I_r(0) - I_g(0) = 0 \tag{5.6}$$

When a moderate forward-bias voltage V_a is applied the recombination current is obtained by applying the Boltzmann factor [6]. Therefore

$$I_r(V_a) = I_r(0)exp\left(\frac{qV_a}{k_B T}\right) \tag{5.7}$$

where T is the absolute temperature, $q = 1.6 \times 10^{-19}$ C is the elementary charge, and k_B is the Boltzmann constant.

The thermal generation current is almost independent of the potential barrier across the junction and determined by the availability of thermally generated minority carriers. Hence

$$I_g(V_a) = I_g(0) \tag{5.8}$$

The *net external current* or the *junction current* is now given by

$$I_j\,(V_a) = I_r(V_a) - I_g(V_a) \tag{5.9}$$

Using Eqs. (5.6) to (5.9) we obtain the relation

$$I_j(V_a) = I_0\left[exp\left(\frac{qV_a}{k_B T}\right) - 1\right] \tag{5.10}$$

where $I_0 = I_g(0)$, is the *saturation current* of the *p-n* junction which is also called the *dark current*. Equation (5.10) is known as the *Shockley equation* [6] that describes the current-voltage behavior of an *ideal p-n* diode.

When the *p-n* junction is illuminated and electrodes are connected across an external load as depicted in Fig. 5.5(a), only a fraction of the light-generated current flows through the external circuit. The potential difference between the *p*-side and *n*-side of the junction will be lowered by the voltage drop across the load, which in turn lowers the barrier potential difference. This increases the recombination current. The voltage drop at the load can be simulated by applying a forward-bias voltage to the junction [6]. Therefore the *light-generated current* I_L is included in Eq. (5.10) by adding it to the junction current. Hence the current through load is given by

$$I(V) = I_L - I_j \tag{5.11}$$

Substituting from Eq. (5.10) in (5.11) we obtain the *characteristic equation* of the ideal solar cell as

$$I(V) = I_L - I_0\left[exp\left(\frac{qV}{k_B T}\right) - 1\right] \tag{5.12}$$

5.5 Solar Cell Parameters

In this section we shall obtain the main parameters used to characterize the performance of solar cells. These include the current-voltage curve, the short circuit current, the open circuit voltage, the maximum power, the fill factor, and the efficiency. The characteristic curve, generated by plotting the current I against the voltage V ($I - V$ curve), is shown in Fig. 5.6. The $I - V$ curve is usually measured under standard conditions [6] where the total irradiation is 1000 Wm^{-2} and the solar radiation spectrum (see chapter 2) resembles the AM 1.5 spectrum (air mass 1.5). The PV cell is maintained at a temperature of 25°C.

The equivalent circuit of the ideal solar cell is depicted in Fig. 5.7(a). The light current or the photocurrent I_L is represented as a current source and the current through the diode in the bypass path is the junction current I_j. From Eq. (5.12), the current through the load I is the difference between the above currents.

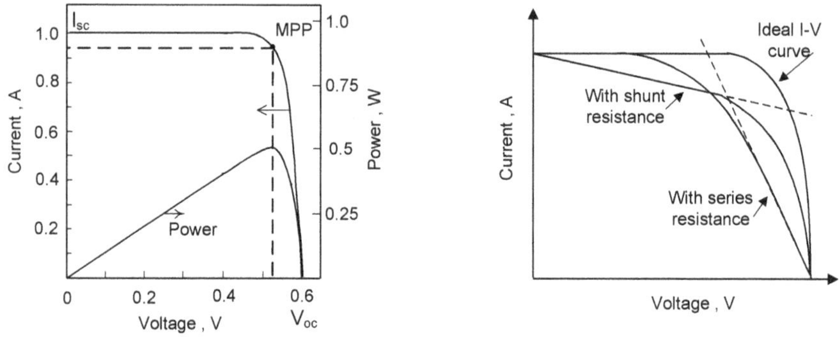

Fig. 5.6 (a) Current-voltage curve for ideal solar cell, (b) Effect of series and shunt resistances

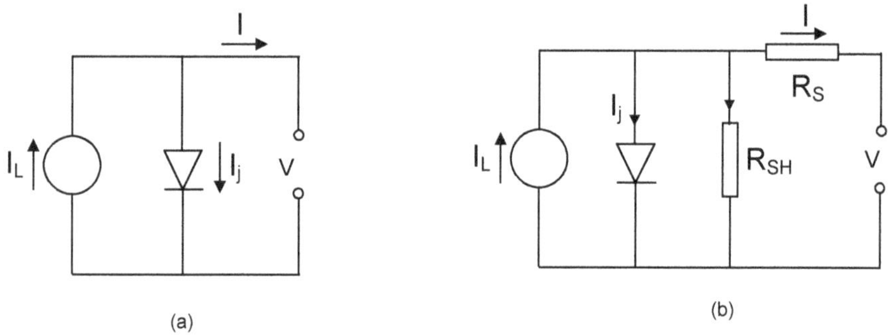

Fig. 5.7 (a) Equivalent circuit of ideal solar cell, (b) Equivalent circuit including series and shunt resistances

5.5.1 *Short circuit current density*

The short circuit current I_{sc} is the current flowing in the circuit in Fig. 5.7(a) when the terminals at the load are short circuited, and the voltage drop across the terminals is zero. This point is indicated on the *I*-axis of the *I-V* characteristic in Fig. 5.6(a). I_{sc} depends on the area of the solar cell. Usually, the area dependence of I_{sc} is removed by the expressing it as current per unit area, which is called the *short circuit current density*.

We observe from Eq. (5.12) that when, $V = 0$,

$$I_{sc} = I_L \tag{5.13}$$

Therefore I_{sc} depends on the light intensity or the photon flux incident on the solar cell which is determined by the spectrum of the incident light. For actual solar cells I_{sc} increases in proportion to the solar radiation as seen in Fig. 5.8(a).

5.5.2 *Open circuit voltage*

The *open circuit voltage* V_{oc} is the voltage across the terminals of the circuit in Fig. 5.7(a) when no current flows through the circuit, and it is the maximum voltage that the cell can deliver. From Eq. (5.12) it follows that V_{oc} can be expressed in the form

$$V_{oc} = \frac{k_B T}{q} ln\left(\frac{I_L}{I_o} + 1\right) \tag{5.14}$$

We note from Eq. (5.14) that V_{oc} depends on the light-generated current density and the saturation current density. The value of V_{oc} increases logarithmically with light intensity [1,6].

5.5.3 *Ideality factor*

The *ideality factor* n is a measure of how closely an actual solar cell follows the ideal characteristic given by Eq. (5.12). It accounts for recombination processes not taken into account in the derivation of the ideal diode equation. The modified solar cell characteristic equation of the *non-ideal* cell takes the form

$$I(V) = I_L - I_0\left[exp\left(\frac{qV}{nk_B T}\right) - 1\right] \tag{5.15}$$

The value of n is obtained from the measured P-V curve of the solar cell.

5.5.4 *Power output of a cell*

Under normal operating conditions, the electrical power output of a solar cell P is given by

$$P = VI \tag{5.16}$$

From the $I - V$ curve in Fig. 5.6(a) we notice that P increases steadily with V while I remains nearly constant. As V approaches V_{oc}, the power drops sharply to zero. There is a point close to V_{oc} where P reaches a maximum. We can obtain this optimum power point by differentiating P with respect to V after substituting for I from Eq. (5.12) in Eq. (5.16). Hence we obtain the following expression for the voltage, V_m at the *maximum power point*

$$V_{oc} - V_m = \frac{k_B T}{q} ln\left[1 + \frac{qV_m}{k_B T}\right] \tag{5.17}$$

Equation (5.17) can be written in the dimensionless form

$$v_{oc} - v_m = ln(1 + v_m) \tag{5.18}$$

The dimensionless voltages are given by

$$v_{oc} = \frac{V_{oc}}{V_{th}} \text{ and } v_m = \frac{V_m}{V_{th}}$$

where $V_{th} = \frac{k_B T}{q}$, is sometimes called the *thermal voltage*.

Equation (5.18) must be solved by a trial-error procedure, which will be illustrated in worked example 5.7. From Eq. (5.12) the current I_m at maximum power is

$$I_m = (I_L + I_o) - I_o exp(v_m) \tag{5.19}$$

Having obtained V_m and I_m from Eqs. (5.18) and (5.19) we can express the maximum power as

$$P_m = I_m V_m \tag{5.20}$$

Differentiating Eq. (5.16) we get

$$\frac{dP}{dV} = V\frac{\partial I}{\partial V} + I\frac{\partial V}{\partial V} \tag{5.21}$$

At the *maximum power point* (MPP) the derivative on the LHS of Eq. (5.21) is zero. Therefore

$$\left[\frac{dI}{dV}\right]_{MPP} = -\frac{I_m}{V_m} \tag{5.22}$$

The maximum power point is indicated in Fig. (5.6) as MPP.

To optimize the operation of a solar power system it is very important to operate it at the maximum power point. This is achieved by a technique call *maximum power point tracking*, or MPPT. The strategies used to operate a PV system at the MPP are categorized as indirect methods and direct methods, which are discussed in detail in Ref.[1,6]. One of the most common indirect methods is to track changes in V_{OC} and adjust V_{MPP} accordingly, because these two voltages are proportional to a good approximation [6]. One of the direct methods of MPPT, called the incremental conductance method, is based on Eq. (5.22).

5.5.5 *Fill factor of a cell*

The fill factor FF is the ratio of the maximum power P_m, to the product of I_{sc} and V_{oc}. Therefore

$$FF = \frac{V_m I_m}{V_{oc} I_{sc}} \tag{5.23}$$

Ideally, the fill-factor is a function of the open circuit voltage, and it can be calculated using the approximate empirical expression [6]

$$FF = \frac{v_{oc} - ln(v_{oc} + 0.72)}{v_{oc} + 1} \tag{5.24}$$

where the *normalized open circuit voltage* is defined as

$$v_{oc} = \frac{V_{oc}}{V_{th}} = \frac{qV_{oc}}{k_B T} \tag{5.25}$$

Notice that FF can be interpreted as the ratio of the areas of the two rectangles, one with sides ($V_m \times I_m$), and the other ($V_{oc} \times I_{sc}$) as indicated in Fig. 5.6(a).

5.5.6 *Conversion efficiency*

The conversion efficiency is the ratio of the maximum power output of a solar cell to the total incident irradiation. The measurement of the conversion efficiency is conducted under standard conditions, where the incident light has the AM1.5 (air mass) solar spectrum, and an intensity of 1000 Wm^{-2}. The cell temperature is maintained at 25°C. The general expression for cell efficiency is

$$\eta = \frac{P_{max}}{I_S A_C} = \frac{V_m I_m}{I_S A_C} = \frac{I_{SC} V_{OC} FF}{I_S A_C} = \frac{J_{SC} V_{OC} FF}{I_S} \tag{5.26}$$

where J_{SC} is the short circuit current density and I_s is the incident solar radiation intensity.

5.5.7 *External quantum efficiency*

The external quantum efficiency $EQE(\lambda)$ is the fraction of the photons incident on a solar cell that creates electron-hole pairs in the cell which are successfully collected. It is dependent on the wavelength of incident radiation. The $EQE(\lambda)$ is obtained by measuring the current I_L through the cell when it is illuminated with monochromatic light of wavelength λ [6]. It is given by

$$EQE(\lambda) = \frac{J_L(\lambda)}{q\psi_\lambda} \tag{5.27}$$

where q is the elementary charge. The spectral photon flux ψ_λ is the number of photons of wavelength λ that are incident per unit time per unit area.

5.5.8 *Thermodynamic efficiency*

The most general efficiency limit of a solar cell is the *thermodynamics efficiency limit* [6]. This is obtained by considering the solar cell as a Carnot engine operated by an ideal heat source. The heat source is an ideal absorber at temperature T_A that receives radiation from a black body at the sun's temperature T_S. The heat sink is at the temperature of the surroundings T_C. The efficiency of the absorber as the heat source of the engine is

$$\eta_h = 1 - \frac{\sigma {T_A}^4}{\sigma {T_S}^4}$$

The efficiency of the Carnot engine is

$$\eta_C = 1 - \frac{T_C}{T_A}$$

The total efficiency is the product of the efficiencies of the absorber and the Carnot heat engine, which can be expressed as

$$\eta_{SC} = \eta_h \eta_C = \left(1 - \frac{T_A{}^4}{T_S{}^4}\right)\left(1 - \frac{T_C}{T_A}\right) \tag{5.28}$$

We obtained a similar efficiency for ideal solar thermal power systems in chapter 4.

5.5.9 *Ultimate conversion efficiency*

The absorber layer of a cell that generates electron-hole pairs is characterized by its band gap energy E_G. Ideally, only light photons with energy equal to or greater than the band gap can create electron-hole pairs in the material. The electrons and holes of a semiconductor tend to occupy energy levels at the bottom of the conduction band and the top of the valence band respectively. Therefore when the energy of the photon is greater than the band gap, the extra energy that the electron-hole pair receives from the photon is released as heat to the lattice by a process called *thermalization* [6].

Consider radiation incident on a solar cell with a band gap energy of E_G. The wavelength of radiation having energy E_G per photon is given by Eq. (5.3) as

$$\lambda_G = \frac{hc}{E_G} \tag{5.29}$$

Only photons with $\lambda \leq \lambda_G$ are absorbed by the cell. The fraction f_{ab} of the *total incident power* that is absorbed by the solar cell can be expressed as

$$f_{ab} = \frac{\int_0^{\lambda_G}(hc/\lambda)\psi_\lambda d\lambda}{\int_0^{\infty}(hc/\lambda)\psi_\lambda d\lambda} \tag{5.30}$$

where ψ_λ is the incident spectral photon flux per unit area per unit time.

The energy in excess of the band gap energy E_G is lost because of thermalization. Hence the fraction of energy absorbed that is delivered as useful energy is

$$f_{use} = \frac{E_G \int_0^{\lambda_G} \psi_\lambda d\lambda}{\int_0^{\lambda_G} (hc/\lambda)\psi_\lambda d\lambda} \tag{5.31}$$

The *ultimate efficiency* [6] is defined as the fraction of the total incident power that is delivered as useful power. This is given by

$$\eta_{ult} = f_{ab} f_{use} = \frac{E_G \int_0^{\lambda_G} \psi_\lambda d\lambda}{\int_0^{\infty} (hc/\lambda)\psi_\lambda d\lambda} \tag{5.32}$$

5.6 Effect of Solar Radiation Level and Cell Temperature

Two of the most important operating conditions for solar cells in practice are the incident solar radiation level and the cell temperature. Incident solar radiation varies widely on all time scales, while the cell temperature is dependent on environmental conditions such as the ambient temperature and the prevailing wind speed. When all other parameters are constant, higher incident solar radiation levels result in larger output currents and therefore higher power outputs as seen in Fig. 5.8(a). The short circuit current density J_{sc} of a PV module is directly proportional to the irradiance [1,6]. Therefore

$$J_{sc} = \lambda I_s \tag{5.33}$$

where λ is the constant of proportionality. Substituting for V_{oc} from Eq. (5.14), and for J_{sc} from Eq. (5.33), in Eq. (5.26) the variation of cell efficiency with irradiation at the standard temperature may be expressed in the simplified form[6].

$$\eta(25°\text{C}) = a\ lnI_s + b \tag{5.34}$$

where a and b are constants.

From Fig. 5.8(b) we see that when the cell temperature increases, V_{oc} decreases significantly, and I_{sc} increases slightly. The effect of

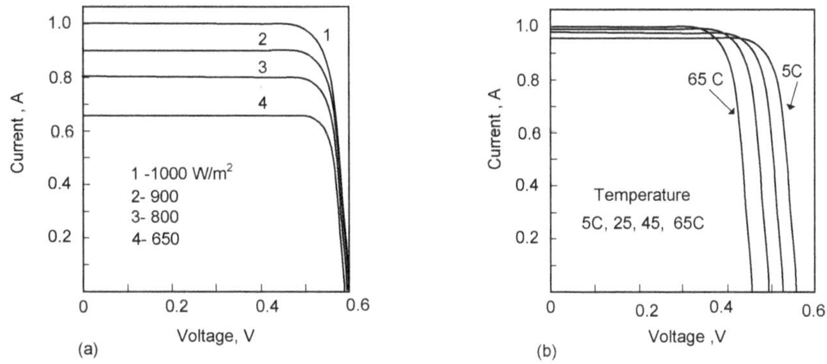

Fig. 5.8 (a) Effect of radiation level on the *I-V* curve, (b) Effect of operating cell temperature

temperature changes of a solar cell on various cell characteristics are expressed in terms of *temperature coefficients* μ that are usually provided by manufacturers in their data sheets. The values of the parameters at the operating temperature T_m can be obtained from the values at the standard temperature T_{st} of 25°C using the following relations.

$$V_{oc}(T_m) = V_{oc,s} + \left(\frac{\partial V_{oc}}{\partial T}\right)\theta_c = V_{oc,s} + \mu_{oc}\theta_c \tag{5.35}$$

$$I_{sc}(T_m) = I_{sc,s} + \left(\frac{\partial I_{sc}}{\partial T}\right)\theta_c = I_{sc,s} + \mu_{sc}\theta_c \tag{5.36}$$

$$P_{mpp}(T_m) = P_{mpp,s} + \left(\frac{\partial P_{mpp}}{\partial T}\right)\theta_c = P_{mpp,s} + \mu_{mpp}\theta_c \tag{5.37}$$

where $\theta_c = (T_m - T_{st})$.

The partial derivatives, denoted by μ in Eqs. (5.35) to (5.37) are the temperature coefficients, usually expressed in the form percent per °C. The effect of temperature changes on cell efficiency may be included in Eq. (5.34) by using a temperature coefficient κ. Hence we have [6]

$$\eta(T_m, I_m) = \eta_s(25^o)[1 + \kappa\theta_c] \tag{5.38}$$

5.7 Thermal Model for a PV Array

The performance of a PV module is strongly effected by its operating temperature and the incident irradiation. The latter can be measured accurately but the former is much harder to evaluate. In order to estimate the temperature manufacturers provide a quantity called the *nominal operating cell temperature* (NOCT). This is the cell temperature at a radiation level of 800 Wm^{-2}, an ambient temperature of 20°C, and a wind speed of 1 ms^{-1}. A *simplified steady state* model to estimate the temperature at an ambient temperature of T_a, and radiation level of G_m is [1,6]

$$T_m = T_a + \frac{(T_{NOCT} - 20)G_m}{800} \tag{5.39}$$

The *Duffie-Beckman* model [1], which includes the effect of wind speed on cell temperature is the following:

$$T_m = T_a + \frac{(T_{NOCT} - 20)G_m}{800}\left(\frac{9.5}{5.7+3.8w}\right)\left(1 - \frac{\eta_{cell}}{\tau\alpha}\right) \tag{5.40}$$

where w is the wind speed over the module, $(\tau\alpha)$ is the transmittance-absorptance product (see chapter 3) of the cell. More detailed CFD models have also been developed to estimate the cell temperature [6].

5.8 Parasitic Resistances

Generally, solar cells have parasitic series and shunt resistances associated with them, as shown in the equivalent circuit in Fig. 5.7(b). Both types of parasitic resistances decrease the fill factor. The major physical factors contributing to the *series resistance* R_S are the bulk resistance of the semiconductor material, the metallic contacts and interconnections, carrier transport through the top layer, and the contact resistance between the metallic contacts and the semiconductor material. The *shunt resistance* R_{SH} is due to non-idealities and impurities of the *p-n* junction.

By applying Ohm's law to different sections of the circuit in Fig. 5.7(b) we can show that the current to the load is given by

$$I = I_L - I_0\left[exp\left(\frac{V+IR_S}{\frac{nk_BT}{q}}\right) - 1\right] - \frac{(V+IR_S)}{R_{SH}} \tag{5.41}$$

The effect of the series resistance and the shunt resistance can be seen in Fig. 5.6(b) where the $I - V$ curve given by Eq. (5.41) is plotted. The two curves sag below the ideal curve, which lowers the fill-factor and the power output of the cell. It is possible to obtain the approximate values of R_S and R_{SH} from the slope (dV/dI) of the *I-V* curve at V_{oc} and I_{sc} respectively.

Usually, the $I - V$ curve of an actual solar cell does not follow Eq. (5.12) due to the factors mentioned above. For design-simulation of PV systems using manufacture's data it is more accurate to use Eq. (5.41) because it takes into account various nonidealities in real PV cells.

However, it is then necessary to know the values of R_S, R_{SH}, the ideality factor n, the light current I_L, and the saturation current I_0. These five parameters have to be estimated using information available in the manufacturer's data sheet, which includes the open-circuit voltage V_{oc}, the short-circuit current I_{sc}, the maximum power conditions, (V_m, I_m), the temperature coefficient of I_{sc}, the temperature coefficient of V_{oc}, and the temperature coefficient of maximum power. Typically, these are measured at the reference conditions of 1000 Wm^{-2} solar radiation, and 25°C cell temperature.

Applying the above conditions to Eq. (5.41) we obtain the following equations [1].

Open circuit voltage:

$$0 = I_L - I_0\left[exp\left(\frac{V_{oc}}{V_{th}}\right) - 1\right] - \frac{(V_{oc})}{R_{SH}} \tag{5.42}$$

Short circuit current:

$$I_{sc} = I_L - I_0\left[exp\left(\frac{I_{sc}R_S}{V_{th}}\right) - 1\right] - \frac{(R_SI_{sc})}{R_{SH}} \tag{5.43}$$

Maximum power point:

$$I_m = I_L - I_0\left[exp\left(\frac{V_m+I_mR_S}{V_{th}}\right) - 1\right] - \frac{(V_m+I_mR_S)}{R_{SH}} \tag{5.44}$$

Applying Eq. (5.22), which is valid at P_{max}:

$$V_m\left[\frac{1}{R_{SH}}+\frac{I_o}{V_{th}}exp\left(\frac{V_m+I_mR_S}{V_{th}}\right)\right]=I_m\left[1+\frac{R_S}{R_{SH}}+\frac{R_SI_o}{V_{th}}exp\left(\frac{V_m+I_mR_S}{V_{th}}\right)\right] \tag{5.45}$$

Temperature coefficient of V_{oc} is given by:

$$\mu_{oc}=\frac{V_{oc}(T_c)-V_{oc}(T_r)}{T_c-T_r} \tag{5.46}$$

The parameter V_{th} in the above equations is given by:

$$V_{th}=\frac{nk_BT}{q}$$

The reference temperature is T_r.

Several numerical techniques have been developed to solve Eqs. (5.42) to (5.46) to estimate the five unknown parameters of the PV cell characteristic Eq. (5.41). One such numerical method is discussed in Ref.[1] using a worked example.

5.9 Multijunction Solar Cells

A number of factors that limit the conversion efficiency of single -layer solar cells were discussed in earlier sections. One of the inherent performance limitations of such cells is the mismatch between the band gap of the cell and energy spectrum of sunlight.

No electron-hole pairs are generated by photons in the portion of the solar spectrum above the wavelength corresponding to the band gap wavelength of the cell, which is about 1.12 μm for Si, and 0.89 μm for GaAs, as seen in Table 5.1. Photons with energy larger than the band gap energy are capable of generating electron-pairs, but the energy in excess of the band gap energy is lost as heat in the thermalization process.

This limitation could be partially overcome by stacking multiple layers of solar cells on top of each other as depicted in Fig. 5.9(a), thus producing a *multijunction* solar cell. Band gaps of the layers decrease from the top layer to the bottom layer. Solar radiation with wavelengths

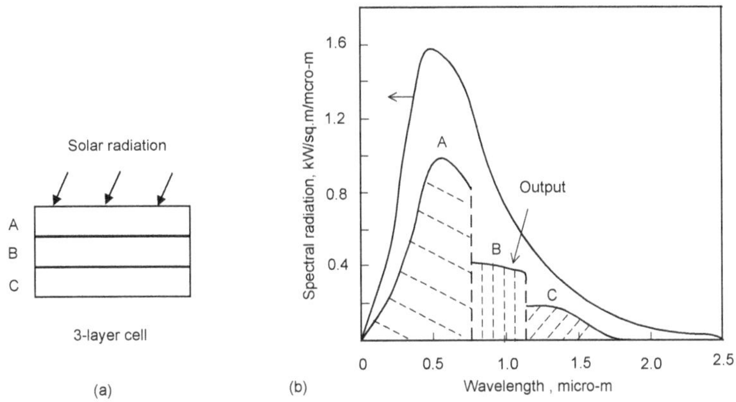

Fig. 5.9 (a) Multijunction solar cell (b) Output of multijunction cells

less than the band gap wavelength λ_a of the top layer A, would be absorbed in it and generate electron-hole pairs. Radiation with wavelengths larger than λ_a will pass through A and would be partially absorbed in layer B, whose band gap wavelength is λ_b. Radiation with wavelengths larger than λ_b will pass through B and be absorbed in C.

The energy absorbed by a multijunction cell with three layers is illustrated schematically in Fig. 5.9(b), using an idealized solar spectrum. The hatched areas in the figure represent the variation of the output with wavelength for the three cells A, B and C. The output and the efficiency of the multijunction cell would be higher than those of the individual cells. However, for this concept to work the layers must be as thin as possible, and this makes the manufacture of such cells more challenging and costly.

5.10 Design of Photovoltaic Systems

The power output of individual solar cells is limited, and therefore they must be connected together to produce enough electricity for most applications. Solar cells may be connected in series, parallel, or both to obtain the desired voltage and current output. A few of the arrangements are shown in Fig. 5.10.

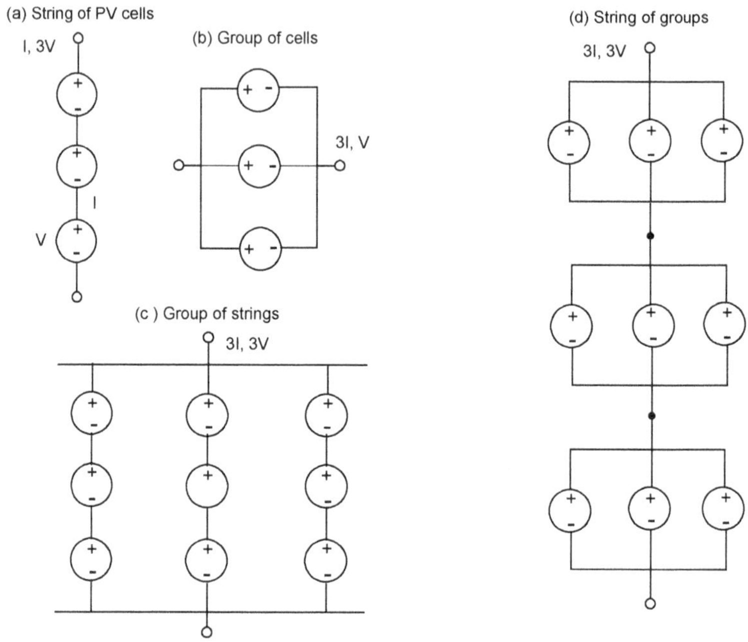

Fig. 5.10 (a) Cells in Series (String), (b) Cells in parallel (Group), (c) Group of strings, (d) String of groups

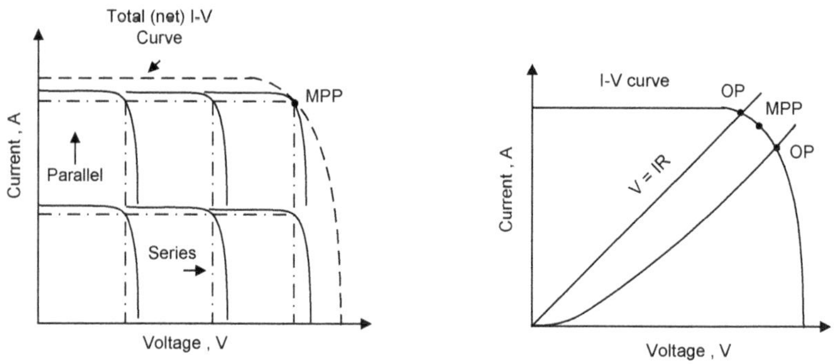

(a) (b)

Fig. 5.11 (a) I-V curve representation of PV circuits: 2 groups (parallel) of 3 strings (series), (b) Loads and operating points (OP)- Maximum power point (MPP)

For boosting the voltage, cells are connected in series as in (a), and to boost the current they are connected in parallel as in (b). Grouping strings together, as in (c) and (d) increases both the voltage and the current. When similar cells are connected in series the voltages are added and the current remains the same. For cells connected in parallel, the currents are added and the voltage remains the same. When two dissimilar cells are connected in series the individual voltages are added and the current remains at the lower value.

The converse is true when two dissimilar cells are connected in parallel, where the currents are added and the voltage remains at the lower value.

Shown in Fig. 5.11(a) is the ideal overall I-V curve of a PV circuit with 2 strings, each with 3 cells in series, arranged in parallel. At the maximum power point (MPP), the current is twice the current for each cell, and the voltage 3 times the individual voltage.

In the following sections we shall discuss the operation of the more common PV power systems used in practice and some basic factors that need to be considered in the design of such systems.

5.10.1 *Direct-coupled PV systems*

This is the simplest type of PV system, and the only components are a PV module, and a load such as an electric motor. An example is a solar attic fan, which operates when the sun shines, and it is the time when it is most needed. Another common application of direct-coupled systems is water pumping. In sunny times, more water is needed and water can be stored at a higher elevation and used at night.

Figure. 5.11(b) shows the operating point when a load is supplied directly by a PV system. For a simple resistive load the $I - V$ curve of the load is a straight line according to Ohm's law. The point of intersection of the load curve and the $I - V$ characteristic of the PV system is the operating point, OP. This is also true for the non-linear load curve shown in Fig. 5.11(b), which is applicable to systems consisting of a motor driven by a PV output.

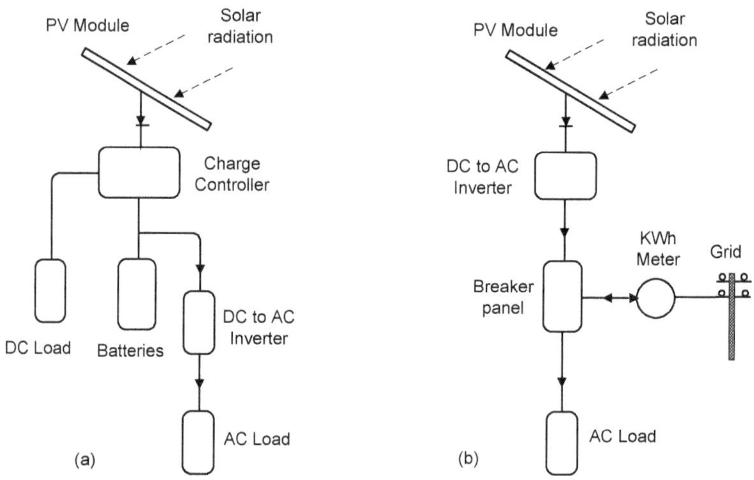

Fig. 5.12 Schematics of (a) Stand-alone PV – power system, (b) Grid-connected PV – power system

5.10.2 *Stand-alone PV systems*

The *stand-alone* solar PV system depicted in Fig. 5.12(a) is suitable for use in remote locations or for applications where power is not available from the local electrical grid. The *direct-current* (DC) generated by the solar array passes through a charge controller that regulates the amount of power that enters the battery pack, where it is stored. The charge controller ensures that the batteries are not over charged or over discharged, because both factors will reduce the life of the battery. In addition, charge controllers help operate the PV system at the maximum power point (MPP), discussed above.

When the batteries are fully charged, the charge controller either disconnects the PV system or short circuits the PV output, avoiding over charging. Moreover, the charge controller prevents over discharging of batteries by disconnecting the load, when necessary. Usually, batteries are sized to supply the loads for several days in case of continuous cloudy periods. The output from the batteries is used for DC applications such as a phone system or a pump. For loads requiring *alternating-current* (AC) power like the lighting system, TVs, power tools, the DC output from the battery pack is first passed through an inverter.

An *inverter* is a device that converts a DC input to an AC output with appropriate voltage and frequency. Inverters can be classified as: (i) stand-alone inverters, (ii) grid-connected or grid-tied inverters, and (iii) bi modal inverters. A stand-alone inverter receives DC power from the PV array and produces a stable constant frequency AC output to meet the AC loads of the system. In this case the inverter acts as a voltage source.

A *grid-connected* inverter needs the grid supply to initiate its operation, and its output voltage and frequency are determined by the conditions of the grid. In this case the inverter acts as a current source, and it will not operate when the grid fails. Bi-modal inverters can function both as stand-alone and grid-tied inverters but their use is rare in practice. All modern inverters have built in maximum power tracking capability.

Inverters can also be classified as: (i) central inverters, (ii) string inverters, and (iii) module inverters. Central inverters convert the entire output of a PV array to AC at one location in the system, and have power ratings in the kW to MW range. String inverters convert power produced by several modules connected in a string (see Fig. 5.10), and their power ratings are from about 500W to several kW. Module inverters or micro inverters are located at each module and they have small power ratings of up to about 500W. Efficiencies of modern PV inverters are about 95 to 97%.

5.10.3 *Grid-connected PV power system*

Grid-connected solar PV systems are more commonly used for domestic and commercial applications where the use of grid-electricity is the norm [4,6]. Such a system is shown schematically in Fig. 5.12(b). The essential components of such a PV system are: (i) PV array, (ii) inverter, (iii) monitoring system, (iv) feed-in meter, (v) grid connection, and (vi) DC and AC cabling.

On sunny days, the DC power generated by the solar array is converted to AC power in the inverter and supplied to the AC loads of the building. No power is drawn from the local electrical grid under these conditions. However, any power more than that required by the loads, is fed to the grid through the energy meter. Under these conditions, the electrical grid functions like a giant storage battery absorbing excess solar-generated energy from various solar installations. When sufficient solar radiation is not available, and during night time, power is drawn from the local grid to supply the required AC loads of the building. This concept is known as *net-metering*, where the consumer is billed on the difference between the power drawn from the grid, and the solar-generated power supplied to the grid. This arrangement is commonly used in domestic PV electric systems.

5.10.4 *Design considerations of stand-alone systems*

In this section we present a simplified procedure, also known as the *heuristic approach*, for the design of a stand-alone PV system [6]. This method does not involve time dependent weather data or their impact on the performance of the system.

A detailed design would essentially be a simulation, that takes into account factors like the dependence of the component efficiencies on the load, time variations in incident solar radiation, and the mode of operation, that is PV-to-load, PV-to-storage, or storage-to-load. The main factors considered in the heuristic approach are the following:

1. The times of operation of the loads and the load currents required.
2. The important energy losses of the system.
3. The average daily solar radiation.
4. Determination of the total solar array current.
5. The optimum arrangement of the modules of the PV array.
6. The size of the required battery that meets the desired reserve time.
7. Selection of a suitable charge controller.

1. The total load current and operation time

To determine the currents required, the load and the operating voltage of the PV system must be fixed. The typical nominal voltages of PV systems are 12V, 24V, or 48V. Once the nominal voltage V_n is known, the average daily ampere-hours can be determined for each load. For a load W watts operating for t_h hours a day, the daily load is Wt_h Watt-hours. Hence the total load in ampere-hours required may be expressed as

$$Ah = \frac{\sum Wt_h}{V_n} \tag{5.47}$$

For AC loads, the energy required is expressed as equivalent DC loads because the PV system produces DC current. The DC equivalent load is obtained by dividing the AC load energy by the efficiency of the inverter, which is typically about 95% for a good inverter. The total ampere-hours is given by Eq. (5.47).

2. The losses of the PV system

Some components of the PV system like the charge controllers and batteries, require energy for their operation. These energy consumptions are considered as system energy losses. These losses are typically about 20 to 30%, and to compensate for these losses the ampere-hours given by Eq. (5.47) is increased by the above amounts.

3. Determination of the average daily solar radiation

The solar energy input to the PV system depends on several factors, most of which were discussed in chapters 2 and 3 on the solar resource and solar thermal systems respectively. Depending on the measured solar radiation data available for a location, different correlations have been developed to estimate the hourly and daily data on a horizontal surface. Form this data the radiation intensity on an inclined surface may be computed.

The tilt of the solar cell module is also an important design parameter, which must be optimized based on the latitude of the location and the usage. The optimum angle for different months of the year, obtained by detailed computer simulation, are listed in Table 5.2. Sun tracking may be used to further improve the performance but this would involve additional costs and maintenance.

4. Determination of the total PV array current

Solar cells are usually characterized with a AM1.5 spectrum, with a total irradiance of 1000 Wm^{-2}. To use the measured characteristics of the solar cell in design a quantity called the *equivalent sun-hours* (*ESH*) is defined [6]. When the *average annual solar radiation* on a surface Q_S is expressed in kWh.m^{-2} per year, the *equivalent sun-hours* (*ESH*) is defined as

$$ESH = \frac{Q_S \times 10^3}{(1000 \times 365.25)} \tag{5.48}$$

The total array current is obtained by dividing the total load expressed in *Ah* (Eq. 5.47) by the *ESH* for the location.

5. Determination of the optimum arrangement of solar modules

Solar modules are manufactured with a range of output powers. The optimum arrangement of modules is to obtain the desired total array current with the minimum number of modules. Several ways to connect modules are shown in Fig. 5.10. For optimum operation, the output voltage of the modules should correspond to the battery voltage. The number of modules in series is determined by dividing the PV system voltage by the module voltage at maximum power. The number of modules in parallel is obtained by dividing the total current required from the PV array by the module current at maximum power.

6. Determination of the battery size

An essential component of a stand-alone PV system is the battery, that supplies the load at night and during cloudy periods with low sunshine. To cater for such situations, the battery should have reserve capacity to meet the required load for a certain period called *days of autonomy*. This number depends on the types of loads being supplied and the general weather of the location. Typically, for telecommunication components the days of autonomy is ten days or more, and for residential applications it's about five days. The capacity of the battery in Ah is calculated by multiplying the total daily DC load of the system including losses by the number of days of recommended reserve time.

Some general losses to be included in PV system design

In sizing PV arrays for grid-connected systems, the general considerations discussed above are applicable. However, there are several practical aspects that need to be included in a conservative design. These are:

1. Manufacturers may have a deviation of about 3% in the rated power output of a module.
2. The operating temperature of a panel will be different from V_{mp} at the standard temperature of 25°C. A correction factor using the temperature coefficient must be applied to the voltage based on the temperature range expected at the location.
3. Dust and dirt deposited on the panel surface will reduce the sunlight absorbed by the cell. In locations with regular rain, the panel surface will be washed by rain water. In dry areas, dust could remain on the surface for a long time if not cleaned regularly. Proximity to busy roads and industrial areas could also effect the performance of the cell adversely. Therefore a correction factor is applied to account for dust and dirt in the design.
4. All cells in a solar array may not perform with the same maximum power output due to manufacturing differences. This mismatch must be accounted for by including a correction factor of about 2%.
5. The power delivered is also reduced due to the resistance of the cables, which should be kept to a minimum. A correction factor must be included to account for such losses.
6. The conversion from DC to AC current in the inverter entails power losses. Modern inverters have efficiencies of about 95% to 97% under factory conditions, but under field conditions these may be reduced to about 93% to 96%. A conservative approach is to assume a power loss of 5%.
7. Solar panels installed on roof tops may have deviations from the optimal tilt conditions desired due to the orientation and slope of the roof. This aspect must be carefully considered in estimating expected power output of the solar panel.

A *worst-case adjustment factor*, including all the above items maybe written in the form:

$$F_{correction} = f_{prod.toler.} \times f_{oper.temp.} \times f_{dirt\&dust} \times f_{mismatch} \times f_{cable\ loss} \times f_{invert.loss} \tag{5.49}$$

The estimation of the above factors must be based on experience with similar systems.

5.11 Detailed Design Method

The heuristic design method introduced above is based on a single parameter called the equivalent sun-hours (*ESH*) at a location. More comprehensive design methods that include the monthly variations of solar radiation and ambient temperature at a location are presented in Ref.[1,7].

We shall discuss one of these methods originally developed by Evens [2]. Considering hourly intervals over a month, the average daily output of a PV panel over a month maybe expressed in the form

$$E = \frac{A \sum \eta_i I_i}{N} \tag{5.50}$$

where A is the panel area, η_i is the hourly panel efficiency, I_i is the hourly insolation, N is the number days of the month, and the summation is over the number of *daylight hours* of the month. It is assumed that the daily weather is like the average weather of the month, but the variations from month to month are large.

The hourly PV cell efficiency η_i is a function of the cell design, the cell temperature T_{ci}, the reference cell temperature T_r, solar radiation intensity I_i, and the ambient temperature T_{ai}. These variables are shown in the schematic diagram in Fig. 5.13. Combining Eqs. (5.34) and (5.38) the panel efficiency may be expressed in the simplified form [7]

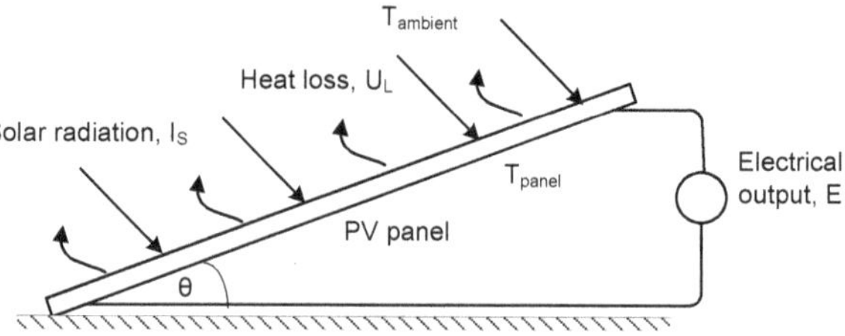

Fig. 5.13 PV panel variables used for detailed design

$$\eta_i = \eta_r[1 - \beta(T_{ci} - T_r) + \mu ln(I_i)] \quad (5.51)$$

η_r is the rated efficiency, β is the temperature coefficient of the cell, and μ is the insolation coefficient. Substituting from Eq. (5.51) in (5.50) we have

$$E = \frac{A\eta_r \sum[1-\beta(T_{ci}-T_r)+\mu ln(I_i)]I_i}{N} \quad (5.52)$$

We now introduce the *average hourly ambient temperature* T_{ai} and the *mean monthly ambient temperature* T_m in Eq. (5.52) to obtain the expression

$$E = \left(\frac{A\eta_r}{N}\right)\sum[I_i - \beta(T_{ci} - T_{ai})I_i - \beta(T_{ai} - T_m)I_i - \beta(T_m - T_r)I_i + \mu I_i ln(I_i)] \quad (5.53)$$

The *monthly average* panel efficiency η may be defined as

$$E = \left(\frac{\eta A}{N}\right)\sum I_i \quad (5.54)$$

From Eqs. (5.53) and (5.54) we get

$$\left(\frac{\eta A}{N}\right)\sum I_i = \left(\frac{A\eta_r}{N}\right)\sum[I_i - \beta(T_{ci} - T_{ai})I_i - \beta(T_{ai} - T_m)I_i - \beta(T_m - T_r)I_i + \mu I_i ln(I_i)] \quad (5.53a)$$

The above equation maybe expressed in the form

$$\eta = \eta_r[1 - \beta(T_c - T_a) - \beta(T_a - T_m) - \beta(T_m - T_r) + \mu ln(I)] \quad (5.55)$$

The average quantities in Eq. (5.55) are obtained by comparing the terms in Eq. (5.53a) with the corresponding terms in Eq. (5.55) as:

$$(T_c - T_a) = \frac{\sum(T_{ci}-T_{ai})I_i}{\sum I_i} \quad (5.56)$$

$$(T_a - T_m) = \frac{\sum(T_{ai}-T_m)I_i}{\sum I_i} \quad (5.57)$$

$$ln(I) = \frac{\sum I_i ln(I_i)}{\sum I_i} \quad (5.58)$$

We shall briefly outline the evaluation of each of the terms in Eq. (5.55). Notice that Eq. (5.56) gives the average difference between the cell temperature and the ambient air for *daylight hours*. This temperature difference is a function of the insolation, heat loss rate from the cell, and the electrical energy output. The overall energy balance for the panel gives (see Fig. 5.13)

$$\alpha\tau I_i = \eta_i I_i + U_l(T_{ci} - T_{ai}) \tag{5.59}$$

The LHS of Eq. (5.59) is the solar radiation absorption rate, the first term on the RHS is the electrical energy output and the second term is the heat loss rate from the panel. The electrical output is relatively small compared to the other terms of Eq. (5.59). Neglecting this term we obtain

$$(T_{ci} - T_{ai}) = \frac{\alpha\tau I_i}{U_l} \tag{5.60}$$

Averaging Eq. (5.60) and comparing with Eq. (5.56) we have

$$\left(\frac{U_l}{\tau\alpha}\right)(T_c - T_a) = \left(\frac{\sum I_i^2}{\sum I_i}\right) \tag{5.61}$$

where T_c and T_a are average temperatures over the summation period.

If the temperature difference in Eq. (5.61) is taken to be the average over a month, then we could correlate the RHS in terms of the monthly clearness index. The clearness index, discussed in chapter 2, is defined as the average radiation on a horizontal surface to the extraterrestrial radiation on a horizontal surface. If its monthly average value is $\overline{K}_T$, then the RHS of Eq. (5.61) maybe written in the form

$$\left(\frac{\sum I_i^2}{\sum I_i}\right) = a_1 + a_2\overline{K}_T \tag{5.62}$$

where the parameters a_1 and a_2 are dependent on the variation of insolation at the location.

The correlation in Eq. (5.62) is usually developed for a panel orientated at the optimal tilt angle θ_m. A correction factor for the deviation of the actual tilt angle θ_a from the optimal value for different months has been obtained by computer simulation. The correction factor C_f which is applied to the monthly average temperature difference $(T_c - T_a)$ is expressed in the form [7]

$$C_f = 1 - 0.000117(\theta_m - \theta_a)^2 \quad (5.63)$$

The optimal tilt angles for different months of the year, expressed as a function of the latitude L, are given in Table 5.2.

Table 5.2 Optimum tilt angle of PV panel (L is latitude)

Month	Jan.	Feb.	Mar.	April	May	June
Opt. Tilt	L+29	L+18	L+3	L-10	L-22	L-25

Month	July	Aug.	Sep.	Oct.	Nov.	Dec.
Opt. Tilt	L-24	L-10	L-2	L+10	L+23	L+30

To determine the term $(T_a - T_m)$ we use Eq. (5.57). It can be written in the form

$$(T_a - T_m) = \frac{\sum T_{ai} I_i}{\sum I_i} - T_m \quad (5.64)$$

We notice that $(T_a - T_m)$ is the difference between the average air temperature during *daylight hours* and the monthly average ambient temperature. This term can be computed by knowing the hourly ambient temperatures for a location. The term $(T_m - T_r)$ in Eq. (5.55) is the difference between the mean monthly temperature and the reference cell temperature, available in manufacture's data.

The term $ln(I)$ has been correlated with $\overline{K}$ using a linear expression like Eq. (5.62). However, the value of μ is assumed to be generally small, so that the last term in Eq. (5.55) can be neglected. The loss coefficient can be estimated by using a thermal model for the PV module, which is available in Ref.[1]. In the absence of sufficient data, a value of 20 Wm^{-2} K^{-1} is recommended.

The monthly energy output is

$$E_m = \eta N_{dm} I_s A \text{ kWh}$$

where η is the monthly mean efficiency from Eq. (5.55), N_{dm} is number of days of the month, I_s is the daily incident solar energy ($kWhm^{-2}day^{-1}$), A is the area of the PV array.

We shall illustrate the application of the detailed design method in worked examples 5.18 and 5.19.

5.12 Efficiency and Cost of Solar PV Cells

The efficiency of a solar PV cell is the fraction of incident solar radiation that is converted to electrical energy, under a set of standard conditions. The data in Table 5.3 shows how the standard cell efficiency has improved over the years. With the current global research efforts, further improvement in solar cell efficiency is likely.

In the USA, the gross cost of solar cells has been decreasing steadily since 2007, with the most dramatic drop occurring between 2014 and 2018, as seen from the data in Table 5.4. The national average cost of solar PV cells in the USA in 2019 was $3.05 /W [11].

The levelized cost of energy (LCOE), discussed in chapter 1, is a commonly used metric to compare the economics of different energy production systems such as solar, wind, coal, and natural gas. It is defined as the ratio of the life-cycle cost of an energy system to the total energy delivered by the system during its life time.

Table 5.3 Improvement in solar cell efficiency, (Data from Ref.[9])

Year	1960	1992	2012	2015
Efficiency	14%	15.8%	17.8%	18.2%

Year	2015	2015	2016	2017
Efficiency	22.5%	22.8%	29.8%	44.5%

Table 5.4 Gross cost ($ per watt) of solar cells in the USA, (Data from Ref.[11])

Year	2014-2nd H	2015-1st H	2015-2nd H	2016-1st H	2016-2nd H
Cost	3.86	3.79	3.69	3.57	3.36

Year	2017-1st H	2017-2nd H	2018-1st H	2018-2nd H
Cost	3.17	3.13	3.12	3.05

Table 5.5 Mean levelized cost (LCOE) of solar PV systems in the USA, in $/MWh, (Data from Ref.[10])

Year	2009	2010	2011	2012	2013	2014
$LCOE_{av}$	359	248	157	125	97.5	79
$LCOE_h$	394	270	166	149	104	86
$LCOE_l$	323	226	148	101	91	72

Year	2015	2016	2017	2018	2019	2020
$LCOE_{av}$	64	55	49.5	43	40	36.5
$LCOE_h$	70	61	53	46	44	42
$LCOE_l$	58	49	48	40	36	31

The mean LCOE values for utility-scale solar PV power in the USA for a period of 12 years from 2009 to 2020 are given in Table 5.5. It is seen that the average LCOE has decreased by about 90% during this period [10]. Improvements in solar cell efficiency would also result in cost reduction of solar PV installations. Globally, the cost of solar PV systems is expected to drop further in the coming years as more commercial plants are deployed. A list of the larger grid-connected solar PV plants operating in different countries is given in Table 5.6.

Table 5.6 Some of the larger commercial solar PV power plants (Data from Wikipedia, Ref.[8])

Country	Location	Coordinates (approx.)	Capacity MW	Land size km^2	Year
India	Bhadla Solar Park	27° N, 72° E	2245	40	2020
China	Huanghe Solar Park	36° N, 95° E	2200	100	2020
India	Pavagad Solar Park	14° N, 77° E	2050	53	2019
Egypt	Benban Solar Park	24° N, 32° E	1650	37	2019
UAE	Noor Abu Dhabi SP	24° N, 55° E	1177	-	2019
Mexico	Villanueva Solar Park	25° N,103° E	828	27.5	2018
Vietnam	Dau Tieng Power	-	600	-	2019
USA	Desert Sunlight Farm	34° N, 115° W	550	16	2015
USA	Topaz Solar Farm	34° N, 120° W	550	19	2014
Spain	Nunez de Balboa	37° N, 6° W	500	10	2020

5.13 Environmental Effects of Solar PV Systems

Several hazardous materials are used in the manufacture of solar PV cells, most of which are used to clean and purify the semiconductor surface. These chemicals include hydrochloric acid, nitric acid, hydrogen fluoride, and acetone [12]. The quantity and type of chemicals used depend on the type of solar cell, the amount of cleaning needed, and the size of the cell.

Thin-film cells contain more toxic materials like gallium arsenide, cadmium-telluride and others, which are not used in the manufacture of traditional silicon PV cells. However, these rare and expensive materials are likely to be recycled than thrown away. The manufacture of single crystal and polycrystalline solar cells is described in Ref.[5].

During operation a solar power plant does not produce any global warming emissions. However, the life-cycle of a solar power plant includes a number of steps like, extraction of ore, processing of various materials, manufacture of cells, transportation, installation, and decommissioning, which produce global warming emissions. For PV cells, the life-cycle emissions are estimated to be between 0.07 to 0.18 kg of carbon dioxide per kWh [12].

5.14 Worked Examples

Example 5.1 The band gap of silicon (Si) is 1.11 ev. Light of intensity 2 mW and wavelength 0.638 μm is incident on a Si solar cell. Calculate (i) the wavelength of radiation whose photons have energies equal to the band gap of Si, (ii) the number of photons incident per second, (iii) the maximum power available, and (iv) the maximum percentage of incident power that can be converted to electricity.

Solution The wavelength of radiation corresponding to the band gap of silicon is given by

(i) $$\lambda_{bg} = \frac{hc}{BG} = \frac{6.625\times10^{-34}\times2.998\times10^{8}}{1.11\times1.6\times10^{-19}} = 1.118\,\mu m$$

(ii) The photon flux n_{ph} of the incident radiation is given by

$$n_{ph}\left[\frac{hc}{\lambda_i}\right] = I = 2 \times 10^{-3}$$

$$\frac{n_{ph} \times 6.625 \times 10^{-34} \times 2.998 \times 10^{8}}{0.638 \times 10^{-6}} = 2 \times 10^{-3}$$

Hence the photon flux is 6.43×10^{15} photons s^{-1}.
Maximum available power from the solar cell is given by

$$P_{max} = \frac{n_{ph}hc}{\lambda_{bg}} = n_{ph}BG$$

$$P_{max} = 6.43 \times 10^{15} \times 1.11 \times 1.6 \times 10^{-19}$$

Hence the maximum power is 1.1412 mW.

(iii) The maximum percentage of power is $\frac{1.142}{2} = 57\%$

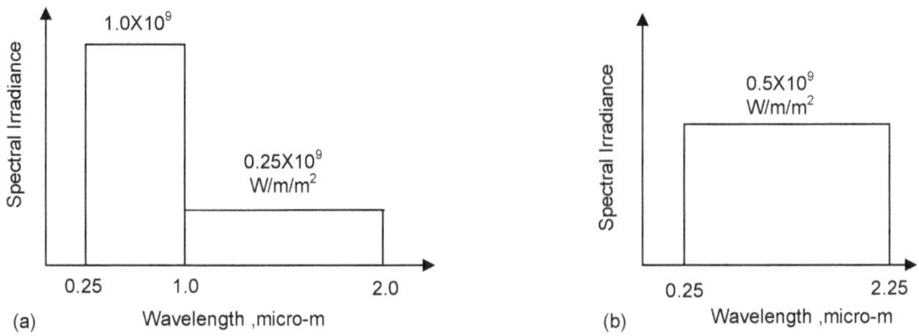

Fig. E5.2 (a) Radiation spectrum for Example 5.2, (b) Radiation spectrum for Example 5.3

Example 5.2 The AM1.5 (air mass 1.5) solar spectrum is approximated by the rectangular form depicted in Fig. E5.2 [6]. Calculate (i) the total radiation, and (ii) the total photon flux of the spectrum.

Solution The incident radiation spectrum is depicted in Fig. E5.2(a). It can expressed as

For $0.25\mu m < \lambda < 1.0\mu m$

$$I_{e\lambda} = 1.0 \times 10^9 \text{ Wm}^{-2}\text{m}^{-1}$$

For $1.0\mu m < \lambda < 2.0\mu m$

$$I_{e\lambda} = 0.25 \times 10^9 \text{ Wm}^{-2}\text{m}^{-1}$$

The total radiation intensity of the spectrum is obtained by integrating the spectrum over the wavelength intervals. Hence we have

$$I_e = \int_{0.25}^{1.0} 1.0 \times 10^9 d\lambda + \int_{1.0}^{2.0} 0.25 \times 10^9 d\lambda$$

$$I_e = 1.0 \times 10^9 \times (1 - 0.25)10^{-6} + 0.25 \times 10^9 \times (2 - 1)10^{-6}$$

Total radiation intensity is $I_e = 10^3$ Wm^{-2}

The spectral photon flux is given by

$$\delta E_\lambda = I_{e\lambda} d\lambda = n_{ph}\left(\frac{hc}{\lambda}\right) d\lambda$$

The total photon flux is given by

$$N_{ph} = \int n_{ph} d\lambda = \int \left(\frac{\lambda}{hc}\right) I_{e\lambda} d\lambda$$

$$N_{ph} = \left(\frac{1}{hc}\right)\left[\int_{0.25}^{1.0} 1.0 \times 10^9 \lambda d\lambda + \int_{1.0}^{2.0} 0.25 \times 10^9 \lambda d\lambda\right]$$

$$N_{ph} = \left(\frac{1}{6.625 \times 10^{-34} \times 3 \times 10^8}\right)\left[\frac{10^9(1^2 - 0.25^2)}{2} + \frac{0.25 \times 10^9(2^2 - 1^2)}{2}\right] \times 10^{-12}$$

The photon flux is $N_{ph} = 4.25 \times 10^{21}$ m^{-2}s^{-1}

Example 5.3 Unpolarized light of wavelength 0.5 μm and intensity 900 Wm^{-2} is incident at right angles on a silicon wafer of thickness 350 μm. The refractive index of Si is 4.29 and the absorption coefficient is 1.11×10^6 m^{-1}. Calculate (i) the fraction of incident light reflected at the surface, (ii) the total number of photons absorbed in Si, and (iii) the photocurrent density generated.

Solution The basic equations governing the reflection and transmission of radiation in semi-transparent media are given in Appendix A1. The reflectivity of the surface for normally incident radiation is given by

$$R = \left(\frac{1-n_2}{1+n_2}\right)^2 = \left(\frac{1-4.29}{1+4.29}\right)^2 = 0.386$$

The intensity of radiation at depth L below the surface is given by

$$I_L = I_o(1 - R)exp(-\alpha L)$$

$$I_L = 900(1 - 0.386)exp(-1.11 \times 10^6 \times 350 \times 10^{-6}) \approx 0$$

Hence the radiation absorbed in the wafer is

$$A = 900(1 - 0.386) = 552.6 \text{ Wm}^{-2}$$

The absorbed photon flux is given by

$$\phi_{ph,a}\left(\frac{hc}{\lambda}\right) = A$$

The absorption rate of photons is

$$\phi_{ph,a} = A\left(\frac{\lambda}{hc}\right) = \frac{552.6 \times 0.5 \times 10^{-6}}{6.625 \times 10^{-34} \times 3 \times 10^8} = 13.9 \times 10^{20} \text{ s}^{-1}\text{m}^{-2}$$

Assume that each absorbed photon generates an electron-hole pair that is collected. Therefore the short circuit current density, which is equal to the light current density by Eq. (5.13) is:

$$J_{sc} = q\phi_{ph,a} = 1.602 \times 10^{-19} \times 13.9 \times 10^{20} = 222.3 \text{ Am}^{-2}$$

Example 5.4 A silicon solar cell of band gap 1.12 ev is uniformly illuminated by monochromatic light of intensity 200 Wm^{-2} and wavelength 0.8 μm. The quantum efficiency at this wavelength is 0.8. Calculate (i) the photon flux, (ii) the light current of the cell if the area of the cell is 5 cm^2.

Solution The photon flux is given by

$$\phi_{ph,a} = I_s\left(\frac{\lambda}{hc}\right)$$

$$\phi_{ph,a} = \frac{200 \times 0.8 \times 10^{-6}}{6.625 \times 10^{-34} \times 3 \times 10^8} = 8.056 \times 10^{20} \text{ s}^{-1}\text{m}^{-2}$$

The quantum efficiency, $EQE = 0.8$ Therefore the electron generation rate is

$$\psi_{el} = EQE \times \phi_{ph,a}$$

$$\psi_{el} = 0.8 \times 8.056 \times 10^{20} = 6.44 \times 10^{20}$$

The light current is

$$I_L = A \times q \times \psi_{el}$$

$$I_L = 5 \times 10^{-4} \times 1.602 \times 10^{-19} \times 6.44 \times 10^{20}$$

Hence the light current is 5.16×10^{-2} A.

Example 5.5 The AM1.5 solar spectrum is roughly-approximated by the rectangular distribution depicted in Fig. E5.2b [6]. Radiation with this spectrum is incident on a solar cell with a band gap of 1.11 ev. The quantum efficiency, $EQE = 1$ Calculate (i) the total irradiation, (ii) the total photon flux, (iii) the amount of incident radiation lost as heat in the cell.

Solution The total solar radiation intensity is given by

$$I_e = \int_{0.25}^{2.25} 0.5 \times 10^9 d\lambda$$

$$I_e = 0.5 \times 10^9 (2.25 - 0.25) 10^{-6} = 1000 \text{ Wm}^{-2}$$

The total photon flux is given by

$$N_{ph} = \int_{0.25}^{2.25} \phi_{ph} d\lambda = \int_{0.25}^{2.25} I_{e\lambda} \left(\frac{\lambda}{hc}\right) d\lambda$$

$$N_{ph} = \left(\frac{1}{hc}\right) \int_{0.25}^{2.25} 0.5 \times 10^9 \lambda d\lambda$$

$$N_{ph} = \left(\frac{0.5\times10^9}{2\times6.625\times10^{-34}\times3\times10^8}\right)(2.25^2 - 0.25^2) \times 10^{-12}$$

The photon flux is $N_{ph} = 6.29 \times 10^{21} \text{m}^{-2}\text{s}^{-1}$

Only photons with wavelength less than or equal to the band-gap wavelength will produce electron-hole pairs in the cell. The total energy of these electrons is

$$Q_{ec} = BG \int_{\lambda_0}^{\lambda_{by}} \phi_{ph}\, d\lambda$$

$$Q_{ec} = \left(\frac{hc}{\lambda_{bg}}\right) \int_{\lambda_0}^{\lambda_{bg}} I_{e\lambda} \left(\frac{\lambda}{hc}\right) d\lambda$$

$$Q_{ec} = \left(\frac{0.5\times 10^{9}}{2\times 1.11\times 10^{-6}}\right)(1.11^2 - 0.25^2)\times 10^{-12} = 263.4 \text{ W}$$

The energy lost as heat is the difference between the incident energy and the above output of the cell.

$$Q_{loss} = 1000 - 263.4 = 736.6 \text{ W}$$

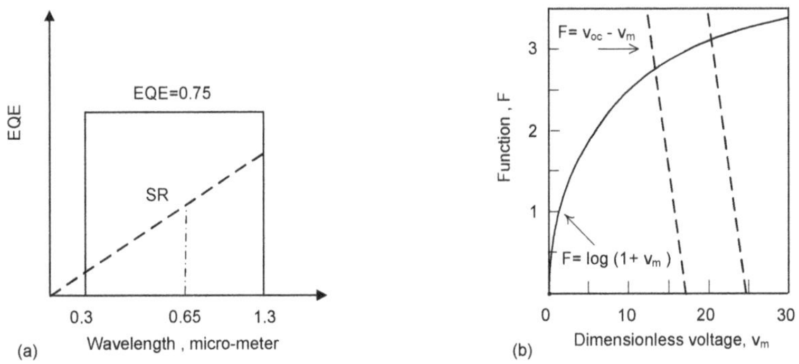

Fig. E5.6 (a) Spectral responsivity and EQE (b) Solution of equation for maximum power point

Example 5.6 Obtain an expression for the spectral responsivity SR of a solar cell, which is defined as the current generated per watt of incident light. The external quantum efficiency (EQE) of the cell under short circuit conditions are shown in Fig. E5.6(a). The total photon flux incident on the cell in the wavelength interval shown is 5.1×10^{21} $m^{-2}s^{-1}$. Calculate (i) the band gap, (ii) the spectral responsivity when the wavelength is 0.65 μm, and (iii) the short circuit current.

Solution From the given quantum efficiency plot we see that the band gap of the material corresponds to a wavelength of 1.3 μm because for larger wavelength, *EQE* is zero. Hence we obtain the band gap as

$$BG = \frac{hc}{\lambda} = \frac{6.625\times10^{-34}\times3\times10^{8}}{1.3\times10^{-6}}\ \mathrm{J}$$

$$BG = 15.28 \times 10^{-20}\ \mathrm{J}$$

$$BG = \frac{15.28\times10^{-20}}{1.6\times10^{-19}} = 0.955\ \mathrm{ev}$$

The *spectral responsivity SR* is defined as the light current generated per watt of incident light. The short circuit current is equal to the light current by Eq. (5.13). Therefore

$$SR(\lambda) = \frac{J_{SC}}{P_{in}(\lambda)} = \frac{qn_e}{n_{ph}E_{ph}} \tag{E5.6.1}$$

$$SR(\lambda) = \frac{qn_e\lambda}{n_{ph}(hc)} = \frac{q\lambda EQE(\lambda)}{hc} \tag{E5.6.2}$$

where the quantum efficiency is defined as the ratio of the electron flux generated to the incident photon flux. Hence

$$EQE(\lambda) = \frac{n_e}{n_{ph}}$$

We notice from Eq. (E5.6.2) that if *EQE* is constant, than *SR* varies linearly with the wavelength.

The spectral responsivity at a wavelength of 0.65 μm is given by

$$SR(0.65) = \frac{q\lambda EQE(\lambda)}{hc} = \frac{1.602\times10^{-19}(C)\times0.65\times10^{-6}(m)\times0.75}{6.625\times10^{-34}(Js)\times3\times10^{8}(ms^{-1})}$$

$$SR(0.65) = 0.393\ \mathrm{AW^{-1}}$$

$$J_{sc} = q \times EQE \times N_{ph}$$

$$J_{sc} = 1.602 \times 10^{-19} \times 0.75 \times 5.1 \times 10^{20}$$

$$J_{sc} = 61.3\ \mathrm{Am^{-2}}$$

Example 5.7 The reverse saturation current of a silicon solar cell is 1.8×10^{-7}A. When illuminated with solar radiation the short circuit current is 5 A. The operating temperature of the cell is 313 K. Calculate (i) the open circuit voltagc, (ii) thc maximum power output, (iii) the fill factor, and (iv) the number of cells needed to supply 275 W of power at 12 V. How must the cells be arranged?

Solution The characteristic equation (5.12) of the ideal solar cell is

$$I = I_L - I_o\left[exp\left(\frac{qV}{k_BT}\right) - 1\right]$$

The short circuit current, when $V = 0$ is

$$I_{SC} = I_L = 5 \text{ A}$$

The open circuit voltage, when $I = 0$ is

$$V_{OC} = \left(\frac{k_BT}{q}\right) ln\left(1 + \frac{I_L}{I_O}\right)$$

$$V_{OC} = \left(\frac{1.381\times10^{-23}\times313}{1.602\times10^{-19}}\right) ln\left(1 + \frac{5}{1.8\times10^{-7}}\right) = 0.462 \text{ V}$$

Define $\gamma = \left(\frac{q}{k_BT}\right) = \frac{1.602\times10^{-19}}{1.381\times10^{-23}\times313} = 37.06 \text{ V}^{-1}$

$\frac{1}{\gamma} = v_{th}$ is called the *thermal voltage.*

The dimensionless open circuit voltage in Eq. (5.18) is given by

$$v_{OC} = 0.462 \times 37.06 = 17.1$$

The dimensionless voltage at the maximum power point is given by the solution of Eq. (5.18)

$$v_{OC} - v_m = ln(1 + v_m) = 17.1 - v_m \quad \text{(E5.7.1)}$$

In Fig. E5.6(b) the graphical solution is shown as the point of intersection of the two functions on RHS and LHS of Eq. (E5.7.1). Two curves of the RHS for the variable parameter v_{OC} are plotted.

Iteration gives the solution of Eq. (E5.7.1) as $v_m = 14.368$.

Hence the voltage at the maximum power point is

$$V_m = \frac{14.368}{37.06} = 0.388 \text{ V}$$

The current at the maximum power point is given by Eq. (5.19) as

$$I_m = (I_L + I_o) - I_o exp(v_m)$$

$$I_m = 5 + 1.8 \times 10^{-7} - 1.8 \times 10^{-7} exp(14.368) = 4.687 \text{ A}$$

The maximum power is

$$P_{max} = V_m I_m = 0.388 \times 4.687 = 1.818 \text{ W}$$

The fill factor is

$$FF = \frac{P_{max}}{V_{oc} I_{sc}} = \frac{1.818}{0.462 \times 5} = 0.787$$

The total number of cells required to supply 275 W is

$$N_c = \frac{275}{1.818} = 152$$

The number of cells in series to obtain a total voltage of 12 V is

$$n_s = \frac{12}{0.388} = 31 \text{ cells}$$

Number of strings in parallel will be

$$n_{st} = \frac{152}{31} = 5$$

Note that we have rounded off the number cells per string and the number of strings. The total power available is 282 W.

Example 5.8 Monochromatic light of wavelength 0.8 μm and intensity 250 Wm^{-2} is incident on a silicon solar cell with a band gap of 1.12 ev and area 4 cm^2. The dark saturation current density is 1.1×10^{-12} A. The cell temperature 30°C. The quantum efficiency is 1 and the ideality factor is 1. Calculate (i) the short circuit current, (ii) open circuit voltage, (iii) maximum power, (iv) the fill factor, and (v) the conversion efficiency.

Solution The band-gap wavelength of silicon is 1.12 μm. Since the wavelength of the given monochromatic radiation is less than the band gap wavelength, and the quantum efficiency is one, all incident photons will produce electron-hole pairs.

The photon flux is given by

$$n_{ph} = \frac{250}{hc/\lambda} = \frac{250 \times 0.8 \times 10^{-6}}{6.625 \times 10^{-34} \times 3 \times 10^{8}} = 1.00 \times 10^{21}\ \text{s}^{-1}\text{m}^{-2}$$

The light current density is given by

$$J_L = qn_{ph} = 1.00 \times 10^{21} \times 1.602 \times 10^{-19} = 161\ \text{Am}^{-2}$$

The characteristic equation (5.12) of the ideal solar cell is

$$I = I_L - I_o\left[exp\left(\frac{qV}{k_BT}\right) - 1\right]$$

The short circuit current, when $V = 0$ is

$$I_{sc} = I_L = J_LA = 161 \times 4 \times 10^{-4} = 0.0644\ \text{A}$$

The open circuit voltage, when $I = 0$ is

$$V_{oc} = \left(\frac{k_BT}{q}\right) ln\left(1 + \frac{I_L}{I_o}\right)$$

$$V_{oc} = \left(\frac{1.381 \times 10^{-23} \times 303}{1.602 \times 10^{-19}}\right) ln\left(1 + \frac{64.4 \times 10^{-3}}{1.1 \times 10^{-12}}\right) = 0.647\ \text{V}$$

Thermal voltage is

$$V_{th} = \frac{k_BT}{q} = \left(\frac{1.381 \times 10^{-23} \times 303}{1.602 \times 10^{-19}}\right) = 0.0261\ \text{V}$$

The dimensionless form of v_{oc} is given by

$$v_{oc} = \frac{0.647}{0.0261} = 24.79$$

The dimensionless voltage at the maximum power point is obtained from the solution of Eq. (5.18).

$$v_{oc} - v_m = ln(1 + v_m) = 24.7 - v_m \qquad \text{(E5.8.1)}$$

The method of solution of Eq. (E5.8.1) was outlined in worked example 5.7.

Hence we obtain the voltage as $v_m = 21.65$

The maximum power voltage is

$$V_m = 21.65 \times 0.0261 = 0.567 \text{ V}$$

The current at the maximum power point is given by Eq. (5.19) as

$$I_m = (I_L + I_o) - I_o exp(v_m)$$

$$I_m = 0.064 + 1.1 \times 10^{-12} - 1.1 \times 10^{-12} exp(21.65) = 0.0612 \text{ A}$$

The maximum power is

$$P_{max} = V_m I_m = 0.567 \times 0.0612 = 0.0347 \text{ W}$$

The fill factor is

$$FF = \frac{P_{max}}{V_{oc} I_{sc}} = \frac{0.0347}{0.064 \times 0.647} = 0.838$$

The energy conversion efficiency is

$$\eta = \frac{0.0347}{250 \times 4 \times 10^{-4}} = 34.7\%$$

Example 5.9 A photovoltaic module has an area of 0.4 × 0.4 m^2. The *I-V* characteristic of the module has been measured under standard conditions. From the measured *I-V* data the following information has been obtained. $V_{oc} = 0.6$ V, $I_{sc} = 41$A. The ideality factor is 1.4. Obtain the (i) the reverse saturation current, (ii) characteristic equation of the module, (iii) the maximum power, (iv) the fill-factor, (v) the energy conversion efficiency, and (vi) the fill-factor using the empirical equation.

Solution The characteristic equation (5.15) of the non-ideal solar cell is

$$I = I_L - I_o \left[exp\left(\frac{qV}{nk_B T}\right) - 1 \right]$$

where $n = 1.4$ is the *ideality factor*.

The short circuit current, when $V = 0$ is

$$I_{sc} = I_L = 41 \text{ A}$$

The open circuit voltage, when $I = 0$ is

$$V_{OC} = \left(\frac{nk_BT}{q}\right) ln\left(1 + \frac{I_L}{I_o}\right) = 0.6 \quad \text{(E5.9.1)}$$

The factor γ is given by

$$\gamma = \left(\frac{q}{nk_BT}\right) = \frac{1.602\times10^{-19}}{1.4\times1.381\times10^{-23}\times298} = 27.83 \text{ V}^{-1}$$

Substituting in Eq. (E5.9.1) we obtain

$$ln\left(1 + \frac{41}{I_o}\right) = 27.83 \times 0.6$$

Hence the dark saturation current,

$$I_o = 2.3027 \times 10^{-6} \text{ A}$$

The characteristic equation is

$$I = 41 - 2.3027 \times 10^{-6}[exp(27.825V) - 1] \quad \text{(E5.9.2)}$$

The dimensionless form of V_{OC} is given by

$$v_{oc} = 27.825 \times 0.6 = 16.695$$

The dimensionless voltage at the maximum power point is given by the solution of Eq. (5.18) as:

$$v_{oc} - v_m = ln(1 + v_m) = 24.7 - v_m \quad \text{(E5.9.3)}$$

The method of solution of Eq. (E5.9.3) is outlined in worked example E5.7. Hence we obtain $v_m = 13.99$.

The maximum power voltage is

$$V_m = \frac{13.99}{27.825} = 0.5027 \text{ V}$$

The current at the maximum power point is given by Eq. (5.19) as

$$I_m = (I_L + I_o) - I_o exp(v_m)$$

$$I_m = 41 + 2.3027 \times 10^{-6} - 2.3027 \times 10^{-6} exp(13.99)$$

$$I_m = 38.26 \text{ A}$$

The maximum power is

$$P_{max} = V_m I_m = 0.5027 \times 38.26 = 19.23 \text{ W}$$

The fill factor is

$$FF = \frac{P_{max}}{V_{oc} I_{sc}} = \frac{19.23}{41 \times 0.6} = 0.78$$

The energy conversion efficiency under standard incidence radiation of 1000 Wm^{-2} is

$$\eta = \frac{19.23}{1000 \times 0.16} = 12\%$$

The empirical equation for the fill-factor is given by Eq. (5.24) as

$$FF' = \frac{v_{oc} - ln(v_{oc} + 0.72)}{v_{oc} + 1} = \frac{16.695 - ln(16.695 + 0.72)}{16.695 + 1} = 0.782$$

Example 5.10 A 0.015 ohm resistor is connected across the solar module described in example 5.9 above. The module is operating under standard conditions. Calculate (i) the power delivered to the resistor, and (ii) the resistance of the load for the module to operate under maximum power conditions.

Solution The non-ideal characteristic equation was obtained as Eq. (E5.9.2). The $I - V$ curve for the resistive load receiving power from the solar cell is a straight line because it follows Ohm's law. Hence its performance characteristic is given by

$$V = IR = 0.015I \tag{E5.10.1}$$

In Fig. 5.11(b) we have plotted the two characteristic equations. Equation (E5.10.1) is a straight line with a slope of 0.015. The point P, where the two curves intersect is the operating point of the system consisting of the solar cell and the load resistance. To obtain the current and voltage at P we need to solve the equation

$$I = 41 - 2.3027 \times 10^{-6}[exp(27.825V) - 1] = \frac{V}{0.015} \quad \text{(E5.10.2)}$$

Equation (E5.10.2) for the voltage, V has to be solved by trial and error because it is a non-linear equation. Hence we obtain the current and voltage at P as:

$$I = 35.31 \text{ A}, \quad V = 0.529 \text{ V}.$$

The power delivered to the resistive load is given by

$$P = VI = 0.529 \times 35.31 = 18.68 \text{ W}$$

In worked example 5.9 we obtained the voltage and current at the maximum power point as 0.5027 V and 38.26 A. This point is indicated as MPP in Fig. 5.11(b). The characteristic curve of the resistive load needs to pass through point MPP for the PV module to delivery maximum power. Using Eq. (E5.10.1) we obtain the required resistance for maximum power as

$$R_{mp} = \frac{0.5027}{38.26} = 0.0131 \text{ ohms}$$

When the incidence radiation level changes, we could obtain maximum power by adjusting load resistance as seen above.

Example 5.11 The parameters of a PV module of area 1.64 m^2, under standard conditions are: $V_{oc} = 66$ V, $I_{sc} = 6.1$ A, $P_{max} = 330$ W, $NOTC = 42$°C, temperature coefficient of $P_{max} = -0.32\%$ per °C, temperature coefficient of $V_{oc} = -0.3\%$ per °C, temperature coefficient of $I_{sc} = 0.06\%$ per °C. The cell is operating with an irradiance of 1000 Wm^{-2} and an ambient temperature of 38°C. Calculate (i) the operating cell temperature, V_{oc}, I_{sc}, P_{max}, and (ii) the change in the fill-factor and the efficiency.

Solution A *simplified steady state* model [1] is used to estimate the cell temperature at an ambient temperature of T_a and radiation level of G_m. Hence we have

$$T_m = T_a + \frac{(T_{NOCT} - 20)G_m}{800}$$

$$T_m = 38 + \frac{(42-20)1000}{800} = 65.5°\text{C}$$

The effect of temperature changes of a solar cell on various cell characteristics are expressed in terms of temperature coefficients provided by manufacturers in their data sheets. Substituting the given temperature coefficients in Eqs. (5.35)–(5.37) we obtain the cell parameters at the operating temperature of 65.5°C as

$$V'_{oc} = 66 - 66\left(\frac{0.3}{100}\right)(65.5 - 25) = 57.98 \text{ V}$$

$$I'_{sc} = 6.1 + 6.1\left(\frac{0.06}{100}\right)(65.5 - 25) = 6.24 \text{ A}$$

$$P'_{max} = 330 - 330\left(\frac{0.32}{100}\right)(65.5 - 25) = 287.2 \text{ W}$$

The fill factor at the operating temperature is

$$FF' = \frac{P'_{max}}{V'_{oc}I'_{sc}} = \frac{287.2}{57.98 \times 6.24} = 0.792$$

The energy conversion efficiency is

$$\eta' = \frac{P'_{max}}{I_S A} = \frac{287.2}{1000 \times 1.64} = 17.5\%$$

The fill factor and the conversion efficiency at the standard conditions are obtained using the same equations as 0.82 and 20.1% respectively.

Example 5.12 A silicon solar cell of area 100 cm^2 is operating with a temperature of 25°C and an irradiation level of 1000 Wm^{-2}. For these conditions, V_{OC} = 600 mV, I_{SC} = 3.3 A. The ideality factor is 1.2. (i) Using the simplified empirical relation for the fill-factor, calculate the maximum conversion efficiency of the cell under the given conditions. (ii) If the operating temperature changes to 55°C, calculate the conversion efficiency of the cell. The temperature coefficients for V_{OC} and I_{SC} are -0.33% and 0.065% per °C respectively.

Solution The characteristic equation (5.15) of the non-ideal solar cell is

$$I = I_L - I_o\left[exp\left(\frac{qV}{nk_BT}\right) - 1\right] \tag{E5.12.1}$$

where $n = 1.2$ is the *ideality factor*.

The short circuit current, when $V = 0$ is

$$I_{sc} = I_L = 3.3 \text{ A}$$

The open circuit voltage, when $I = 0$ is

$$V_{oc} = \left(\frac{nk_BT}{q}\right) ln\left(1 + \frac{I_L}{I_o}\right) = 0.6 \text{ V}$$

The thermal voltage is given by

$$v_{th} = \left(\frac{nk_BT}{q}\right) = \frac{1.2\times1.381\times10^{-23}\times298}{1.602\times10^{-19}} = 0.030826 \text{ V}$$

Substituting in Eq. (E5.12.1) we obtain

$$0.6 = 0.030826 ln\left(1 + \frac{3.3}{I_o}\right) \text{ V}$$

Hence the dark saturation current is

$$I_o = 1.1625 \times 10^{-8} \text{ A}$$

The dimensionless form of V_{oc} is given by Eq. (5.18) as

$$v_{oc} = \frac{0.6}{0.030826} = 19.46$$

The empirical equation for the fill-factor is Eq. (5.24)

$$FF' = \frac{v_{oc} - ln(v_{oc} + 0.72)}{v_{oc} + 1} = \frac{19.46 - ln(19.46 + 0.72)}{19.46 + 1} = 0.804$$

The maximum power is given by

$$P_{max} = V_{oc} I_{sc} FF = 0.6 \times 3.3 \times 0.804 = 1.59 \text{ W}$$

The energy conversion efficiency is

$$\eta = \frac{P_{max}}{I_s A} = \frac{1.59}{1000\times100\times10^{-4}} = 15.9\%$$

The voltage and current at the new operating temperature of 55°C is given by Eqs. (5.35)–(5.37) as

$$V'_{oc} = 0.6 - 0.6\left(\frac{0.33}{100}\right)(55 - 25) = 0.54 \text{ V}$$

$$I'_{sc} = 3.3 + 3.3\left(\frac{0.065}{100}\right)(55 - 25) = 3.364 \text{ A}$$

$$v'_{oc} = \left(\frac{qV'_{oc}}{nk_BT}\right) = \frac{1.602\times10^{-19}\times0.54}{1.2\times1.381\times10^{-23}\times328} = 15.9$$

The empirical equation for the fill-factor is Eq. (5.24).

$$FF' = \frac{v_{oc} - ln(v_{oc}+0.72)}{v_{oc}+1} = \frac{15.9 - ln(15.9+0.72)}{15.9+1} = 0.774$$

The maximum power is given by

$$P'_{max} = V'_{oc}I'_{sc}FF = 0.54 \times 3.364 \times 0.774 = 1.406 \text{ W}$$

The energy conversion efficiency is

$$\eta' = \frac{P'_{max}}{I_S A} = \frac{1.406}{1000\times100\times10^{-4}} = 14.06\%$$

Example 5.13 A silicon solar cell of area 100 cm^2 is operating at a temperature of 298 K and an irradiation level of 1000 Wm^{-2}. For these conditions, $V_{OC} = 600$ mV, $I_{SC} = 3.3$ A. The ideality factor is 1.2. The cell has a series resistance of 0.1 ohms and a shunt resistance of 3 ohms. Obtain the *I-V* characteristic of the cell.

Solution The characteristic equation (5.41) of the non-ideal solar cell including series and shunt resistances is

$$I = I_L - I_o\left[exp\left[\frac{q(V+IR_S)}{nk_BT}\right] - 1\right] - \frac{(V+IR_S)}{R_{sh}} \qquad \text{(E5.13.1)}$$

where $n = 1.2$ is the *ideality factor*.

The light current I_L and the reverse saturated current I_o are unknown. We substitute the short circuit and open circuit conditions in Eq. (E5.13.1) to obtain the following relations.

The short circuit current, when $V = 0$

$$I_{sc} = I_L - I_o\left[exp\left[\frac{q(I_{SC}R_S)}{nk_BT}\right] - 1\right] - \frac{(I_{SC}R_S)}{R_{sh}} \qquad \text{(E5.13.2)}$$

The open circuit voltage, when $I = 0$

$$I_L = I_o\left[exp\left[\frac{q(V_{OC})}{nk_BT}\right] - 1\right] + \frac{V_{OC}}{R_{sh}} \qquad \text{(E5.13.3)}$$

Define the thermal voltage as

$$v_{th} = \left(\frac{nk_BT}{q}\right) = \frac{1.2\times1.381\times10^{-23}\times298}{1.602\times10^{-19}} = 0.03083 \text{ V}$$

Substituting the given numerical values and conditions:
$V_{OC} = 600$ mV, $I_{SC} = 3.3$ A, $n = 2$, $R_{sh} = 3\Omega$ and $R_s = 0.1\Omega$ in Eqs. (E5.13.2) and (E5.13.3) we obtain the following simultaneous equations

$$I_L - 4.459\times10^4 I_o = 3.41$$

$$I_L - 2.838\times10^8 I_o = 0.2$$

Solve the above two equations simultaneously to get I_L and I_o. Hence we have

$$I_L = 3.409 \text{ A} \quad \text{and} \quad I_o = 1.1309\times10^{-8} \text{ A}$$

These values are then substituted in Eq. (E5.13.1) to obtain the non-ideal *I-V* characteristic of the cell including parasitic resistances. This gives

$$I = 3.409 - 1.1309\times10^{-8}[exp[32.44(V + 0.1I)] - 1] - \frac{(V+0.1I)}{3.0} \qquad \text{(E5.13.4)}$$

Points on the *I-V* curve have to be obtained by solving the above non-linear equation iteratively. For each chosen value of V on the *I-V* curve we obtain the value of I by solving Eq. (E5.13.4) by trial and error. Hence we locate the maximum power point on the *I-V* curve.

Example 5.14 A multijunction junction solar cell is made using three layers of semiconductors as shown schematically in Fig. 5.9(a). The band gaps of the cells A, B and C are 1.86 ev, 1.44 ev, and 0.65 ev respectively. The cell is tested under a simulated solar spectrum, which can be approximated by the rectangular shape depicted in Fig. E5.2(b). The external quantum efficiency of each layer is 0.8. Calculate the short circuit currents, and the power outputs of the cells A, B and C.

Solution The operation of the multijunction cell was described in section 5.9. The photon flux is given by

$$\phi_{ph} d\lambda = \frac{I_{e\lambda}\lambda d\lambda}{hc}$$

The number of the electron-hole pairs generated by the photon flux is

$$e_\lambda d\lambda = \frac{EQE(\lambda) I_{e\lambda}\lambda d\lambda}{hc} \quad \text{(E5.14.1)}$$

The resulting short circuit current is given by

$$J_{sc} = \int_{\lambda_o}^{\lambda_G} \frac{qEQE(\lambda) I_{e\lambda}\lambda d\lambda}{hc} \quad \text{(E5.14.2)}$$

The band gap wavelength of layer A is given by

$$\lambda_g = \frac{hc}{BG} = \frac{6.625 \times 10^{-34} \times 2.998 \times 10^{8}}{1.86 \times 1.6 \times 10^{-19}} = 0.667\ \mu m$$

Similarly, the band gap wavelengths of cells B and C are 0.862 μm and 1.91 μm. Note that the band gap wavelengths of all layers are within the range ($0.25\ \mu m < \lambda < 2.25\ \mu m$) of the rectangular spectrum.

The integration in Eq. (E5.14.2) can be performed conveniently because of the rectangular shape of the radiation spectrum in Fig. E5.2(b). Hence we have

$$J_{sc,A} = \frac{1.602 \times 10^{-19} \times 0.8}{6.625 \times 10^{-34} \times 2.998 \times 10^{8}} \int_{0.25}^{0.667} 0.5 \times 10^{9} \lambda d\lambda$$

$$J_{sc,A} = 6.45 \times 10^{5} \times 0.0956 \times 10^{-12} = 0.617 \times 10^{-7}\ \text{Am}^{-2}$$

Similarly, for layers B and C,

$$J_{sc,B} = 0.481 \times 10^{-7}\ \text{Am}^{-2}$$

$$J_{sc,C} = 4.68 \times 10^{-7}\ \text{Am}^{-2}$$

The power output of a cell is given by

$$E = BG \int_{\lambda_o}^{\lambda_G} \frac{EQE(\lambda) I_{e\lambda} \lambda d\lambda}{hc} = \int_{\lambda_o}^{\lambda_G} \frac{EQE(\lambda) I_{e\lambda} \lambda d\lambda}{\lambda_{BG}}$$

$$E_A = \frac{0.8}{0.667 \times 10^{-6}} \int_{0.25}^{0.667} 0.5 \times 10^9 \lambda d\lambda$$

$$E_A = 1.2 \times 10^6 \times 1.0 \times 10^9 \times 0.0955 \times 10^{-12} = 115 \text{ Wm}^{-2}.$$

Similarly for layers B and C,

$$E_B = 69 \text{ Wm}^{-2}, \qquad E_C = 304 \text{ Wm}^{-2}$$

Total power output is 488 Wm^{-2}. The total incident energy is 1000 Wm^{-2}. Hence the overall efficiency is 48.8%.

Example 5.15 In a direct PV system the output is supplied to a load resistance of 20 Ω. The light current and the saturation current of the PV module are 13.6A and 0.008A respectively. The thermal voltage,

$$\left(\frac{nK_BT}{q}\right) = 23.6 \text{ V},$$

and the series resistance is 0.9 Ω. The shunt resistance is very large. Calculate the power delivered to the load.

Solution The characteristics of a solar panel and a resistive load are depicted in Fig. 5.11(b). The operating point of the system is given by intersection of these the two curves. The PV characteristic of a solar cell with a series resistance is shown in Fig. 5.6(b). The $I - V$ characteristic of the module is obtained from Eq. (5.41) by substituting $R_{SH} \to \infty$. Hence we get

$$I = I_L - I_o \left[exp \left[\frac{q(V + IR_S)}{nk_BT} \right] - 1 \right]$$

$$I = 13.6 - 0.008 \left[exp \left[\frac{(20 + 0.9)I}{23.6} \right] - 1 \right] \qquad \text{(E5.15.1)}$$

For the resistive load

$$V = 20I \qquad \text{(E5.15.2)}$$

Substituting for V from Eq. (E.15.2) in (E5.15.1) we have

$$13.6 - I = 0.008exp\left[\frac{(20+0.9)I}{23.6}\right] = 0.008exp(0.886I)$$

We solve the above equation by trial and error.

Hence we get $I = 7.5$ A.

The power input to the resistance is

$$P = I^2R = 20 \times 7.5^2 = 1125 \text{ W}$$

Example 5.16 A home at a location with 5 hours of equivalent sun-hours (*ESH*) has a total daily load of 1820 Wh. The PV panels to be installed at the home have an output of 50 W at 12 V. The system voltage is 24 V, and the voltage and capacity of the batteries are 6 V and 125 Ah respectively. The battery system is to have 3 days of autonomy and a discharge limit of 0.7. The charge controller has an efficiency of 96%. Using the heuristic approach specify the numbers of the PV panels and batteries required. How should the panels be arranged?

Solution A schematic diagram of a stand-alone PV system is shown in Fig. 5.12(a). The average daily load is 1820 Wh and the number of equivalent sun-hours is 5. The efficiency of the charge controller is 0.96. Hence the required wattage of the panel is

$$E = \frac{1820}{5\times0.96} = 379 \text{ W}$$

The number of panels required is

$$n_p = \frac{379}{50} = 7.58$$

The system voltage is 24 V and the PV module is 12 V, therefore the number of panels in series per string is 2. Assuming the total number of panels as 8, the number of strings in parallel is 4.

The current in a panel is $\frac{50}{12} = 4.16$ A. Current in a string is 4.16 A. Total current is:

$$I_{tot} = 4 \times 4.16 = 16.64 \text{ A}$$

The daily load is 1820 Wh. For 3 days of autonomy, and a discharge limit of 0.7, the energy to be stored in the batteries is:

$$E_b = 3 \times \frac{1820}{0.7} = 7800 \text{ Wh}.$$

The batteries used are 6 V and125 Ah. Therefore the number of batteries required is:

$$n_b = \frac{7800}{(6 \times 125)} = 10.4$$

The system voltage is 24 V, and the battery voltage is 6V. The number of batteries in series is, $\frac{24}{6} = 4$. The number of batteries in parallel is:

$$\frac{10.4}{4} = 2.6$$

Therefore the number of batteries in parallel is 3.

Table E5.17.1 Capacities and run times of equipment

Equipment	Capacity (W)	Run time (day) (h)	Run time (night) (h)
Lighting (DC)	30	2	4
Refrigerator (DC)	200	1	1
Water pump (DC)	500	1.5	0.5
Computer (AC)	250	4	0
Printer (AC)	175	0.5	0
Kettle (AC)	1200	0.25	0.25

Equipment	Load (day) (Wh)	Load (night) (Wh)	Load (total) (Wh)
Lighting (DC)	60	120	180 (DC)
Refrigerator (DC)	200	200	400 (DC)
Water pump (DC)	750	250	1000 (DC)
Computer (AC)	1000	0	1000 (AC)
Printer (AC)	87.5	0	87.5 (AC)
Kettle (AC)	300	300	600 (AC)

Example 5.17 The capacities and run times of the DC and AC equipment used in a small house are listed in Table E5.17.1. The location receives 5.5 equivalent sun-hours (*ESH*) per day. (a) Calculate the daily daytime and nighttime DC and AC loads. (b) Assuming that all loads are DC, calculate the number of 50W PV panels required if the distribution system efficiency is 96%, and the battery system used is 75% efficient. (c) Calculate the size of the inverter required. (d) If an inverter with an efficiency of 88% is used, calculate the daily energy required from the PV panel.

Solution (a) The day loads and night loads are given by:

$$L = capacity \times run\ time$$

The computed loads for day and night are given in Table E5.17.1 above. The total load is the sum of the day and night loads.

(b) If all the loads are DC then the total daily load is 3267.5 Wh. If no distinction is made between the combined efficiencies for daytime and nighttime loads, then the total PV load is

$$\frac{3267.5}{0.96 \times 0.75} = 4538.2 \text{ Wh.}$$

The watts required from the PV system is

$$\frac{4538.2}{5.5} = 825.13 \text{ W.}$$

The number of PV panels required is

$$\frac{825.13}{50} = 17 \text{ panels.}$$

(c) The maximum AC load is

$$(250 + 175 + 1200) = 1625 \text{ W}$$

This is the minimum size of the inverter required.

(d) Consider the DC and AC loads separately, with no distinction between daytime and nighttime loads. Then the total DC load to be supplied is

$$\frac{1580}{0.96 \times 0.75} = 2194.4 \text{ Wh.}$$

The total AC load to be supplied is

$$\frac{1687.5}{0.96 \times 0.88 \times 0.75} = 2663.35 \text{ Wh.}$$

Therefore total load to be supplied by the PV system is

$$(2192.4 + 2663.35) = 4855.75 \text{ Wh.}$$

The watts required is

$$\frac{4855.75}{5.5} = 882.86 \text{ W.}$$

The number of 50W PV panels required is

$$\frac{882.86}{50} = 18$$

Example 5.18 A PV system of area 20 m^2 faces south and is inclined at 40°, at a location with a latitude of 40° N. The following design data are available:

For the PV system: $U_l = 20$ Wm^{2}K^{-1}, $\alpha = 0.8$, $\tau = 0.96$, $\eta_r = 0.15$, $\beta = 0.0045$ C^{-1}

For the location: $\bar{K}_T = 0.505$, $a_1 = 0.219$, $a_2 = 0.832$, $I_s = 5.2$ kWh m^{-2} per day, $T_m = 16.2$°C, $T_r = 20$°C.

Calculate the total output of the system for the month of June.

Solution For the month of June, we obtain the optimum tilt angle from data in Table 5.2 as

$$\theta_{opt} = L - 25 = 40 - 25 = 15^{\circ}$$

$$\theta_{opt} - \theta_{act} = 15 - 40 = -25^{\circ}$$

Hence the tilt correction factor is given by Eq. (5.63) as:

$$C_f = 1 - 0.000117\left(\theta_{opt} - \theta_{act}\right)^2 = 0.927$$

Equation (5.61) for energy balance gives:

$$\left(\frac{U_l}{\tau\alpha}\right)(T_c - T_a) = \left(\frac{\sum I_i^2}{\sum I_i}\right) = a_1 + a_2\bar{K}_T$$

$$a_1 + a_2\bar{K}_T = 0.219 + 0.832 \times 0.55 = 0.676$$

Substituting numerical data in the two equations above we get

$$(T_c - T_a) = \frac{0.676 \times 0.96 \times 0.8 \times 0.927}{0.02} = 24.06^{\circ}\text{C}$$

Assume that

$$T_a - T_m = 3 \text{ K}, \qquad T_m - T_r = 16.2 - 20 = -3.8 \text{ K}.$$

The monthly mean efficiency is given by Eq. (5.55) as:

$$\eta = \eta_r[1 - \beta(T_c - T_a) - \beta(T_a - T_m) - \beta(T_m - T_r) + \mu ln(I)]$$

Substituting numerical values and neglecting the last term in the above equation we have

$$\eta = 0.15[1 - 0.0045(24.06 + 3 - 3.8)] = 0.134$$

Hence the energy output for the month of June is

$$E_m = \eta N_{dm} I_s A$$

$$E_{June} = 0.134 \times 30 \times 5.2 \times 20 = 418 \text{ kWh}$$

Note that in the actual design the correction factor given in Eq. (5.49) has to be applied to obtain the final estimate of the power.

Example 5.19 A south- facing PV array consisting of 45 panels, each of area 1.24 m^2, is installed at a location with a latitude of 40° N. The tilt angle is 35°.

Data for a panel are as follows: rated efficiency = 18%, rated temperature = 25°C, temperature coefficient of efficiency, $\beta = 0.0037$ C^{-1}, heat loss coefficient = 20 Wm^{-2}K^{-1}, $\tau\alpha = 0.84$.

Design data for the location are: $\mu \approx 0$, $(T_a - T_m) = 3$°C, $a_1 = 0.219$, $a_2 = 0.832$.

Weather data for the location for three months are tabulated below. Estimate the monthly energy output of the system.

Variable	January	March	June
$\overline{K}_T$	0.36	0.46	0.542
Daily Insolation, MJ m^{-2}	4.85	12.18	23.4
T_a °C	-2.4	2.5	21.2
Optimum tilt, deg.	L+29	L+3	L-25

Solution The tilt angle for January is

$$\theta_{opt} = L + 29 = 40 + 29 = 69^{\circ}$$

$$\theta_{opt} - \theta_{act} = 69 - 35 = 34^{\circ}$$

The tilt correction factor is given by Eq. (5.61) as:

$$C_f = 1 - 0.000117(34)^2 = 0.865$$

The energy balance Eq. (5.61) gives:

$$\left(\frac{U_l}{\tau\alpha}\right)(T_c - T_a) = \left(\frac{\sum I_i^2}{\sum I_i}\right) = a_1 + a_2\overline{K}_T$$

$$a_1 + a_2\overline{K}_T = 0.219 + 0.832 \times 0.36 = 0.518$$

$$(T_c - T_a) = \frac{0.518 \times 0.84 \times 0.865}{0.02} = 18.81^{\circ}\text{C}$$

Assume that

$$T_a - T_m = 3\text{ K}, \qquad T_m - T_r = -5.4 - 25 = -30.4\text{ K}.$$

The monthly mean efficiency is given by Eq. (5.55) as

$$\eta = \eta_r[1 - \beta(T_c - T_a) - \beta(T_a - T_m) - \beta(T_m - T_r) + \mu ln(I)]$$

Substituting numerical values and neglecting the last term we have

$$\eta = 0.18 \times [1 - 0.0037(18.81 + 3 - 30.4)] = 0.186$$

Hence the energy output for the month of January is

$$E = AH_m\eta N_{dm} = 45 \times 1.24 \times 4.85 \times 0.186 \times 31 = 1560\text{ MJ}$$

Using the same procedure for the months of March and June we obtain the following results.

For March: $\theta_{opt} = 43^o$, $C_f = 0.999$, $(T_c - T_a) = 25.24$ K,

$(T_a - T_m) = 3$ K, $(T_m - T_r) = -25.5$ K, $\eta = 0.178$,

$E = 3750.3$ MJ

For June: $\theta_{opt} = 15^o$, $C_f = 0.953$, $(T_c - T_a) = 26.8$ K,

$(T_a - T_m) = 3$ K, $(T_m - T_r) = -6.8$ K, $\eta = 0.165$,

$E = 6463.3$ MJ

Example 5.20 The annual energy output of a solar PV system of area 60 m^2 has been estimated to be 10580 kWh using detailed monthly insolation data. The rated capacity of the system is 9.8 kW and the life time is 25 years. The initial costs are as follows: solar panels = 1.5 \$ / watt, inverter = 0.45 \$ / watt, hardware = 0.4 \$ / watt, labor = 0.55 \$ / watt, professional services = 0.25 \$ / watt. Assume a minimum rate of return of 0.05. Calculate (i) the capacity factor, and (ii) the levelized cost of energy.

Solution (i) The maximum possible output of the system is given by

$$P_{max} = 9.8 \times 24 \times 365 = 85848 \text{ kWh}$$

Actual estimated output, $P_{act} = 10580$ kWh

The capacity factor is given by

$$CF = \frac{P_{act}}{P_{max}} = \frac{10580}{85848} = 0.12$$

(ii) The total initial cost of the system is given by

$$C_{in} = (1.5 + 0.45 + 0.4 + 0.55 + 0.25) \times 9800 = \$\,30870$$

The annualized cost is

$$A = P\left[\frac{i(1+i)^N}{(1+i)^N - 1}\right] = 30870\left[\frac{0.05 \times 1.05^{25}}{1.05^{25} - 1}\right] = 2190.3$$

Additional annul costs, dependent on location of plant, include maintenance costs, expected replacement costs, prevailing incentive schemes, taxes and others. Neglecting these costs we obtain the levelized cost of energy, *LCOE* as

$$LCOE = \frac{annualized\ cost}{annual\ energy\ output} = \frac{2190.3}{10580} = 0.207\ \$\ \text{per kWh}$$

Number of tools for cost evaluation of engineering projects are presented in Chapter 1.

Problems

P5.1 Monochromatic light of wavelength 0.72 μm and intensity 2.5 mW is incident on a GaAs solar cell with a band gap energy of 1.4 ev. Calculate (i) the wavelength of radiation with photon energy equal to the band gap of GaAs, (ii) energy per photon of incident light, (iii) the photon flux, (iv) the maximum power available, and (v) the maximum percentage of incident power that can be converted to electricity. [*Answers*: (i) 0.887 μm, (ii) 2.76×10^{-19} J, (iii) 9.057×10^{15} s^{-1}, (iv) 2.029 mW, (v) 81%]

P5.2 Unpolarized light of intensity 800 Wm^{-2} and wavelength 0.4 μm is incident at right angles on a silicon wafer of thickness 250 μm. The absorption coefficient of Si is 1.11×10^6 m^{-1} and the refractive index is 4.29. Calculate (i) the fraction of incident light reflected at the surface, (ii) the total number of photons absorbed in Si, and (iii) the photocurrent density generated.
[*Answers*: (i) 0.386, (ii) 9.89×10^{20} m^{-2}s^{-1}, (iii) 158.4 Am^{-2}]

P5.3 Monochromatic light of wavelength 0.75 μm and intensity 280 Wm^{-2} is incident uniformly on a silicon solar cell of area 8 cm^2 and band gap 1.12 ev. The quantum efficiency at this wavelength is 0.75. Calculate (i) the photon flux, (ii) the electric power generated, and (iii) the short circuit current of the cell.
[*Answers*: (i) 10.57×10^{20}s^{-1}m^{-2}, (ii) 0.114 W, (iii) 10.16×10^{-2} A]

P5.4 Monochromatic light of intensity 300 Wm^{-2} and wavelength 0.65 μm is incident on a silicon solar cell with a band gap of 1.12 ev and area 8 cm^2. The dark saturation current density is 1.1×10^{-11} A. The cell temperature is 25°C. The quantum efficiency is 0.9 and the ideality factor is 1. Calculate (i) the short circuit current, (ii) open circuit voltage, (iii) maximum power, (iv) the fill factor, and (v) the conversion efficiency.
[*Answers*: (i) 0.113 A, (ii) 0.592 V, (iii) 0.0553 W, (iv) 0.827, (v) 23%]

P 5.5 From the measured *I-V* data of a PV module of area 0.16 m^2, the following information has been obtained, $V_{OC} = 0.7$ V, $I_{SC} = 42$A. The cell temperature is 25°C and the ideality factor is 1.2. A 0.02 ohm resistor is connected across the PV module. (a) Obtain the reverse saturation current, and the characteristic equation of the module. (b) Calculate the power delivered to the resistor.
[*Answers*: (a) 5.66×10^{-9}A, (b) 21.35W]

P5.6 The following data has been obtained for a silicon solar cell operating at a temperature of 30°C and an insolation level of 900 Wm^{-2}: $V_{OC} = 0.58$V, $I_{SC} = 3.1$ A, ideality factor = 1.2. The area of the cell 0.01 m^2. (i) Using the simplified empirical relation for the fill-factor, calculate the maximum conversion efficiency of the cell under the given conditions. (ii) If the operating temperature changes to 50°C, calculate the conversion efficiency of the cell. The temperature coefficients for V_{OC} and I_{SC} are -0.35% and 0.07% per °C respectively.
[*Answers*: (i) 15.95%, (ii) 14.6%]

P5.7 A silicon solar cell of area of 100 cm^2 is operating at a temperature of 303 K and an irradiation level of 1000 Wm^{-2}. For these conditions, $V_{OC} = 0.55$ V, $I_{SC} = 3.2$ A. The ideality factor is 1.4.

The cell has a series resistance of 0.12 Ohms and the shunt resistance is very large. (a) Find (i) the saturation current, (ii) the light current, and (iii) the maximum power conditions. (b) If the series resistance is zero and the ideality factor is one, calculate the maximum power conditions.
[*Answers*: (a) (i) 9.63×10^{-7}A, (ii) 3.23 A, (iii) 1.935 A, 0.284 V; (b) 3.03 A, 0.473 V]

P5.8 A stand-alone PV system is to be installed in a small house with the DC and AC equipment listed in Table P5.8. The capacities and the day and night run times of the equipment used are tabulated. The location receives 5.5 equivalent sun-hours (*ESH*) per day.

The battery system is to have 3 days of autonomy and a discharge limit of 0.7. The system voltage is 24 V, and the voltage and capacity of the batteries used are 6V and 125Ah respectively.

(a) Calculate the daily daytime and nighttime DC and AC loads. (b) Assuming that all loads are DC, calculate the number of 50 W, 12 V PV panels required if the distribution system efficiency is 95%, and the battery system used is 75% efficient. (c) Calculate the size of the inverter required.

(d) If an inverter with an efficiency of 85% is used, find the number of PV panels required. How should the PV panels be arranged? (e) Calculate the number of batteries required. (f) Draw a block diagram to show the energy flows from the incident solar radiation to the various loads. [*Answers*: (b) 108, (c) 1600 W, (d) 118, (e) 50]

Table P5.8 Capacities and run times of equipment

Equipment	Capacity (W)	Run time (day) (h)	Run time (night) (h)
Lighting (DC)	40	3	5
Refrigerator (DC)	250	2	1.5
Water pump (DC)	600	1.5	0.5
Computer (AC)	225	5	2
Printer (AC)	175	1.0	1.0
Kettle (AC)	1200	0.25	0.25

P5.9 A south-facing PV system of area 20 m^2 is inclined at 30° at a location with a latitude of 35° N. The following data for August are available: $U_l = 20$ Wm^{2}K^{-1}, $\alpha = 0.85$, $\tau = 0.95$, $\eta_r = 0.14$, $\beta = 0.0043$ C^{-1}, $\bar{K}_T = 0.54$, $a_1 = 0.219$, $a_2 = 0.832$, $\mu = 0$, $I_s =$ 22.2 MJ m^{-2} per day, $T_a = 23.2$°C, $T_r = 25$°C. Calculate the total output of the system for the month of August. [*Answer*: 1720 MJ]

P5.10 A 750 kW solar PV power plant costs \$ 1.1 million to install. The yearly operating and maintenance cost is \$ 10,500. Assume a capacity factor of 0.17 for the location and a discount rate of 8%. The life-time of the plant is 25 years. (a) Calculate the levelized cost of energy for this system. (b) Perform a sensitivity evaluation by changing (i) the discount rate by ±15%, and (ii) the capacity factor by ±15%. [*Answer*: (a) $101.7 per MWh]

Reference

1. Duffie, John A., Beckman, William A. with Blair, Nathan., *Solar Engineering of Thermal Processes, Photovoltaics and Wind, Fifth Edition*, John Wiley & Sons Inc. New Jersey, 2020.
2. Evans, D.L., Simplified method for predicting photovoltaic array output, *Solar Energy*, Vol. 27, No. 6, (1981), 555-560.
3. Hersch, Paul and Zweibel, Kenneth., *Basic Photovoltaic Principles and Methods*, *SERI/SP*-290-1448, Technical Information Office, Solar Energy Research Institute, Golden, Colorado,1982.
4. Jenkins, Nick., and Ekanayake, Janaka., *Renewable Energy Engineering*, Cambridge University Press, Cambridge, 2017.
5. Kreith, Frank and Krumdieck, Susan., *Principles of Sustainable Energy Systems*, CRC Press, Taylor & Francis Group, New York, 2015.
6. Smets Arno HM, Jager, Klaus, Isabella, Olindo, van Seaaij, Rane ACMM and Zeman, Miro., *Solar Energy: The physics and engineering of photovoltaic conversion, technologies and systems*, UIT Cambridge Ltd, Cambridge, 2016.
7. Vanek, Francis M., Albright, Louis D. and Angenent, Largus T., *Energy Systems Engineering: Evaluation and Implementation*, McGraw-Hill Book Company, New York, 2012.
8. 'List of Photovoltaic Power Stations, Wikipedia', [https://en.wikipedia.org/wiki/List of_photovoltaic power stations].
9. http://news.energysage.com/solar-panel-efficiency-cost-over-time/.
10. http://www.lazard.com/media/450773/lazard-levelized-cost-of-energy-version-120-vfinal.pdf.
11. http//news.energysage.com/how-much-does-the-average-solar-panel—installation-cost-in-the-u-s-/.
12. http://www.ucsusa.org/clean-energy/renewable-energy/environmental-impacts.

Chapter 6

Wind Power Generation

6.1 Introduction

The movement of air across the surface of the earth, caused by areas of high pressure and low pressure, is called wind. Incoming solar radiation interacts with the atmosphere and the surface of the earth to create a thermal engine, absorbing heat at a higher temperature, and releasing heat at a lower temperature. This temperature difference drives the winds, maintaining circulation of the atmosphere against frictional dissipation [1,5].

Waves in water bodies like lakes, seas and oceans are most caused by wind. Friction between the wind and the water surface creates wind-driven waves or surface waves. The kinetic energy of wind can be converted to useful mechanical energy with the aid of wind turbines, which in turn can drive generators to produce electricity.

Historically, mechanical devices that enabled the use of wind energy for human purposes existed in Persia around 900-1000 AD [5]. In the seventh century, early versions of vertical axis windmills were used in Afghanistan for milling grain and drawing water. Horizontal axis windmills were developed in Northwest Europe, in the eleventh and twelfth centuries, mainly to mill grain. During the period from the twelfth to the nineteenth centuries, the Dutch built many windmills to move water. Until the industrial revolution, that started in the eighteenth century, wind remained the major source of energy in Europe. With the advent of steam power in the 1800s, wind energy receded in importance, but by that time wind turbine design in Europe had reached a high level of sophistication.

Later attempts to use wind energy for the generation of electricity was motivated by the availability of small electrical generators. These small wind power systems were used mainly in remote locations that were not served by the local electricity grid. In the late 1960s large wind turbines for the generation of electricity re-emerged as an important alternative due to concerns over the environmental effects of fossil fuels. This development has continued up to the present time, and the installation of large wind energy systems has grown rapidly. A detailed account of the historical development of wind energy is available in Ref.[5].

6.2 Types of Wind Turbines

There are two basic types of wind turbines called horizontal axis wind turbines (HAWT) and vertical axis wind turbines (VAWT). The size of wind turbines varies widely depending on the application. Small wind mills are used in remote areas to pump water, and wind turbines with an electricity generating capacity of about 10 kW are in use to power single homes. The largest wind turbines in operation have power ratings of up to 8 MW, and larger wind turbines are in development. A list of the largest onshore wind power systems in operation is given in Table 6.1[7].

6.2.1 *Horizontal axis wind turbines*

The most common type of modern wind turbine, shown schematically in Fig. 6.1(a), is the horizontal axis wind turbine (HAWT). Although HAWTs with one, two, three and four blades have been designed and tested, the most widely installed turbines have three blades. Blades are attached to a central hub which is connected to a drive shaft that transmits power from the turbine to the electric generator, through a gear train. The gear train increases the speed of the shaft to match the required electric generator speed. The generator and the gear box are housed in an enclosure called the *nacelle* that sits atop the turbine tower. The nacelle is rotated about a vertical axis by a 'yaw mechanism' to ensure that the turbine faces the wind at all times. Power is transmitted to a transformer

Table 6.1 Some of the largest onshore wind farms (Data from Ref.[7])

Installation	Country	State	Location	Rating (MW)
Gansu Wind Farm	China	Gansu	40°N	7975
Atla Wind Energy Center	USA	California	40°N 118E	1548
Muppandal Wind Farm	India	Tamil Nadu	40°N 118E	1500
Jaisalmer Wind Park	India	Rajasthan	27°N 70E	1064
Shepherds Flat Wind Farm	USA	Oregon	46°N 120W	845
Roscoe Wind Farm	USA	Texas	32°N 100W	781.5
Horse Hollow Wind Energy Center	USA	Texas	32°N 100W	735.5
Capricorn Ridge Wind Farm	USA	Texas	32°N 100W	662.5
Fantanele-Cogealac Wind Farm	Romania	Constanta	44°N 29W	600
Fowler Ridge Wind Farm	USA	Indiana	40°N 87W	599.8

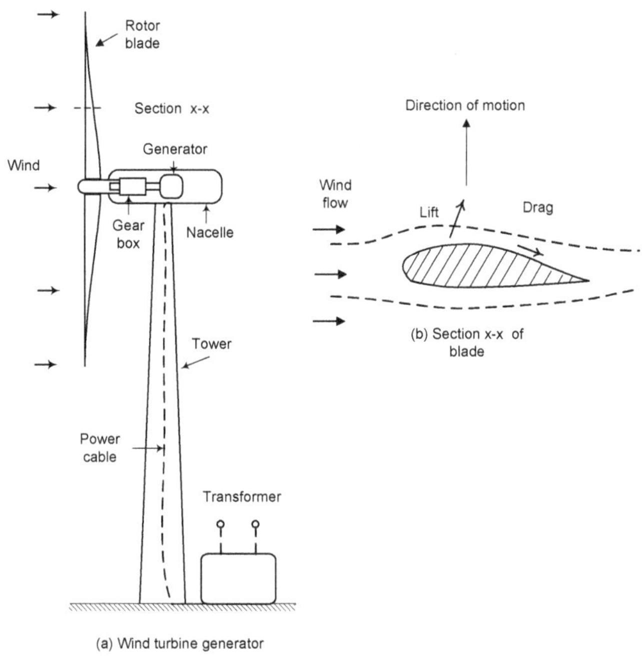

Fig. 6.1 (a) Horizontal axis wind turbine-generator, (b) Forces on rotor blade

at the base of the tower through a cable. When possible, controls and monitors are installed inside the base of the tower because it is difficult to access the nacelle.

Since a tower produces turbulence behind it, the turbine is pointed upwind of the tower, and located a considerable distance in front. The main advantages of HAWTs are: (i) The tall tower base allows access to stronger winds in sites with high wind shear or wind speed gradient. In some sites wind speed can change by 20% with every ten meters change in elevation. (ii) Since the blades are facing the wind continuously, they capture the available power with high efficiency.

However, HAWTs do have some disadvantages. These are: (i) The tall tower must support the heavy weight of the blades, the gear box, and the generator, and during operation the tower is subject to strong bending moments. (ii) A yaw control mechanism is required to turn the blades toward the wind. (iii) Breaking is required during high winds to prevent the turbine from spinning out of control and damaging itself.

Blade cross sections of modern HAWTs are shaped like airplane wings as seen in Fig. 6.1(b). One surface of the section is more curved than the other. As air flows over the curved side the streamline are squeezed, and consequently the air speed is higher over this surface. The increased speed results in a lower pressure on the surface. The pressure on the flatter surface remains nearly unchanged. The difference in pressure between the two surfaces generates a net force perpendicular to the relative wind flow direction, called the *lift force*. The force normal to the lift force is called the *drag force*. The component of the lift force acting in the direction of rotation of the blade produces the torque that drives the shaft of the turbine.

6.2.2 *Vertical axis wind turbines*

Two common designs of vertical axis wind turbines (VAWT) are depicted in Fig. 6.2. The Savonius turbine [1,5] shown in Fig. 6.2(a) is one of the simplest wind turbines. Its design can be visualized as a hollow cylinder cut in half along the axis, and displaced along the diameter. This produces an s-shaped cross section, and two open

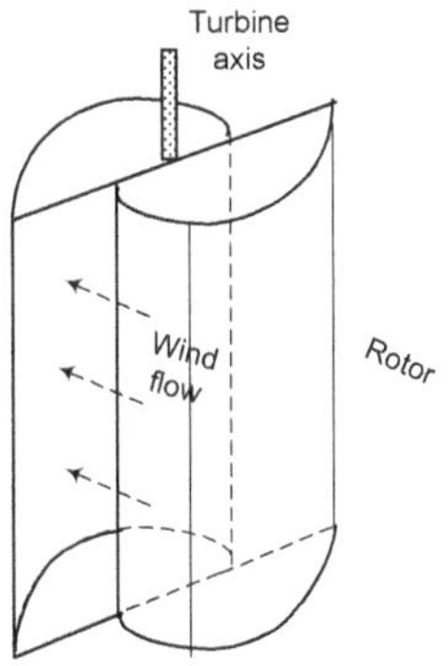

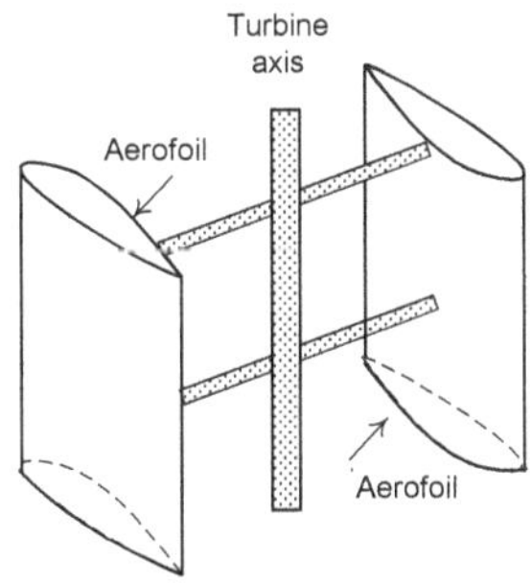

(b) Darrieus Wind Turbine

(a) Savonius Wind Turbine

Fig. 6.2 Vertical axis wind mills, (a) Savonius rotor, (b) Darrieus rotor

rectangular sides face the wind. The turbine is mounted symmetrically on a vertical axis parallel to the height. Wind enters through one of the open sides and exits through the other. This is called a *drag-type* device because the force causing it to rotate about the axis is in the same direction as the air flow through it. Many variations of this basic design have been built and tested. Savonius turbines have reached efficiencies up to about 30%. These turbines are useful whenever cost and reliability are more important than efficiency. For this reason, most wind anemometers are Savonius turbines.

A common VAWT with airfoil shaped blades is the Darrieus wind turbine [5], shown in Fig. 6.2(b). When the turbine is rotating, the air foils move forward through the air in a circular path. For a fixed wind direction, the wind speed relative to a blade changes continuously as it rotates about the vertical axis. Hence the lift force on the blade, and the resulting driving torque on the shaft, are variable. When the turbine is stationary, no net rotational force arises even with high wind speeds, and therefore the turbine is not self-starting. Unlike, HAWTs, the power extraction system including the gear train and the generator of VAWTs can be located at the bottom of the vertical shaft.

Some of the advantages of VAWTs are: (i) they can generate power in any wind direction, (ii) a strong tower is not needed because the gearbox and generator are placed on the ground, (iii) easy to transport and install compared to HAWTs, and (iv) maintenance costs are generally lower.

The disadvantages of VAWTs are: (i) their efficiency is low compared to HAWTs, (ii) they are usually mounted close to the ground, where wind speeds are generally lower, (iii) they are not self-starting, so require some external starting torque, and (iv) may require guy wires to hold it up vertically.

6.3 Power Available in the Wind

Consider the air in an imaginary cylinder of cross sectional area A, equal to area swept by the wind turbine, as shown in Fig. 6.1(a) and Fig. 6.9(a). If the average air speed in the axial direction is U, then the mass flow rate of air is

$$\dot{m} = A\rho U \tag{6.1}$$

where ρ is the density of air.

The kinetic energy *per unit mass* of air is given by

$$E_{ke} = \frac{1}{2}U^2 \tag{6.2}$$

The *power* available in the wind is the product of the mass flow rate and the kinetic energy per unit mass. Hence we have

$$P_{wind} = \dot{m}E_{ke} = \frac{1}{2}A\rho U^3 \tag{6.3}$$

For a wind turbine with blades of radius R, the above expression becomes

$$P_{wind} = \frac{1}{2}\rho\pi R^2 U^3 \tag{6.4}$$

We observe form Eq. (6.4) that the wind power available to a turbine can be increased by having larger blades, and locating the turbine at the top of a tall tower where the average wind speed is higher. However, this would increase the cost of manufacturing the turbine, and constructing a stronger tower that could withstand the larger forces involved [3].

6.4 Power Output Curve of a Wind Turbine

The power available in the wind at any speed is given by Eq. (6.4). However, only a fraction of this power could be extracted by a wind turbine. The ratio of the power output of the turbine rotor to the power available in the wind is called the *power coefficient*. It is defined as

$$C_p = \frac{P_{rotor}}{\frac{1}{2}\rho A U^3} \tag{6.5}$$

The power output of a turbine-generator system (see Fig. 6.1a) can be written in the form

$$P_{out}(U) = \frac{1}{2}\rho A \eta C_p(U) U^3 \tag{6.6}$$

where the effficiency η of the turbine-generator is defined as the ratio of the electric generator power output to the turbine rotor power [5].

The power output curve of a wind turbine, $P_w(U)$ is a plot of the variation of rotor power output with wind speed. A typical power curve [6] for a large turbine is depicted in Fig. 6.3. Usually power curves of this type are based on test data, and are supplied by turbine manufacturers.

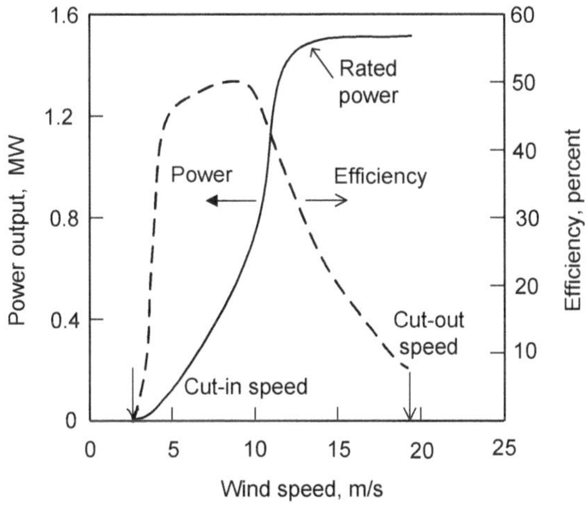

Fig. 6.3 Power output and efficiency of a 1.5 MW wind turbine, (Data from Ref.[6])

Until the *cut-in* wind speed is reached no power could be extracted because of the losses and the inertia of the system. Beyond the cut-in speed the power output increases steadily with wind speed.

Beyond the rated speed, the turbine power output is maintained constant at the rated power by installing appropriate control stratagies. The *cut-out* speed is the wind speed beyond which the cost of increasing the structural strength needed to deal with higher speeds,exceeds the cost of the additional energy produced. Once the cut-out speed is reached the turbine is shut down to prevent damage to the system.

The efficiency of the turbine rises sharply, initially, and then drops off more gradually, reaching a maximum close to the rated conditions of the turbine, as seen in Fig. 6.3. Ideally, the efficiency of the turbine is maximum at the rated power point.

The yearly energy output of a wind turbine is obtained by combining the power curve of the turbine, that is, the variation of the power output with wind speed, and the yearly wind speed distribution at the location. Due to the variability of wind speed, a turbine's annual energy production is not equal to the 'nameplate rating' multiplied by the number of hours in a year, which is the maximum possible yearly energy production.

The ratio of actual yearly energy output of a turbine to the maximum possible output is called the *capacity factor CF* of the turbine. It is seen that *CF* depends both on the design of the turbine as well as the wind conditions at the location.

6.5 The Wind Resource

Uncertainities associated with the design of a fossil fuel power plant are mainly the availability and price fluctuations of fuel. However, the feasibility of generating wind power at a given location depends critically on the average wind speed at the site over the year, which varies from year to year, season to season, day to day, and with the height above the ground. Two of the important issues that need to be addressed before a wind power system is installed are: where should the wind turbine be located, and how much energy be can generated over a year? The feasibility study required to answer these two questions is called '*wind resource assessment*'.

The average wind speed at a location depends on three main factors, amongst others. These are: (i) Roughness of the terrain,which is dependendent on the nature of the surrounding area, such as, the presence of a body of water or forest cover. Forests increase roughness causing more turbulance and mixing of air, and the wind speed tends to decrease as a result. (ii) The second factor is the topography of the terrain. The presence of hills or mountains tends to increase the wind speed as the top of the hill is approached due to the compression of the stream lines. (iii) The third factor is the sheltering caused by obstacles like buildings and houses in the neighborhood, which tends to decrease the wind speed [5]. The mathematical models used to predict the wind resource at a location are presented below.

6.5.1 *Wind generation*

Uneven heating of the atmosphere and the rotation of the earth are the main causes of wind. For instance, the equatorial regions of the earth on average absorb more solar radiation than the poles. Wind transfers some of this energy from the equator to the poles. As the wind moves it experiences a force perpendicular to the direction of motion due to the Coriolis acceleration, which is directed to the right in the Northern Hemisphere [4,5]. The distance of the air stream from the axis of rotation of the earth decreases as air moves toward the pole. Therefore in order to conserve angular momentum, the air speed in the west to east direction increases.

These combined effects result in circulation cells in the atmosphere, which effect the wind conditions close to the earth's surface. The pressure gradient of the air in the vertical direction is balanced by the downward gravitational force. Hence winds blow predominantly in the horizontal direction. The different physical processes governing wind generation is discussed in detail in Ref.[5], where a simple physical model is developed for the mechanics of wind motion.

6.5.2 *Atmospheric pressure and temperature*

We shall now develop a simple model to estimate pressure and temperature changes in the atmosphere. Consider a quesient vertical air column of area δA and height z. For the equilibrium of a small element of air of thickness δz, the weight of the element is balanced by the net vertical pressure force. Hence we have

$$\delta A \delta P + \delta A \rho(z) g \delta z = 0 \tag{6.7}$$

where the pressure varies from P to $(P + \delta P)$ over the height δz. The density at height z is ρ and the acceleration due to gravity is g. From Eq. (6.7) we obtain the differential equation for the pressure variation of the atmosphere as

$$\frac{dP}{dz} = -\rho(z) g \tag{6.8}$$

Assuming that atmospheric air behaves like an ideal gas we write the equation of state as

$$P = \rho R_a T \tag{6.9}$$

where R_a is the gas constant for air, and T is the absolute temperature. From Eqs. (6.8) and (6.9) we obtain the differential equation

$$\frac{dP}{dz} = -\left(\frac{Pg}{R_a T}\right) \tag{6.10}$$

Equation (6.10) can be solved for different assumed conditions of the atmosphere.

For an isothermal atmosphere, $(T = constant)$, the solution of Eq. (6.10) is

$$P(z) = P_o exp\left(\frac{-Pgz}{R_a T}\right)$$

where P_o is the pressure at $z = 0$.

If the atmosphere is assumed to be *adiabatic*, then the process law is

$$\frac{P^{(1-1/\gamma)}}{T} = constant$$

where $\gamma = c_p/c_v$, is the ratio of specific capacities of air. Substituting for P in Eq. (6.10) from the above relation we obtain the linear temperature variation given by (see worked example 6.3)

$$\frac{dT}{dz} = -\frac{g}{c_p} = 0.0098 \tag{6.11}$$

From Eq. (6.11) we note that the temperature decreases by about 1°C with every 100 m change in elevation.

The atmosphere has three stability conditions called, stable, unstable and neutral states depending on temperature variations [5]. The atmospheric conditions are neutral when the temperature does not change with height due to adiabatic cooling. This state is accompanied with high winds. When the air temperature increases with altitude the atmosphere is in a stable state, and the wind speed increases with height under these conditions. When the air temperature at ground level is higher then the temperature at higher altitides, the atmosphere is in an unstable state. The atmosphere becomes turbulent under these conditions, with warm air rising and cold air sinking.

6.5.3 *Variation of wind speed with height*

One of the important inputs to the design and location of wind turbines is the variation of wind speed with height at a location. This information is often not readily available. There are two methods available in fluid mechanics for predicting the variation of fluid velocity in boundary layer flow. These are called: (i) the *log law* velocity profile, and (ii) the *power law* velocity profile. We will now apply these methods to the atmospheric boundary layer.

Logarithmic velocity profile

The shear stress in the boundary layer close to a surface is given by the Prandtl mixing length theory as [4]

$$\tau = \rho l^2 \left(\frac{\partial U}{\partial z}\right)^2 \tag{6.12}$$

where U is the horizontal velocity at height z from the surface, and l is the mixing length. Hence we have

$$\frac{\partial U}{\partial z} = \frac{1}{l}\left(\frac{\tau}{\rho}\right)^{1/2} = \frac{U^*}{l} \tag{6.13}$$

where U^* is the friction velocity. If we assume that the surface is smooth, then $l = kz$, and $k = 0.4$, the Karman's constant. Substituting in Eq. (6.13) we have

$$\frac{\partial U}{\partial z} = \frac{U^*}{kz} \tag{6.14}$$

Integrating Eq. (6.14) the *logarithmic wind velocity profile* is obtained as

$$U(z) = \frac{U^*}{k} ln\left(\frac{z}{z_o}\right) \tag{6.15}$$

The integration of Eq. (6.14) is from a lower height z_o because natural surfaces are never smooth. The value z_o for a few types of terrain are listed in Table 6.2 [5]. Wind speed $U(z_r)$, measured at a reference height z_r, can be used to obtain the wind speed at any other height z by recasting Eq. (6.15) in the form

$$\frac{U(z)}{U(z_r)} = \frac{ln(z/z_o)}{ln(z_r/z_o)} \tag{6.16}$$

Table 6.2 Surface roughness length for different terrain types (Data from Ref.[5])

Terrain description	z_o (mm)
Very smooth ice, mud	0.01
Calm open sea	0.20
Snow surface	3.0
Lawn grass	8.0
Rough pasture	10.0
Fallow field	30.0
Crops	50.0
Few trees	100.0
Many trees, few buildings	250.0
Forest and woodland	500.0

Power law velocity profile

The power law velocity profile assumes that the wind velocity is proportional to a fractional power α of the height z. Hence we have

$$\frac{U(z)}{U(z_r)} = \left(\frac{z}{z_r}\right)^{\alpha} \tag{6.17}$$

where z_r is the reference height where the wind velocity is $U(z_r)$.

For flow over flat plates, $\alpha = 1/7$. The value of α depends on a number of conditions including, elevation, time of day, season, nature of terrain, wind speed, temperature [5]. The variation of wind speed with height for three values of α is shown in Fig. 6.4. The rate of increase of wind speed is high initially and becomes more gradual at higher elevations. It is desirable to locate the turbine rotor at a larger height both due to the high wind speeds available and also due to the more gradual change in wind speed.

The exponent of the power-law has been correlated with the parameter z_o of the log-law velocity profile. One such correlation, due to Counihan is [5]

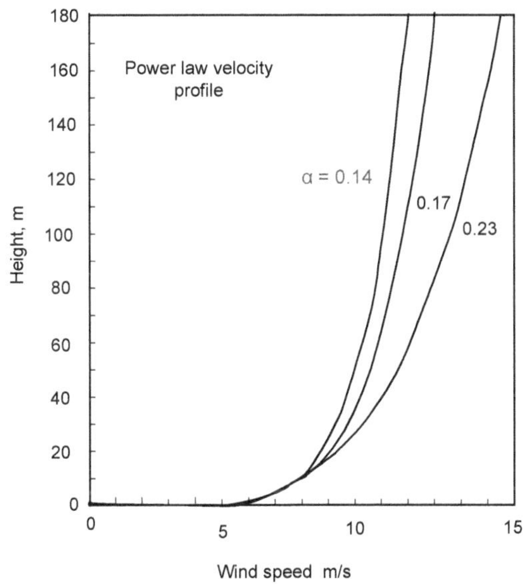

Fig. 6.4 Power law wind velocity profile

$$\alpha = 0.096Log_{10}z_o + 0.016(Log_{10}z_o)^2 + 0.24 \quad (6.18)$$

for 0.001 m $< z_o <$ 10 m, where z_o is the surface roughness in m, which is listed in Table 6.2 for a number terrain types.

6.6 Wind Speed Data Analysis

The statististical approach used for long-term prediction of the energy output of a turbine, relies on the fequency distribution of the wind speed over a year. This frequency distribution is made up of the number of hours of a year when the wind speed falls in an interval between two specified values, for instance, 6 ms^{-1} and 6.5 ms^{-1}. In addition, measured wind data is used to generate a distribution of wind direction that is often displayed as a 'wind rose' [6]. If measured data on average wind speed over a long period like a year is available, then a direct method could be used to estimate the expected output of a turbine. However, if such detailed data are not available, then theoretical wind-speed probability distributions have to be relied upon.

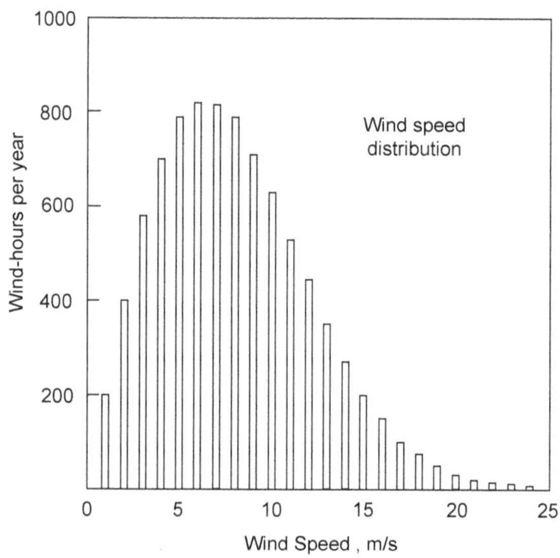

Fig. 6.5 Typical hourly wind speed distribution over a year

We shall first consider the direct method where a series of N wind speed measurements are available. Each measurement U_i is the average wind speed over a time interval Δt, which is commonly one hour. A typical hourly averaged wind speed distribution over a year is shown graphically in Fig. 6.5.

The following statistical parameters may be obtained from this series of observations.

(i) The long-term average wind speed $\bar{U}$ over the entire period of data collection, which is often one year, is given by

$$\bar{U} = \frac{1}{N}\sum_1^N U_i \tag{6.19}$$

(ii) The standard deviation of the wind speed σ_U is

$$\sigma_u = \left[\frac{1}{N-1}\sum_1^N (U_i - \bar{U})^2\right]^{1/2} \tag{6.20}$$

From Eqs. (6.19) and (6.20) we get

$$\sigma_u = \left[\frac{1}{N-1}\sum_1^N U_i^2 - N\bar{U}^2\right]^{1/2} \tag{6.20a}$$

(iii) The average wind power avilable at a location may be obtained by combining the expression for instantaneous power of the wind, given by Eq. (6.3), and the measured wind speed distribution. Hence we have the average wind power available $\bar{P}$ over the data collection period as

$$\bar{P} = \frac{A\rho}{2N}\sum_1^N U_i^3 \tag{6.21}$$

(iv) The average wind turbine power output $\bar{P}_w$ is

$$\bar{P}_w = \frac{1}{N}\sum_1^N P_w(U_i) \tag{6.22}$$

where P_w is the power output of turbine given by the power curve (see Fig. 6.3).

6.6.1 *The bin-method*

The *bin-method* is similar to the direct method outlined above and requires detailed measured data on wind speeds over a long period of time, like an year, at a location. The measured wind speeds are separated into bins, which are essentially wind speed intervals. For instance, a bin of width 1 ms^{-1} centered at a wind speed of 10 ms^{-1}, will include all measured wind speeds between 9.5ms^{-1} and 10.5 ms^{-1}. The graphical representation of the bins is similar to the histogram shown in Fig. 6.5.

Let the wind speed at the middle of bin i be u_i and the number of measurements in bin i be n_i. If there are N_B bins in all, then the total number of measurements is given by

$$N = \sum_1^{N_B} n_i \tag{6.23}$$

The *average wind speed* over the measurement period is given by

$$\bar{U} = \frac{1}{N}\sum_1^{N_B} u_i n_i = \sum_1^{N_B} u_i f_i \tag{6.24}$$

where $f_i = \frac{n_i}{N}$, is the frequency of occurrence for bin i.

The standard deviation of the velocity distribution is given by

$$\sigma_W{}^2 = \frac{1}{N-1}\sum_1^{N_B} (u_i - \bar{U})^2 n_i \tag{6.25}$$

From Eqs. (6.24) and (6.25) we get

$$\sigma_W{}^2 = \frac{N}{N-1}\left[\sum_1^{N_B} u_i{}^2 f_i - \bar{U}^2\right] \tag{6.25a}$$

The average wind power avilable at the location may be obtained by combining the expression for instantaneous power of the wind, given by Eq. (6.3), and the characteristic wind speed for bin i. Hence we have

$$\bar{P} = \frac{A\rho}{2N}\sum_1^{N_B} u_i{}^3 n_i = \frac{A\rho}{2}\sum_1^{N_B} u_i{}^3 f_i \tag{6.26}$$

where $f_i = \frac{n_i}{N}$ is the frequency of bin i.

The average power output of the wind turbine is given by

$$\bar{P}_W = \frac{1}{N}\sum_1^{N_B} P_W(u_i)\, n_i = \sum_1^{N_B} P_W(u_i)\, f_i \tag{6.27}$$

6.6.2 *Statistical analysis of wind data*

Often detailed data, of the type described in the previous section, on the variation of wind speed at a location may not be available. In such a situation, we can use statistical analysis to determine the wind speed distribution at the location, and hence the wind energy potential.

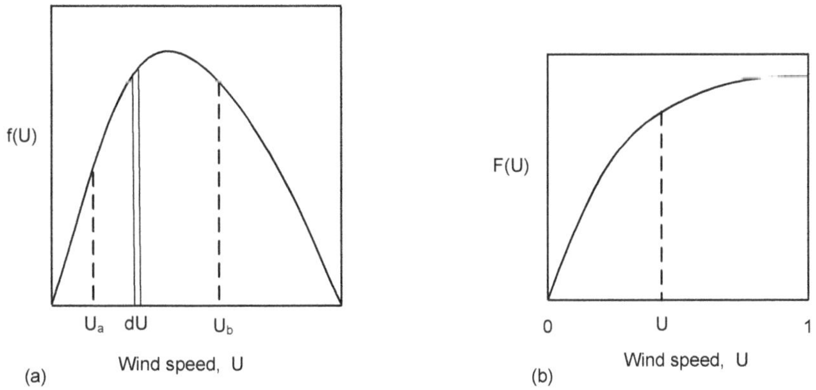

Fig. 6.6 (a) Probability density function, (b) Cumulative probability distribution

Probability density function

In statistical analysis the likelihood of occurance of a particular value of a random variable, such as the wind speed, is described by its probability density function (PDF), $f(U)$. A typical PDF is shown in Fig. 6.6(a). The probabilty of occurance of a wind speed between U_a and U_b is given by

$$f(U_a \to U_b) = \int_{U_a}^{U_b} f(U)\, dU \tag{6.28}$$

The total area under the probability density curve is

$$\int_0^\infty f(U) dU = 1 \tag{6.29}$$

The *mean wind speed*, $\bar{U}$ is given by

$$\bar{U} = \int_0^\infty U f(U)\, dU \tag{6.30}$$

The *standard deviation* of the wind speed is

$$\sigma_U = \left[\int_0^{\infty} (U - \bar{U})^2 f(U) dU\right]^{1/2} \tag{6.31}$$

The *mean wind power* available is

$$\bar{P} = \frac{A\rho}{2} \int_0^{\infty} U^3 f(U)\, dU \tag{6.32}$$

Cumulative distribution function

The cumulative distribution function (CDF), $F(U)$ represents the probability or the fraction of time during which the wind speed is smaller than or equal to a given wind speed, U. It is obtained by integrating the PDF given by Eq. (6.28) from zero to U. Hence we have

$$F(U) = \int_0^{U} f(U') dU' \tag{6.33}$$

6.6.3 *Commonly used probability distribution functions*

The two probability distributions used commonly for the analysis of wind speed data are: (i) the Weibull distribution, and (ii) the Rayleigh distribution. The Weibull distribution uses two parameters, called the *scale factor* and the *shape factor*, to predict the wind speed distribution. The Rayleigh distribution, which is a special case of the Weibull distribution, uses one parameter which is the average wind speed. These distributions are called skew distributions because they are defined only for values greater than zero.

Weibull distribution

The Weibull distribution, shown in Fig. 6.7, involves two parameters k and c, which are the shape factor and the scale factor respectively. The mathematical forms of the Weibull PDF, f_W and the CDF, F_W are:

$$f_W(U) = \left(\frac{k}{c}\right)\left(\frac{U}{c}\right)^{k-1} exp\left[-\left(\frac{U}{c}\right)^k\right] \tag{6.34}$$

$$F_W(U) = 1 - exp\left[-\left(\frac{U}{c}\right)^k\right] \tag{6.35}$$

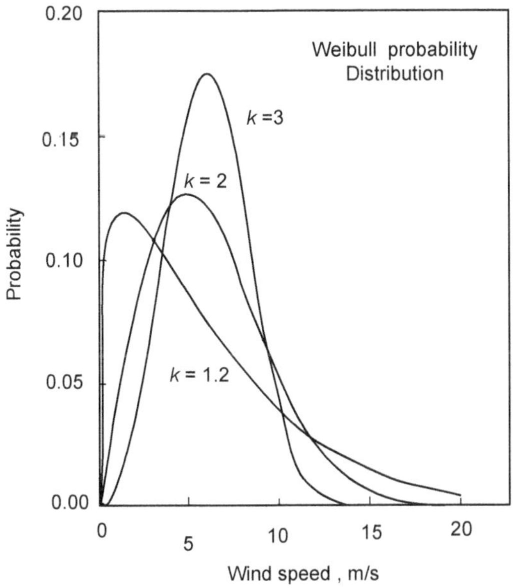

Fig. 6.7 Weibull PDF for a few different values of *k*.

The *average wind speed* $\bar{U}$ is obtained by substituting for $f(U)$ in Eq. (6.30). Hence we have

$$\bar{U} = \int_0^\infty U\left(\frac{k}{c}\right)\left(\frac{U}{c}\right)^{k-1} exp\left[-\left(\frac{U}{c}\right)^k\right] dU \tag{6.36}$$

Integration of Eq. (6.36) gives the average wind speed as

$$\bar{U} = c\Gamma\left(1 + \frac{1}{k}\right) \tag{6.37}$$

The standard deviation is given by Eq. (6.31) as

$$\sigma^2 = \bar{U}^2\left[\frac{\Gamma(1+2/k)}{\Gamma^2(1+1/k)} - 1\right] \tag{6.38}$$

where Γ is the Gamma function which has the integral form

$$\Gamma(x) = \int_0^\infty e^{-t}t^{x-1}dt$$

The value of the gamma function at x is returned in the Matlab software pakage by using the expression: $y = gamma(x)$.

Two approximate analytical expressions that relate the mean and the standard deviation to the parameters k and c of the Weibull PDF are given in Ref.[5]. These are valid for $1 \leq k \leq 10$

$$k = \left(\frac{\sigma_U}{\bar{U}}\right)^{-1.086} \tag{6.39}$$

$$c = \frac{\bar{U}}{\Gamma(1+1/k)} \tag{6.40}$$

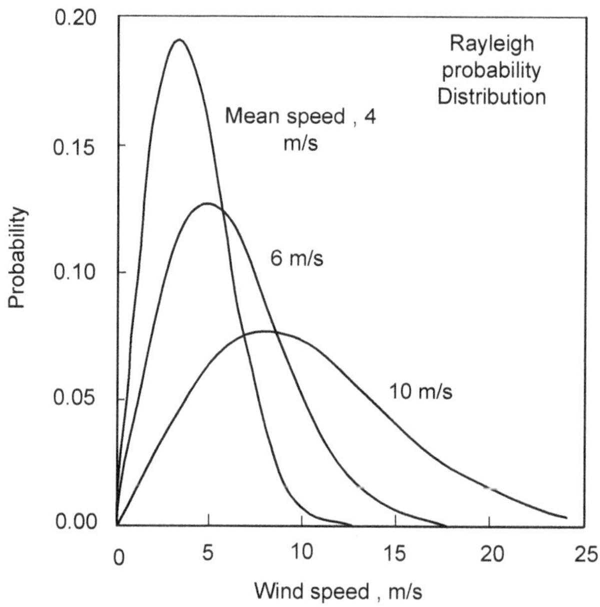

Fig. 6.8 Rayleigh PDF for a few different mean wind speeds

Rayleigh distribution

The Rayleigh distribution, depicted in Fig. 6.8, is simpler because it requires only one parameter, which is the average wind speed $\bar{U}$. When $k = 2$ the Weibull distribution reduces to the Rayleigh distribution. The expressions for the Rayleigh PDF, $f_R(U)$ and CDF, $F_R(U)$ are given by

$$f_R(U) = \frac{\pi}{2}\left(\frac{U}{\bar{U}^2}\right) exp\left[-\frac{\pi}{4}\left(\frac{U}{\bar{U}}\right)^2\right] \tag{6.41}$$

$$F_R(U) = 1 - exp\left[-\frac{\pi}{4}\left(\frac{U}{\bar{U}}\right)^2\right] \tag{6.42}$$

We notice that the Rayleigh PDF is skewd to the left, which means that lower wind speeds are more likely than higher wind speeds. Typically the PDFs are determined at a desired height that is usually the hub height of the wind turbine.

6.7 Performance of Wind Turbines

The long-term energy production of a wind turbine may be obtained using the power curve of the turbine and the local wind speed distribution. In this section we shall first develop an idealized model to predict the power output of a wind turbine.

6.7.1 *The actuator disk model*

The *actuator disk model* [3-5] is an idealized model of the wind turbine which gives the maximum power that could be extracted from the power available in the wind. Depicted in Fig. 6.9 is a column of wind in steady flow from left to right across the actuator disk. The air speed decreases in the flow direction as momentum is extracted by the disk. Due to mass conservation, the area of the wind column increases.

The following assumptions are made in applying the conservation equations of mass, energy,and momentum to the system:

(i) air flow is steady, homogeneous, incompressible, and inviscid, (ii) static pressure far upstream and downstream are equal, (iii) air velocity on either side of the actuator disk are equal, (iv) the disk has an infinite number of infinitely thin blades, (v) velocity and thrust on the disk are uniform, and (vi) no rotation of the wake flow.

Applying the conservation equations of mass, energy and momentum to the control volume shown in Fig. 6.9 we obtain the following equations. The air velocity U and the pressure p at different sections are indicated in Fig. 6.9.

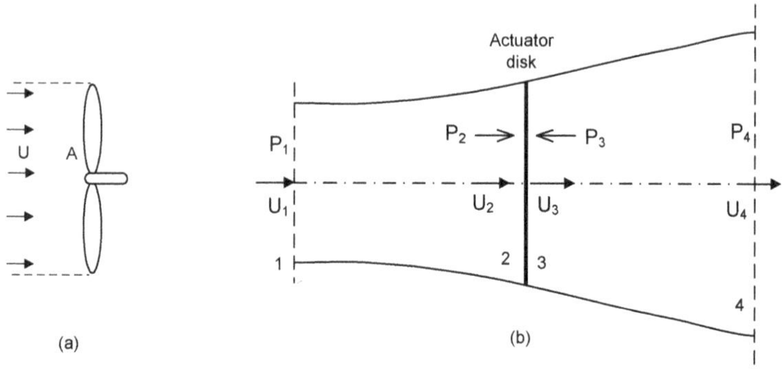

Fig. 6.9 (a) Available energy in wind, (b) Actuator disk model

Mass conservation equation:

$$\dot{m}_a = A\rho U_2 = A\rho U_3 \tag{6.43}$$

where $\dot{m}_a$ is the mass flow rate and ρ is the density.

Bernoulli's equation between sections 1 and 2:

$$\frac{p_1}{\rho} + \frac{U_1{}^2}{2} = \frac{p_2}{\rho} + \frac{U_2{}^2}{2} \tag{6.44}$$

Bernoulli's equation between sections 3 and 4:

$$\frac{p_3}{\rho} + \frac{U_3{}^2}{2} = \frac{p_4}{\rho} + \frac{U_4{}^2}{2} \tag{6.45}$$

From assumption (ii):

$$p_1 = p_4 \tag{6.46}$$

From assumption (iii):

$$u_2 = u_3 \tag{6.47}$$

Adding the two sides of Eqs. (6.44) and (6.45) and applying conditions in Eqs. (6.46) and (6.47) we obtain

$$\frac{p_3}{\rho} + \frac{U_1{}^2}{2} = \frac{p_2}{\rho} + \frac{U_4{}^2}{2} \tag{6.48}$$

Applying the momentum equation to control volume 1-4 we obtain the following expression for the *thrust on the fluid* by the disk T_f.

$$T_f = \dot{m}_a(U_4 - U_1) \tag{6.49}$$

The thrust *on the disk* T_d is given by

$$T_d = A(p_2 - p_3) \tag{6.50}$$

From Newton's third law it follows that

$$T_d = -T_f \tag{6.51}$$

Substituting from Eqs. (6.49) and (6.50) in Eq. (6.51) we have

$$A(p_2 - p_3) = \dot{m}_a(U_1 - U_4) \tag{6.52}$$

Substitute for $\dot{m}_a$ from Eq. (6.43) and for $(p_2 - p_3)$ from Eq. (6.48) in Eq. (6.52) to obtain

$$U_2(U_1 - U_4) = \frac{{U_1}^2 - {U_4}^2}{2} \tag{6.53}$$

$$U_2 = \frac{U_1 + U_4}{2} \tag{6.54}$$

We define a parameter, called the *axial induction factor* a as

$$a = \frac{U_1 - U_2}{U_1} \tag{6.55}$$

$$U_2 = (1 - a)U_1 \tag{6.56}$$

Note that the factor a is the fractional decrease in the free stream wind speed at the disk cross section. Substituting from Eq. (6.56) in Eq. (6.54) the air speed can now be expressed in the form

$$U_4 = (1 - 2a)U_1 \tag{6.57}$$

Substtituting the above expressions in Eq. (6.49) we obtain the thrust *on the disk* as

$$T_d = \dot{m}_a(U_1 - U_4) = 2A\rho {U_1}^2 a(1 - a) \tag{6.58}$$

The power extracted is given by

$$P_e = T_d U_2 = 2A\rho U_1{}^3 a(1-a)^2 \tag{6.59}$$

The ratio of the power extracted by the wind turbine to the power available in the wind is called the *power coefficient* or coefficient of performance C_p. Therefore

$$C_p(U_1) = \frac{P_e}{P_{wind}} = \frac{P_e}{\frac{1}{2}A\rho U_1{}^3} \tag{6.60}$$

$$C_p(U_1) = \frac{2A\rho U_1{}^3 a(1-a)^2}{\frac{1}{2}A\rho U_1{}^3} = 4a(1-a)^2 \tag{6.61}$$

Similarly, the *thrust coefficient* C_T is defined as the ratio of the thrust force to the dynamic force. Therefore

$$C_T(U_1) = \frac{T_d}{F_{dynamic}} = \frac{T_d}{\frac{1}{2}A\rho U_1{}^2} \tag{6.62}$$

$$C_T(U_1) = \frac{2A\rho U_1{}^2 a(1-a)}{\frac{1}{2}A\rho U_1{}^2} = 4a(1-a) \tag{6.63}$$

In Eqs. (6.58) and (6.59) U_1 is a constant. The parameter a is treated as a variable, and we seek its value for which P_e and T_d are maximum. Differentiating the RHS of Eqs. (6.58) and (6.59) with respect to a we obtain the following expressions for the maximum thrust and power.

$$T_{d,max} = \frac{1}{2}A\rho U_1{}^2 \tag{6.64}$$

$$P_{e,max} = \left(\frac{16}{27}\right)\left(\frac{1}{2}A\rho U_1{}^3\right) = 0.592\left(\frac{1}{2}A\rho U_1{}^3\right) \tag{6.65}$$

From Eqs. (6.65) and (6.3) it follows that the *maximum power extractable* is about 59.2% of the available power in the wind. This limit is called the *Betz limit*. From Eq. (6.61) it follows that the maximum value of the power coefficient is 0.592.

The power coefficient is also a function of how fast the tips of the turbine blades are rotating compared to the incoming wind speed. The ratio of the speed of the blade tips to the wind speed is called the *tip speed ratio TSR*. Hence we have

$$TSR = \frac{\omega R}{U_1} = \lambda \tag{6.66}$$

where ω is the rotational speed of the rotor and R is the radius of the blade.

A *torque coefficient* C_Q may be defined as the torque generated by the rotor Q to the maximum possible torque Q_{max}.

$$C_Q = \frac{Q}{Q_{max}} \tag{6.67}$$

The maximum torque would occur if the maximum thrust generated by the rotor $T_{d,max}$ is applied at the tip of the rotor. Hence we have

$$Q_{max} = T_{d,max}R = \frac{1}{2}A\rho {U_1}^2 R \tag{6.68}$$

From Eqs. (6.66) and (6.68) it follows that

$$Q_{max} = \frac{1}{2}A\rho {U_1}^2 R = \frac{1}{2}A\rho {U_1}^3 \left(\frac{\lambda}{\omega}\right) \tag{6.69}$$

Substituting from Eqs. (6.66), (6.67) and (6.68) in Eq. (6.69) we get

$$Q_{max} = \frac{Q}{C_Q} = \left(\frac{\lambda}{\omega}\right)\left(\frac{P_e}{C_p}\right) \tag{6.70}$$

The power output of the turbine is given by

$$P_e = Q\omega \tag{6.71}$$

From Eqs. (6.70) and (6.71) we have

$$C_p = C_Q\lambda \tag{6.72}$$

The maximum C_p predicted by the actuator disk model is decreased due to: (i) wake rotation behind the rotor, (ii) finite number of blades, and (iii) the aerodynamic drag [5].

The analysis presented above has been extended to include wake rotation by considering the energy imparted to an annular stream tube in the wake flow. This rotation is in the opposite direction to the rotor because of angular momentum conservation. The energy extracted by the rotor is decreased due to energy carried by wake rotation, which is taken into account in turbine design by introducing a parameter a', called the *angular induction factor*. This is given by [5]

$$a' = \frac{\omega}{2\Omega} \tag{6.73}$$

where ω is the angular speed of the air in the wake and, Ω is the angular speed of the rotor.

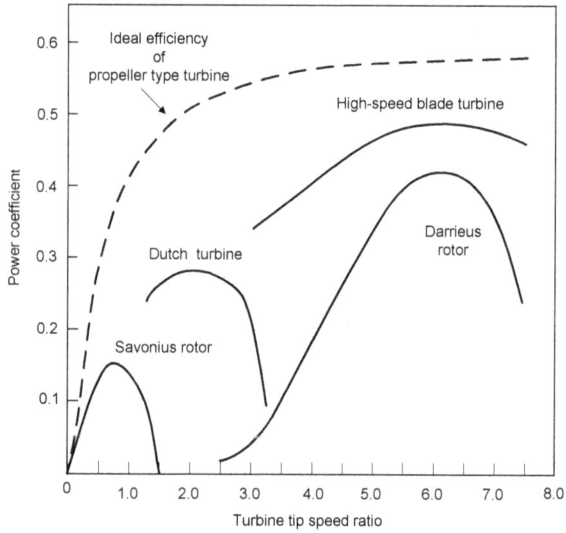

Fig. 6.10 Power coefficient of different wind turbine designs (data from Ref.[4])

The variation of the power coefficient C_p with tip speed ratio TSR of different commercial wind turbines is shown in Fig. 6.10. For the ideal turbine based on the actuator disk model, shown at the top of the figure, C_p increases with TSR until the Betz limit of 0.59 is reached. For modern high-speed horizontal axis turbines, C_p increases with TSR and reaches a maximum value of about 0.5. For the vertical axis turbine, the Darrieus rotor, the curve shows a sharp peak at about 0.4 to 0.45. The Dutch turbine and the vertical axis Savonius rotor have relatively low values of C_p.

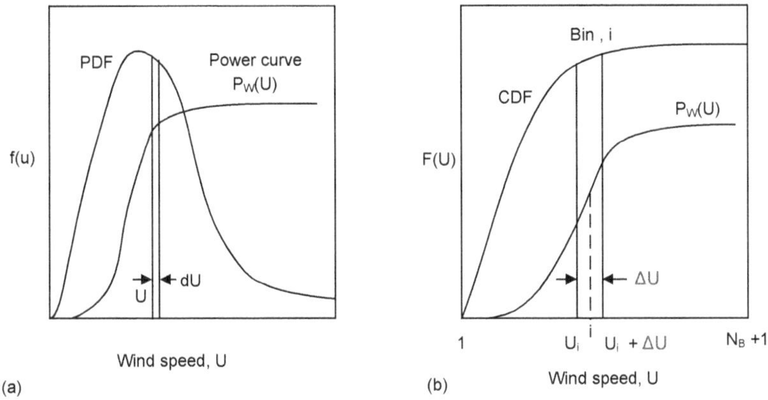

Fig. 6.11 Typical hourly wind speed distribution over a year. (a) PDF, (b) CDF

6.7.2 *Estimation wind power production*

The main information required to estimate the yearly energy production of a wind turbine are: (i) the PDF, $f(U)$ of the wind speed at the location, and (ii) the power output curve of the turbine, $P_w(U)$. A typical turbine power output curve is shown in Fig. 6.11. The average wind power output is given by

$$\bar{P}_w = \int_0^\infty P_w(U)\, f(U) dU \tag{6.74}$$

The power output of the turbine may be expressed in terms of the power coefficient $C_p(U)$ using Eq. (6.60). Hence we obtain

$$P_w(U) = \frac{1}{2}\rho A \eta C_p(U) U^3 \tag{6.75}$$

Substituting from Eq. (6.75) in Eq. (6.74) we have

$$\bar{P}_w = \frac{1}{2}\rho A \eta \int_0^\infty C_p(U)\, U^3 f(U) dU \tag{6.76}$$

Equation (6.76) can be used to estimate the average power output of the turbine by substituting the expressions for the power coefficient of the turbine and the appropriate wind speed PDF, such as the Weibull distribution. The integration of the resulting expression may be done numerically.

Moreover, the cumulative probability distribution, $F(U)$ can be used to obtain the total power output of the turbine by a procedure similar to the bin method described in section 6.6.1. Consider the bin i which is bound by wind speeds from U_i and $(U_i + \Delta U)$ as shown in Fig. 6.11(b).The average wind speed of the bin is $(U_i + \Delta U/2)$. Then the number of hours included in bin i is given by

$$N_i = [F(U_i + \Delta U) - F(U_i)]N_t \tag{6.77}$$

where N_t is the total number of hours. The power output of the turbine at the average wind speed of bin i is

$$P_{wi} = P_w(U_i + \Delta U/2) \tag{6.78}$$

Substituting for P_w from Eq. (6.75) in Eq. (6.78) we get

$$P_{wi} = \frac{1}{2}\rho A\eta C_p(U_i + \Delta U/2)(U_i + \Delta U/2)^3 \tag{6.79}$$

The total power output over the period N_t is given by

$$\bar{P}_w = \sum_1^{N_B} P_{wi}N_i \tag{6.80}$$

$$\bar{P}_w = \sum_1^{N_B} P_w(U_i + \Delta U/2)[F(U_i + \Delta U) - F(U_i)]N_t \tag{6.81}$$

The application of the above method will be illustrated in worked examples 6.6 to 6.13.

6.8 Aerodynamic Design of Wind Turbines

The actuator disk model developed above is an idealized model which assumes that the rotor of the turbine has infinite number of infinitely thin blades. It gives the Betz limit as the maximum possible fraction of the power in the wind that could be extracted. Two additional parameters that emerged from the model are the axial induction factor a, and the angular induction factor a'. Actual wind turbines have a finite number of blades, which is usually three in the case of large modern turbines. Such a turbine is depicted in Fig. 6.1(a).

In this section we will use the *blade element theory* to develop a more realistic aerodynamic model that takes into account the details of the flow over the turbine blade, and the resulting aerodynamic forces.

6.8.1 *Blade element theory*

In the blade element model, the blade is divided into several elements, each with a small radial thickness. As the radial location changes, the tangential speed of the blade element increases because it is proportional to the radius. Each element will therefore experience a slightly different air flow over it.

Moreover, the shape of the blade element changes with its radial location. Two key assumptions made in developing the model: (i) there is no aerodynamic interaction between different blade elements, and (ii) the only forces acting on the blade elements are the lift and drag forces, which are determined by the lift and drag coefficients.

A typical cross section of a blade element is shown in Fig. 6.12(b). It can be visualized as the plan view of section x-x of the vertical blade shown in Fig. 6.1(a), which is rotating in the counter clockwise direction. The rounded front edge of the *airfoil section* of the blade is called the

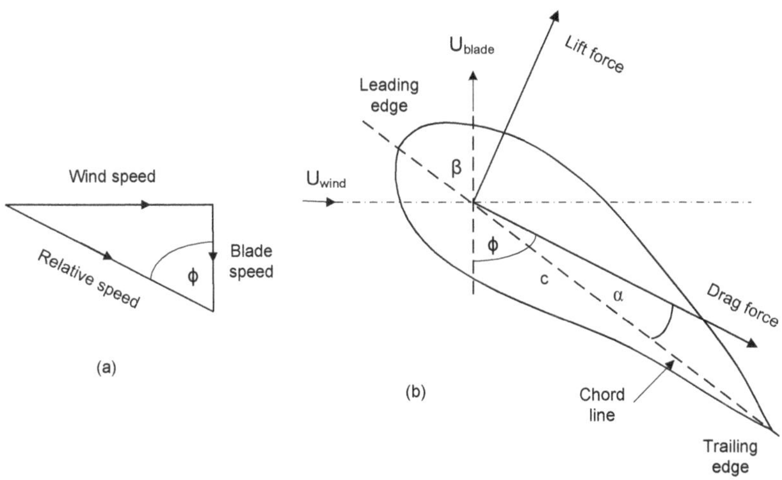

Fig. 6.12 Air flow and forces on blade, (a) velocity triangle, (b) forces on blade section

leading edge, while the sharper rear edge is the *trailing edge*. The line joining these two edges is the *chord line*, and the distance from the leading to the trailing edge is called the *chord length*.

The air flow over the airfoil section generates two net forces called the *lift force* and the *drag force*. Since the blade is moving with respect to the upstream air flow, the above forces depend on the *relative motion* of the air. The velocity triangle shown Fig. 6.12(a), gives the magnitude and direction of the relative air velocity. The speed of the wind at the blade, including the *axial induction* factor a is (see Eq. 6.56)

$$V_w = (1 - a)U \tag{6.82}$$

where U is the upstream wind speed.

The tangential speed of the blade, including the *angular induction factor* a' is [5] (see Eq. 6.73)

$$V_b = \Omega r + \frac{\omega r}{2} = \Omega r(1 + a') \tag{6.83}$$

where r is the radial distance of the blade section from the axis of rotation, and Ω is the rotational speed of the blade. From the velocity triangle it follows that

$$tan\phi = \frac{V_w}{V_b} = \frac{U(1-a)}{\Omega r(1+\acute{a})} = \frac{(1-a)}{(1+\acute{a})\lambda_r} \tag{6.84}$$

The *local tip speed ratio* is defined as

$$\lambda_r = \frac{\Omega r}{U} = \frac{\Omega R}{U}\left(\frac{r}{R}\right) = \lambda\left(\frac{r}{R}\right) \tag{6.85}$$

where λ is the blade tip speed ratio.

The air *speed relative* to the blade may be written as

$$W = \frac{U(1-a)}{sin\phi} \tag{6.86}$$

Notice that the angle ϕ changes from blade element to blade element as the radius changes.

The angle α between the direction of the relative air speed and the chord line of the section is called the *angle of attack*. The angle β

between the blade chord line and the plane of rotation of the blade is called the *section pitch angle*.

From Fig. 6.12(b),

$$\phi = \alpha + \beta \tag{6.87}$$

The *section twist* is defined as

$$\theta_t = \beta - \beta_0 \tag{6.88}$$

where β_0 is the blade pitch at the tip.

Forces on a blade element

Consider a blade element of thickness dr at a radial distance r from the axis of rotation of the turbine, as depicted in Fig. 6.12(b). Let the lift and drag forces on the element be dL and dD respectively, which by definition are perpendicular and parallel to the direction of *relative air flow* over the element. Then the resultant forces dF_t in the tangential direction and dF_a in the axial direction are:

$$dF_t = dLsin\phi - dDcos\phi \tag{6.89}$$

$$dF_a = dLcos\phi + dDsin\phi \tag{6.90}$$

The lift and drag forces may be written as [5]:

$$dL = \frac{1}{2}C_L\rho W^2 cdr \tag{6.91}$$

$$dD = \frac{1}{2}C_D\rho W^2 cdr \tag{6.92}$$

where C_L and C_D are the *lift and drag coefficients*, c is the chord length of the blade section, and ρ is the density of air.

Substituting from Eqs. (6.91) and (6.92) in Eqs. (6.89) and (6.90) we obtain

$$dF_t = \frac{1}{2}\rho W^2(C_L sin\phi - C_D cos\phi)cdr \tag{6.93}$$

$$dF_a = \frac{1}{2}\rho W^2(C_L cos\phi + C_D sin\phi)cdr \tag{6.94}$$

The *torque* on the turbine due to the tangential force on the blade elements at radius r is given by

$$dT = \frac{N_B}{2}\rho W^2(C_L sin\phi - C_D cos\phi)crdr \tag{6.95}$$

where N_B is the number of blades of the turbine.

The *total axial force* on the turbine due to the forces on blade elements at radius r is

$$dF_{at} = \frac{N_B}{2}\rho W^2(C_L cos\phi + C_D sin\phi)cdr \tag{6.96}$$

Substituting from Eq. (6.86) in Eqs. (6.95) and (6.96) we have

$$dT = \frac{N_B \rho U^2(1-a)^2}{2sin\phi^2}(C_L sin\phi - C_D cos\phi)crdr \tag{6.97}$$

$$dF_{at} = \frac{N_B \rho U^2(1-a)^2}{2sin\phi^2}(C_L cos\phi + C_D sin\phi)cdr \tag{6.98}$$

Power output of the turbine

The power output of the turbine is given by

$$P = \int_{r_h}^{R} \Omega dT \tag{6.99}$$

where r_h is the turbine hub radius and R is tip radius.

Substituting from Eq. (6.97) in Eq. (6.99) we have

$$P = N_B \rho U^2 \int_{r_h}^{R} \frac{(1-a)^2}{2sin\phi^2}(C_L sin\phi - C_D cos\phi)cr\Omega dr \tag{6.100}$$

The power coefficient is defined by

$$C_P = \frac{P}{P_{wind}} = \frac{P}{\frac{1}{2}\rho\pi R^2 U^3} \tag{6.101}$$

Substituting from Eqs. (6.100) and (6.84) in Eq. (6.101) we obtain the overall power coefficient as

$$C_p = \left(\frac{N_B}{\pi R^2}\right)\int_{r_h}^{R} \frac{(1-a)^3}{(1+a')tan\phi sin\phi^2}(C_L sin\phi - C_D cos\phi)cdr \tag{6.102}$$

The above equations can be used to estimate the power output of a turbine rotor. The main steps of the design method are as follows:

1. Choose the desired tip speed based on the application, such as electricity generation, water pumping and others.
2. Choose the number of blades.
3. Select the airfoil section, and obtain data like c, C_L and C_D.
4. Create an ideal blade shape by dividing it into several sections, typically 10 to 20.
5. Evaluate the performance based on the blade element theory (BET) developed above.

The solution of the above equations of BET requires the estimation of the axial induction factor and the radial induction factor that vary with flow conditions.

An iterative approach has to be used to obtain these quantities. For this purpose two important equations need to be derived using the momentum theory, which are available in Ref.[5]. The derivation is beyond the scope of this chapter. However, we shall use the equations to illustrate the design procedure. They may be expressed in the simplified form:

$$C_L = \frac{4sin\phi(cos\phi - \lambda_r sin\phi)}{\sigma_1(sin\phi + \lambda_r cos\phi)} \tag{6.103}$$

where the *solidity ratio* is given by

$$\sigma_1 = \frac{N_B c}{2\pi r}$$

The axial and radial induction factors are obtained from the following relations:

$$\frac{1}{a'} = \left(\frac{4cos\phi}{\sigma_1 C_L} - 1\right) \tag{6.104}$$

$$\frac{a}{a'} = \frac{\lambda_r}{tan\phi} \tag{6.105}$$

We shall illustrate the application of the above equations for estimating the performance of turbine rotors in worked examples 6.16 and 6.17.

A typical curve of power coefficient C_p versus tip speed ratio (TSR) obtained using the blade element method is shown in Fig. 6.13(a). C_p increases steadily with the TSR and reaches a maximum value of around 0.5 when TSR is about 9. The theoretical maximum value of C_p is 0.592, which is the Betz limit. The performance curves for actual turbines, shown in Fig. 6.10, display similar trends. This curve is an important input to the control system of a turbine to operate it at the maximum power point under varying wind conditions.

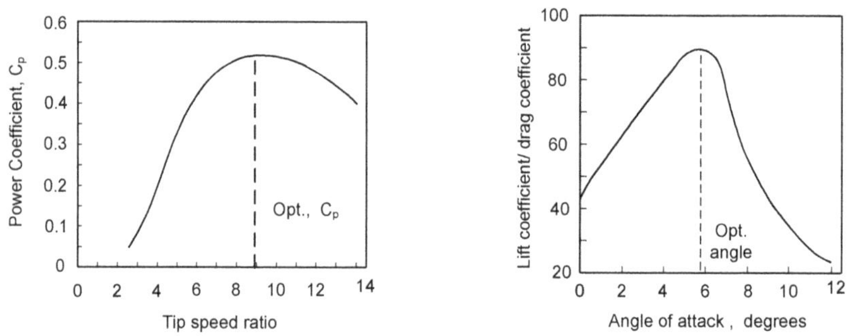

Fig. 6.13 (a) Power coefficient versus tip speed ratio, (b) C_{lift}/C_{drag} versus angle of attack

The cross sectional shapes of modern turbine blades are called airfoils. Each airfoil section (see Fig. 6.12) has a characteristic lift coefficient and a drag coefficient, whose ratio is a function of the angle of attack (AOA) as shown in Fig. 6.13(b). An important condition for the extraction of maximum power from the wind is to maintain the AOA at the optimum value, for which the ratio of lift to drag coefficient is a maximum. This value is about 6° for the airfoil in Fig. 6.13(b). It is not possible to satisfy this condition for a turbine blade if the chords of the different cross sections of the blade are oriented in the same direction.

The reason for this can be clarified by considering three typical blade cross sections, one near the root, one near the middle, and one close to the tip. Since the rotational speed of the rotor is constant, the blade speeds at the above three cross sections are proportional to the respective radii. The velocity triangles for the three sections are shown in Fig. 6.14(a).

Assuming the wind velocity to be same at the three sections, we notice that the relative air velocity at the sections are different both in magnitude and direction ($V_{tip} > V_{mid.} > V_{root}$). If the angle of attack α, which is the angle between the relative air speed and thc chord of the air foil section, is to be kept at the same optimum value for the three sections, the blade has to be twisted towards the plane of rotation as seen in Fig. 6.14(a). This is the case with modern large wind turbine blades. However, the blades of small wind mills usually have a uniform cross section along the length. Deviation from the optimum AOA also occurs due to changes in thc wind speed which is managed by the turbine control system.

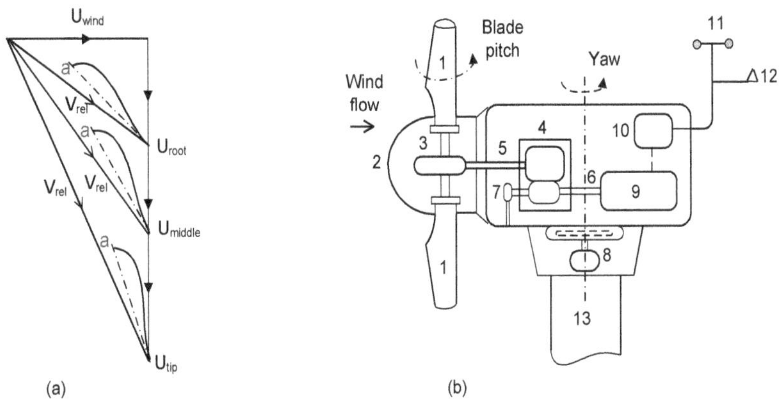

Fig. 6.14(a) Velocity triangles for different points on blade.
Fig. 6.14(b) Components of a turbine: 1-blade, 2-nose of hub, 3-blade pitch control mechanism, 4-gear box, 5-low speed shaft, 6-high speed shaft, 7-brake, 8-yaw control mechanism, 9-electric generator, 10-control system, 11-cup anemometer, 12-wind vane, 13-turbine tower

6.9 Turbine Components and Control

The main components of a modern horizontal axis wind turbine are shown in the simplified schematic diagram in Fig. 6.14(b). Wind blows from the left and the nose (2) of the turbine faces the wind. Such an arrangement is called an up wind design. An alternative arrangement is the down wind design where the turbine is placed behind the tower. The disadvantages of the latter design outweigh the advantages and is therefore not commonly used [5].

The blades (1) are attached to the aerodynamically-shaped nose, through which the blade rotation is transmitted to the low-speed shaft (5), connected to the gear box (4). The high-speed output shaft (6) of the gear box is connected to the electric generator (9) that produces the electrical energy output. A brake (7) is used to shut down the system in an emergency. The blade pitch control mechanism (3) responds to signals from the controller to maintain the pitch angle at the desired value. Both electrical-motor driven systems and hydraulic actuators are used to adjust the blade pitch.

The cup anemometer (11) measures the prevailing wind speed, and the wind vane (12) measures the wind direction. This data is transmitted to the system controller (10). The wind turbine is rotated about a vertical axis to always face the wind by the yaw control mechanism (8). In modern large turbines, this is essentially a ring-gear system driven by an electric motor.

In smaller wind turbines the normal force on the wind vane due to misalignment from the wind direction is adequate to move the turbine about the vertical axis to face the wind. All components described above are located in an enclosure called the nacelle that sits a atop of the turbine tower (13), through which the electrical cables pass from the generator to the transformer located on the ground.

Two types of control concepts are used in commercial wind turbines to control the rotor speed [4]. The control variable in both cases is the blade pitch angle, which is the angle between the blade chord line and the plane of rotation of the blade (see Fig. 6.12). In the first scheme, the pitch of the blades is varied to maintain the rotor speed constant. In the second control scheme (VSVP) the blade pitch and the rotor speed are both varied. The mechanical design and the electric generator design are both simpler for constant speed turbine systems.

The VSVP turbine systems are more complicated in design but have the advantage of producing a larger power output compared to fixed speed turbine systems. Feathering the turbine blades is a procedure to stop the rotor during emergency shutdowns, or whenever the wind speed exceeds the cut-out speed. The blades are oriented along the wind direction during feathering.

Smaller turbines usually don't have pitch control, and a method called *stall regulation* is used to control the speed of the turbine. The principle can be visualized by referring to the velocity triangles in Fig. 6.14(a). Assume that the blades are of a simple design with no twist. Then the angle of attack α decreases from the root to the tip of the blade. We notice that when the wind speed increases and the rotational speed remains the same, the angle of attack α increases at all three sections.

When α increases beyond a threshold value, called the *stall angle*, air flow over the upper surface of the airfoil (see Fig. 6.1b) separates from the surface. This phenomenon is called *stall*, and at this stage the lift force decreases dramatically and the drag force increases. Therefore the speed of the blade drops.

For blades with a constant twist angle throughout its length, stall occurs first at the root, where the increase in α is largest and moves towards the tip as the wind speed increases further. Therefore stall regulation offers a passive method for controlling the turbine speed, although it is less predictable.

6.10 Wind Farms

A *wind farm* is a collection of wind turbines located within a region used to generate electricity. They can be very large in extent covering hundreds of square killometers. Turbines in an *offshore wind farm* are located in a body of water.

Wind farms require fairly consistant wind flow, blowing at the right speed over the area of the wind farm. Once an appropriate site is selected, it is important to place the turbines in such a way as to maximize the amount of power received from the wind. Typically they are spaced about 3 to 10 rotor diameters apart. The placement is usually a compromise between the land cost of the farm, and the potential loss of power due to blockage effects. Increasing the spacing between the turbines in the downwind direction reduces the power losses caused by wake effects. This would, however, decrese the number of turbines that could be accomodated in the wind farm.

A potential advantage of a large wind farm is the reduction in intermittancy of the power output [2]. The power output of a single wind turbine can be intermittent because the wind speed can fluctauate widely about the rated wind speed. However, turbines located over a geographically diverse large land area are unlikely to experience similar fluctuations in wind speed at the same time. Therefore their combined power output would be less intermittant.

Offshore wind farms address some of the limitations of land-based wind farms like the space for wind turbines. Moreover, turbines located offshore have access to a much better wind resource. However, the cost of locating wind turbines in deep water is more costly and technologically challanging.

The construction methods used in offshore wind farms are called bottom-fixed foundations, and floating wind turbines. Botton-fixed foundations include gravity, monopile, tripod, and jacket foundations [2].

Gravity foundations are made of precast concrete and are used in water that is about 30 m to 50 m deep. Monopile foundations are used in water depths of about 20 m to 30 m. They are made of steel and are driven 30 m into the sea floor. Tripod foundations are made of steel and their feet are drilled into the sea floor. Jacket flundations have steel beams welded together and their feet secured to the sea floor. The water depths are from 50 m to 60 m.

Floating wind turbines can be located in any depth of water. They have floating platforms that are secured to the sea floor by wires.

6.11 Environmental Effects of Wind Energy Systems

The land area required by wind turbine installations varies substantially depending on the site. Turbines placed in flat areas typically use more land than those located in hilly areas. However, these do not occupy all of this land because they must be spaced approximately 5 to 10 rotor diameters apart. In the USA a new utility-scale wind power facility, which is typically about 2 MW capacity, uses between 12 and 57 hectare per MW of power output capacity [9]. Only about 0.4 hectare per MW is distributed permanently and less than 1.4 hectare per MW is distributed during construction. Thus, the turbines themselves and the surrounding infrastructure occupy a small portion of the total area of a wind facility. A significant fraction of the land in a wind farm can be used for a variety of other productive purposes, including livestock grazing, agriculture, highways, and hiking.

Many recent research studies found evidence of bird and bat deaths from collisions with wind turbines and due to changes in air pressure caused by the spinning turbines, as well as from habitat disruption. Offshore wind turbines can have similar impacts on marine birds, but as with onshore wind turbines bird deaths associated with offshore wind turbines are minimal. Wind farms located offshore will also impact fish and other marine wildlife [9].

Sound and visual impact are the two main public health and community concerns associated with operating wind turbines. Most of the sound generated by wind turbines is caused by the movement of turbine blades through the air. There is also mechanical sound generated by the turbine itself. Overall sound levels depend on turbine design and wind speed. Depending on the location and the lighting conditions, wind turbines can create an effect known as 'shadow flicker'. With careful siting, planting trees, or installing window awnings, this annoying condition can be minimized [9].

Wind energy systems do not produce any global warming emissions during operation. However, the mean life-cycle global warming potential of offshore and onshore wind turbines are assessed to be 0.12 and 0.11 kg of carbon dioxide per kWh respectively [9].

6.12 Cost Evaluation of Wind Power Systems

The general methods for cost evaluation of energy systems were discusses in section 1.5 in chapter 1. In this section we will briefly outline their application to wind energy systems. The *levelized cost of energy* is the metric that is often used to compare the economic viability of different energy systems. It is given by the expression [4]

$$LCOE = \frac{FCR \times ICC}{AEP_{net}} + AOE \qquad (6.106)$$

where

FCR	the fixed charge rate used to annualize the initial cost
ICC	the initial capital cost of the turbine and system
CF	capacity factor
P_{rated}	rated turbine power (kW)
AEP_{net}	net annual energy production (kWh/year)

$AEP_{net} = CF \times P_{rated} \times 8760$

LLC	the land lease cost (\$/kWh)
$O\&M$	the levelized operating and maintenance cost (\$/year)
LRC	the levelized replacement and overhaul cost (\$/year)
AOE	annual operating expenses (\$/kWh)

$$AOE = LLC + \frac{(O\&M + LRC)}{AEP_{net}}$$

We shall illustrate the application of the costing method in worked example 6.20.

Published data [8] on the variation of LCOE in the USA over a period of 10 years is listed in Table 6.3. The high and low values are based on the assumptions made in the computation. The average LCOE has dropped steadily, with the decrease becoming more gradual during last five years.

Table 6.3 Mean levelized cost $(LCOE)_{av}$, $(LCOE)_h$, $(LCOE)_l$ of wind power systems in the USA, in \$/MWh. (Data from Ref.[8])

Year	2009	2010	2011	2012	2013	2014
$(LCOE)_{av}$	135	123.5	71	71.5	70	59
$(LCOE)_h$	169	148	92	95	95	81
$(LCOE)_l$	101	99	50	48	45	37

Year	2015	2016	2017	2018	2019	2020
$(LCOE)_{av}$	54.5	47	45	42.5	41	40
$(LCOE)_h$	77	62	60	56	54	54
$(LCOE)_l$	32	32	30	29	28	26

6.13 Worked Examples

Example 6.1 A wind turbine with a rotor diameter of 15 m is installed at a location with an average annual wind speed of 8.5 ms^{-1}. The conversion efficiency of the turbine-generator system is 38%. Estimate the annual energy output of the system based on the average wind speed. The mean density of air is 1.225 kgm^{-3}.

Solution The *power* available in the wind is the product of the mass flow rate and the kinetic energy per unit mass. Hence we have

$$P_{wind} = \dot{m}E_{ke} = \frac{1}{2}A\rho U^3$$

For a wind turbine with blades of radius *R*, the above expression becomes

$$P_{wind} = \frac{1}{2}\rho\pi R^2 U^3$$

Based on the *average annual* wind speed, the average power in the wind is

$$\bar{P}_{wind} = \frac{1}{2} \times 1.225 \times \pi \times 7.5^2 \times 8.5^3 = 66.47 \text{ kW}$$

The average power output of the system is

$$\bar{P}_{out} = 0.38 \times 66.47 = 25.25 \text{ kW}$$

The average annual energy output is given by

$$\bar{E}_{out} = 25.25 \times 3600 \times 24 \times 365 \times 10^{-6} = 796.3 \text{ GJ}$$

This is only a first estimate because taking the cube of the average wind speed is strictly not correct. The wind speed distribution used to obtain the average wind speed needs to be used to estimate the annual energy output. Such a detailed analysis is performed in worked example 6.7.

Example 6.2 A three-bladed wind turbine with a rotor diameter of 35 m produces a rated power output of 350 kW at a wind speed of 12 ms^{-1}. The height of the tower is 45 m. The tip speed ratio of the rotor is 4.8 and the generator speed is 1800 rpm. The thrust coefficient is 0.52. The mean density of air is 1.225 kgm$^{-3.}$ Calculate (i) the rotational speed of the rotor, (ii) the tip speed, (iii) the gear ratio of the gear box between the rotor and the generator, (iv) the power in the wind, (v) the power coefficient, and (vi) the overturning moment acting on the tower at the base.

Solution Using the definition of the tip speed ratio, the rotor speed can be expressed as

$$\omega_{ro} = \frac{\lambda U_W}{R_r} = \frac{4.8\times 12}{17.5} = 3.29 \text{ rad.s}^{-1}$$

The rotational speed is

$$N_r = \frac{\omega_{ro}}{2\pi} = \frac{3.29\times 60}{2\pi} = 31.43 \text{ rpm}$$

The tip speed is given by

$$V_{tip} = \omega_{ro} \times R_r = 3.29 \times 17.5 = 57.57 \text{ ms}^{-1}$$

The gear ratio is given by (see Fig. 6.14b)

$$GR = \frac{N_{gen}}{N_r} = \frac{1800}{31.43} = 57.27$$

Power in the wind is

$$P_{wind} = \frac{1}{2}\rho\pi R^2 U^3 = \frac{1.225\times\pi\times 17.5^2\times 12^3}{2} = 1018.3 \text{ kW}$$

The power coefficient of the turbine is

$$C_p = \frac{P_{out}}{P_{wind}} = \frac{350}{1018.3} = 0.34$$

The thrust coefficient of the turbine is given by Eq. (6.62) as

$$C_T(U_1) = \frac{T_d}{F_{dynamic}} = \frac{T_d}{\frac{1}{2}A\rho {U_1}^2}$$

Hence we have

$$T_d = 0.54 \times \left(\frac{1}{2} \times \pi \times 17.5^2 \times 1.225 \times 12^2\right) = 45.82 \text{ kN}$$

The moment at the base of the tower is given by

$$M_{tb} = T_d H = 45.82 \times 45 = 2062 \text{ kNm}$$

Example 6.3 (i) For an adiabatic atmosphere, show that the temperature varies linearly with elevation.
(ii) (a) Wind speeds (ms^{-1}) measured at two heights of 12 m and 55 m at a location are listed in Table E6.3.1. Obtain the value of α assuming that the variation of wind speed with height is given by

$$\frac{U(z_1)}{U(z_2)} = \left(\frac{z_1}{z_2}\right)^{\alpha}$$

(b) If the wind speed varies with height according to the relation below, obtain the value of z_o.

$$\frac{U(z)}{U(z_r)} = \frac{ln(z/z_o)}{ln(z_r/z_o)}$$

Check the validity of the correlation in Eq. (6.18).

Table E6.3.1 Wind speed U (ms^{-1}) at two heights

z = 12 m	U = 5.0	9.5	7.0	6.3	8.1	7.6	5.6
z = 55m	U = 6.2	11.8	8.7	7.8	10.0	9.4	7.0

Solution (i) The condition for an adaibatic atmosphere is:

$$PV^{\gamma} = c \tag{E6.3.1}$$

where P and V are the pressure and volume respectively, and $\gamma = \frac{c_p}{c_v}$.
Treating air as an ideal gas:

$$PV = mR_aT \tag{E6.3.2}$$

$$c_p - c_v = R$$

$$\frac{\gamma - 1}{\gamma} = \frac{R_a}{c_p}$$

From Eqs. (E6.3.1) and (E6.3.2) we get

$$\frac{P^{\frac{\gamma-1}{\gamma}}}{T} = constant \tag{E6.3.3}$$

Differentiating Eq. (E6.3.3) we have

$$\frac{(\gamma-1)}{\gamma}\left(\frac{dP}{P}\right)=\frac{dT}{T}$$

The pressure variation of the atmosphere with height is given by Eq. (6.10) as:

$$\frac{dP}{dz}=-\left(\frac{Pg}{R_aT}\right)$$

From the above two equation we obtain

$$dT=\frac{g(\gamma-1)dz}{\gamma R_a}=-\frac{g}{c_p}dz$$

Substituting $c_p = 1003$ JkgK^{-1} for air, and $g = 9.81$ ms^{-1} we get

$$\frac{dT}{dz}=0.0098 \text{ Km}^{-1}$$

Therefore the temperature varies linearly with elevation.

(ii) (a) Applying the power-law wind velocity distribution to the two heights we have

$$\frac{U(55)}{U(12)}=\left(\frac{55}{12}\right)^{\alpha}=4.583^{\alpha}$$

$$U(55)=4.583^{\alpha}U(12)$$

We plot a graph of $U(55)$ versus $U(12)$ and obtain the slope by linear regression. This could be done by using a software such as MS Exel. Hence we have

$$4.583^{\alpha}=1.234$$

Therefore $$\alpha=0.138$$

(b) Assuming a log variation of wind speed we obtain

$$\frac{U(55)}{U(12)}=\frac{ln(55/z_o)}{ln(12/z_o)}$$

$$U(55)=\frac{ln(55/z_o)}{ln(12/z_o)}U(12)$$

We plot $U(55)$ vs $U(12)$ and obtain the slope by linear regression.

Hence we have

$$1.234 = \frac{ln(55/z_o)}{ln(12/z_o)}$$

The above equation gives

$$z_o = 0.0179 \text{ m}$$

From the data in Table 6.2, this roughness value corresponds approximately to a fallow field terrain.

Counihan [4] obtaind the following correlation to predicted value of α.

$$\alpha = 0.096Log_{10}z_o + 0.016(Log_{10}z_o)^2 + 0.24$$

for 0.001 m < z_o < 10 m, where z_o is the surface roughness in m, which is listed in Table 6.2 for a number terrain types. Substituting $z_o = 0.0179$ in the correlation we get $\alpha = 0.121$. In part (a) we obtained $\alpha = 0.138$.

Example 6.4 The measured wind speed at a height of 10 m at a location with a few trees is 6 ms^{-1}. (i) Using the power law method and the log law method, calculate the wind speed at a height of 50 m. (ii) If all the trees at the location are removed, calculate the wind speed at a height of 50 m.

Solution For a terrain with a few trees, $z_o = 0.1$ m, from data in Table 6.2. The log-law wind speed distribution gives

$$\frac{U(50)}{U(10)} = \frac{ln(50/z_o)}{ln(10/z_o)}$$

The wind speed at a height of 50 m is

$$U(50) = \frac{ln(50/0.1)}{ln(10/0.1)} \times 6 = 8.1 \text{ ms}^{-1}$$

We use the following correlation to predicted value of α.

$$\alpha = 0.096Log_{10}z_o + 0.016(Log_{10}z_o)^2 + 0.24 \qquad \text{(E6.4.1)}$$

Substituting $z_o = 0.1$ m, we get $\alpha = 0.16$

Applying the power law speed distribution we get

$$\frac{U(50)}{U(10)} = \left(\frac{50}{10}\right)^{\alpha} = 5^{0.16}$$

$$U(50) = 6 \times 5^{0.16} = 7.76 \text{ ms}^{-1}$$

(ii) When the trees are removed, the data in Table 6.2 gives, $z_o = 0.01$ m. The correlation in Eq. (E6.4.1) gives, $\alpha = 0.112$.

We first use the two terrain conditions to determine the speed at 10 m when the trees are removed. The relevant equation is

$$U(z) = \frac{U^*}{k} ln\left(\frac{z}{z_o}\right)$$

Assume that $\left(\frac{U^*}{k}\right)$ is constant. Then we have

With trees: $U_1(10) = \frac{U^*}{k} ln\left(\frac{10}{0.1}\right) = \frac{U^*}{k} \times 4.605 = 6$

With trees removed: $U_2(10) = \frac{U^*}{k} ln\left(\frac{10}{0.01}\right) = \frac{U^*}{k} \times 6.908$

Hence the wind speed at 10 m with trees removed is 9 ms^{-1}.

Using the log-law we obtain the speed at 50 m as

$$U(50) = \frac{ln(50/0.01)}{ln(10/0.01)} \times 9 = 11.1 \text{ ms}^{-1}$$

Applying the power law we have

$$U_2(50) = U_2(10) \times 5^{0.112} = 9 \times 5^{0.112} = 10.77 \text{ ms}^{-1}$$

Example 6.5 The hub height of a wind turbine is 60 m and the rotor diameter is 30 m. It is installed at a location with a few trees. Obtain the ratio of the power available in the wind at the highest point of the rotor to the lowest point of the rotor.

Solution The power available in the wind is given by Eq. (6.4)

$$P_{wind} = \frac{1}{2}\rho\pi R^2 U^3 \quad \text{(E6.5.1)}$$

The height at the top of the rotor is 90 m and at the bottom of the rotor is 30 m. Let the wind speeds at the two heights be U_1 and U_2. Neglecting any changes in air density and applying Eq. (E6.5.1) at the two heights we have

$$\frac{P_{w1}}{P_{w2}} = \frac{{U_1}^3}{{U_2}^3} \tag{E6.5.2}$$

For a terrain with a few trees, we obtain from Table 6.2, $z_o = 0.1$ m

We apply the log law to obtain the ratio of the wind speeds at the two heights. Hence we have

$$\frac{U(90)}{U(30)} = \frac{ln(90/z_o)}{ln(30/z_o)} = \frac{ln900}{ln300} = 1.1926 \tag{E6.5.3}$$

Substituting in Eq. (E6.5.2) from Eq. (E6.5.3) we obtain the ratio of the power available at the two heights as

$$\frac{P_{w1}}{P_{w2}} = \frac{{U_1}^3}{{U_2}^3} = 1.1926^3 = 1.696$$

Example 6.6 A wind turbine rated at 1.2 MW has a rotor with a radius of 3 m. The rated wind speed is 14 ms^{-1}. The cut-in and cut-out speeds are 4 ms^{-1} and 24 ms^{-1} respectively. The turbine is located at a site with an average wind speed of 10 ms^{-1}. The wind speed distribution is well represented by the Rayleigh PDF. Calculate (i) the number of hours per year that the wind speed is below the cut-in speed, (ii) the number of hours per year that the speed exceeds the cut-out speed, and (iii) the yearly energy production when the turbine is running at the rated power.

Solution We shall use the Rayleigh cumulative distribution to calculate the required number of hours. The Rayleigh *cumulative frequency* distribution is given by

$$F_R(U) = 1 - exp\left[-\frac{\pi}{4}\left(\frac{U}{\bar{U}}\right)^2\right]$$

The fraction of the number of hours that the wind speed is below the cut-in speed is given by

$$F_R(0 \rightarrow 4) = 1 - exp\left[-\frac{\pi}{4}\left(\frac{4}{10}\right)^2\right] = 0.118$$

Hence the number of hours is,

$$N = 8760 \times 0.118 = 1034 \text{ hours.}$$

The fraction of the number of hours the wind speed exceeds the cut-out speed is given by

$$F_R(24 \to \infty) = 1 - \left\{1 - exp\left[-\frac{\pi}{4}\left(\frac{24}{10}\right)^2\right]\right\} = 0.0108$$

Hence the number of hours is

$$N = 8760 \times 0.0108 = 94.6 \text{ hours.}$$

The turbine operates at the rated power between the rated speed and the cut-out speed. The fraction of the number of hours the wind speed is between these limits is

$$F_R(14 \to 24) = exp\left[-\frac{\pi}{4}\left(\frac{14}{10}\right)^2\right] - exp\left[-\frac{\pi}{4}\left(\frac{24}{10}\right)^2\right] = 0.2037$$

Hence the number of hours is,

$$N = 8760 \times 0.2037 = 1784 \text{ hours.}$$

The energy produced during this period is

$$1.2 \times 1784 = 2140.9 \text{ MWh}$$

Fig. E6.7.1 Wind speed bin- data

Bin No.	1	2	3	4	5	6	7
U_{min}, ms^{-1}	0	3	4	5	6	7	8
U_{max}, ms^{-1}	3	4	5	6	7	8	9
$U_{ave.}$, ms^{-1}	1.5	3.5	4.5	5.5	6.5	7.5	8.5
Hours	781	536	673	806	838	831	833
Frequency	8.92	6.12	7.68	9.20	9.57	9.49	9.51

Bin No.	8	9	10	11	12	13
U_{min}, ms^{-1}	9	10	11	12	13	21
U_{max}, ms^{-1}	10	11	12	13	21	-
U_{ave}, ms^{-1}	9.5	10.5	11.5	12.5	17	-
Hours	683	574	474	367	1358	2
Frequency	7.8	6.6	5.41	4.19	15.5	0.02

Example 6.7 A horizontal axis windmill has a rated output of 1500 kW and the rotor diameter is 69 m. It is operating at a location with the annual wind speed data listed in Table E6.7.1. The density of air is 1.15 kgm^{-3}. Thirteen wind speed bins are given in columns 1-13. The average wind speed of a bin is in row 4. Measured values of the number of hours during which the wind speed falls within the speed interval of the bin is given in row 5. Calculate (i) the frequency of occurrence for each bin, (ii) the average annual wind speed, (iii) the standard deviation of the wind speed, (iv) the energy available in the wind annually, and (v) the annual energy production of a wind turbine with a mean system efficiency of 19%.

Solution From the data in Table E6.7.1, the frequency of occurrence of a bin is given by

$$f_i = \frac{n_i}{8760}$$

for example, bin 10,

$$f_{10} = \frac{474}{8760} = 5.41\ \%$$

The rest of the frequencies are listed in row 6. The *average annual* wind speed is given by

$$\bar{U} = \sum_1^{13} f_i U_{bi} = (1.5 \times 8.92 + 3.5 \times 6.12 + 4.5 \times 7.68 + \cdots)/100$$

$$\bar{U} = 8.56\ \text{ms}^{-1}$$

The standard deviation is given by

$$\sigma_w{}^2 = \frac{N}{N-1}\left[\sum_1^{N_B} u_i{}^2 f_i - \bar{U}^2\right]$$

$$\sigma_w{}^2 = \frac{13}{12}\left[(8.92 \times 1.5^2 + 6.12 \times 3.5^2 + 7.68 \times 4.5^2 \ldots)/100 - 8.56^2\right]$$

$$\sigma_w{}^2 = 23.04$$

The standard deviation is 4.8 ms^{-1}.

The swept area of the rotor is

$$A_t = \frac{\pi}{4}D^2 = \frac{\pi}{4}69^2 = 3739.2 \text{ m}^2$$

The energy available in the wind is given by

$$\bar{P} = \frac{A\rho}{2N}\sum_1^{N_B} u_i{}^3 n_i = \frac{A\rho}{2}\sum_1^{N_B} u_i{}^3 f_i$$

$$\bar{P} = \frac{1.15\times 3739.2}{2}[(8.92 \times 1.5^3 + 6.12 \times 3.5^3 + 7.68 \times 4.5^3 \ldots)/100]$$

$$\bar{P} = 2.64 \times 10^3 \text{ kW}$$

Hence the *average* power of the wind is 2640 kW. The average output of the wind turbine is

$$0.19 \times 2640 = 501.6 \text{ kW}.$$

Example 6.8 (a) Using the wind speed bin data given in Table E6.7.1, obtain the Rayleigh PDF that fits the average wind speed. Hence calculate (i) the average power available in the wind, and (ii) the annual energy production of a wind turbine with a mean efficiency of 19%.
(b) Use the Rayleigh PDF with the same average wind speed directly to obtain the average power of the wind at the location.

Solution (a) The Rayleigh *cumulative frequency* distribution is given by

$$F_R(U) = 1 - exp\left[-\frac{\pi}{4}\left(\frac{U}{\bar{U}}\right)^2\right]$$

We shall now use the above *cumulative distribution* to estimate the numbers of hours in the bins listed in Table E6.7.1.

Let the wind speeds of the boundary of bin n be $U_{min,n}$ and $U_{max,n}$. Then the number of hours per year in bin n is given by

$$\delta N_n = N_{dy}\left[F_R\left(U_{max,n}\right) - F_R\left(U_{min,n}\right)\right]$$

The frequency of occurrence for bin n is given by

$$f_{rn} = exp\left[\frac{\pi}{4}\left(\frac{U_{min}}{\bar{U}}\right)^2\right] - exp\left[-\frac{\pi}{4}\left(\frac{U_{max}}{\bar{U}}\right)^2\right]$$

The mean wind speed of the Rayleigh distribution is equal to that of the measured speed distribution. Hence we have

$$\bar{U} = 8.56 \text{ ms}^{-1}$$

Consider bin number 5 as an example to illustrate the procedure.

$$U_{min} = 6 \text{ ms}^{-1}, U_{max} = 7 \text{ ms}^{-1}$$

The frequency for bin 5 is

$$f_{rn} = exp\left[-\frac{\pi}{4}\left(\frac{6}{8.56}\right)^2\right] - exp\left[-\frac{\pi}{4}\left(\frac{7}{8.56}\right)^2\right] = 8.85\%$$

Using the same equation we compute the frequencies of all the bins, which are listed in the Table E6.8.1 below.

Table E6.8.1 Bin data obtained using the Rayleigh PDF

Bin no.	1	2	3	4	5	6	7
f_n %	9.19	6.56	7.74	8.50	8.85	8.87	8.39

Bin no.	8	9	10	11	12	13
f_n %	7.73	6.90	5.98	5.03	15.48	-

The annual average power available in the wind is given by

$$\bar{P} = \frac{A\rho}{2N}\sum_1^{N_B} u_i^3 n_i = \frac{A\rho}{2}\sum_1^{N_B} u_i^3 f_i$$

$$\bar{P} = \frac{1.15 \times 3739.2}{2}[(9.19 \times 1.5^3 + 6.56 \times 3.5^3 + 7.74 \times 4.5^3 \ldots)/100]$$

$$\bar{P} = 2.6456 \times 10^3 \text{ kW}$$

Hence the *average* power of the wind is 2645.6 kW.

The average output of the wind turbine is

$$0.19 \times 2645.6 = 502.7 \text{ kW}.$$

(b) The Rayleigh PDF is:

$$f_R(U) = \frac{\pi}{2}\left(\frac{U}{\bar{U}^2}\right) exp\left[-\frac{\pi}{4}\left(\frac{U}{\bar{U}}\right)^2\right]$$

The average power of the wind is given by

$$\bar{P}_W = \frac{\rho A}{2}\int_0^\infty U^3 f_R(U)dU$$

$$\bar{P}_W = \frac{\rho A}{2}\int_0^\infty \frac{\pi}{2}\left(\frac{U}{\bar{U}^2}\right) exp\left[-\frac{\pi}{4}\left(\frac{U}{\bar{U}}\right)^2\right] U^3 dU \qquad \text{(E6.8.1)}$$

Substituting numerical values we get

$$\bar{P}_W = \frac{1.15\times 3739.2}{2}\int_0^\infty \frac{\pi}{2}\left(\frac{U}{8.56^2}\right) exp\left[-\frac{\pi}{4}\left(\frac{U}{8.56}\right)^2\right] U^3 dU$$

The integration of the RHS of the above equation is done numerically using the trapezoidal rule. Hence we get annual average power in the wind as 2575.6 kW.

The integral in Eq. (E6.8.1) has an analytical expression, which is available in Ref.[5]. Using it we obtain the average power in the wind as

$$\bar{P}_{wa} = \frac{3}{4}\rho D^2 \bar{U}^3$$

$$\bar{P}_{wa} = 0.75 \times 1.15 \times 69^2 \times 8.56^3 = 2575.6 \text{ kW}$$

The average wind power obtained using the measured bin frequencies and bin- average speeds in example E6.7 is 2640 kW.

Example 6.9 Using the average wind speed of the measured speed distribution given in Table E6.7.1, obtain the Weibull PDF that fits the data. Hence calculate (i) the energy available in the wind annually, and (ii) the annual energy production of a wind turbine with a mean efficiency of 19%.

Solution The Weibull *cumulative frequency* distribution is given by

$$F_W(U) = 1 - exp\left[-\left(\frac{U}{c}\right)^k\right]$$

Two approximate analytical expressions that relate the mean and the standard deviation to the parameters k and c of the Weibull distribution are given in Ref.[5]. They are valid for $1 \leq k \leq 10$.

$$k = \left(\frac{\sigma_U}{\bar{U}}\right)^{-1.086} \qquad c = \frac{\bar{U}}{\Gamma(1+1/k)}$$

where Γ is the gamma function.

We shall now use the above cumulative distribution to estimate the number of hours in the bins listed in Table E 6.7.1. Let the wind speeds of the boundary of bin n be $U_{min,n}$ and $U_{max,n}$. Then the number of hours per year in bin n is given by

$$\delta N_n = N_{dy}\left[F_W\left(U_{max,n}\right) - F_W\left(U_{min,n}\right)\right]$$

The frequency is given by

$$f_{rn} = exp\left[-\left(\frac{U_{min}}{c}\right)^k\right] - exp\left[-\left(\frac{U_{max}}{c}\right)^k\right]$$

The mean wind speed and the standard deviation of the Weibull distribution are equal to corresponding parameters of the measured speed distribution given in Table E6.7.1. Hence we have

$$\bar{U} = 8.56 \text{ ms}^{-1} \text{ and } \sigma_U = 4.8 \text{ ms}^{-1}$$

Hence we obtain the parameters of the Weibull distribution as:

$$k = \left(\frac{\sigma_U}{\bar{U}}\right)^{-1.086} = \left(\frac{4.8}{8.56}\right)^{-1.086} = 1.874$$

$$c = \frac{\bar{U}}{\Gamma(1+1/k)} = 9.64$$

Consider bin number 5 as an example to illustrate the procedure.

$$U_{min} = 6 \text{ ms}^{-1}, \qquad U_{max} = 7 \text{ ms}^{-1}$$

The frequency for bin 5 is

$$f_{rn} = exp\left[-\left(\frac{6}{9.64}\right)^{1.874}\right] - exp\left[-\left(\frac{7}{9.64}\right)^{1.874}\right] = 8.53\%$$

Using the same equation we compute the frequencies of all the bins listed in Table E6.9.1 below.

Table E6.9.1 Bin data obtained using the Weibull PDF

Bin no.	1	2	3	4	5	6	7
f_n %	10.6	6.88	7.83	8.37	8.53	8.34	7.89

Bin no.	8	9	10	11	12	13
f_n %	7.25	6.48	5.64	4.80	16.03	-

The *average power* available in the wind is given by

$$\bar{P} = \frac{A\rho}{2N}\sum_1^{N_B} u_i^{\,3} n_i = \frac{A\rho}{2}\sum_1^{N_B} u_i^{\,3} f_i$$

$$\bar{P} = \frac{1.15 \times 3739.2}{2}[(10.6 \times 1.5^3 + 6.88 \times 3.5^3 + 7.83 \times 4.5^3 \,...)/\,100\,]$$

$$\bar{P} = 2.652 \times 10^3 \text{ kW}$$

Hence the *average power* of the wind is 2652 kW.

The average output of the wind turbine is

$$0.19 \times 2652.0 = 503.9 \text{ kW}.$$

Example 6.10 A wind turbine with a rated output of 1500 kW is installed at a location having the annual wind speed distribution given in Table E6.7.1. The diameter of the rotor is 69 m and the density of air is 1.15 kgm^{-3}. The actual power curve of the turbine is well approximated by the following analytical form:

For $0 < U < 2.5$ ms^{-1} $\quad P(U) = 0$ kW

For $2.5 \leq U < 12.5$ ms^{-1} $\quad P(U) = 1.5(U - 2.5)^3$ kW

For $12.5 \leq U < 22$ ms^{-1} $\quad P(U) = 1500$ kW

For $22 \geq U$ ms^{-1} $\quad P(U) = 0$ kW

Use the measured wind speed data in Table E6.7.1 to calculate (i) the annual energy output of the turbine, (ii) average efficiency of the turbine, (iii) the variation of the power coefficient C_p with wind speed, and (iv) the capacity factor.

Solution The area of the rotor is

$$A = \frac{\pi}{4}D^2 = \frac{\pi}{4} \times 69^2 = 3739.3 \text{ m}^2$$

The mean turbine power output is given by Eq. (6.74) as:

$$\bar{P}_w = \int_0^{\infty} P_w(U)\, f(U)dU$$

The above variations are depicted graphically in Fig. (6.11).

We express the above equation in terms of the bin frequencies, and the power output at the bin-averaged speeds. Hence we have

$$\bar{P}_w = \sum_0^{2.5} P_{wi} f_i + \sum_{2.5}^{12.5} P_{wi} f_i + \sum_{12.5}^{22} P_{wi} f_i + \sum_{22}^{\infty} P_{wi} f_i$$

$$\bar{P}_w = \sum_0^{2.5} 0 \times f_i + \sum_{2.5}^{12.5} 1.5(U_i - 2.5)^3 \times f_i + \sum_{12.5}^{22} 1500 \times f_i + \sum_{22}^{\infty} 0 \times f_i$$

Substituting the average wind speeds and the frequencies listed in Table E6.7.1 in the above equation we obtain the *average power output* as 507.85 kW.

In worked example 6.7 the annual average power available in the wind was obtained as 2640 kW.

The average efficiency of the turbine is given by

$$\eta = \frac{\text{average power output}}{\text{average power in the wind}} = \frac{507.85}{2640} = 19.2\%$$

The *power coefficient*, C_p is the ratio of the power output of the turbine to the power in the wind at any wind speed.

It is given by the expression

$$C_p = \frac{P(U)}{\frac{1}{2}\rho A U^3}$$

Consider the different ranges of the turbine power curve given above.

For $0 < U < 2.5$ ms^{-1}, $P(U) = 0$ kW, $C_p = 0$

For $2.5 \leq U < 12.5$ ms^{-1}, $P(U) = 1.5(U - 2.5)^3$ kW,

$$C_p = \frac{1.5(U-2.5)^3}{\frac{1}{2}\rho A U^3} = \frac{1.5\times10^3}{0.5\times1.15\times3739.3}\left(1 - \frac{2.5}{U}\right)^3 = 0.698\left(1 - \frac{2.5}{U}\right)^3$$

For $12.5 \leq U < 22$ ms^{-1}, $P(U) = 1500$ kW,

$$C_p = \frac{1500}{\frac{1}{2}\rho A U^3} = \frac{1500}{2.15\times10^3\times U^3}$$

For $22 \geq U$ ms^{-1}, $P(U) = 0$ kW, $C_p = 0$

We see that the C_p increases from zero for $U \leq 2.5$ ms^{-1} to a maximum of 0.357 at $U = 12.5$ ms^{-1}. It then decreases steadily to 0.065 at $U = 22$ ms^{-1}.

The capacity factor is given by

$$CF = \frac{annual\ energy\ output(kWh)}{energy\ output\ at\ rated\ power(kWh)} = \frac{507.85\times8760}{1500\times8760} = 0.338$$

Example 6.11 Using the Rayleigh distribution for the wind speed at the location in example 6.8 with same wind turbine as in example 6.10, calculate (i) the annual energy output of the turbine, (ii) the average efficiency of the turbine, and (iii) the capacity factor. Compare the solutions obtained using the bin method and direct numerical integration of the power curve for the Rayleigh PDF.

Solution We obtained the Rayleigh frequency distribution listed in Tables E6.8.1 above. The average power output of the turbine is given by

$$\bar{P}_w = \Sigma_0^{2.5} 0 \times f_i + \Sigma_{2.5}^{12.5} 0.768U_i{}^3 \times f_i + \Sigma_{12.5}^{22} 1500 \times f_i + \Sigma_{22}^{\infty} 0 \times f_i$$

Substituting the Rayleigh frequency distribution given in Table E6.8.1 in the above equation the turbine power output is obtained as 522.3 kW. The average power in the wind for the Rayleigh distribution is 2645.6 kW.

The average efficiency of the turbine is given by

$$\eta = \frac{average\ power\ output}{average\ power\ in\ the\ wind} = \frac{522.3}{2645.6} = 19.7\%$$

The capacity factor is given by

$$CF = \frac{annual\ energy\ output(kWh)}{energy\ output\ at\ rated\ power(kWh)} = \frac{522.3\times8760}{1500\times8760} = 0.348$$

Example 6.12 An ideal wind turbine has a constant power coefficient equal to β times the Betz limit of, $C_{pb} = 16/27$. The wind speed distribution follows a Rayleigh PDF. (i) Obtain an expression for the average power output of the turbine. (ii) Calculate the annual energy output of the ideal machine for a mean wind speed of 6.5 ms^{-1} and rotor radius of 9.5 m. The air density is 1.225 kgm^{-3}.

Solution The power output at a wind speed U is given by Eq. (6.6) as:

$$P_w = \frac{1}{2}\rho A\eta C_p(U)U^3$$

For the ideal turbine considered, the efficiency $\eta = 1$ and the power coefficient, $C_p = \beta C_{pb}$. Notice that $\beta = 1$ for the Betz limit. The wind speed probability follows a Rayleigh distribution. The average power output is given by Eq. (6.76) as:

$$\bar{P}_w = \int_0^\infty P_w(U) f_R(U) dU = \frac{1}{2}\rho A\eta \int_0^\infty C_p(U)U^3 f_R(U)\, dU \qquad \text{(E6.12.1)}$$

For a constant power coefficient we have

$$\bar{P}_w = \frac{1}{2}\rho A\eta\beta C_{pb} \int_0^\infty U^3 f_R(U) dU \qquad \text{(E6.12.2)}$$

The Rayleigh PDF is given by

$$f_R(U) = \frac{\pi}{2}\left(\frac{U}{\bar{U}^2}\right) exp\left[-\frac{\pi}{4}\left(\frac{U}{\bar{U}}\right)^2\right] \qquad \text{(E6.12.3)}$$

Substituting from Eq. (E6.12.3) in (E6.12.2) we get

$$\bar{P}_w = \frac{\pi}{4}\rho A\eta\beta C_{pb}\int_0^\infty \left(\frac{U^4}{\bar{U}^2}\right) exp\left[-\frac{\pi}{4}\left(\frac{U}{\bar{U}}\right)^2\right] dU \quad \text{(E6.12.4)}$$

For an *ideal* machine

$$\eta = 1, \quad \beta = 1, \quad \text{and} \quad C_{pb} = \frac{16}{27}$$

An analytical expression for the integral on the RHS of Eq. (E6.12.4) is available in Ref.[5]. Using it we get

$$\bar{P}_w = \frac{3C_{pb}}{4}\rho D^2\bar{U}^3$$

Substituting numerical data in the above equation we have

$$\bar{P}_w = 0.75 \times \left(\frac{16}{27}\right) \times 1.225 \times 19^2 \times 6.5^3 = 53.98 \text{ kW}$$

The annual energy production is given by

$$E_y = 8760\bar{P}_w = 8760 \times 53.98 = 472.8 \text{ MWh}$$

Example 6.13 A wind turbine is to be installed at a location with an average wind speed of 5.6 ms^{-1}. The rated power and speed of the turbine are 10 kW and 10 ms^{-1}. The power curve $P_w(U)$ versus wind speed U of the turbine is approximated by the form:

$$P_w(U) = 0 \text{ kW} \quad \text{for} \quad 0 < U \le 4.0 \text{ ms}^{-1}$$

$$P_w(U) = 10\left(\frac{U}{6} - \frac{2}{3}\right) \text{kW} \quad \text{for} \quad 4.0 < U \le 10.0 \text{ ms}^{-1}$$

$$P_w(U) = 10 \text{ kW} \quad \text{for} \quad 10.0 < U \le 20.0 \text{ ms}^{-1}$$

$$P_w(U) = 0 \text{ kW} \quad \text{for} \quad U \ge 20.0 \text{ ms}^{-1}$$

Using the Rayleigh PDF for the wind speed distribution, calculate (i) the annual energy output, and (ii) the capacity factor.

Solution The average power output is obtain using the Rayleigh wind speed distribution. Therefore

$$\bar{P}_w = \int_0^\infty P_w(U) f_R(U)\, dU \quad \text{(E6.13.1)}$$

The Rayleigh distribution is given by

$$f_R(U) = \frac{\pi}{2}\left(\frac{U}{\bar{U}^2}\right) exp\left[-\frac{\pi}{4}\left(\frac{U}{\bar{U}}\right)^2\right]$$

The integration of the Eq. (E6.13.1) is carried out considering the different ranges of wind speed for which the power curve equations are given above. Hence we have

For $0 < U \leq 4.0$ $\quad \bar{P}_{w1} = \int_0^4 0 \times f_R(U)\, dU = 0$

For $4.0 < U \leq 10.0$ $\quad \bar{P}_{w2} = \int_4^{10} 10\left(\frac{U}{6} - \frac{2}{3}\right) \times f_R(U)\, dU$

For $10.0 < U \leq 20.0$ $\quad \bar{P}_{w3} = \int_{10}^{20} 10 \times f_R(U)\, dU$

For $20.0 < U \leq \infty$ $\quad \bar{P}_{w4} = \int_{20}^{\infty} 0 \times f_R(U)\, dU = 0$

The average wind speed is, $\bar{U} = 5.6\ \mathrm{ms}^{-1}$.

Substituting the Rayleigh PDF in the power curve equations above we have

$$\bar{P}_w = \left(\frac{10}{6}\right)\int_4^{10}\left(\frac{\pi U^2}{2\times 5.6^2}\right) exp\left(-\frac{\pi}{4}\left(\frac{U}{5.6}\right)^2\right) dU - \left(\frac{20}{3}\right)\left[exp\left(-\frac{\pi}{4}\left(\frac{4}{5.6}\right)^2\right) - exp\left(-\frac{\pi}{4}\left(\frac{10}{5.6}\right)^2\right)\right] + 10\left[exp\left(-\frac{\pi}{4}\left(\frac{10}{5.6}\right)^2\right) - exp\left(-\frac{\pi}{4}\left(\frac{20}{5.6}\right)^2\right)\right]$$

The integration of the first term on the RHS of the above equation is performed numerically using the trapezoidal rule. Hence the total power output is

$$\bar{P}_w = 7.079 - 3.9208 + 0.8168 = 3.975\ \mathrm{kW}$$

The rated power is 10 kW. Therefore the capacity factor,

$$CF = \frac{3.975\times 8760}{10\times 8760} = 0.39$$

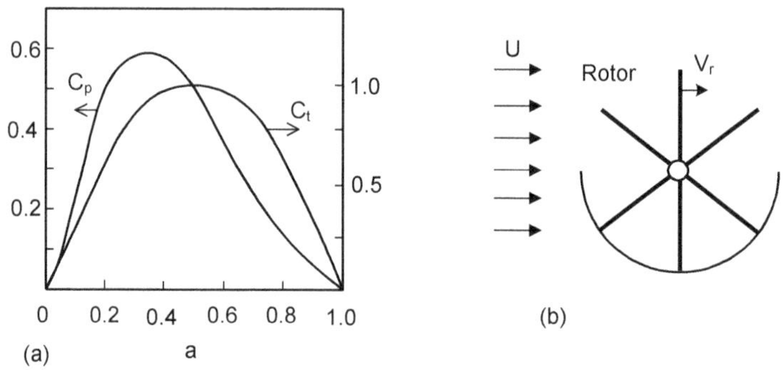

Fig. E6.14 (a) Variation of C_p and C_t with a (b) Vertical axis turbine model

Example 6.14 A wind turbine with a rotor radius of 4m has an axial induction factor of 0.4. The wind speed is 6 ms^{-1} and the air density is 1.225 kgm^{-3}. Using the actuator disk model, calculate (i) the power output, (ii) the thrust, (iii) the power coefficient, and (iv) the thrust coefficient. Derive an expression for the power coefficient in terms of the thrust coefficient. Hence obtain the upper limit of the power coefficient.

Solution The area of the rotor is

$$A = \pi R^2 = \pi \times 4^2 = 50.26 \text{ m}^2$$

The actuator disk model gives the power output as (Eq.6.59):

$$P_e = 2A\rho {U_1}^3 a(1-a)^2$$

$$P_e = 2 \times 50.26 \times 1.225 \times 6^3 \times 0.4 \times (1-0.4)^2 = 3830 \text{ W}$$

The actuator disk model gives the thrust as (Eq. 6.58):

$$T_a = 2A\rho {U_1}^2 a(1-a)$$

$$T_a = 2 \times 50.26 \times 1.225 \times 6^2 \times 0.4(1-0.4) = 1064 \text{ N}$$

The power in the wind is given by

$$P_w = \frac{1}{2} A\rho {U_1}^3 = \frac{1}{2} \times 50.26 \times 1.225 \times 6^3 = 6649.4 \text{ W}$$

The power coefficient is given by

$$C_p = \frac{P_e}{P_w} = \frac{3830}{6649.4} = 0.576$$

Note that the maximum power coefficient is the Betz limit of 0.59 which occurs when $a = \frac{1}{3}$.

The thrust coefficient is given by

$$C_t = \frac{T_a}{0.5A\rho {U_1}^2} = \frac{1064}{0.5\times50.26\times1.225\times6^2} = 0.96$$

The variation of C_p and C_t with the axial induction factor is shown in Fig. E6.14(a).

The power coefficient and the thrust coefficient are given by Eqs. (6.61) and (6.63) respectively as:

$$C_p = 4a(1-a)^2$$

$$C_t = 4a(1-a)$$

Eliminating a between the above equations we obtain the relation

$${C_p}^2 - C_pC_t + {C_t}^3/4 = 0$$

Hence we obtain

$$C_p = \frac{C_t}{2}\left[1 \pm (1-C_t)^{1/2}\right]$$

The upper limit of C_t in the above equation is 1, and at this point the value of C_p is 0.5 and $a = 0.5$. Beyond this upper limit the Betz theory is invalid because U_4 in Eq. (6.57) would be negative if $a > 0.5$.

Example 6.15 An idealized model of a vertical axis wind mill is shown in Fig. E6.14(b). The free stream wind speed is U and the steady rotational speed of the rotor is ω. Obtain an expression for the power coefficient of this turbine. What is the maximum value of the power coefficient?

Solution A schematic diagram of the ideal vertical axis turbine is shown in Fig. E6.14(b). Wind strikes one half of the turbine and drags the blade in the same direction. Such a device is called a drag device because the blades move in the same direction as the wind. The Savonius turbine shown in Fig. 6.2(a) is a real vertical axis turbine working on the same principle.

Let r be the mean radius of the rotor and ω the angular velocity. The relative speed of the wind with respect to the rotor is

$$V_r = U - \omega r$$

The drag force on the rotor is given by [5]

$$F_d = \frac{1}{2} A C_d \rho V_r^{\,2} = \frac{1}{2} A C_d \rho (U - \omega r)^2$$

where A is the drag surface area and C_d is the three-dimensional drag coefficient. For a square plate, $C_d = 1.1$. The torque experienced by the rotor is

$$M_r = F_d r = \frac{1}{2} A C_d \rho (U - \omega r)^2 r$$

The rotor power output is

$$P_t = M_r \omega = \frac{1}{2} A C_d \rho (U - \omega r)^2 r \omega$$

$$P_t = \frac{1}{2} A C_d \rho U^3 \left(1 - \frac{\omega r}{U}\right)^2 \left(\frac{\omega r}{U}\right)$$

The ratio of the rotor surface speed to the wind speed is defined as the speed ratio, $\lambda = \frac{\omega r}{U}$. Hence the power output of the turbine may be expressed as

$$P_t = \frac{1}{2} A C_d \rho U^3 \lambda (1 - \lambda)^2$$

The power of the wind, based on an area equal to $2A$ is given by

$$P_w = \frac{1}{2} \rho (2A) U^3$$

Hence the power coefficient of the turbine may be expressed as [5]

$$C_p = \frac{P_t}{P_w} = \frac{1}{2} C_d \lambda (1 - \lambda)^2$$

The power coefficient is zero when $\lambda = 0$ when there is no motion and when $\lambda = 1$ when there is no relative motion between the wind and the rotor surface. It has a maximum value when $\lambda = \frac{1}{3}$, which is obtained by differentiating the above expression. Assuming $C_d = 1.1$ for a flat plate, the maximum power coefficient is 0.08. Note that this value is significantly lower than the Betz limit of 0.59 obtained from the actuator disk model.

Example 6.16 A three-bladed wind turbine is designed for the following conditions: (i) free stream wind speed is 8 ms^{-1}, (ii) rotational speed is 7.5 rpm, (iii) blades have a constant cross section and no twist, length is 40 m, chord is 1.0 m, lift coefficient is 1.2, drag coefficient is 0.08, and (iv) density of air is 1.2 kgm^{-3}. Estimate the total torque developed, the power output, and the thrust force at the top of the tower under the design conditions given.

Solution For the purpose of illustrating the main steps of the blade element design procedure, we shall make several simplifying assumptions. These assumptions could be relaxed in a more detailed design calculation which would require the use of a computerized procedure.

Assume that the lift coefficient, the drag coefficient, and the blade pitch angle are fixed along the length of the blade. The axial induction factor and the radial induction factor are neglected. The velocity triangle is shown in Fig. 6.12(a). At different radial sections of the blade it is affected by the changing tangential speed.

Consider the aerodynamic forces acting on a small section of thickness δr of the blade at a radial distance r from the axis of the turbine. The blade section experiences a lift force normal to the direction of the relative wind velocity and a drag force along the direction of relative wind velocity.

To illustrate the numerical procedure consider the blade section at a distance of 20 m from the axis. The velocity triangle is depicted in Fig. 6.12(a). We obtain the following values from the geometry.

$$tan\phi = \frac{U_w}{r\omega} = \frac{8}{20\times0.785} = 0.5095, \qquad \phi = 27^\circ$$

$$V_r = \frac{U_w}{sin\phi} = \frac{8}{sin27} = 17.62\ \mathrm{ms^{-1}}$$

The lift force on the blade section is given by Eq. (6.91) as:

$$\delta L_r = \frac{1}{2}\rho V_r^{\ 2} C_L c \delta r$$

$$\delta L_r = \frac{1}{2} \times 1.2 \times 17.62^2 \times 1.2 \times 1 \times \delta r = 223.5\delta r$$

The drag force on the blade section is given by Eq. (6.92) as:

$$\delta D_r = \frac{1}{2}\rho V_r^{\ 2} C_D c \delta r$$

$$\delta D_r = \frac{1}{2} \times 1.2 \times 17.62^2 \times 0.08 \times 1 \times \delta r = 14.9\delta r$$

The forces in the tangential and axial directions at the section are obtained by resolving the above forces in these directions (see Fig. 6.12). Hence we have

$$\delta F_{\theta,r} = \delta L_r sin\phi - \delta D_r cos\phi$$

$$\delta F_{a,r} = \delta L_r cos\phi + \delta D_r sin\phi$$

Substituting numerical values we get

$$\delta F_{\theta,r} = (223.5sin27 - 14.9cos27)\delta r = 88.2\delta r$$

$$\delta F_{a,r} = (223.5cos27 + 14.9sin27)\delta r = 205.9\delta r$$

The torque due to the tangential force at $r = 20$ m is given by

$$\delta \Gamma_r = \delta F_{\theta,r} r \delta r = 88.2 \times 20\delta r = 1764\delta r$$

To compute the total torque and the total axial force due to wind forces the blade is divided into 4 radial sections as indicated in Table E6.16.1. At each radial node the torque and the axial force are computed as outlined above. To obtain the total torque and axial force these values are integrated using the trapezoidal rule.

Table E6.16.1 Results of computation

Radius, m	0	10	20	30	40
V_r ms^{-1}	-	11.2	17.6	24.9	32.4
$\Gamma_r \times 10^3$ Nm/m	0	0.603	1.764	3.456	5.514
$F_a \times 10^3$ N/m	0	0.0677	0.2061	0.4317	0.7457

The total torque developed is

$$M = 3 \times 85.81 = 257.4 \text{ kNm}$$

The total axial force is

$$F_a = 3 \times 10.78 = 32.34 \text{ kN}$$

The power output is given by

$$P_w = M\omega = 257.4 \times 7.5 \times \frac{2\pi}{60} = 202 \text{ kW}$$

Example 6.17 A wind turbine rotor has two blades of length 13 m. The tip speed ratio is 8. For the blade section at radius equal to 7.15 m, calculate (i) angle of relative wind, (ii) angle of attack, (iii) lift and drag coefficients, (iv) the axial and radial induction factors, and (v) local contribution to the power coefficient. The geometrical properties of the blades are given in the table below. The pitch at the blade tip is -1.97°.

The lift and drag coefficients of the airfoil are well represented by the following equations [5].

For $\alpha < 21^\circ$

$$C_L = 0.42625 + 0.11628\alpha - 6.3997 \times 10^{-4}\alpha^2 - 8.712 \times 10^{-5}\alpha^3 - 4.2576 \times 10^{-6}\alpha^4$$

For $\alpha > 21^\circ$ $\quad C_L = 0.95$

For all angles

$$C_D = 0.011954 + 1.9972 \times 10^{-4}\alpha + 1.0332 \times 10^{-4}\alpha^2$$

Table E6.17.1 Characteristics of turbine blade at different radial sections

Radius, m	0.95	1.95	3.25	4.55	5.85
r/R	0.05	0.15	0.25	0.35	0.45
Chord, m	1.0	1.0	1.0	1.0	0.87
Twist, deg.	13.0	11.0	9.0	7.0	5.0

Radius, m	7.15	8.45	9.75	11.05	12.35
r/R	0.55	0.65	0.75	0.85	0.95
Chord, m	0.72	0.61	0.54	0.47	0.42
Twist, deg.	3.4	2.2	1.40	0.7	0.2

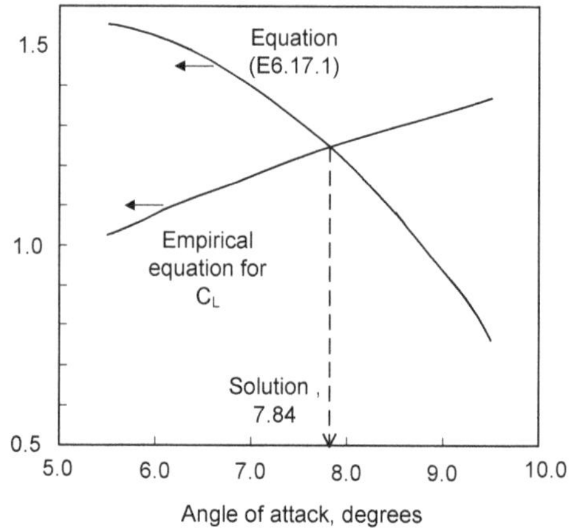

Fig. E6.17.1 Solution of Eq. (E6.17.1) and Eq. (6.17.5)

Solution To determine the aerodynamic torque developed by the blades we need to compute the induction factors a and a'. In Example E6.16 we ignored these factors. The equations needed for the design of a turbine blade are given in section 6.8, which are summarized below [5].

$$C_L = \frac{4sin\phi(cos\phi-\lambda_r sin\phi)}{\sigma_1(sin\phi+\lambda_r cos\phi)} \tag{E6.17.1}$$

where the *solidity ratio* is given by

$$\sigma_1 = \frac{N_B c}{2\pi r}$$

$$\frac{1}{a'} = \left(\frac{4cos\phi}{\sigma_1 C_L} - 1\right) \tag{E6.17.2}$$

$$\frac{a}{a'} = \frac{\lambda_r}{tan\phi} \tag{E6.17.3}$$

In addition, we require the empirical equations relating C_L and C_D to the angle of attack α. These are given above.

The angle of the relative wind ϕ is related to α by the equation (see Fig. 6.12).

$$\phi = \alpha + \beta$$

The *section twist* is defined as

$$\theta_t = \beta - \beta_0$$

where β_0 is the blade pitch at the tip.

This set of equations must be solved iteratively. We start with an assumed value of α, and obtain the ϕ using the two equations above, and the given value of θ_t at the section. For the blade section at a radius of 7.15 m we have

$$N_B = 2, \quad \theta_t = 3.4, \quad \beta_0 = -1.97, \quad r = 7.15, \quad c = 0.72,$$

$$\sigma_1 = \frac{N_B c}{2\pi r} = \frac{2\times 0.72}{2\pi\times 7.15} = 0.0321, \quad \lambda_r = \frac{\lambda r}{R} = \frac{8\times 7.15}{13} = 4.4$$

$$C_L = \frac{4 sin\phi(cos\phi - 4.4 sin\phi)}{0.0321(sin\phi + 4.4 cos\phi)} \tag{E6.17.4}$$

$$\phi = \alpha + 3.4 - 1.97 = \alpha + 1.43$$

$$C_L = 0.42625 + 0.11628\alpha - 6.3997\times 10^{-4}\alpha^2 - 8.712\times 10^{-5}\alpha^3 - 4.2576\times 10^{-6}\alpha^4 \tag{E6.17.5}$$

$$C_D = 0.011954 + 1.9972\times 10^{-4}\alpha + 1.0332\times 10^{-4}\alpha^2 \tag{E6.17.6}$$

We solve Eqs. (E6.17.4) and (E6.17.5) simultaneously by initially guessing the value of α and adjusting it until the values of C_L given by the expressions on the right hand sides of the above equations are equal.

The variation of the two expressions with α is shown graphically in Fig. E6.17.1 The point of intersection of the two curves gives an approximate value which can be refined using an iterative approach. The final results are:

$$\alpha = 7.84 \quad \phi = 9.27 \quad C_L = 1.242 \quad \text{and} \quad C_D = 0.199$$

$$\frac{1}{a'} = \left(\frac{4cos\phi}{\sigma_1 C_L} - 1\right) = \frac{4cos9.27}{0.0321\times 1.242} - 1 = 98.02,$$

$$a' = 0.0102$$

$$\frac{a}{a'} = \frac{\lambda_r}{tan\phi} = \frac{4.4}{tan9.27} = 26.96,$$

$$a = 0.275$$

The power coefficient at the section is given by Eq. (6.102) as

$$\frac{\delta C_p}{\delta r} = \left(\frac{N_B}{\pi R^2}\right) \frac{(1-a)^3}{(1+a')tan\phi sin\phi^2} (C_L sin\phi - C_D cos\phi)c$$

Substituting numerical data we get

$$\frac{\delta C_p}{\delta r} = \frac{2(1-0.275)^3}{13^2\pi(1+0.0102)tan9.27sin9.27^2}(1.242sin9.27 - 0.199cos9.27)0.72 = 8.805 \times 10^{-4}$$

Hence the contribution to power coefficient per unit radius at the location is 8.805×10^{-4} per meter.

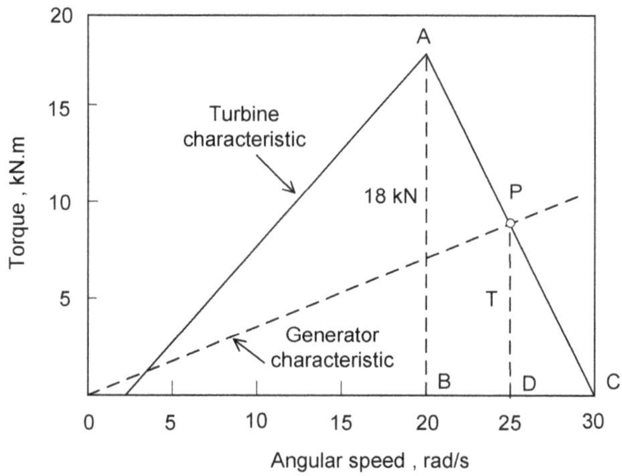

Fig. E6.18.1 Turbine and generator characteristics

Example 6.18 The torque-speed characteristics of a wind turbine is approximated by the triangular form shown in Fig. E6.18.1. It drives an electric generator whose power output is proportional to the square of rotational speed. When the speed is 300 rpm, the power output of the generator is 360 kW. Calculate the power output of the generator when it is driven by the wind turbine.

Solution The torque–speed characteristic of the turbine is shown in Fig. E6.18.1. It has a triangular form with the maximum torque of 18 kNm at an angular speed of 20 rad. s^{-1}. Th generator characteristic is given as

$$P_g(kW) = a\omega^2 = T_g\omega \tag{E6.18.1}$$

Hence torque-speed characteristic of the generator is

$$T_g = a\omega$$

Using the given condition,

$$P_g = 360 \text{ kW at } \omega = \left(\frac{2\pi \times 300}{60}\right) = 31.42 \text{ rad. s}^{-1}.$$

Substituting in Eq. (E6.18.1) we obtain the condition

$$a = 0.365 \text{ kNm.s}$$

Hence the torque-speed characteristic of the generator is

$$T_g = 0.365\omega$$

The torque-speed characteristics of the turbine and the generator intersect at P, which is the stable operating point of the system. Let the torque and angular speed at P be T and ω respectively. Considering the two triangles PDC and ABC of the turbine characteristic we get

$$\frac{T}{(30-\omega)} = \frac{18}{(30-20)}$$

$$T = 1.8(30 - \omega)$$

The generator characteristic at point P gives

$$T = 0.365\omega$$

From two equations above we obtain

$$\omega = 25 \text{ rad. s}^{-1} \quad \text{and} \quad T = 9.125 \text{ kNm}$$

Hence the power output of the system is

$$P_g = 9.125 \times 25 = 228 \text{ kW}$$

The maximum power produced by the turbine at A is

$$P_{t,max} = 18 \times 20 = 360 \text{ kW}$$

If the generator characteristic could be changed to pass through point A, maximum available power could be extracted.

Example 6.19 The moment of inertia of a wind turbine rotor is 4.8×10^6 kgm^2. It is operating under steady conditions with a rotational speed of 22 rpm and a power of 1600 kW. If the connection to the electrical generator fails suddenly and the break fails to operate, how long does it take for the rotational speed to double? Assume that the aerodynamic forces on the turbine are unchanged.

Solution Under steady operation the electrical resisting torque is balanced by the aerodynamic torque developed by the rotor due to wind flow over the blades. An example is seen in Fig. E6.18.1. The aerodynamic torque is related to the power by the equation

$$T_a \omega_t = P = 1600 \times 10^3$$

$$T_a \times \left(\frac{22 \times 2 \times \pi}{60}\right) = 1600 \times 10^3$$

Therefore the torque produced by the rotor is 694.5×10^3 Nm.

When the connection to the electrical generator is lost there is no resisting electric torque on the turbine shaft. Hence it will be accelerated by the aerodynamic torque that is still being generated by the wind. We apply Newton's second law to the rotor to obtain the equation

$$I_r \left(\frac{d\omega}{dt}\right) = T_a$$

Integrating the above equation we have

$$\Delta t = \frac{I_r(\omega_f - \omega_o)}{T_a}$$

The time taken for the rotor speed to double is

$$\Delta t = \frac{I_r(\omega_f - \omega_o)}{T_a} = \frac{4.8 \times 10^6 \times 2.304}{694.5 \times 10^3} = 15.9 \text{ s}$$

Example 6.20 A windmill with a capacity factor of 0.32 is rated at 4 MW. The cost of the turbine is 1300 \$ per kW and the annual maintenance and operations cost is $86,100.0 Balance of the system (BOS) costs $1,865,000.0 The levelized replacement cost is $133,000.0 per year. Assume a rate of return of 6% for the project. The life time of the turbine is 20 years. Calculate the levelized cost of energy for this turbine.

Solution The annual energy production is:

$$E_y = CF \times rated\ power \times 8760 \text{ kWh}$$

$$E_y = 0.32 \times 4000 \times 8760 = 11{,}212{,}800.00 \text{ kWh}$$

Initial capital cost ICC includes the turbine cost and the balance of system (BOS) cost.

$$ICC = (1300 \times 4000) + 1865000 = \$\,7{,}065{,}000.00$$

The rate of return is 6%. The capital recovery factor is given by

$$CRF = \frac{i(1+i)^N}{(1+i)^N - 1} = \frac{0.06 \times 1.06^{20}}{1.06^{20} - 1} = 0.08718$$

Hence the annualized initial capital cost is:

$$C_{ia} = 7{,}065{,}000.00 \times 0.08718 = \$615{,}926.70$$

Annual operating, maintenance, and replacement cost is:

$$C_{O\&M} = 86{,}100.00 + 133{,}000.00 = \$219{,}100.00$$

Total annualize cost is:

$$C_{it} = 615{,}926.70 + 219{,}100.00 = \$\,835{,}026.70$$

The levelized cost of energy is

$$LCOE = \frac{835{,}026.70}{11{,}212{,}800} = 0.0745 \text{ \$ per kWh}$$

More details on the computation of the levelized cost of energy are given in Chapter 1.

Problems

P6.1 A wind turbine with a rotor diameter of 10 m is installed at a site with a wind speed of 5.8 ms^{-1}. The efficiency of the turbine-generator system is 38%. The mean density of air is 1.15 kgm^{-3}. Calculate (i) the power in the wind, (ii) power output of the turbine-generator, and (iii) the power output of a Betz-limit turbine under the above conditions. [*Answers*: (i) 8.81 kW, (ii) 3.35 kW, (iii) 5.2 kW]

P6.2 The hub height of a HAWT is 95 m and the blade diameter is 80 m. The wind speed at the hub height is 11.5 ms^{-1} and the output of the system is 2 MW. The rotational speed of the turbine is 15 rpm, and the overall efficiency of the generator system is 0.94. The density of air is 1.225 kg m^{-3}. Calculate (i) the power coefficient, (ii) the tip speed ratio, (iii) the turbine torque, and (iv) the torque coefficient.
[*Answers*: (i) 0.45, (ii) 5.46, (iii) 1.36 MNm, (iv) 0.082]

P6.3 The rotor diameter of a HAWT is 50 m and the hub height is 65 m. The efficiency of the gearbox and the generator is 95% and the gear ratio of the gearbox is 1:30. The wind speed at hub height is 10 ms^{-1} and tip speed ratio is 8. The thrust coefficient is 0.54 and the power coefficient is 0.33. The density of air is 1.25 kgm^{-3}. Calculate (i) the rotational speed and torque on the generator shaft, (ii) the power output of the system, (iii) the torque coefficient, and (iv) the overturning moment acting on the tower at the base. [*Answers*: (i) 916.7 rpm, 4 kNm, (ii) 384.7 kW, (iii) 0.041, (iv) 4.31 MNm]

P6.4 At a wind turbine site with a few trees, the measured wind speed at a height of 10 m is 8 ms^{-1}. Calculate the wind speed at a height of 60 m using the power-law method and the log-law method.
[*Answers*: 10.65 ms^{-1}, 11.1 ms^{-1}]

P6.5 A wind turbine with a rated power output of 5 kW is located at a site whose annual wind speed distribution is given in Table P6.5 in a bin format. The rotor diameter is 8 m and the density of air is 1.15 kgm^{-3}. Calculate (i) the average annual wind speed at the location, (ii) the

standard deviation of the wind speed, (iii) the annual average power of the wind, and (iv) the maximum average output of a wind turbine at the location.
[*Answers*: (i) 5.56 ms^{-1}, (ii) 2.93 ms^{-1}, (iii) 9.15 kW, (iv) 5.4 kW]

Table P6.5 Annual wind speed bin-data

Bin Number	U_{min} ms^{-1}	U_{max} ms^{-1}	U ave. ms^{-1}	Hours/ year	Frequency %
1	0	0	0	82	0.94
2	0	1	0.5	204	2.33
3	1	2	1.5	498	5.68
4	2	3	2.5	808	9.22
5	3	4	3.5	1210	13.81
6	4	5	4.5	1254	14.31
7	5	6	5.5	1247	14.24
8	6	7	6.5	1025	11.70
9	7	8	7.5	710	8.11
10	8	9	8.5	550	6.28
11	9	10	9.5	442	5.05
12	10	11	10.5	329	3.76
13	11	12	11.5	220	2.51
14	12	13	12.5	125	1.43
15	13	14	13.5	53	0.61
16	14	∞	-	3	0.03

P6.6 A wind turbine with rated power output of 5 kW is located at a site with the annual wind speed distribution given in Table P6.5. The rotor diameter is 8 m and the density of air is 1.15 kgm^{-3}. Using (a) the Rayleigh PDF, (b) the Weibull PDF, calculate (i) the annual average power of the wind, and (ii) the maximum average output of a wind turbine at the location.
[*Answers*: (a) (i) 9.78 kW, (ii) 5.77 kW, (b) (i) 10.87 kW, (ii) 6.4 kW]

P6.7 A wind turbine with a rated output 5 kW is installed at a location with the annual wind speed distribution given in Table P6.5. The diameter of the rotor is 8 m and the density of air is 1.15 kgm^{-3}. The actual power curve of the turbine is approximated by the analytical form given by:

For $0 < U < 2.5$ ms^{-1} $P(U) = 0$ kW
For $2.5 \leq U < 8$ ms^{-1} $P(U) = 0.03(U - 2.5)^3$ kW
For $8 \leq U < 14$ ms^{-1} $P(U) = 5$ kW
For $14 \geq U$ ms^{-1} $P(U) = 0$ kW

Using (a) the given bin data, and (b) the Rayleigh PDF, calculate (i) the annual average power output of the turbine, (ii) average efficiency of the turbine, and (iii) the capacity factor. [*Answers*: (a) (i) 1.66 kW, (ii) 18%, (iii) 0.33, (b) (i)1.65kW, (ii) 16.9%, (iii) 0.33]

P6.8 The blade characteristics of a two-bladed wind turbine at different radial distances are given in Table E6.17.1. The pitch at the blade tip is -1.97°. The lift and drag coefficients of the airfoil are well-represented by the following equations:

For $\alpha < 21°$

$$C_L = 0.42625 + 0.11628\alpha - 6.3997 \times 10^{-4}\alpha^2 - 8.712 \times 10^{-5}\alpha^3 - 4.2576 \times 10^{-6}\alpha^4$$

For $\alpha > 21°$ $C_L = 0.95$. For all angles

$$C_D = 0.011954 + 1.9972 \times 10^{-4}\alpha + 1.0332 \times 10^{-4}\alpha^2$$

Calculate for the given radial sections of the blade: (i) the angle of attack, (ii) the induction factors, (iii) the lift and drag coefficients.

Hence calculate the power coefficient by adding the contributions from the different radial sections. [*Answers*: At $r = 3.25$ m, $\alpha = 13.45°$, $a = 0.222$, $a' = 0.041$, $C_L = 1.523$, $C_D = 0.033$]

P6.9 The torque-speed characteristic, $T_t\ (kNm)\ vs\ \omega\ (rad.s^{-1})$ of a wind turbine can be approximated by the triangular form below:

$0 \leq \omega < 2.5$ $T_t = 0$

$2.5 \leq \omega \leq 20$ $T_t = \left(\frac{16}{17.5}\right)(\omega - 2.5)$

$20 \leq \omega \leq 30$ $T_t = \left(\frac{16}{10}\right)(30 - \omega)$

$\omega > 30$ $T_t = 0$

The turbine is connected to a load that absorbs power according to the equation

$$P_l = 12\omega \text{ kW}$$

Calculate (i) the power delivered by the turbine under steady conditions, and (ii) the maximum power available.
[*Answers*: (i) 270 kW, (ii) 320 kW]

P6.10 A wind turbine system has a rated output of 3 MW and a capacity factor of 0.32. The unit cost of the turbine is 1200 $ per kW and the annual operating and maintenance cost is $82,200. Balance of the system (BOS) costs $1,725,000. The replacement cost is $120,000 per year. Assume a rate of return of 7% for the project. The life time is 20 years. Calculate the levelized cost of energy for this system.
[*Answer*: $0.0837 per kWh]

References

1. Duffie, John A., Beckman, William A. with Blair, Nathan, *Solar Engineering of Thermal Processes, Photovoltaics and Wind, Fifth Edition*, John Wiley & Sons Inc. New Jersey, 2020.
2. Jacobson, Mark Z, 100% *Clean, Renewable Energy and Storage for Everything*, Cambridge University Press, Cambridge, U.K., 2021.
3. Jenkins, Nick., and Ekanayake, Janaka., Renewable Energy Engineering, Cambridge University Press, Cambridge, U.K. 2017.
4. Kreith, Frank and Krumdieck, Susan, *Principles of Sustainable Energy Systems*, CRC Press, Taylor & Francis Group, New York, 2015.
5. Manwell, J.F., McGowan, J.G., and Rogers, A.L., *Wind Energy Explained: Theory, Design, and Applications*, 2nd Edition, Wiley & Sons, West Sussex, 2010.
6. Vanek, Francis M., Albright, Louis D. and Angenent, Largus T., *Energy Systems Engineering: Evaluation and Implementation*, McGraw-Hill Book Company, New York, 2012.
7. 'List of onshore wind farms,Wikipedia' [https://en.wikipedia.org/wiki/List of onshore wind farms].
8. http://www.lazard.com/media/450773/lazard-levelized-cost-of-energy-version-120-vfinal.pdf.
9. http://ucsusa.org/clean-energy/renewable-energy/environmental-impacts.

Chapter 7

Energy Storage Systems

7.1 Introduction

The last decade has seen a rapid increase in the number roof-top and utility-scale solar PV installations, and large wind farms. The integration of these systems with existing electricity grids has posed several new challenges to electrical-grid operators due to the intermittency of solar and wind resources. Moreover, there is often a mismatch between the availability of the resource and electricity demand. A unique characteristic of solar and wind energy is their intermittency on every time scale. Several ways have been proposed to deal with this time-variability of renewable energy supply, which include the following [3]:

(i) Interconnect geographically dispersed, technologically diverse renewable generation types, such as wind, solar, wave, and tidal to smooth out daily supply variability. For example, solar power generation is highest at midday, and wind is often strongest at night and early morning. The combined solar-wind resource has lower variability than either individual source.

(ii) Use renewable energy sources with storage potential, such as, hydroelectric, geothermal, and biomass to fill energy deficits between electricity demand and intermittent solar and wind generation.

(iii) Use demand-side management to shift flexible loads to periods with high solar and wind availability, and away from times when they are low.

(iv) Store excess solar and wind power, which would otherwise be curtailed, for later use at times when generation is not sufficient to meet the load. Some energy storage technology types include pumped-hydro, compressed air, electrochemical batteries, flywheels, and hydrogen [2,3].

Large-scale deployment of solar and wind energy systems, and their integration into the existing energy infrastrucuture has changed the electrial energy demand curve drastically. As more solar and wind generating systems are integrated into the grid, their peak energy contribution may even exceed the demand, and therefore the overall solar and wind generating capacity may have to be curtailed to prevent the grid from being over loaded.

Storing excess solar-generated and wind-generated electricity for use during periods of high demand is an ideal solution. However, the cost of storing electricity on a megawatt-scale is high, and the technology to achieve it is not well developed at the present time. There is much on-going research on large-scale storage of electricity, because a breakthrough would facilitate even a total switch to solar and wind sources in the not too distant future [3]. In the next section we shall present the different energy storage technologies that are being developed.

7.2 Energy Storage Technologies

Energy storage technologies (EST) can be broadly divided into five major categories. These are: mechanical, electrical, chemical, electro-chemical and thermal storage systems [6]. These energy storage types include the following technologies:

- Mechanical energy storage: (a) compressed air storage systems, (b) flywheel kinetic energy systems, (c) pumped-hydroelectric systems.
- Electrochemical energy storage: (a) secondary batteries, (b) flow batteries.
- Thermochemical energy storage: (a) hydrogen production, (b) hydrogen fuel cells.
- Electrical energy storage: (a) capacitors, (b) super-capacitors.
- Thermal energy storage: (a) sensible heat systems, (b) phase change systems.

Details of these different energy storage technologies are presented below.

7.2.1 *Characteristics of energy storage systems*

Several parameters are used to characterize the design and performance of energy storage systems. These include the following: depth of discharge (DOD), self discharge time (SDT), storage size, efficiency of storage, cycle life, specific energy, energy density [6].

Depth of discharge (DOD)

Depth of discharge is the percentage of total stored energy in a storage device that is utilized during operation. For example, if 20 percent of stored energy in a battery must be left behind to maintain its proper functioning and its design life, then the DOD is 80 percent.

Self discharge time (SDT)

Self discharge time is the time required for a fully charged self-standing storage device to reach a particular DOD. The SDT varies widely, from minutes for flywheels to several days for thermal storage systems.

Storage size

Storage size characterizes the energy capacity of a storage device, which is usually measured in kWh. The power of a storage technology is the rate of energy flows to and from the system.

Efficiency of storage

Efficiency of a storage system is the ratio of the energy output to energy input. The efficiency accounts for losses during the energy input and output phases and the storage period. Moreover, all components needed at the interconnections to operate the energy storage system must be considered in computing the overall efficiency.

For example, a compressed air system used to store electrical energy output from a wind turbine, incorporates an electric motor-compressor and a turbine-generator to absorb and discharge electrical energy (see Fig. 7.1). Therefore the overall efficiency of the storage system has to include the efficiencies of all the above subcomponents. In the case of a thermal energy storage system, the heat losses during the energy storage period must be included in the efficiency. Energy efficiency of a few storage technologies is listed in Table 7.1.

Cycle-life

Cycle-life is the number of charge/discharge cycles that a storage system can undergo while maintaining its design performance. This is usually a function of the DOD for some devices. For instance, over discharging batteries can adversely affect their cycle life. However, thermal storage systems and flywheel storage systems can undergo many charge/discharge cycles with minimum effect on performance.

Specific energy

Specific energy of a device is the storage capacity per unit mass, usually expressed in MJ kg^{-1} or $kWhkg^{-1}$. It is an indication of the weight of the device. For transport applications, such as cars and trucks, a high specific energy is desirable.

Energy density

Energy density of a device is storage capacity per unit volume. It gives an indication of the space required to house the storage system.

In the following sections we will discuss the design and operation of practical systems belonging to the various energy storage technologies listed in section 7.1.

7.3 Compressed Air Storage Systems (CAS)

A typical compressed air energy storage (CAS) system is depicted schematically in Fig. 7.1. Several variations of this basic design have been proposed and studied. In this hybrid design atmospheric air enters a two-stage compressor which is run by an electric motor through a clutch. The motor receives electrical power from a wind power system or a solar PV plant whose energy output is to be stored for later use. The air leaving the low-pressure compressor is passed through an intercooler to lower its temperature. Air is further compressed in a high-pressure compressor. Some designs incorporate an additional cooler after this compressor, called an aftercooler. The intercooler and the aftercooler help to enhance the overall efficiency of the compression process. High pressure air enters the storage reservoir to complete the charging phase.

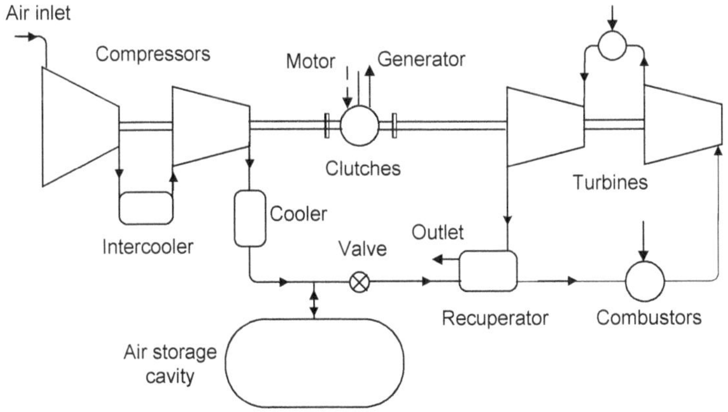

Fig. 7.1 Compressed air energy storage (CAS) system

In the discharge phase, compressed air from the storage reservoir first passes through a recuperator where it is heated at constant pressure by hot gases leaving the low pressure turbine. It then enters a combustor supplied with fuel, such as, natural gas or biogas from an external source. The need to use a fossil fuel combustor is a drawback of this storage system, but it improves the overall efficiency of the system significantly. Two of the existing CAS systems use this hybrid design [1,7]. Using high pressure air directly from the storage cavity to operate the turbine results in a low overall efficiency for the system.

Hot gases leaving the combustor expands through the high pressure turbine. It then passes through a second combustor before expanding through the low pressure turbine. The two turbines are connected to a common shaft. During the discharge process the clutch on the turbine side is engaged to run the generator that delivers electrical power. Usually, the air flow rates during the charge and discharge phases are not equal.

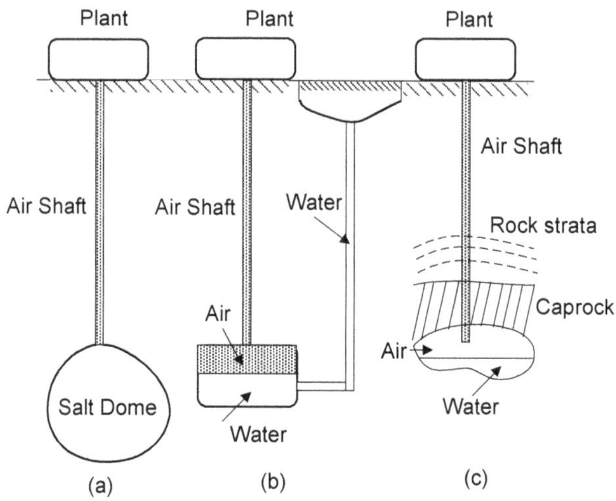

Fig. 7.2 Schematic of compressed air storage (CAS) cavities

7.3.1 *Air storage devices*

For small CAS systems, the storage device could be a metal tank. However, for large-scale storage of compressed air a proven method is the use of underground cavities, some of which are shown schematically in Fig. 7.2. There are three methods of producing suitable cavities for air storage. In the first method, depicted in Fig. 7.2(a), fresh water is pumped into natural salt formations to dissolve the salt and remove the saturated water. Such cavities have been used to store oil and gas, and are widely available [6,7].

In the second method, which is more expensive, a cavity is produced artificially in hard rock. These cavities can be operated at constant pressure by having a column of water to balance the air pressure in the storage cavity as shown in Fig. 7.2(b). In the third option, shown in Fig. 7.2(c), air is pumped into an underground aquifer with a suitable domed caprock to replace the water and create the space for air storage. This is a slow process, but once the air space is formed, acceptable air charge and discharge rates are possible because air is less viscous than water.

7.3.2 *Analysis of CAS systems*

We analyzed the performance of a conventional gas turbine cycle in section 4.5 in chapter 4. One of the main differences between a CAS system and a gas turbine plant is that in the latter a large fraction of the turbine output work is used to run the compressor while in a CAS system the entire turbine output is available for electricity generation.

We shall use some of the equations obtained in section 4.5 to develop a simplified analysis of the charging and discharging phases of the CAS system. The power input to a compressor P_c is given by

$$P_c = m_{ai}c_{pi}T_{ai}\left[\left(\frac{P_2}{P_{ai}}\right)^{\frac{\gamma-1}{\gamma}} - 1\right] \tag{7.1}$$

where m_{ai} is the air flow rate, c_{pi} is the specific heat capacity of inlet air, γ is the ratio of the specific heat capacities of air, T_{ai} is the inlet air temperature, P_{ai} is the inlet air pressure, and P_2 is the air pressure at the outlet of the compressor. Equation (7.1) is applicable to the low pressure compressor and the high compressor. The heat removal processes in the intercooler and the aftercooler occur at constant pressure.

Ideally, the heat transfer to air in the recuperator and the combustor occurs at constant pressure. The expansion processes in the two turbines are assumed to be reversible adiabatic processes. The power output of the high pressure turbine, P_{th} is given by

$$P_{th} = m_{f1}c_{p1}T_{f1}\left[1 - \left(\frac{P_{12}}{P_1}\right)^{\frac{(\gamma_1-1)}{\gamma_1}}\right] \tag{7.2}$$

where m_{f1} is the fluid flow rate through the high pressure turbine, c_{p1} is the specific heat capacity of the fluid, γ_1 is the ratio of the specific heat capacities, T_{a1} is the inlet fluid temperature, P_1 is the inlet pressure, and P_{12} is the outlet pressure. Similarly, the power output of the low pressure turbine is given by

$$P_{tl} = m_{f2}c_{p2}T_{f2}\left[1 - \left(\frac{P_o}{P_{12}}\right)^{\frac{(\gamma_2-1)}{\gamma_2}}\right] \tag{7.3}$$

The symbols in Eq. (7.3) have the same meaning as the corresponding symbols for the high pressure turbine given above. The inlet and outlet pressures for the low pressure turbine are P_{12} and P_o respectively.

The fluid flow rate through the turbine is the sum of the air flow rate m_a and the fuel flow rate m_{fu}. Therefore

$$m_f = m_a + m_{fu} \tag{7.4}$$

The fuel-air ratio (m_{fu}/m_a) is approximately constant [7].

The total power output is given by

$$P_{tot} = (P_{th} + P_{tl}) \tag{7.5}$$

The analysis of the storage reservoir is done using the unsteady mass and energy flow equations [8]. These are given respectively by

$$\frac{dM_s}{dt} = m_i - m_o \tag{7.6}$$

$$\frac{dE_s}{dt} = m_i h_i - m_o h_o \tag{7.7}$$

where $M_s(t)$ and $E_s(t)$ are the total mass and internal energy of air in the storage reservoir at any time t. The inflow and outflow rates and specific enthalpies are m_i, m_o and h_i, h_o respectively.

Depending on the type of storage reservoir used, a CAS system may be operated in a number of different ways [7]. A constant volume storage system (Figs. 7.2a and c) can be operated with (i) constant inlet air pressure to the turbine, and (ii) variable inlet pressure to the turbine, equal to the discharge pressure of the storage reservoir.

To maintain the turbine inlet pressure constant, as in mode (i) above, the air leaving the storage reservoir is throttled. Although this mode of operation incurs energy losses due to throttling, the turbine can be operated at the optimum inlet pressure and hence with a high efficiency. This mode of operation, however, requires a larger storage volume per unit power output. Two of the existing CAS systems operate with a constant turbine inlet pressure of about 45-46 bar, and the storage pressure varying between about 70 bar and 46 bar [1,7].

When the turbine inlet pressure is variable, as in mode (ii), there are no inlet losses but the turbine efficiency is adversely affected. When a water compensated reservoir (Fig.7.2b) is used the storage pressure remains constant during the entire operation. Thc losscs arc a minimum under these conditions and it improves the system efficiency.

One of the parameters for assessing the storage requirements of a CAS system is the amount of electrical energy delivered per unit storage volume, which may be expressed as [7]

$$E_G = \eta_m \eta_g \int_0^t m_a W_m dt \tag{7.8}$$

where η_m and η_g are the efficiencies of the turbine and the generator, and W_m is the total mechanical energy production per unit mass during the period of discharge t.

We shall illustrate the application of the above analysis of CAS systems in worked examples 7.1 to 7.4.

7.4 Flywheel Energy Storage Systems

A flywheel is a body that stores kinetic energy by virtue of its rotation about an axis [2,9]. Early flywheels, in the form of large metal wheels, were used in steam engines and other reciprocating machines to smooth their power output. The use of flywheels for electrical energy storage is a more recent development. This was brought about by improvements in bearing design, rotor materials, electronic power converters, and other auxiliary components like vacuum systems.

The main components of a modern flywheel energy system are depicted in Fig. 7.3. It consists of a spinning rotor which is made of steel or a high-strength composite material. Low speed rotors spinning at speeds below about 6000 rpm are made of steel, and have conventional bearings.

High speed flywheels with rotor speeds on the order of 50,000 rpm are made of composite materials. They have magnetically levitated bearings that reduce friction and increase operational life. The flywheel is housed in a containment maintained at a vacuum pressure to reduce

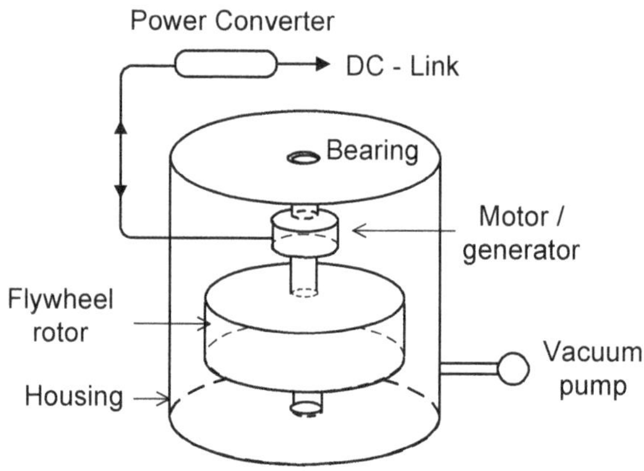

Fig. 7.3 Flywheel energy storage system

aerodynamic losses and rotor stresses. The electric power interface consists of a motor/generator, a power electronics convertor, and a power controller.

As a flywheel absorbs energy from an external source its rotational speed increases steadily due to the output torque of the motor acting on it. During the energy storage phase the flywheel will continue to rotate at the maximum speed, if friction losses can be completely eliminated. This, however, is not practically possible, but an important design goal of a flywheel system is to have an acceptably low self discharge rate. During the discharge phase the flywheel delivers electric power by transferring its kinetic energy to a generator.

Flywheel storage systems have very high charge and discharge cycles and a life of about 20 years. They have a high efficiency and a fast response time. The disadvantages of flywheels are the high cost, low storage capacity, and relatively high self-discharge rates.

7.4.1 *Analysis of flywheel storage systems*

The energy stored in a flywheel rotor is a function of the rotor shape and material. The rotational kinetic energy is given by

$$E_{ke} = \frac{1}{2} I \omega^2 \tag{7.9}$$

where I is the moment of inertia of the flywheel and ω is the angular velocity.

The useful energy stored in a flywheel E_{st} is determined by the design maximum and minimum angular speeds ω_{max} and ω_{min} respectively. This is given by

$$E_{st} = \frac{1}{2} I({\omega_{max}}^2 - {\omega_{min}}^2) \tag{7.10}$$

For a solid cylinder of radius r and mass m, the moment inertia is

$$I = \frac{1}{2} mr^2 \tag{7.11}$$

For a hollow cylinder of outer radius r_o, inner radius r_i, and mass m, the moment of inertia is

$$I = \frac{1}{2} m({r_o}^2 + {r_i}^2) \tag{7.12}$$

From Eqs. (7.10) and (7.11) it is seen that the energy stored in a flywheel is directly proportional to its mass and varies as the square of the rotational speed. Therefore for maximum energy storage a flywheel should run at the highest possible speed. The maximum speed at which a flywheel could operate is determined by the ultimate strength of the rotor material.

We shall use a simplified analysis to obtain the dependence of the maximum stress on other properties of the flywheel. For this purpose we consider a thin cylinder of mass m and mean radius r as the rotor. The moment of inertia is given by

$$I = mr^2 = \rho V r^2 \tag{7.13}$$

where V is the volume and ρ is the density. The kinetic energy is given by Eq. (7.9) as

$$E_k = \frac{1}{2} mr^2\omega^2 = \frac{1}{2} \rho V r^2 \omega^2 \tag{7.14}$$

The circumferential stress or the hoop stress of the flywheel is obtained by considering the equilibrium of an elemental angular section of the thin cylinder. Hence we get the hoop stress as

$$\sigma_\theta = \rho r^2 \omega^2 \tag{7.15}$$

From Eqs. (7.14) and (7.15) we obtain the kinetic energy per unit mass of the flywheel as

$$E_{km} = \frac{1}{2}\left(\frac{\sigma_\theta}{\rho}\right) \tag{7.16}$$

From Eq. (7.16) we note that to build a flywheel with high kinetic energy per unit mass, a material with high yield stress and low density is required. Lightweight composite materials have these properties.

Similarly, the kinetic energy per unit volume is

$$E_{kV} = \frac{1}{2}\sigma_\theta \tag{7.17}$$

A flywheel with high kinetic energy per unit volume should have a high yield stress as seen from Eq. (7.17). Again, composite materials satisfy this requirement.

The above expressions for specific energy density can be generalized to other rotor geometries by introducing a shape factor K. These are given by

$$\frac{E_k}{m} = K\frac{\sigma_{max}}{\rho} \tag{7.18}$$

$$\frac{E_k}{V} = K\sigma_{max} \tag{7.19}$$

where σ_{max} is the maximum stress.

The following are shape factors for the most common flywheel geometries:

$$K_{thin\,rim} = 0.5,\ K_{thick\,rim} = 0.305,\ K_{solid\,disk} = 0.606.$$

The applications of flywheel systems are illustrated in worked examples 7.5 to 7.7.

7.5 Pumped Hydroelectric Energy Storage (PHES)

Pumped hydroelectric energy storage is a system used for load balancing in electric power networks. It is one of the oldest, large scale energy storage systems that has been deployed in a number of countries. In PHES systems energy is stored in the form of gravitational potential energy of water by pumping water from a lower elevation to a higher elevation as shown schematically in Fig. 7.4.

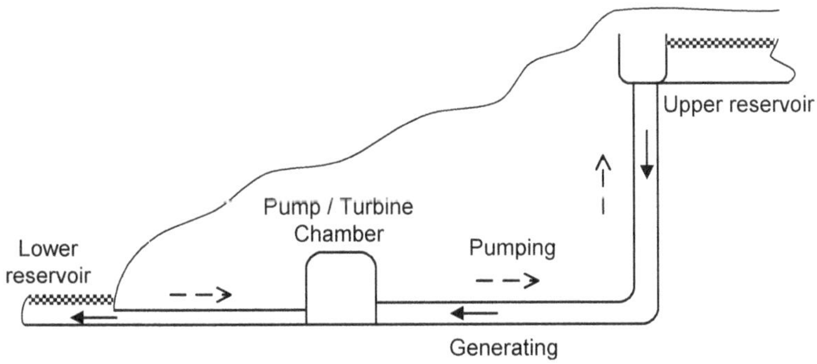

Fig. 7.4 Pumped- hydroelectric energy storage system

In conventional applications, electrical power is used for pumping water during off-peak hours. During periods of high electricity demand, the process is reversed and water is run through a turbine -generator system to supply power.

The pumping and generation phases of the cycle involve energy losses, and therefore the system is a net consumer of energy. Also there is water loss due to evaporation during the storage phase which could be several hours. However, if the upper reservoir collects significant rainfall or has a river feeding it, then the plant could produce more energy than it consumes. With modern turbines and pumps, PHES could achieve round-trip efficiencies of about 80% to 85%.

The power output and capacity of a PHES is a function of the elevation difference between the water reservoirs, the water flow rate, and the efficiencies of the pump and the turbine. The maximum water flow rate is determined by the size of the reservoir and the desired storage time. Depending on the allowable depth of discharge, a fraction of water available is maintained as a reserve.

The average power input during pumping is given by

$$P_p = \frac{m_p g H_p}{\eta_p} \tag{7.20}$$

where m_p the average flow rate of water, g is the acceleration due to gravity, H_p is the average difference in elevation, and η_p is the overall efficiency of the motor-pump and the piping network.

The average power delivered during generation is given by

$$P_g = m_g g H_g \eta_g \tag{7.21}$$

where m_g the average flow rate of water, H_g is the average difference in elevation, and η_g is the overall efficiency of the turbine-generator system, including the losses in the piping network.

The main advantages of PHES are: (i) mature technology, capable of storing large quantities of energy, (ii) very low operating costs, (iii) overall high efficiency of about 70% to 80%, (iv) power output of the turbine-generators could be adjusted rapidly.

The main disadvantages are: (i) high capital cost, (ii) potential sites with desired conditions are not readily available, (iii) high environmental impacts.

PHES systems involve most of the features and components of conventional hydroelectric power systems [4]. These include turbines, pipe networks and controls, reservoirs, dams, and others. To avoid repetition, we shall discuss some these components in more detail in chapter 9 on hydropower systems. Worked examples 7.8 to 7.10 illustrate the application of PHES systems.

7.6 Battery Storage Systems

Batteries are devices that store chemical energy and convert it to electrical energy when an external conduction path is provided. Batteries that must be discarded once they are fully discharged are called *primary* batteries. They cannot be recharged. *Secondary* batteries, on the other hand, are rechargeable. Once discharged they can be recharged by connecting them to an external power source. Electrical energy supplied to a secondary battery is stored as chemical energy and later delivered to a load on demand as electricity.

The operation of batteries is based on electrochemical principles and processes [6]. The chemical reaction in a battery involves the flow of electrons from one material to another, which are known as *electrodes*. The electrodes are immersed in an *electrolyte* through which charged ions flow to balance the flow of electrons. How a battery works, how much energy it can store, and its voltage are dependent on the chemical reactions between the anode, the cathode, and the electrolyte. We shall illustrate these characteristics of batteries using lead-acid and lithium-ion batteries as examples.

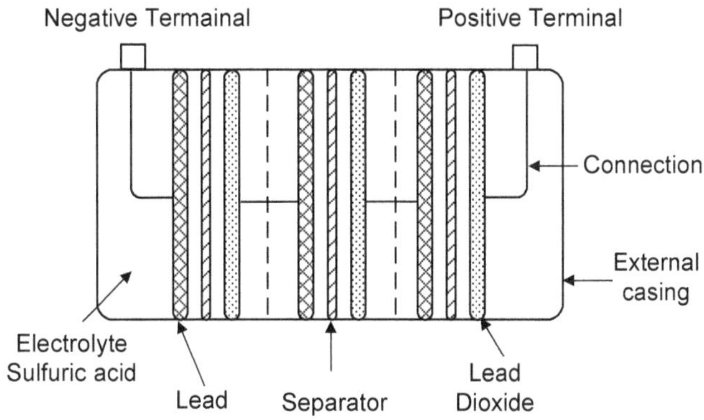

Fig. 7.5 Lead – acid battery

7.6.1 *Lead-acid batteries*

Lead-acid batteries have been used in transportation applications for many decades. They are heavy and must be charged often for many applications. In automotive applications they can provide a high current for a short period to start an engine.

The main components of a lead-acid battery are shown schematically in Fig. 7.5. Each cell of the battery has a sponge lead anode (negative electrode) and a lead oxide cathode (positive electrode). The electrodes are separated by porous membranes and immersed in dilute sulfuric acid solution, which is the electrolyte. Aqueous sulfuric acid dissociates into positive hydrogen ions and negative sulfate ions. This is represented by the chemical equation:

$$2H_2SO_4 \rightarrow 4H^+ + 2SO_4{}^{2-} \tag{7.22}$$

At the anode, lead is converted to lead sulfate releasing two electrons to the external circuit. These electrons constitute the electric current which delivers power to the external load. The chemical equation for this *oxidation* reaction is:

$$Pb(s) + SO_4{}^{2-}(aq) \rightarrow PbSO_4(s) + 2e^- \tag{7.23}$$

According to Eq. (7.23) two moles of electrons are produced for each mole of sulfate ions and sponge lead.

At the cathode the following *reduction* reaction occurs. The hydrogen and sulfate ions from the solution react with lead oxide and the electrons from the external circuit to produce lead sulfate and water. This is given by the chemical equation:

$$PbO_2(s) + 4H^+(aq) + SO_4{}^{2-}(aq) + 2e^- \leftrightarrow PbSO_4(s) + 2H_2O(l) \tag{7.24}$$

It is seen that at both electrodes lead is converted to lead sulfate, and sulfuric acid in the solution is consumed. Hydrogen ions combine with oxygen to form water. Therefore as the battery discharges the sulfuric acid solution becomes more dilute.

The net cell reaction may be represented by the equation:

$$Pb(s) + PbO_2(s) + 2H_2SO_4 \leftrightarrow 2PbSO_4(s) + 2H_2O(l) \tag{7.25}$$

The reaction from left to right occurs when the battery is discharging through a load, and the right to left reaction occurs when the battery is being recharged. It is done by connecting the battery to an external power source. The electron flow is then reversed, and water is consumed to produce sulfate and hydrogen ions from lead sulfate on the electrodes. Once recharged the battery can deliver electric current on demand.

The advantages of lead acid batteries are low cost and high power density. However, they have a short cycle life of about 1500 deep cycles. Therefore lead-acid batteries are not used in large-scale energy storage applications.

7.6.2 *Lithium-ion batteries*

Primary batteries that use pure metallic lithium are not rechargeable because lithium is unstable. In rechargeable lithium batteries the electrodes are made of lithium compounds. The cathode is made of a metal oxide such as lithium cobalt oxide ($LiCoO_2$) attached to a current collector made of aluminum foil. The anode uses layered graphitic carbon with a copper current collector. The electrolyte is a lithium salt such as $LiClO_4$ or $LiBF_4$ dissolved in an organic solvent like ether.

During discharge, an oxidation reaction at the anode produces positively charged lithium ions and negatively charged electrons. The chemical equation for this reaction is [6]:

$$Li_xC_6 \leftrightarrow 6C + xLi^+ + xe^- \tag{7.26}$$

The above equation is expressed in moles, with x as a coefficient.

The oxidation reaction may also produce uncharged material that remains at the anode. Lithium ions move through the electrolyte, and electrons move through the external circuit. They recombine at the cathode, together with the cathode material in a reduction reaction. This is given by the equation:

$$Li_{1-x}CoO_2 + xLi^+ + xe^- \leftrightarrow LiCoO_2 \tag{7.27}$$

The overall reaction is represented by the equation:

$$LiCoO_2 + 6C \leftrightarrow Li_{1-x}CoO_2 + Li_xC_6 \tag{7.28}$$

Advantages of lithium-ion batteries are: high energy density, light weight, low maintenance, low self discharge rate.

Disadvantages of lithium-ion batteries are: limited shelf-life due to slow decay of capacity, protective circuit is needed to avoid overcharging and over-discharging, if damaged the electrolyte, the anode and the cathode can undergo combustion, at present costly to manufacture. Lithium-ion batteries are used in electric/ hybrid vehicles, cell phones and laptop computers.

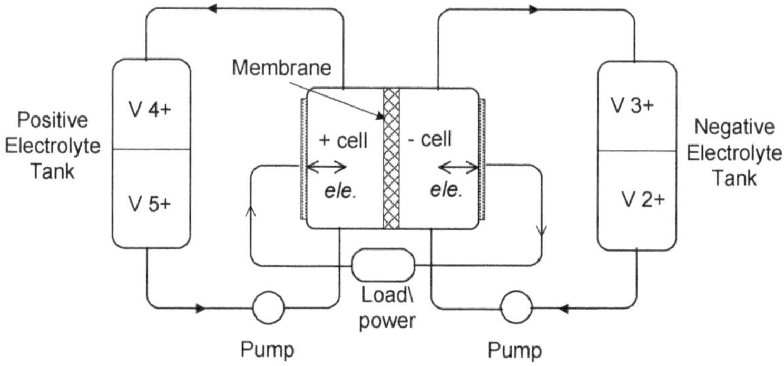

Fig. 7.6 Flow battery

7.6.3 *Flow batteries*

A flow battery is a novel battery in which electrolytes are pumped through a cell [6]. The most common battery of this type, depicted in Fig. 7.6, is the vanadium *redox* (reduction-oxidation) or VR flow battery. The electrolytes used in VR batteries are created by electrolytically dissolving vanadium pentoxide (V_2O_5) in sulfuric acid (H_2SO_4). The electrolytes are contained in two tanks and are pumped through the positive and negative cells during charging and discharging.

During the discharge cycle V_{2+} is oxidized to V_{3+} in the negative cell and an electron is released to the external circuit to do work. In the positive half cell V_{5+} in the form of VO_{2+} accepts an electron from the external circuit and is reduced to V_{4+} in the form of VO_2. Hydrogen $(H+)$ ions are exchanged between the two cells to maintain charge neutrality. The hydrogen ions diffuse through the membrane separating the cells. Charged vanadium species and water can also diffuse across the membrane.

The reaction at the cathode is represented by:

$$VO_2 + 2H^+ + e^- \leftrightarrow VO^{2+} + H_2O \tag{7.29}$$

The reaction at the anode is:

$$V^{2+} \leftrightarrow V^{3+} + e^- \tag{7.30}$$

The complete reaction is given by:

$$VO_2 + 2H^+ + V^{2+} \leftrightarrow VO^{2+} + H_2O + V^{3+} \quad (7.31)$$

For charge neutrality the transfer of an electron between the electrodes must be accompanied by the transfer of a hydrogen ion across the membrane between the cells.

The main advantages of the vanadium flow battery (VFB) are:
(i) the capacity can be increased simply by having larger electrolyte storage tanks, (ii) it can be completely discharged with no detrimental effects on performance, (iii) if the electrolytes mix accidentally, the battery does not suffer permanent damage, (iv) the electrolyte is aqueous and inherently safe and non-flammable, (v) VFBs can undergo 12,000 to 14000 charge/discharge cycles.

Some of the disadvantages of VFBs are:
(i) relatively poor energy density compared to other storage batteries, (ii) the battery is heavy and therefore suitable only for stationary applications, (iii) relatively high toxicity of oxides of vanadium.

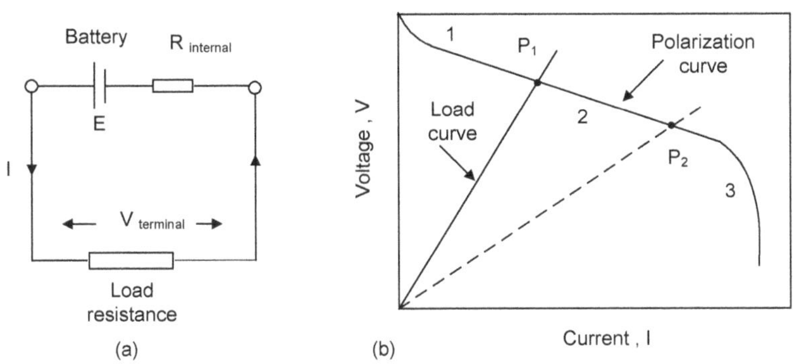

Fig. 7.7 (a) Battery with resistive load, (b) Polarization curve of battery

7.6.4 *Battery characteristics*

In this section we shall present a few important design characteristics of batteries. Figure 7.7(a) shows a battery connected to an external resistive load R_{load}. When the battery is not charging or discharging the voltage across the terminals is called the *open-circuit* voltage, and is equal to the *electromotive force* (emf) of the cell. However, as the cell delivers current to the external circuit the terminal voltage drops as shown in Fig. 7.7(b).

The variation of the terminal voltage with current is called the *polarization curve*. The voltage drop depicted by the curved section 1 of the graph is due to *activation losses* at the electrodes. The voltage loss in section 2 is due to *internal ohmic resistances* of the cell. The sharp voltage drop shown in section 3 is due to the slowness of the various transport processes in the cell.

The variation of voltage with current for the load resistance R_{load} follows Ohms law, and the resulting graph is a straight line as shown in Fig. 7.7(b). The intersection of this curve with the polarization curve gives the *operating point* of the system. As the load resistance is increased the operating point moves from P_1 to P_2. For most practical purposes, the idealized polarization curve is assumed to be linear, and its slope is the *internal resistance* R_{int} of the cell. Hence at the operating point we have

$$V = IR_{load} = E - IR_{int} \tag{7.32}$$

where E thc opcn-circuit voltage or the emf of the cell.

When cells are connected in series, the voltage of the battery is the sum of the individual voltages. When identical cells are connected in parallel, the total current is the sum of individual currents, and the voltage is that of a single cell.

The *capacity* of a battery is the amount of electric charge it can deliver at the *rated voltage*. It is measured in units of amp-hour (Ah). For example, a battery rated at 100 Ah can deliver 5A over a period of 20 hours at room temperature. The fraction of the stored energy that a battery can deliver depends on several factors, including battery chemistry, the rate at which charge is delivered (current), the required terminal voltage, the storage period, the ambient temperature, and others.

Figure 7.8(a) shows that the charge-discharge curves are generally steady except close to the fully charged and fully discharged conditions. A fundamental characteristic of a battery is the variation of the terminal

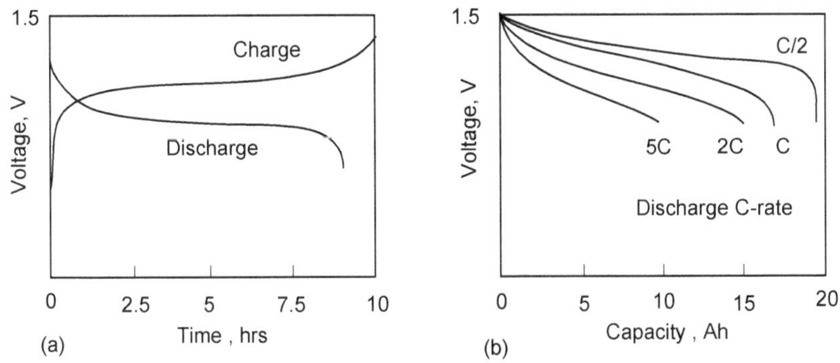

Fig. 7.8 (a) Charge-discharge of battery, (b) discharge C-rate of battery

voltage with time $V_t(t)$, during discharging at constant current, which is shown in Fig. 7.8(a). The area under the curve is proportional to the total energy delivered. For an ideal battery, the terminal voltage is constant, and the area under the curve is a rectangle. Different batteries have different $V_t(t)$ curves.

A battery's charge and discharge rates are controlled by the C-rating of the battery. The rated capacity (Ah) is generally expressed as the 1-C rate. This means that a battery with a rated capacity of 10 Ah should be able to provide 10 A for one hour. The same battery if discharged at a C rating of 0.5-C will deliver 5 A over 2 hours, and if discharged at a 2-C rate will provide 20 A for 0.5 hours. The discharge time T_{dis} (h) is related to the C-rating C_r (h^{-1}) of a battery by the equation

$$T_{dis} = \frac{1}{C_r} \tag{7.33}$$

The discharge current I_{dis}, the rated capacity E_r (Ah), and the C-rating are related by the equation

$$C_r = \frac{I_{dis}}{E_r} \tag{7.34}$$

The variation of terminal voltage with capacity for different C-ratings of a battery is shown in Fig. 7.8(b).

The change in capacity of rechargeable lead-acid batteries at different discharge rates t_a may be obtained from Peukert's law, which has the mathematical form:

$$t_a = t_r \left(\frac{C_r}{t_r I_a}\right)^k \tag{7.35}$$

where t_r is the rated discharge time [h], C_r is the rated capacity [Ah], I_a is the actual discharge current [A], t_a is the actual discharge time [h], and k is the Peukert constant. For an ideal battery, the capacity is independent of the current and $k = 1$. For a flooded lead-acid battery, $k = 1.25 - 1.4$.

Peukert's law becomes a key factor for batteries used in electric cars, where batteries rated for discharge times of about 20 hours are used for a much shorter discharge time of about 1 hour. In this case, the discharge currents are much higher than the rated discharge currents, which results in significant internal resistance power dissipation.

7.7 Fuel Cells

An indirect method for storing solar and wind energy is the production and storage of hydrogen, which may be supplied to fuel cells to generate electricity. Hydrogen can be produced by electrolyzing water [6]. In this section we shall discuss the principles of operation of fuel cells.

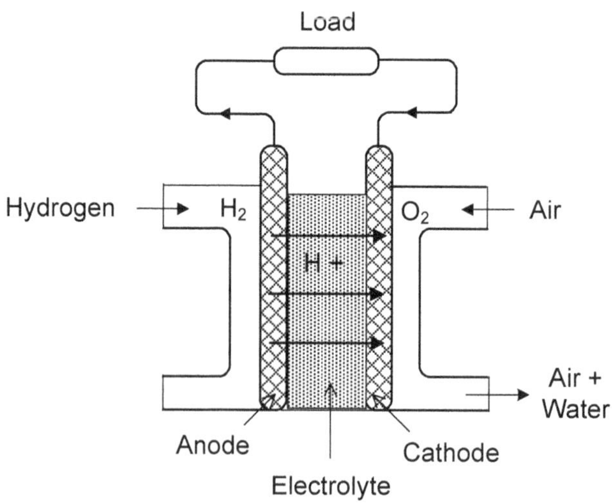

Fig. 7.9 Hydrogen fuel cell

A fuel cell, like a battery, is a device that converts the chemical energy of a fuel directly to electricity. However, unlike a battery it does not store chemical energy. To generate electricity a fuel cell needs a steady supply of a fuel such as hydrogen. Fuel cells are being used in space applications, and in commercial and industrial buildings as primary and backup power sources. Many companies are developing fuel cells for application in transport vehicles.

A hydrogen fuel cell is shown schematically in Fig. 7.9. It consists of two electrodes separated by an electrolyte, which transmits ions but is impervious to electrons. The fuel, which is hydrogen, is supplied to the anode while oxygen or air flows into the cathode. The anode has a porous catalyst that promotes the dissociation of hydrogen into hydrogen ions and electrons. The chemical equation for the reaction is:

$$H_2 \rightarrow 2H^+ + 2e^- \tag{7.36}$$

The electrons flow to the cathode through the external circuit doing work. The concentration gradient causes H^+ to diffuse through the electrolyte towards the cathode. At the cathode, the electrons from the external circuit, and the hydrogen ions migrating through the electrolyte, react with the oxygen supplied to the cathode to produce water. The platinum catalyst on the cathode promotes this reaction. The chemical equation for the reaction is:

$$\frac{1}{2}O_2 + 2H^+ + 2e^- \rightarrow H_2O \tag{7.37}$$

The net equation for the reactions in the fuel cell is:

$$H_2 + \frac{1}{2}O_2 \rightarrow H_2O \tag{7.38}$$

The amount of electric current available to the external circuit depends on the chemical activity and mass flow rate of the fuel.

7.7.1 *Analysis of fuel cells*

In this section we shall outline a thermodynamic analysis of fuel cells to predict its ideal open circuit voltage and current. The same concepts are applicable to batteries in general. The fuel cell (Fig. 7.9) may be treated as steady flow device (SFD) with fuel and oxidizer entering and products of the reaction leaving the control volume. Assume that mass and heat exchanges occur with an external ambient at constant temperature and pressure. Under these conditions, the maximum work output of SFD is the change in the total Gibb's free energy [5,8]. Hence we have

$$W_{max} = -\Delta G = G_1 - G_2 \tag{7.39}$$

where G is the Gibbs's free energy, and subscripts 1 and 2 represent inlet and outlet states. The Gibb's free energy is given by the expression [5]

$$G = H - T_o S \tag{7.40}$$

The constant ambient temperature is T_o, S is the entropy, and H is the enthalpy.

Applying the energy equation to the SFD we obtain

$$Q = H_2 - H_1 + W_{max} \tag{7.41}$$

where H is the enthalpy of the flow stream and Q is the heat input.

From Eqs. (7.39)–(7.41) it follows that the heat transfer is

$$Q = T_o(S_2 - S_1) \tag{7.42}$$

The fuel cell does not function like a typical heat engine, whose efficiency is defined as the ratio of the net work output to the heat input.

Therefore, the ideal efficiency of a fuel cell ε is usually defined as the ratio of the maximum work output to the change in total enthalpy. Therefore

$$\varepsilon = \frac{W_{max}}{H_2 - H_1} = \frac{G_2 - G_1}{H_2 - H_1} \tag{7.43}$$

The ideal external electrical work done by the fuel cell is given by the product of the charge Q_e carried by the electrons through the circuit and the terminal emf or voltage V of the cell.

Assume that at the anode reactants are consumed at the rate of n_a kmole *per unit time*. Let the number of electrons released per kmole of reactant be j. Then the electron flow expressed in kmoles of electrons per unit time is $n_a j$.

The number of electrons per kmole of electrons is the Avagadro number given by $N_A = 6.022169 \times 10^{26}$. An electron has a charge of 1.6022×10^{-19} Coulomb/electron. Therefore the charge transported by a kmole of electrons is

$$F = 1.6022 \times 10^{-19} N_A = 96.6487 \times 10^6 \text{ C per kmol}$$

The quantity F is called the *Faraday constant* which can be expressed in the form,

$$F = 96{,}487 \text{ kJ/V.kmol of electrons.}$$

The electric current I [amps] can be written as

$$I = jn_a F \tag{7.44}$$

The instantaneous power delivered [kJ] by the cell is

$$P = IV = jn_a FV \tag{7.45}$$

The maximum power [W] from the cell is given by Eq. (7.39) as

$$P_{max} = n_a W_{max} = -\Delta G n_a \tag{7.46}$$

where ΔG is the change in Gibb's free energy per *kmol of reactant*.

From Eqs. (7.45) and (7.46) we obtain the *ideal terminal voltage* of the cell as

$$V_c = -\frac{\Delta G}{jF} \tag{7.47}$$

Note that V_c is the ideal cell voltage. Internal resistance and other effects diminish this value in actual fuel cells. The above analysis may be used to determine the ideal properties of batteries [6].

In practical fuel cells, the ideal terminal voltage is not achieved due to several reasons. The difference between the ideal voltage and the actual voltage is known as the *overvoltage* or the *polarization*. As the current delivered by the cell increases, there is a significant drop in the terminal voltage as shown in Fig. 7.10. The initial sharp drop in voltage is called *activation* polarization. This is due to the energy absorbed to overcome the activation energy barrier at the electrodes.

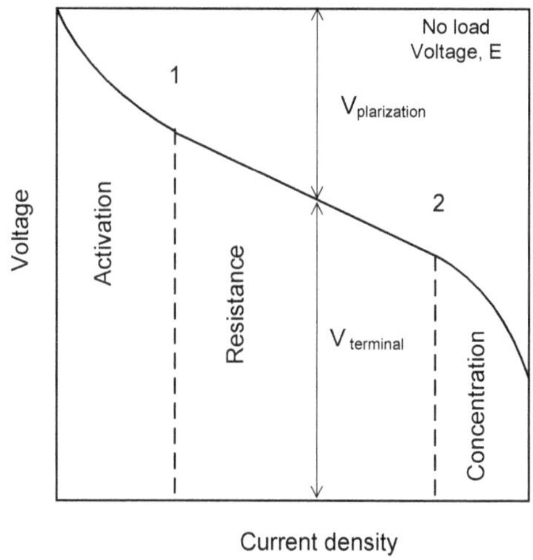

Fig. 7.10 Polarization curve of fuel cells

In the middle section of the curve, which is linear, the voltage drop is due to the internal resistance of the cell. This is composed of the electrode resistance, the bulk electrolyte resistance, and the interface contact resistance. The loss due to *resistance* polarization can be reduced (i) by proper shape of electrodes, (ii) by reducing the gap between electrodes, and (iii) by using concentrated electrolyte.

The sharp loss seen in the last section of the curve is called *concentration* polarization. This occurs mainly due to slowness of the various transport processes within the cell. All three polarization losses are reduced by operating the cell at a high temperature.

The losses in a fuel cell are accounted for by defining a performance index called the *voltage efficiency* η_v. It can be expressed as:

$$\eta_v = \frac{V}{V_c} \tag{7.48}$$

where V is the operating voltage and V_c is the ideal open circuit voltage (emf). The ideal cell voltage is given by Eq. (7.47).

The overall performance of a fuel cell is expressed in terms of an efficiency that is analogues to the *thermal efficiency* of a heat engine.

The heat input in the case of a fuel cell is the change in enthalpy of the fuel ΔH. The ideal energy output per kmol of fuel is given by

$$\Delta G = nFV_c \tag{7.49}$$

where n is the number of kmols of electrons produced per kmol of fuel. Hence the efficiency is

$$\eta = \frac{\Delta G}{\Delta H} \tag{7.50}$$

Six different types of fuel cells have been developed [6]. These are: (1) proton exchange membrane fuel cells, (2) direct methanol fuel cells, (3) molten carbonate fuel cells, (4) alkaline fuel cells, (5) phosphoric acid fuel cells, (6) solid oxide fuel cells. These types of fuel cells differ in operating temperature, efficiency, and cost. A detailed discussion of the characteristics of different fuel cells is available in Ref.[6].

7.8 Supercapacitor Energy Storage Systems

Capacitors, unlike batteries, store energy in an electrostatic field rather than in chemical form. A basic capacitor consists of a dielectric or an electrolyte separating two plates carrying opposite charges. It is charged by applying an external voltage across the plates, which makes the positive and negative charges to migrate to plates with the opposite polarity. The potential difference V between the plates of a capacitor is given by

$$Q = CV \tag{7.51}$$

where Q is the charge carried by the plates and C is the *capacitance*.

The capacitance depends on the area of the plates A, the distance between the plates d, the permittivity of free space ε_o, and the dielectric constant of the material between the plates ε_r. It may be expressed as

$$C = \frac{A\varepsilon_o\varepsilon_r}{d} \tag{7.52}$$

The energy stored in a capacitor U is given by

$$U = \frac{1}{2} QV \tag{7.53}$$

From Eqs. (7.51)-(7.53) it follows that

$$U = \frac{1}{2} CV^2 = \frac{1}{2} \frac{A\varepsilon_o \varepsilon_r}{d} V^2 \tag{7.54}$$

Equation (7.54) shows that for high energy storage in a capacitor, the dielectric constant and the plate area should be as large as possible, and the plate separation should be as small as possible.

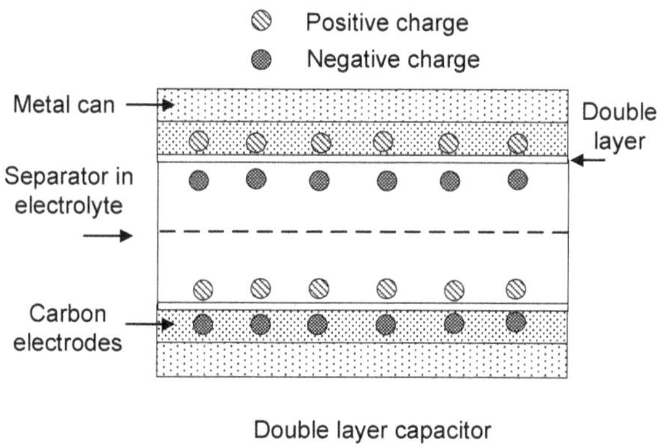

Fig. 7.11 Supercapacitor

Supercapacitors or ultracapacitors are advanced capacitors designed to improve their energy storage potential. These are also called *electric double layer capacitors* (EDLC) and they look much like regular batteries [10]. They have double layer construction consisting of two carbon electrodes immersed in an organic electrolyte as shown schematically in Fig. 7.11.

During charging, the applied volage causes electrically charged ions in the electrolyte to migrate towards the electrodes with opposite polarity. This produces two separately charged layers called Helmholtz double layers as seen in Fig. 7.11. The two electrodes form a series circuit of two individual capacitors C_1 and C_2. The total capacitance C_{tot} is given by

$$\frac{1}{C_{tot}} = \frac{1}{C_1} + \frac{1}{C_2} \tag{7.55}$$

Although similar to batteries, the EDLCs rely on electrostatic action, and no chemical reactions are involved. These capacitors use the double-layer effect to store electric energy, and they do not have a solid dielectric to separate the charges as in conventional capacitors. As a result EDLCs have much higher capacitance values because of the very large area of the activated carbon electrodes and the extremely thin double-layer distance on the order of 0.3 to 0.8 nm.

The main advantages of EDLCs are:
(i) cell voltage is determined by the application and not by the cell chemistry, (ii) very high cell voltages are possible, (iii) high power available, (iv) simple charging method, (v) rapid charge and discharge compared to batteries, (vi) they cannot be overcharged, (vii) long life cycle of over 5×10^6 cycles, (viii) long operational life of about 10 to 20 years [10].

The disadvantages are:
(i) power is available only for a short duration, (ii) low capacity, (iii) relatively low energy density, (iv) high self discharge rate compared to batteries.

7.9 Thermal Energy Storage Systems

In this section we shall discuss different types of thermal energy storage technologies used in renewable energy systems. These include: (i) sensible heat storage, (ii) phase change heat storage systems, and (iii) thermochemical storage systems.

7.9.1 *Sensible heat storage*

Thermal energy storage systems have been used extensively in several solar thermal applications. The most well-established of these are in domestic hot water heating systems and industrial process heating systems. We discussed their design and operation in chapter 3 where several design options were presented.

In chapter 4 on solar thermal power generation, we introduced the solar pond as a sensible heat storage system for low temperature power generation. In the case of high temperature solar thermal power systems, using trough concentrators and solar power towers, the widely used heat storage facility is the two-tank system. Few optional designs of such systems together with the different types of storage and heat removal fluids were discussed in chapter 4. The above devices are called sensible heat storage systems because heat is stored as sensible heat of a fluid.

The quantity of heat stored per unit volume of fluid when the storage temperature increases from T_1 to T_2 is given by

$$\Delta Q = \int_{T_1}^{T_2} \rho C_v dT \tag{7.56}$$

where C_v and ρ are the specific heat capacity at constant volume, and the density of the heat storage medium, respectively. These properties are temperature dependent, but for engineering design calculations their values averaged over the temperature range of operation may be used.

7.9.2 *Packed bed thermal storage*

All solar heating and power generation systems considered in chapters 3 and 4 involved liquids as heat removal and storage media. However, for solar home heating and solar drying applications the heat removal fluid is usually air. Such systems use granular materials like pebbles or rocks, as the energy storage medium. The solid material is placed in a container and supported by a screen, with openings for air entry and exit. Air flows through the intestacies between the pebbles. Some of the advantages of the system are: high heat transfer coefficient between the air and the pebbles, and thermal stratification in the flow direction.

Unlike liquid based storage systems, the pebble bed system does not allow simultaneous heat storage and removal. During the heat storage phase, the pebbles near the entrance are heated first and the temperature of the air drops to the initial bed temperature in a short distance. As time progresses this temperature distribution moves towards the exit of the bed. When fully charged the bed has a uniform temperature. To recover heat the flow direction is reversed with the cold air entering the bed from the opposite end.

We shall develop a simplified model to predict the long term performance of a pebble bed storage system.

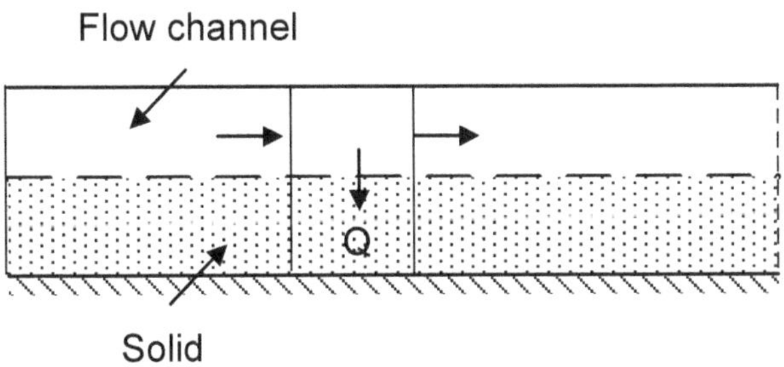

Fig. 7.12 Model of pebble bed storage system

An idealized physical model of a pebble bed is shown in Fig. 7.12 where the pebbles are represented by a solid slab with air flowing in a channel above it. The heat exchange area between the pebbles and air is represented by an interface. In the storage phase, air transfers heat to the slab across the interface.

Following assumptions are made in deriving the governing energy equations for air and the slab: the energy stored in air is negligible, the axial heat conduction in air and the bed are negligible, all physical properties are constant, no heat is lost to the surroundings.

Consider a small control volume of length δx at a distance x from the entrance. The air temperature change across the control volume is δT_a and the bed temperature at this location is T_b.

The energy equation for air is

$$mc_{pa}\delta T_a = -h_v(T_a - T_b)A\delta x \tag{7.57}$$

where m is the mass flow rate of air, c_{pa} is the specific heat capacity. The bed cross sectional area is A, and the volumetric heat transfer coefficient between the pebbles and the air is h_v. The latter is obtained by multiplying the usual heat transfer coefficient by the bed particulate surface area per unit bed volume. The LHS of Eq. (7.57) is the increase in energy of the air and the RHS is energy transferred from the air to the bed, both over a time interval δt.

The energy equation for the pebbles is

$$\rho_b(1-\epsilon)A\delta x c_b \delta T_b = h_v(T_a - T_b)A\delta x \delta t \tag{7.58}$$

The density, void fraction, and specific heat capacity of the bed are ρ_b, ϵ and c_b respectively. The LHS of Eq. (7.58) is the increase in energy of the bed, and the RHS is energy transferred from air, both over a time interval δt.

The energy equations (7.57) and (7.58) lead to the following partial differential equations:

$$mc_{pa}\frac{\partial T_a}{\partial x} = -Ah_v(T_a - T_b) \tag{7.59}$$

$$\rho_b(1-\epsilon)c_b\frac{\partial T_b}{\partial t} = h_v(T_a - T_b) \tag{7.60}$$

Both analytical and numerical methods have been developed to solve the differential Eqs. (7.59) and (7.60). For long-term performance evaluation of packed bed storage systems numerical solution of the above equations is recommended [1]. We shall illustrate the numerical procedure in the worked examples to follow in this chapter.

7.9.3 *Phase change thermal storage*

Phase change thermal energy storage systems use a material that under goes change of phase, typically from a solid to a liquid when it stores energy[1]. The process usually occurs at a constant temperature, which is the melting point of the material. However, there would be some subcooling and superheating accompanying the energy transfer from the heating fluid to the phase change medium (PCM).

Some of the desirable properties of a phase change storage material area: high latent heat per unit volume, reversible over many cycles without serious degradation, limited subcooling and superheating, and reasonable cost of material.

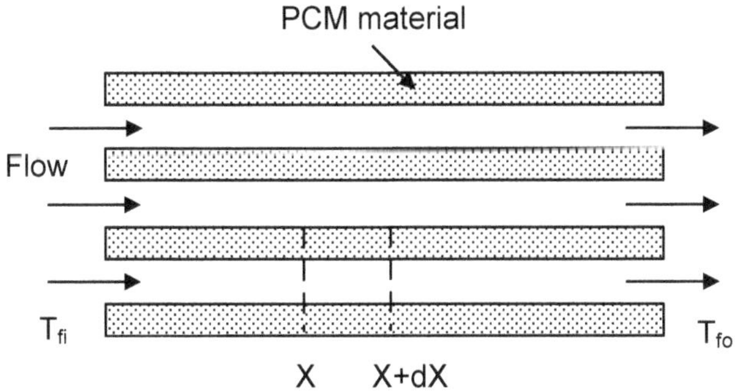

Fig. 7.13 Model of PCM storage system

A schematic view of a phase change storage system is depicted in Fig. 7.13. The phase change material is contained in thin containers in the form of slabs with flow passages for air located between them. The energy stored in a phase change material per unit mass Q_s when it is heated from a temperature T_1 to T_2 is given by

$$Q_s = c_s(T_m - T_1) + L + c_l(T_2 - T_m) \tag{7.61}$$

where c_s and c_l are respectively the specific heat capacities of the solid and the liquid, L is the latent heat, and T_m is the phase change temperature of the material.

A material that has been used in house-heating applications is Glauber's salt [1]. The reaction accompanying heat addition to the material may be represented by the equation

$$Na_2SO_4.10H_2O + Q \leftrightarrow Na_2SO_4 + 10H_2O \tag{7.62}$$

The reaction proceeds from left to right during energy addition, and in the opposite direction when energy is extracted.

Other materials that have been studied for energy storage are paraffin waxes. They can be obtained with various melting points, and as mixtures with ranges of melting points.

Ice is a phase change substance that has most of the desirable properties mentioned above. Building air conditioning plants use ice storage systems to exploit the differences in electricity prices between daytime and night, which is common in most countries. Ice is produced during off-peak hours when electricity costs are low, and melted to cool building air during periods when energy costs are high.

7.10 Comparison of Different Energy Storage Technologies

Several important characteristics of energy storage systems are compared in Table 7.1. These characteristics are discussed in section 7.1.1. The data has been obtained by considering the average values available in published sources [2].

Table 7.1 Characteristics of storage technologies (Data from Ref.[2])

Technology	Power	Energy	Resp. time	Efficiency
Comp. air	110-290 MW	1.16-3 GWh	12 min	90%
Flywheel	5 kW-90 kW	5-200 kWh	12 min	80-95%
Pumped Hydro	100 MW-2 GW	0.4-20 GWh	12 min	70-80%
Batteries	100 W-100 MW	1 kWh-200 MWh	Seconds	60-80%
Supercapacitors	2 kW-1 MW	1 Wh-1 kWh	10^{-3} Sec.	80-95%
H_2 - Fuel cells	1 kW-1 MW	20 kWh-7 MWh	-	59%

Technology	Discharge Time	Life time, Years	Daily self discharge, %	Maturity
Comp. air	1-24 hr.	20-40	Small	Proven
Flywheel	ms-15min	15	High	Mature
Pumped Hydro	1-24 hr.	40-60	Very small	Mature
Batteries	sec-hours	5-15	0.1-0.3	Mature
Supercapacitors	ms-hours	20	20-40	Proven

7.11 Worked Examples

Example 7.1 Air at a pressure of 98 kPa and temperature 27°C, enters a two stage compressor at the rate of 9.2 m^3 per min. It is compressed to 336 kPa in the low pressure compressor and enters an intercooler where it is cooled to the initial temperature before it enters the high pressure compressor. Air leaves the high pressure compressor at a pressure of

1150 kPa. The compression processes in the two compressors are polytropic with an index of 1.35. Neglect any changes in kinetic energy. Calculate (i) the work input with and without the intercooler, (ii) heat removal rate in the intercooler, and (iii) the exit air temperature with and without the intercooler.

Solution Assume that air is an ideal gas. The mass flow rate of air is given by

$$\dot{m} = \dot{V}\rho = \dot{V}\left(\frac{P}{RT}\right)$$

$$\dot{m} = \left(\frac{9.2}{60}\right)\left(\frac{98}{0.287\times 300}\right) = 0.174 \text{ kg s}^{-1}$$

Without intercooler and single-stage compression
The total power input is given by [5,6]

$$W_{in} = \dot{m}RT\left(\frac{n}{n-1}\right)\left[\left(\frac{P_2}{P_1}\right)^{\left(\frac{n-1}{n}\right)} - 1\right]$$

$$W_{tot} = 0.174 \times 0.287 \times 300\left(\frac{1.35}{0.35}\right)\left[\left(\frac{1150}{98}\right)^{\left(\frac{1.35-1}{1.35}\right)} - 1\right] = 51.56 \text{ kW}$$

The exit air temperature is given by

$$\frac{T_2}{300} = \left(\frac{1150}{98}\right)^{\left(\frac{1.35-1}{1.35}\right)} = (11.73)^{0.259},\ T_2 = 567.7 \text{ K}$$

With intercooler and two-stage compression
Power input to the low pressure stage is

$$W_{lp} = \dot{m}RT\left(\frac{n}{n-1}\right)\left[\left(\frac{P_2}{P_1}\right)^{\left(\frac{n-1}{n}\right)} - 1\right]$$

$$W_{lp} = 0.174 \times 0.287 \times 300\left(\frac{1.35}{0.35}\right)\left[\left(\frac{336}{98}\right)^{\left(\frac{1.35-1}{1.35}\right)} - 1\right] = 21.72 \text{ kW}$$

Similarly, for the high pressure stage

$$W_{hp} = 0.174 \times 0.287 \times 300\left(\frac{1.35}{0.35}\right)\left[\left(\frac{1150}{336}\right)^{\left(\frac{1.35-1}{1.35}\right)} - 1\right] = 21.68 \text{ kW}$$

Total power input with intercooler is 43.4 kW
The exit temperature from low pressure stage is given by

$$\frac{T_2}{300} = \left(\frac{336}{98}\right)^{0.259}, \quad T_2 = 412.77 \text{ K}$$

Heat removal rate in the intercooler is

$$Q_{ic} = \dot{m}c_p(T_2 - T_1)$$

$$Q_{ic} = 0.174 \times 1.005(412.77 - 300) = 19.72 \text{ kW}$$

Exit temperature from the high pressure stage is

$$\frac{T_2}{300} = \left(\frac{1150}{336}\right)^{0.259}, \quad T_2 = 412.6 \text{ K}$$

Example 7.2 In a compressed air storage system, air from the storage tank at 42 bar and 30°C enters a combustor where it is heated at constant pressure to 410°C. Air expands to 7.8 bar in the high pressure turbine and enters a second combustor where it is heated to 350°C. In the low pressure turbine air expands to 1.2 kPa. The polytropic index for expansions in the two turbines is 1.38. For unit mass of air leaving the storage tank, calculate (i) the work output of the two turbines, (ii) the heat input in the two combustors, and (iii) the exit temperature of the air. Assume that air is an ideal gas with constant properties.

Solution As air discharges from the storage tank to the combustor, the pressure in the tank drops continuously. We shall consider the processes undergone by unit mass of air during this discharge phase.

Heat input in the first combustor under the given conditions is

$$Q_{com1} = \dot{m}c_p(T_2 - T_1)$$

$$Q_{com1} = 1.0 \times 1.005(410 - 30) = 381.9 \text{ kWkg}^{-1}$$

The work output in the high pressure turbine is [5,6]

$$W_{out} = \dot{m}RT\left(\frac{n}{n-1}\right)\left[1 - \left(\frac{P_1}{P_2}\right)^{\left(\frac{n-1}{n}\right)}\right]$$

$$W_{ohp} = 1.0 \times 0.287 \times 683 \left(\frac{1.38}{1.38-1}\right)\left[1 - \left(\frac{7.8}{42}\right)^{\left(\frac{1.38-1}{1.38}\right)}\right]$$

$$W_{ohp} = 263.8 \text{ kWkg}^{-1}$$

Exit temperature from the high pressure turbine is

$$\frac{T_2}{683} = \left(\frac{7.8}{42}\right)^{\left(\frac{1.38-1}{1.38}\right)}, \quad T_2 = 429.9 \text{ K}$$

Heat input in the second combustor is

$$Q_{com2} = \dot{m}c_p(T_2 - T_1)$$

$$Q_{com2} = 1.0 \times 1.005(623 - 429.9) = 194.06 \text{ kWkg}^{-1}$$

The work output in the low pressure turbine is

$$W_{out} = \dot{m}RT\left(\frac{n}{n-1}\right)\left[1 - \left(\frac{P_1}{P_2}\right)^{\left(\frac{n-1}{n}\right)}\right]$$

$$W_{olp} = 1.0 \times 0.287 \times 623 \left(\frac{1.38}{1.38-1}\right)\left[1 - \left(\frac{1.2}{7.8}\right)^{\left(\frac{1.38-1}{1.38}\right)}\right]$$

$$W_{olp} = 261.25 \text{ kWkg}^{-1}$$

For the expansion in the low pressure turbine

$$\frac{T_2}{623} = \left(\frac{1.2}{7.8}\right)^{\left(\frac{1.38-1}{1.38}\right)}$$

Hence the exit temperature of the air $T_2 = 372.3\ K$

Example 7.3 A rigid compressed air storage tank of volume 10 m^3 contains air at 40°C and 900 kPa. Calculate the mass, pressure and temperature of the air in the tank, after two thirds the original mass of air has been discharged to a turbine. Neglect any heat transfer between the air in the tank and the surroundings. Assume that air is an ideal gas with constant properties.

Solution Let the mass of air in the tank at time t after the discharge process has commenced be M, and the temperature and pressure of the air in the tank be T and P respectively. Let the mass of air discharged during a small time interval δt be δm.

We now apply the unsteady mass and energy balance equations for a control volume surrounding the tank. Assume that air is an ideal gas with constant properties, and neglect any heat transfer between the air and the surroundings.

Mass balance equation is:

$$\frac{dM}{dt} = -\dot{m} \tag{E7.3.1}$$

Energy balance equation is:

$$\frac{dE}{dt} = -\dot{m}h \tag{E7.3.2}$$

The internal energy E of the air in the tank, and the specific enthalpy h of the air in the discharge port are given by

$$E = Mc_vT \quad \text{and} \quad h = c_pT \tag{E7.3.3}$$

Manipulating Eqs. (E7.3.1), (E7.3.2) and (E7.3.3) we obtain the differential equation

$$M\frac{dT}{dt} = (\gamma - 1)T\frac{dM}{dt} \tag{E7.3.4}$$

where, the ratio of the specific heat capacities for air is given by

$$\gamma = c_p/c_v = 1.4.$$

The solution of Eq. (E7.3.4) gives

$$\frac{T}{T_i} = \left(\frac{M}{M_i}\right)^{\gamma-1} \tag{E7.3.5}$$

where M_i is the initial mass of air, T_i is the initial temperature of air in the tank.

Substituting the ideal gas equation in Eq. (E7.3.5) we have

$$\frac{T}{T_i} = \left(\frac{P}{P_i}\right)^{\frac{\gamma-1}{\gamma}} \tag{E7.3.6}$$

Substituting the given numerical values in Eq. (E7.3.5) we get

$$\frac{T}{313} = \left(\frac{1}{3}\right)^{\gamma-1} = \left(\frac{1}{3}\right)^{0.4}$$

The temperature of the air is 201.69 K.

Substituting numerical values in Eq. (E7.3.6) we have

$$\frac{201.69}{313} = \left(\frac{P}{900}\right)^{\frac{\gamma-1}{\gamma}} = \left(\frac{P}{900}\right)^{0.2857}$$

The pressure of the air in the tank is 193.28 kPa.

Apply the ideal gas equation to the air in the tank. Hence we have

$$PV_t = MRT$$

$$193.28 \times 10 = 0.287 \times 201.69M$$

The mass of air in the tank is 33.4 kg.

Example 7.4 Compressed air, initially stored at 850 kPa and 40°C, is to be used to power an adiabatic air turbine operating between the discharge pressure of the storage tank and the ambient pressure of 100 kPa. The turbine produces an average output of 2.5 kW over a period of 12 min. with a mean efficiency of 45 percent. The tank pressure at the end of this period is 150 kPa. Calculate the volume of the required storage tank. Neglect any heat transfer between the air and the surroundings, and assume that air is ideal gas with constant properties.

Solution A compressed air storage system is shown schematically in Fig. 7.1. Consider the situation at time t after the discharge process has commenced. The mass, temperature, and pressure in the tank are related by the Eqs. (E7.3.5) and (E7.3.6). Hence we get

$$M = M_i\left(\frac{p}{p_i}\right)^{1/\gamma} \tag{E7.4.1}$$

Mass of air entering the turbine δm during the time interval δt is obtained by differentiating Eq. (E7.4.1) as:

$$\delta m = -\delta M = -\left(\frac{M_i}{\gamma p_i^{1/\gamma}}\right) p^{\frac{1-\gamma}{\gamma}} \delta p \qquad \text{(E7.4.2)}$$

Work output of the *adiabatic* turbine per unit mass of air is [5]

$$W_{out} = RT\left(\frac{\gamma}{\gamma-1}\right)\left[1-\left(\frac{P_o}{P}\right)^{\left(\frac{\gamma-1}{\gamma}\right)}\right] \qquad \text{(E7.4.3)}$$

where p and p_o are the inlet and outlet pressures respectively.

It follows from Eqs. (E7.4.2) and (E7.4.3) that the work output during the small time interval δt is

$$\delta W_{out} = \delta m RT\left(\frac{\gamma}{\gamma-1}\right)\left[1-\left(\frac{P_o}{P}\right)^{\left(\frac{\gamma-1}{\gamma}\right)}\right] \qquad \text{(E7.4.4)}$$

Substituting from Eq. (E7.4.2) in Eq. (E7.4.4) we have

$$\delta W_{out} = -RT\left(\frac{\gamma}{\gamma-1}\right)\left[1-\left(\frac{P_o}{P}\right)^{\left(\frac{\gamma-1}{\gamma}\right)}\right]\left(\frac{M_i}{\gamma p_i^{1/\gamma}}\right) p^{\frac{1-\gamma}{\gamma}} \delta p \qquad \text{(E.7.4.5)}$$

The total energy output when the pressure in the tank decreases from p_i to p_f is obtained by integrating Eq. (E7.4.5) over this pressure interval.

Substituting for T from Eqs. (E7.3.5) in Eq. (E7.4.5) we obtain the following simplified expression for the total energy output

$$E_{out} = \frac{-RM_iT_i}{(\gamma-1)p_i}\int_{p_i}^{p_f}\left[1-\left(\frac{P_o}{P}\right)^{\frac{\gamma-1}{\gamma}}\right]dp \qquad \text{(E7.4.6)}$$

Integrating Eq. (E7.4.6) we get

$$E_{out} = -\frac{V_t p_o}{\gamma-1}\left[\frac{p_f}{p_o}-\frac{p_i}{p_o}-\gamma\left(\frac{p_f}{p_o}\right)^{\frac{1}{\gamma}}+\gamma\left(\frac{p_i}{p_o}\right)^{\frac{1}{\gamma}}\right]$$

Substituting numerical data in the above equation we have

$$E_{out} = -\frac{100V_t}{0.4}\left[\frac{150}{100}-\frac{850}{100}-1.4(1.5)^{\frac{1}{1.4}}+1.4(8.5)^{\frac{1}{1.4}}\right] = 604.37V_t$$

The actual output of the turbine is

$$E_{act} = \eta E_{out} = 0.45 \times 604.3V_t$$

where η is the turbine efficiency.

The required energy output is

$$E_{req} = 2.5 \times 12 \times 60 = 1800 \text{ kJ}$$

The required tank volume is obtained equating E_{act} and E_{req}. Hence we get

$$V_t = \frac{1800}{0.45 \times 604.3} = 6.62 \text{ m}^3$$

Example 7.5 A flywheel made of steel consists of a solid cylinder of diameter 1.2 m and thickness 10 cm, integral with an outer cylindrical rim of diameter 1.4 m and thickness 16 cm. During energy storage, a constant motor torque acting on the flywheel increases its rotational speed from 500 rpm to 4000 rpm. The energy output of the motor during this period is 35 MJ. Calculate (i) the moment inertia of the flywheel, (ii) the energy stored in the flywheel, and (iii) efficiency of energy storage.

Solution A flywheel energy storage system is shown schematically in Fig. 7.3.

Mass of the inner cylinder is

$$m = \frac{\pi}{4} D^2 L\rho = \frac{\pi}{4} \times 1.2^2 \times 0.1 \times 7700 = 870.85 \text{ kg}$$

Moment of the inertia of the inner cylinder is given by Eq. (7.11) as:

$$I = \frac{1}{2} mr^2 = \frac{1}{2} \times 870.85 \times 0.6^2 = 156.75 \text{ kgm}^2$$

Mass of the outer annular cylinder is

$$m = \frac{\pi}{4}\left({D_o}^2 - {D_i}^2\right) L\rho$$

$$m = \frac{\pi}{4}(1.4^2 - 1.2^2) \times 0.16 \times 7700 = 503.16 \text{ kg}$$

Moment of inertia of the outer cylinder is given by Eq. (7.12) as:

$$I = \frac{m}{2}({r_o}^2 + {r_i}^2)$$

$$I = \frac{503.16}{2}(0.7^2 + 0.6^2) = 213.84 \text{ kgm}^2$$

Total moment inertia of the flywheel about its axis is

$$I_f = 156.75 + 213.84 = 370.6 \text{ kgm}^2$$

The total energy stored in the flywheel is given by Eq. (7.10) as:

$$E_{max} = \frac{I_f}{2}({\omega_{max}}^2 - {\omega_{min}}^2)$$

where the maximum and minimum angular speeds are

$$\omega_{max} = 418.87 \text{ rad.s}^{-1} \text{ and } \omega_{min} = 52.36 \text{ rad. s}^{-1}$$

Hence we have

$$E_{max} = \frac{370.6}{2}(418.87^2 - 52.36^2) = 32 \text{ MJ}$$

The work done by the motor torque of 35 MJ is equal to the energy supplied to the flywheel. Hence the efficiency of energy storage is

$$\eta = \frac{32}{35} = 91.4\%$$

The difference between the energy supplied and the energy stored is due to losses in the flywheel system.

Example 7.6 A steel flywheel with a moment of inertia of 260 kgm^2 is acted upon by a constant motor torque of 175 Nm. It experiences a frictional torque given by, $2.6 \times 10^{-3}\omega$ Nm, where ω is the angular speed in rad. s^{-1}.

(a) Calculate the time taken for the rotational speed to reach the maximum value of 4000 rpm. (b) Calculate the self discharge time to 80% of the maximum energy stored if the frictional torque has the same form. (c) At the end of the self discharge period, the flywheel discharges energy to a generator and experiences a constant load torque of 200 Nm. Calculate the time to discharge 90% of the maximum energy, if the frictional torque remains unchanged.

Solution The equation of motion of the flywheel is given by

$$I\frac{d\omega}{dt} = T_m - B\omega \tag{E7.6.1}$$

where T_m is the motor torque and $B\omega$ is the frictional torque.

The solution of the differential equation (E7.6.1) is given by

$$\frac{\lambda-\omega_2}{\lambda-\omega_1} = e^{-a(t_2-t_1)} \tag{E7.6.2}$$

where $\lambda = \frac{T_m}{B}$ and $a = \frac{B}{I}$. The initial and final angular speeds are ω_1 and ω_2 respectively. Substituting given numerical data we get

$$\lambda = \frac{T_m}{B} = \frac{175}{2.6\times10^{-3}} = 67.31 \times 10^3 \text{ s}^{-1}$$

$$a = \frac{B}{I} = \frac{2.6\times10^{-3}}{260} = 10^{-5} \text{ s}^{-1}$$

The initial and final speeds are $\omega_1 = 0$, and $\omega_2 = 418.8$ rad. s^{-1} respectively.

(a) Substituting these values in Eq. (E7.6.2) we obtain the time to reach the maximum speed of 418.8 rad. s^{-1} as 10.4 minutes.

(b) During the storage period, the flywheel looses energy due to the frictional torque acting on it. There is no motor torque, therefore $T_m = 0$ and $\lambda = 0$.

When its energy reaches 80% of the maximum energy we have

$$0.8 \times \frac{I}{2}{\omega_{max}}^2 = \frac{I}{2}{\omega_3}^2$$

$$\omega_3 = 0.8^{1/2} \times 418.8 = 374.65$$

Substituting in Eq. (E7.6.2) we obtain

$$\frac{\omega_4}{\omega_3} = e^{-a(t_2-t_1)} = \frac{374.65}{418.8}$$

Hence we obtain the self discharge period as 3.09 hours.

(c) During the energy discharge period the governing equation of motion is

$$I\frac{d\omega}{dt} = -T_g - B\omega \qquad \text{(E7.6.3)}$$

where T_g is the generator torque.

The solution of the differential equation (E 7.6.3) is given by

$$\frac{\lambda+\omega_2}{\lambda+\omega_1} = e^{-a(t_2-t_1)} \qquad \text{(E7.6.4)}$$

where $\lambda = \frac{T_g}{B}$ and $a = \frac{B}{I}$. Substituting numerical values we get

$$\lambda = \frac{200}{2.6\times10^{-3}} = 76.92 \times 10^3$$

and

$$a = \frac{2.6\times10^{-3}}{260} = 10^{-5}$$

The initial and final speeds are ω_1 and ω_2 respectively.

When energy of the flywheel reaches 10% of the maximum energy we have

$$0.1 \times \frac{I}{2}{\omega_{max}}^2 = \frac{I}{2}{\omega_2}^2$$

$$\omega_2 = 0.1^{1/2} \times 418.8 = 132.46 \text{ rad. s}^{-1}$$

Substituting in Eq. (E7.6.4) we get

$$ln\left(\frac{76920+132.46}{76920+374.65}\right) = -10^{-5}(t_2 - t_1)$$

Hence the time to discharge to 10% of the maximum energy is 5.2 minutes.

Example 7.7 A cylindrical steel flywheel has a diameter of 1.8 m and length of 0.3 m. The ultimate stress for steel is 690 MNm^{-2}, and the density is 7700 kgm^{-3}. (a) Calculate the maximum possible kinetic energy and speed of the flywheel. (b) The flywheel is acted upon by a constant motor torque of 1800 Nm during the energy storage phase. Calculate the time taken to store 80% of the maximum possible energy, neglecting frictional effects. (c) The flywheel stores energy for 4.5 hours. Calculate the self discharge that occurs during this period if the frictional torque is given by 0.011ω Nm, where ω is the angular speed in rad. s^{-1}.

(d) Calculate the time taken to discharge 80% of the available energy to a generator that produces a constant load torque of 1600 Nm. Neglect frictional effects.

Solution (a) Mass of the flywheel is

$$m = \frac{\pi}{4} D^2 L\rho = \frac{\pi}{4} \times 1.8^2 \times 0.3 \times 7700 = 5878.2 \text{ kg}$$

Maximum possible energy storage is given by Eq. (7.18) as

$$\frac{E_{km}}{m} = K\frac{\sigma_{max}}{\rho} = 0.606 \times \frac{690\times10^6}{7700}$$

Hence $$E_{km} = 319.2 \times 10^6 \text{ J}$$

The moment of inertial of the flywheel is

$$I = \frac{1}{2} mr^2 = \frac{1}{2} \times 5878.2 \times 0.9^2 = 2380.67 \text{ kgm}^2$$

The maximum speed is given by

$$\frac{1}{2} I\omega_{max}{}^2 = \frac{2380.67\omega_{max}{}^2}{2} = 319 \times 10^6$$

Hence the maximum speed is 517.68 rad. s^{-1}

(b) When 80% of the maximum energy is stored the rated speed is

$$\omega_r = 0.8^{1/2} \times 517.68 = 463.0 \text{ rad. s}^{-1}$$

The time to reach this speed is given by

$$I(\omega_2 - \omega_1) = T_m \Delta t$$

$$1800\Delta t = 2380.67 \times (463.0 - 0)$$

Hence the time is 10.2 mts.

(c) The flywheel self discharges energy due to the action of the frictional torque. The governing equation is

$$I\frac{d\omega}{dt} = -B\omega$$

The solution of the above equation is

$$\omega_2 = \omega_1 e^{-at}$$

where

$$a = \frac{B}{I} = \frac{0.011}{2380.76} = 4.62 \times 10^{-6} \text{ s}^{-1}$$

Substituting numerical data in the above equation the speed at the end of 4.5 h is obtained as

$$\omega_2 = 463.0e^{-4.62\times10^{-6}\times4.5\times3600} = 429.6 \text{ rad. s}^{-1}$$

(d) The speed of the flywheel when 80% of the energy available has been discharged to the generator is

$$\omega_f = 0.2^{1/2} \times 429.6 = 192.12 \text{ rad. s}^{-1}$$

The time to reach this speed is given by

$$2380.67 \times (429.6 - 192.12) = 1600t_1$$

Hence the time is 5.89 mts.

Example 7.8 The difference in elevation from head water to tail water of a hydroelectric power system is 55 m, and the water flow rate through the turbine is 5.2 m^3s^{-1}. The rotational speed of the turbine shaft is 190 rpm and the torque in the shaft is 120 kNm. The output of the electric generator is 2200 kW. The energy losses in the pipe network and fittings are given by, $E_{fr} = 4.8Q^2$ kW, where Q is the flow rate in m^3s^{-1}. Calculate (i) the maximum power available to the turbine-generator system, (ii) the power losses in the piping system, (iii) the losses and efficiency of the turbine, and (iv) the losses and efficiency of the generator.

Solution A hydroelectric power system is shown schematically in Fig. 7.4. The maximum power available is given by

$$P_{av} = H\rho gQ$$

$$P_{av} = 55 \times 10^3 \times 9.81 \times 5.2 = 2805.7 \text{ kW}$$

Power loss due to friction is

$$E_{fr} = 4.8Q^2 = 4.8 \times 5.2^2 = 129.8 \text{ kW}$$

Power available to the turbine-generator system is

$$P_{t-g} = 2805.05 - 129.8 = 2675.86 \text{ kW}$$

The turbine power is

$$P_t = 120 \times 10^3 \times \frac{190 \times 2\pi}{60} = 2387.6 \text{ kW}$$

Power loss in the turbine is

$$P_{tl} = 2675.86 - 2387.6 = 288.24 \text{ kW}$$

Power loss in the generator is

$$P_{gl} = 2387.6 - 2200 = 187.6 \text{ kW}$$

Efficiency of the turbine is

$$\eta_{tur} = \frac{2387.6}{2675.86} = 89\%$$

Efficiency of the generator is

$$\eta_{gen} = \frac{2200}{2387.6} = 92\%$$

Example 7.9 The difference in elevation between two reservoirs of a pump storage system is 30 m. It includes 1600 m of piping of diameter 300 mm and some fittings. The pump used has the following characteristics:

$$H_p = -0.0014Q^2 + 0.017Q + 45.335$$

where H_p is the pump head in m and Q is the water flow rate in liter s^{-1}.

The equivalent frictional head loss of the pipe-network h_{fr} (m) per meter of pipe can be expressed in the form

$$h_{fr} = 3.819 \times 10^{-3} V^2$$

where V is the water flow velocity in ms^{-1}.

Calculate (i) the water flow rate through the system, (ii) the pump head, (iii) the power input to the pump if the efficiency is 74%, and (iii) the electrical energy input if the efficiency of the motor is 93%.

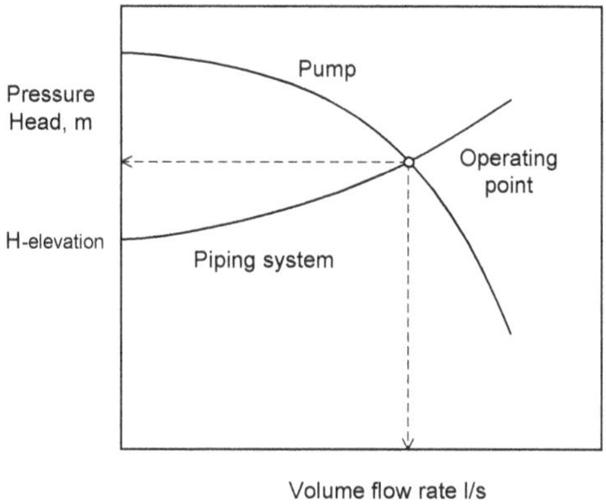

Fig. E7.9.1 Pump characteristic and system characteristic

Solution A schematic of a pump storage system is shown in Fig. 7.4. The energy equation for the pumping system, neglecting inlet and outlet kinetic energy is

$$z_l + H_p = z_h + H_{fr} \tag{E7.9.1}$$

where z_l and z_h are the elevations of the lower and upper reservoirs. The pump head is H_p, and H_{fr} is the frictional head loss. Friction losses per unit length in the piping network is given by

$$h_{fr} = cv^2$$

Equation (E7.9.1) may be written as

$$H_p = z_h - z_l + Lh_{fr}$$

Substituting the given expressions in the above equation we have

$$-0.0014Q^2 + 0.017Q + 45.335 = 30 + 1600 \times 3.819 \times 10^{-3}V^2 \tag{E7.9.2}$$

The flow velocity can be expressed in the form

$$V = \frac{4Q}{\pi D^2} = \frac{4\times10^{-3}Q}{\pi\times0.3^2} = 14.147 \times 10^{-3}Q$$

where the volume flow rate Q is in ls^{-1}.

Substituting in Eq. (E7.9.2) we get

$$-0.0014Q^2 + 0.017Q + 45.335 = 30 + 1.223 \times 10^{-3}Q^2 \quad \text{(E4.9.3)}$$

The LHS of Eq. (E7.9.3) is the *pump curve* and the RHS is called the *system curve*. These curves are plotted in Fig. E7.9.1. The point of intersection of the two curves gives the *operating point* of the pumping system. These curves are very useful in the design of pumping systems because they could be used to simulate various scenarios. For instance, the partial closing of a valve would increase frictional loss and hence move the system curve upwards. The pump curve is unaffected. On the other hand, using two pumps in series or parallel would alter the overall pump curve.

Equation (E7.9.3) may be simplified as

$$Q^2 - 6.481Q - 5846.36 = 0$$

The positive root of the above equation gives the flow rate as 79.77 ls^{-1}. Hence the pump head is 37.78 m.

The power input by the pump is

$$P_p = H_p\rho gQ = 37.8 \times 10^3 \times 9.81 \times 79.77 \times 10^{-3} = 29.6 \text{ kW}$$

The power input to the pump from the motor is

$$P_{pi} = \frac{29.6}{0.74} = 40 \text{ kW}$$

Power input to the motor is

$$P_m = \frac{40}{0.94} = 42.57 \text{ kW}$$

Example 7.10 The difference in elevation between the lower and upper reservoirs of a pumped-hydro system is 100 m. The pump is driven by a motor which receives an input of 30 MW for a period of 5 hours. The efficiencies of the motor and pump are 94% and 78% respectively. Friction losses in the pumping system is 6% of the elevation head.

During the 8-hour discharge period the stored water is passed through a turbine with an efficiency of 84%. The losses in the piping network is 6% of the available head. The efficiency of the generator is 92%. Calculate (i) the volume of water stored, (ii) energy of the water stored, (iii) the average power output of the turbine, (iv) the average electrical power output, and (v) the round trip efficiency of the system.

Solution Pressure head to be supplied by the pump includes the elevation head and the friction head. Hence we have

$$H_p = 100 + 0.06 \times 100 = 106 \text{ m}$$

The power supplied by the electric motor for pumping water is given by

$$H_p \rho g \dot{Q}_w = P_m \eta_m \eta_p$$

$$106 \times 10^3 \times 9.81 \dot{Q}_w = 30 \times 10^6 \times 0.94 \times 0.78$$

where $\dot{Q}_w$ is the volume flow rate of water in m^3s^{-1}.

Therefore the water flow rate is 21.15 m^3s^{-1}. The total volume of water stored during the pumping period of 5 hr. is 380.75×10^3 m^3.

Total energy of stored water is

$$E_{ws} = V_{ws} H_{el} \rho g$$

$$E_{ws} = 380.75 \times 10^3 \times 100 \times 10^3 \times 9.81 = 373.515 \times 10^3 \text{ MJ}$$

Average water flow rate through the turbine is

$$\dot{Q}_{tu} = \frac{380.75 \times 10^3}{8 \times 3600} = 13.22 \text{ m}^3\text{s}^{-1}$$

The head available to the turbine is given by

$$H_{tu} = 100 - 0.06 \times 100 = 94 \text{ m}$$

The average power output of the turbine is given by

$$W_{tu} = \dot{Q}_{tu} H_{tu} \rho g = 13.22 \times 94 \times 10^3 \times 9.81 = 12.19 \text{ MW}$$

The average electrical power output is

$$W_{el} = 12.19 \times 0.84 \times 0.92 = 9.42 \text{ MW}$$

The round trip efficiency of the system is given by

$$\eta_{rt} = \frac{9.42 \times 8}{30 \times 5} = 50.2\%$$

Note that well-designed pumped-hydro systems have efficiencies of about 75%.

Example 7.11 A lead acid battery is connected to a variable external resistance. When the resistance is 3.0 Ω the current flow is 3.833 A. When the external resistance is increased to 4.0 Ω the current flow is 2.949 A. Calculate (i) the emf of the battery, (ii) the internal resistance of the battery, (iii) the power generated by the battery, and (iv) the heat generated in the resistance.

Solution An external variable resistance is connected to the battery as shown in Fig. 7.7a. Apply Ohm's law to the battery to obtain the battery curve. Hence we get

$$V = E - Ir \tag{E7.11.1}$$

where E is the emf and r is the internal resistance of the battery.

Applying Ohm's law to the external load resistance we get the load curve as

$$V = IR \tag{E7.11.2}$$

From Eqs. (E7.11.1) and (E7.11.2) the operating point (see Fig. 7.7b) is obtained as

$$R + r = \frac{E}{I} \tag{E7.11.3}$$

Substituting the two given current and resistance conditions in Eq. (E7.11.3) we get

$$r + 3 = 0.2609E$$

$$r + 4 = 0.3391E$$

Solving the above equations the internal resistance and emf are obtained as

$$r = 0.336\Omega \qquad E = 12.787 \text{ V}$$

The power generated in the battery for condition 1 is

$$P_{e1} = EI = 12.787 \times 3.833 = 49.01 \text{ W}$$

Power dissipated in the resistance is

$$P_{r1} = RI^2 = 3 \times 3.833^2 = 44.07 \text{ W}$$

The power generated in the battery for condition 2 is

$$P_{e2} = EI = 12.787 \times 2.949 = 37.7 \text{ W}$$

Power dissipated in the resistance is

$$P_{r2} = RI^2 = 4 \times 2.949^2 = 34.786 \text{ W}$$

Example 7.12 (a) Three identical batteries with an emf of 12.4 V and internal resistance of 0.14 Ω are connected in series to a load resistance of 8 Ω. Calculate (i) the terminal voltage and the current flow through the resistance, (ii) power supplied to the resistance, and (b) If the three batteries in (a) are connected in parallel to the load, calculate (i) the terminal voltage, (ii) the current, and the power supplied to the load.

Solution (a) When three batteries are connected in series with the load resistance, the equivalent emf and internal resistance are given by:

$$E_{eq} = E_1 + E_2 + E_3 = 3 \times 12.4 = 37.2 \text{ V}$$

$$r_{eq} = r_1 + r_2 + r_3 = 3 \times 0.14 = 0.42 \ \Omega$$

The battery with these equivalent properties is connected to the external load resistance as shown in Fig. 7.7a.

The current flowing through the load is

$$I = \frac{E_{eq}}{r_{eq}+R} = \frac{37.2}{0.42+8} = 4.418 \text{ A}$$

The terminal voltage is

$$V = IR = 8 \times 4.418 = 35.34 \text{ W}$$

Power dissipated in the resistor is

$$P_l = I^2R = 4.418^2 \times 8 = 156.15 \text{ W}$$

(b) When the batteries are connected in parallel, the equivalent internal resistance and emf are given by

$$\frac{1}{r_{eq}} = \frac{1}{r_1} + \frac{1}{r_2} + \frac{1}{r_3}$$

$$\frac{E_{eq}}{r_{eq}} = \frac{E_1}{r_1} + \frac{E_2}{r_2} + \frac{E_3}{r_3}$$

Substituting numerical data in the above equations we get

$$\frac{1}{r_{eq}} = \frac{1}{0.14} + \frac{1}{0.14} + \frac{1}{0.14} = \frac{3}{0.14}$$

$$\frac{E_{eq}}{r_{eq}} = \frac{12.4}{0.14} + \frac{12.4}{0.14} + \frac{12.4}{0.14} = \frac{3\times12.4}{0.14}$$

Therefore

$$r_{eq} = 0.0467\ \Omega, \qquad E_{eq} = 12.4 \text{ V}$$

The battery with these equivalent properties is connected to the external load resistance as shown in Fig. 7.7a.

The current flowing through the load is

$$I = \frac{E_{eq}}{r_{eq}+R} = \frac{12.4}{0.0467+8} = 1.54 \text{ A}$$

The terminal voltage is

$$V = IR = 8 \times 1.54 = 12.33 \text{ V}$$

Power dissipated in the resistor is

$$P_l = I^2R = 1.54^2 \times 8 = 18.97 \text{ W}$$

Example 7.13 (a) Obtain an expression for the change in capacity of a battery when the discharged current is changed.

(b) A lead acid battery has a rated capacity of 100 Ah and the rated discharge time is 20 h. What is the C-rating of the battery?

Calculate the time to fully discharge the battery and the C-rating if the discharge rate is (i) 10 A, and (ii) 20 A. What is the capacity under the above discharge rates?

Solution (a) The change in capacity of rechargeable lead-acid batteries at different discharge rates t_a may be obtained from Peukert's law, which has the mathematical form:

$$t_a = t_r \left(\frac{C_r}{t_r I_a}\right)^k \qquad \text{(E7.13.1)}$$

where t_r is the rated discharge time [h], C_r is the rated capacity [Ah], I_a is the actual discharge current [A], t_a actual discharge time [h], and k is the Peukert constant. The actual capacity of the battery is given by

$$C_a = I_a t_a$$

The rated capacity is given by

$$C_r = I_r t_r$$

Multiplying Eq. (E7.13.1) by I_a/C_r we get

$$\frac{I_a t_a}{C_r} = \left(\frac{I_a t_r}{C_r}\right)\left(\frac{C_r}{t_r I_a}\right)^k = \left(\frac{C_r}{t_r I_a}\right)^{k-1}$$

$$\frac{C_a}{C_r} = \left(\frac{C_r}{t_r I_a}\right)^{k-1} = \left(\frac{I_r}{I_a}\right)^{k-1} \qquad \text{(E7.13.2)}$$

(b) The rated discharge time is 20 h. Therefore the C-rating is $1/20 = 0.05$ h^{-1}.

The rated capacity is 100 Ah. Hence the rated current is 5 A.

(i) When the discharge rate is 10 A, the time to discharge is 10 h. Hence the C-rating is 0.1. Assume that $k = 1.2$. The capacity is given by Eq. (E7.13.2) as

$$\frac{C_a}{100} = \left(\frac{5}{10}\right)^{k-1} = \frac{1}{2^{0.2}}$$

The actual capacity is 87 Ah.

(ii) When the discharge rate is 20 A, the time to discharge is 5 h. The C-rating is $1/5 = 0.2$. The capacity is given by

$$\frac{C_a}{100} = \left(\frac{5}{20}\right)^{k-1} = \frac{1}{4^{0.2}}$$

The actual capacity is 75.8 Ah.

Example 7.14 An ideal fuel cell using hydrogen and oxygen operates at 25°C Calculate (i) ideal cell voltage, (ii) the first law efficiency, and (iii) the heat transfer per kmol.

Solution The half reaction generating electrons is

$$H_2 \rightarrow 2H + 2e^-$$

The overall chemical reaction equation is

$$H_2 + \frac{1}{2}O_2 \rightarrow H_2O(l)$$

The steady flow work output is given by Eq. (7.40) as:

$$W_{rev} = \Delta H - T\Delta S$$

$$W_{rev} = \Sigma_P(n_i\Delta h_i{}^o - Tn_is_i{}^o) - \Sigma_R(n_i\Delta h_i{}^o - Tn_is_i{}^o) \quad \text{(E7.14.1)}$$

where *P* and *R* represent products and reactants of the chemical reaction. The enthalpy of formation of a constituent of the reaction (reactants and products) is Δh^o, and the absolute entropy of a constituent is s^o. (see section 10.4.3)

The maximum work is the change in the Gibb's function Δg_n.

The following data are obtained from Ref. [8].

For liquid water $\Delta h^o = 285830$ kJkmol^{-1},

For hydrogen and oxygen, $\Delta h^o = 0$.

The absolute entropy, s^o (kJkmol^{-1}K^{-1}) values are:

H_2O - $s^o = 69.95$, O_2 - $s^o = 205.04$, H_2 - $s^o = 130.57$

Substituting numerical data in Eq. (E7.14.1) we have

$$\Delta g_n = -285830 - (273 + 25)\left(69.95 - 130.57 - \frac{1}{2} \times 205.04\right)$$

$$\Delta g_n = -237214.3$$

The ideal cell voltage is given by Eq. (7.47) as:

$$V_i = -\frac{\Delta g_n}{96487 \times N_e} = \frac{237214.3}{96487 \times 2} = 1.23 \text{ V}$$

The first law efficiency is given by Eq. (7.50) as:

$$\eta_1 = \frac{\Delta g_n}{\Delta h^o} = \frac{237214.3}{285830} = 0.83$$

The heat transfer (kJkmol^{-1}) is given by Eq. (7.41) as:

$$Q_l = -\Delta H + W_{rev} = -T\Delta S$$

$$Q_l = -T\Delta S = -298\left(69.95 - 130.57 - \frac{1}{2} \times 205.04\right) = 48615.7$$

Example 7.15 A fuel cell uses gaseous propane (C_3H_8) and stoichiometric amount of oxygen at 25°C, and the pressure is 1 bar. The products CO_2 and liquid water leave at the anode and cathode respectively. Calculate (i) maximum work output, (ii) the half-cell reactions, (iii) the ideal voltage, and (iv) the ideal efficiency.

Solution The chemical equation for the reaction is

$$C_3H_8 + 5O_2(g) \rightarrow 3CO_2(g) + 4H_2O(l)$$

The steady flow work output is given by

$$W_{rev} = \Delta H - T\Delta S = \Delta G \tag{E7.15.1}$$

The maximum work is the change in the Gibb's function.
The following data for molar enthalpy are obtained from Ref.[8].

For liquid water $\Delta h^o = 285830$ kJkmol^{-1}, For gaseous propane, $\Delta h^o = 103850$ kJkmol^{-1}, For carbon dioxide, $\Delta h^o = 393520$ kJkmol^{-1}, For oxygen, $\Delta h^o = 0$.
The absolute molar entropy (kJkmol^{-1}K^{-1}) values are:
CO_2 - $s^o = 213.685$, H_2O - $s^o = 69.92$, O_2 - $s^o = 205.04$,
C_3H_8 - $s^o = 269.91$. (see section 10.4.3)

The change in enthalpy (kJkmol^{-1}) is

$$\Delta H = 3(-393520) + 4(-285830) - 1(-103850) = -2220030.0$$

The change in entropy (kJkmol^{-1}K^{-1}) is

$$\Delta S = 3(213.685) + 4(69.92) - 1(269.91) - 5(205.04) = -374.38$$

Substituting the data in Eq. (E7.15.1) we have

$$W_{rev} = \Delta H - T\Delta S = \Delta G = \Delta g_n$$

$$W_{rev} = -2220030.0 - (298)(-374.38)$$

$$W_{rev} = -2108464.8 \text{ kJkmol}^{-1}$$

The half- reaction producing electrons is

$$C_3H_8 + 6H_2O \rightarrow 3CO_2 + 20H^+ + 20e^-$$

The number of electrons released per kmol of reactant C_3H_8 is, $N_e = 20$.
The ideal cell voltage is given by

$$V_i = -\frac{\Delta g_n}{96487 \times N_e} = \frac{2108430}{96487 \times 20} = 1.093 \text{ V}$$

The first law efficiency is

$$\eta_1 = \frac{\Delta g_n}{\Delta h^o} = \frac{2108430}{2220030} = 0.95$$

The heat transfer (kJkmol^{-1}) is given by Eq. (7.41) as:

$$Q_l = -\Delta H + W_{rev} = -T\Delta S$$

$$Q_l = -T\Delta S = -298(-374.38) = 111565.24$$

Example 7.16 A bank of fuel cells using methane (CH_4) and oxygen operates at 25°C and 1 bar pressure. The voltage of each cell is 60% of the ideal voltage and the current is 1.25 A. Of the fuel supplied only 80% takes part in electrical energy production. Calculate (i) the fuel flow rate, (ii) the power output, and (iii) heat transfer rate of each cell.

Solution The chemical equation for the reaction is

$$CH_4 + 2O_2(g) \rightarrow CO_2(g) + 2H_2O(l)$$

The steady flow work output is given by

$$W_{rev} = \Delta H - T\Delta S = \Delta G$$

$$W_{rev} = \sum_{P-R} n_i \Delta h_i{}^o - T \sum_{P-R} n_i s_i{}^o = -\Delta g_n \qquad \text{(E7.16.1)}$$

where *P* and *R* represent products and reactants of the chemical reaction. The maximum work is the change in the Gibb's function.
The following data are obtained from Ref. [8].
For liquid water $\Delta h^o = -285830$ kJkmol^{-1}, For gaseous methane $\Delta h^o = -74850$ kJkmol^{-1}, For liquid water $\Delta h^o = 285830$ kJkmol^{-1}, For carbon dioxide $\Delta h^o = -393520$ kJkmol^{-1}, For oxygen $\Delta h^o = 0$.
The absolute entropy (kJkmol^{-1}K^{-1}) values are:
CO_2 - $s^o = 213.685$, H_2O - $s^o = 69.92$, O_2 - $s^o = 205.04$,
CH_4 - $s^o = 186.16$.

The change in enthalpy (kJkmol^{-1}) is

$$\Delta H = (-393520) + 2(-285830) - 1(-74850) = -890330$$

The change in entropy (kJkmol^{-1}K^{-1}) is

$$\Delta S = (213.685) + 2(69.92) - 1(186.16) - 2(205.04) = -242.715$$

Substituting data in Eq. (E7.16.1) we have

$$\Delta g_n = -890330.0 - (298)(-242.715)$$

$$\Delta g_n = -818000.93 \text{ kJkmol}^{-1}$$

The half-reaction producing electrons is

$$CH_4 + 2H_2O \rightarrow CO_2 + 8H^+ + 8e^-$$

The number of electrons released per kmol of reactant CII_4 is, $N_e = 8$. The ideal cell voltage is given by

$$V_i = -\frac{\Delta g_n}{96487 \times N_e} = \frac{818000.93}{96487 \times 8} = 1.06 \text{ V}$$

The actual voltage is

$$V_a = 0.6 \times 1.06 = 0.636 \text{ V}$$

The actual current is 1.25 A. Hence the power output is

$$P = 1.25 \times 0636 = 0.795 \text{ W}$$

The current is also given by

$$I = n_f F j = 96.487 \times 10^6 \times 8 \times n_f = 1.25 \text{ A}$$

where n_f is the number moles of CH_4 reacting per unit time.

Hence the fuel participating in the reaction is $n_f = 1.619 \times 10^{-9}$ kmol per s.

Actual rate of fuel input is

$$n_a = \frac{n_f}{0.8} = 2.024 \times 10^{-9} \text{ kmol per s}$$

Applying the first law to the actual system we have

$$Q_{in} = H_{out} - H_{in} + W_{out}$$

$$Q_{in} = -(1.619 \times 10^{-9} \times 890330 \times 10^3) + 0.795$$

Hence the rate of heat loss is

$$Q_{out} = 1.44 - 0.795 = 0.646 \text{ W}$$

Example 7.17 Each plate of a parallel plate capacitor has an area of 0.6 m^2 and the spacing between the plates is 3.0 mm. The medium between the plates is uniform and has a permittivity of 4.1. (a) Calculate the capacitance of the capacitor, the charge stored and the energy stored in the capacitor when a voltage of 40 V is applied between its terminals. (b) If 4 such capacitors are connected in parallel and a voltage of 60 V is applied to the terminals, calculate the charge and the energy stored. (c) If the 4 capacitors are connected in series, calculate the charge and energy stored.

Solution (a) The capacitance is given by

$$C = \frac{\varepsilon_0 \varepsilon_r A}{d}$$

$$C = \frac{8.84\times10^{-12}\times4.1\times0.6}{3\times10^{-3}} = 7.248 \times 10^{-9}\ \text{F}$$

The charge stored in the capacitor is given by

$$Q = CV = 7.248 \times 10^{-9} \times 40 = 2.899 \times 10^{-7}\ \text{C}$$

The energy stored in the capacitor is given by

$$U = \frac{1}{2}QV = \frac{1}{2} \times 2.899 \times 10^{-7} \times 40 = 5.798 \times 10^{-6}\ \text{J}$$

(b) When capacitors are connected in parallel the equivalent capacitance is the sum of the individual values.

$$C_{eq} = \sum C_i$$

Hence we have

$$C_{eq} = 4 \times 7.248 \times 10^{-9} = 28.99 \times 10^{-9}\ \text{F}$$

The voltage across each capacitor is the same.

The charge stored in the capacitor is given by

$$Q = C_{eq}V = 28.99 \times 10^{-9} \times 60 = 1.739 \times 10^{-6}\ \text{C}$$

The energy stored in the 4 capacitors is

$$U = \frac{1}{2}QV = \frac{1}{2} \times 1.739 \times 10^{-6} \times 60 = 5.218 \times 10^{-5}\ \text{J}$$

(c) When capacitors are connected in series the equivalent capacitance is given by

$$\frac{1}{C_{eq}} = \sum \frac{1}{C_i}$$

$$\frac{1}{C_{eq}} = \frac{4}{7.248\times10^{-9}}$$

$$C_{eq} = 0.25 \times 7.248 \times 10^{-9} = 1.812 \times 10^{-9}\ \text{F}$$

The voltage across capacitors in series is the sum of the voltages across the individual capacitors. The charge stored is the same.

The charge stored in the capacitor is given by

$$Q = C_{eq}V = 1.812 \times 10^{-9} \times 60 = 1.087 \times 10^{-7}\ \text{C}$$

The energy stored in the 4 capacitors is

$$U = \frac{1}{2}QV = \frac{1}{2} \times 1.087 \times 10^{-7} \times 60 = 3.26 \times 10^{-6}\ \text{J}$$

Example 7.18 The potential difference between the terminals of a 10 μF capacitor is 50 V. The terminals of a 2.5 μF uncharged capacitor are connected to those of the charged capacitor. Calculate (i) the initial energy stored in the larger capacitor, (ii) the potential difference between the common terminals of the two capacitors after they are connected, and (iii) the total energy stored in the two capacitors after they are connected.

Solution (i) The initial energy stored in the larger capacitor is

$$U_i = \frac{1}{2}CV^2 = \frac{1}{2} \times 10 \times 10^{-6} \times 50^2 = 0.0125\ \text{J}$$

When the terminals of two capacitors are connected, their terminal voltages are equal.

$$V_{1f} = V_{2f} = V_f \tag{E7.18.1}$$

The initial charge in the large capacitor is now distributed between the two capacitors. This is due to charge conservation. Therefore

$$Q_{1f} + Q_{2f} = Q_{1i}$$

$$C_1V_{1f} + C_2V_{f2} = C_1V_{1i}$$

Substituting from Eq. (E7.18.1) we get

$$12.5V_f = 10 \times 50$$

Hence the final voltage of the two capacitors is 40 V.

The total energy of the two capacitors is

$$U_f = \frac{1}{2}C_1{V_f}^2 + \frac{1}{2}C_2{V_f}^2$$

$$U_f = \frac{1}{2} \times 10 \times 10^{-6}{V_f}^2 + \frac{1}{2} \times 2.5 \times 10^{-6}{V_f}^2$$

$$U_f = \frac{1}{2} \times 12.5 \times 10^{-6} \times 40^2 = 0.01 \text{ J}$$

The total energy of the two capacitors is 0.01 J

Example 7.19 A thermal storage system consists of an insulated cylindrical tank holding 7850 kg of water at an initial temperature of 30°C. Hot water at a constant temperature of 94°C flows through a heat exchanger immersed in the tank at the rate of 0.75 kgs^{-1}. The effectiveness of the heat exchanger is 0.92 and the (UA)–product for heat loss from the tank is 14 WK^{-1}. The average ambient temperature during the 5 hour period of heat supply is 24°C.

The tank stores heat for a period of 8 hours during which time the average ambient temperature is 15°C. During the 6 hour discharge period heat is supplied to a process through a second heat exchanger immersed in the tank. The effectiveness of this heat exchanger is 0.9 and the water from the process enters it at a constant temperature of 21°C with a mass flow rate of 0.8 kgs^{-1}. The average ambient temperature during the discharge period is 18°C.

Calculate (i) the water temperature in the tank at the end of the heat supply period, (ii) the temperature just before heat discharge to the process begins, (iii) the total energy loss when the tank is storing the heat supplied, and (iv) the temperature of the tank at the end of the discharge period.

Solution Consider the charging phase of the system when hot water flows into the tank through the heat exchanger. The governing energy is

$$Mc\frac{dT_s}{dt} = \epsilon_1 \dot{m}c\left(T_{fi} - T_s\right) - (AU)_t(T_s - T_a) \qquad \text{(E7.19.1)}$$

where T_s is the average water temperature at time t. The water inlet temperature and the ambient temperature are T_{fi} and T_a respectively.

The term on the LHS of Eq. (E7.19.1) is the rate of energy storage in the tank, the first term on the RHS is the rate of heat transfer from the heat exchanger to the tank, and the second term is the rate of heat loss to the surroundings. We assume that the water in the tank is fully mixed.

It is convenient to express all temperatures in Eq. (E7.19.1) with reference to the constant ambient temperature of 24°C. Hence Eq. (E7.19.1) becomes

$$Mc\frac{d\theta_s}{dt} = \epsilon_1 \dot{m}c\left(\theta_{fi} - \theta_s\right) - (UA)_t\theta_s \qquad \text{(E7.19.2)}$$

where $\theta_s = T_s - T_a$ and $\theta_{fi} = T_{fi} - T_a$

Substituting numerical values in Eq. (E7.19.2) we get

$$(7850 \times 4.19)\frac{d\theta_s}{dt} = 0.92 \times 0.75 \times 4.19\left(\theta_{fi} - \theta_s\right) - 14 \times 10^{-3}\theta_s$$

$$\frac{d\theta_s}{dt} = 8.79 \times 10^{-5}(70 - \theta_s) - 4.25 \times 10^{-7}\theta_s$$

$$\frac{d\theta_s}{(69.67 - \theta_s)} = 8.833 \times 10^{-5} dt$$

The solution of the above differential equation gives the temperature at the end of the energy supply period as:

$$\frac{69.67 - \theta_{sf}}{69.67 - 6} = e^{-8.833 \times 10^{-5} \times 5 \times 3600} = 0.2039$$

$$\theta_{sf} = 56.69°\text{C}$$

Hence the storage tank temperature at the end of heat supply is

$$T_{sf} = \theta_{sf} + 24 = 56.69 + 24 = 80.7°\text{C}.$$

We now consider the storage phase when the water looses heat to the surroundings. The governing energy equation is

$$Mc\frac{dT_s}{dt} = -(AU)_t(T_s - T_a) \qquad \text{(E7.19.3)}$$

Using the constant ambient temperature of 15°C as the reference temperature we rewrite Eq. (E7.19.3) as

$$Mc\frac{d\theta_s}{dt} = -(UA)_t\theta_s$$

Substituting numerical values we get

$$(7850 \times 4.19)\frac{d\theta_s}{dt} = -14 \times 10^{-3}\theta_s$$

$$\frac{d\theta_s}{dt} = -4.25 \times 10^{-7}\theta_s \qquad \text{(E7.19.4)}$$

The initial tank temperature is $\theta_{ti} = (80.7 - 15) = 65.7$ °C.

The solution of Eq. (E7.19.4) gives the temperature at the end of the storage period as:

$$\frac{\theta_{s2}}{65.7} = e^{-4.25\times10^{-7}\times8\times3600} = 0.9878$$

Hence the temperature at the end of 8 hours is

$$T_{s2} = \theta_{s2} + 15 = 64.89 + 15 = 79.9\text{°C}.$$

The heat lost during the heat storage period is

$$Q_{loss} = 7850 \times 4.19 \times (80.7 - 79.9) = 26.3 \times 10^3\text{kJ}$$

Consider the 6 hour energy supply period. The governing energy equation is

$$Mc\frac{dT_s}{dt} = -\epsilon_2\dot{m}c(T_s - T_l) - (AU)_t(T_s - T_a) \qquad \text{(E7.19.5)}$$

where T_l is the process water inlet temperature to the tank.

We express the temperatures in Eq. (E7.19.5) with reference to the constant ambient temperature of 18°C. Hence Eq. (E7.19.5) becomes

$$Mc\frac{d\theta_s}{dt} = -\epsilon_2\dot{m}c(\theta_s - \theta_l) - (UA)_t\theta_s$$

Substituting numerical values we get

$$(7850 \times 4.19)\frac{d\theta_s}{dt} = -0.90 \times 0.8 \times 4.19(\theta_s - \theta_l) - 14 \times 10^{-3}\theta_s$$

$$\frac{d\theta_s}{dt} = -9.172 \times 10^{-5}(\theta_s - 3) - 4.25 \times 10^{-7}\theta_s$$

$$\frac{d\theta_s}{(\theta_s - 2.986)} = -9.214 \times 10^{-5}dt$$

The solution of the above differential equation gives the temperature at the end of the discharge period as:

$$\frac{\theta_{s2} - 2.986}{61.9 - 2.986} = e^{-9.214\times10^{-5}\times6\times3600} = 0.1366$$

Hence the storage temperature at the end of 6 hours is

$$T_{s2} = \theta_{s2} + 18 = 11.3 + 18 = 29.03^{\circ}\text{C}.$$

Example 7.20 Develop a numerical procedure to predict the air temperature and bed temperature variation of a packed bed thermal storage system. The initial temperature distribution of the bed and the inlet air temperature are known.

Solution An idealized model of a packed bed storage system is shown in Fig. 7.12. The differential equations governing temperatures of the air and the storage bed are given by Eqs. (7.59) and (7.60) as

$$mc_{pa}\frac{\partial T_a}{\partial x} = -Ah_v(T_a - T_b) \tag{E7.20.1}$$

$$\rho_b(1-\epsilon)c_b\frac{\partial T_b}{\partial t} = h_v(T_a - T_b) \tag{E7.20.2}$$

To simulate the performance of the storage system, the bed is divided into N finite sections, where the bed temperatures are assumed uniform.

Consider the section i of the bed whose temperature at time t is T_{bi} and the air temperature at the entrance and exit of the section are T_{ai} and $T_{a(i+1)}$ respectively.

The air temperature variation in section i is given by

$$mc_{pa}\frac{\partial T_a}{\partial x} = -Ah_v(T_a - T_{bi}) \quad \text{(E7.20.3)}$$

The solution of Eq. (E7.20.3) can be expressed as

$$\frac{T_{ao}-T_{bi}}{T_{ai}-T_{bi}} = e^{-\lambda\Delta x} \quad \text{(E7.20.4)}$$

where $\lambda = \frac{Ah_v}{mC_p}$, and $\Delta x = \frac{L}{N}$ is the length of a section.

The heat flow from the air to the bed section may expressed as

$$\Delta Q_i = mc_{pa}(T_{ai} - T_{ao}) \quad \text{(E7.20.5)}$$

From Eqs. (E7.20.4) and (E7.20.5) we obtain the effectiveness of heat transfer as

$$\eta = \left(1 - e^{-\lambda\Delta x}\right)$$

Hence we have

$$\Delta Q_i = mc_{pa}\eta(T_{ai} - T_{bi}) \quad \text{(E7.20.6)}$$

Substituting the expression for heat transfer given by Eq. (E7.20.6) in Eq. (E7.20.2) we get

$$A\rho_b(1-\epsilon)c_b\frac{dT_{bi}}{dt} = mc_{pa}\eta(T_{ai} - T_{bi}) \quad \text{(E7.20.7)}$$

Equation (E7.20.7) is a set of N ordinary differential equations for the bed temperatures.

The finite difference forms of Eq. (E7.20.7) and Eq. (E7.20.3) are

$$A\rho_b(1-\epsilon)c_b\frac{(T'_{bi}-T_{bi})}{\Delta t} = mc_{pa}\eta[T_{ai} - (T'_{bi} + T_{bi})/2] \quad \text{(E7.20.8)}$$

$$mc_{pa}\left(T_{ai} - T_{a(i+1)}\right) = mc_{pa}\eta(T_{ai} - T_{bi}) \quad \text{(E7.20.9)}$$

With all the initial bed temperatures known, the numerical procedure starts at node 1 where the inlet air temperature is known. A new bed temperature for the next time step, T'_{bi} is calculated from Eq. (E7.20.8), and an outlet air temperature from Eq. (E7.20.9). This air temperature is the inlet temperature to node 2. The above procedure is continued until the last node at the exit of the bed is reached.

Problems

P7.1 An adiabatic air turbine is operated between a compressed air tank, initially at 800 kPa and 50°C, and the ambient pressure of 100 kPa. The average output of the turbine is 2.5 kW over a period of 15 min. with a mean efficiency of 55 percent. The tank pressure at the end of this period is 175 kPa. Calculate the volume of the required storage tank. [*Answer*: 7.57 m^3]

P7.2 The upper reservoir of a pumped-hydro system is 150 m above the lower reservoir. The pump is driven by a motor which receives an average input of 40 MW for a period of 6 hours. The efficiencies of the motor and the pump are 95% and 78% respectively. Friction loss in the pumping system is 7% of the elevation head. During the 9-hour discharge period the stored water is passed through a turbine with an efficiency of 85%. The loss in the piping network is 7% of the available head. The efficiency of the generator is 94%. Calculate (i) the volume of water stored, (ii) average electrical power output of the generator, and (iii) the round trip efficiency of the system.
[*Answers*: (i) 4.066×10^5 m^3, (ii) 13.72 MW, (iii) 51.45%]

P7.3 The moment of inertia of a steel flywheel is 280 kgm^2. It is acted upon by a constant motor torque of 185 Nm and experiences a frictional torque given by, $2.6 \times 10^{-3}\omega$ Nm, where ω is the angular speed in rad. s^{-1}. (i) Calculate the time taken for the rotational speed to reach the maximum value of 5000 rpm. (ii) Calculate the self discharge time to 70% of the maximum energy stored if the frictional torque has the same form. (iii) At the end of the self discharge period, the flywheel discharges energy to a generator and experiences a load torque of 220 Nm. Calculate the time to discharge 80% of the maximum stored energy, if the frictional torque remains unchanged.
[*Answers*: (i) 13.26 min., (ii) 5.33 hrs., (iii) 4.31 min.]

P7.4 The voltages of three batteries are respectively, 2 V, 1.8 V and 1.5 V, and their internal resistances are respectively 0.6 Ω, 0.7 Ω and 0.9 Ω. These batteries are connected in series across a load resistance of 4 Ω. (i) Calculate the terminal voltage, the current output, and the power supplied to the resistance. (ii) Calculate the corresponding quantities if the batteries are connected in parallel.
[*Answers*: (i) 3.42 V, 0.86 A, 2.92 W, (ii) 1.69 V, 0.424 A, 0.716 W]

P7.5 The rated capacity and the rated discharge time of a battery are 150 Ah and 10 h respectively. What is the rated C-value of the battery? Calculate the time to fully discharge the battery and the C-rating if the discharge rate is (i) 10 A, and (ii) 20 A. What is the capacity under the above discharge rates? [*Answers*: 0.1 C, (i) 15 h, 0.067 C, 162.7 A.h, (ii) 7.5 h, 0.133 C, 141.6 Ah]

P7.6 An ideal fuel cell using gaseous acetylene (C_2H_4) and oxygen operates at 25°C and 1 bar pressure. Calculate (i) ideal cell voltage, (ii) the first law efficiency, and (iii) the heat transfer.
[*Answers*: (i) 1.28 V, (ii) 95%, (iii) 64,451.0 kJkmol^{-1}]

P7.7 Bank of fuel cells operating at 25°C and 1 bar uses gaseous hydrogen and oxygen as reactants. The actual voltage of the cell is 60 percent of the ideal voltage when the current is 1.20 A. Twenty percent of the fuel passes through the cell without reacting and is wasted. Calculate (i) the fuel flow rate, (ii) the power output, and (iii) the heat transfer rate. [*Answers*: (i) 7.77×10^{-9} kmols^{-1}, (ii) 0.817 W, (iii) 0.7 W]

P7.8 The two plates of a parallel plate capacitor are spaced 3.2 mm apart and the medium between has a permittivity of 3.8. The plate areas are 1.2 m^2 and the voltage between them is 60 V. (i) Calculate the energy stored. (ii) If 3 such capacitors are connected in parallel, and a voltage of 80 V is applied to the terminals, calculate the energy stored. (iii) If 3 such capacitors are connected in series, and a voltage of 80V is applied to the terminals calculate the energy stored.
[*Answers*: (i) 2.26×10^{-5} J, (ii) 12.08×10^{-5} J, (iii) 1.34×10^{-5} J]

P7.9 A thermal storage system consists of an insulated cylindrical tank holding 6900 kg of water at an initial temperature of 40°C. Hot water at a constant temperature of 88°C flows through a heat exchanger immersed in the tank at the rate of 1.2 kgs^{-1}. The effectiveness of the heat exchanger is 0.75, and the (UA)–product for heat loss from the tank is 16 WK^{-1}. The average ambient temperature is 24°C. (i) Calculate the time taken for the water temperature in the tank to reach 80°C. (ii) Heat is removed from the tank at a constant rate of 55 kW until the water temperature is 40°C. For how long can the heat supply be maintained? Neglect heat losses to the surroundings during this period.
[*Answers:* (i) 3.86 h, (ii) 5.84 h]

P7.10 The length and cross sectional area of a pebble-bed thermal storage system are 2.0 m and 15 m^2 respectively. The pebbles have a density of 1360 kgm^{-3} and a specific heat capacity of 0.88 $kJkg^{-1}K^{-1}$. The void fraction of the bed is 0.48. The mass flow rate of air through the bed is 0.93 kgs^{-1} and the mean specific heat of air is 1.12 $kJkg^{-1}K^{-1}$. The volumetric heat transfer coefficient has been estimated as 2.045 $kWm^{-3}K^{-1}$. The bed is initially at a uniform temperature of 29°C. Hot air enters the bed at a constant temperature of 65°C. Obtain the temperature distribution of the bed at different times from 1 h to 5 h during the charging period. Neglect any heat losses to the surroundings. Use the numerical procedure developed in example E7.20.

References

1. Duffie, John A., Beckman, William A. with Blair, Nathan, *Solar Engineering of Thermal Processes, Photovoltaics and Wind, Fifth Edition*, John Wiley & Sons Inc. New Jersey, 2020.
2. Ibrahim, H, Ilinca, A, and Perron, J., Energy storage systems-Characteristics and comparisons, *Renewable and Sustainable Energy Reviews*, 12 (2008) 1221-1250.
3. Jacobson, Mark Z, 100% *Clean, Renewable Energy and Storage for Everything*, Cambridge University Press, Cambridge, U.K., 2021.
4. Jenkins, Nick., and Ekanayake, Janaka., *Renewable Energy Engineering*, Cambridge University Press, Cambridge, U.K. 2017.
5. Jones, J.B., and Hawkins, G.A., *Engineering Thermodynamics: An Introductory Textbook*, John Wiley & Sons, Inc. New Jersey, 1986.

6. Kreith, Frank and Krumdieck, Susan, *Principles of Sustainable Energy Systems*, CRC Press, Taylor & Francis Group, New York, 2015.
7. Succar, Sumir, and Williams, Robert H, *Compressed Air Energy Storage: Theory, Resources, and Applications for Wind Power.* Princeton Environmental Institute (PEI), Princeton University, April 2008.
8. Wark, Kenneth, JR, *Advanced Thermodynamics for Engineers,* McGraw-Hill Inc., New York, 1995.
9. Amiyar, Mustafa E and Pullen, Keith R, *A Review of Flywheel Energy Storage System Technologies and Applications*, MDPI, applied science, http://res.mdpi.com/dattachment/applsci/applsci-07-00286/article_deploy/applsci-07-00286.pdf
10. http://www.mpoweruk.com

Chapter 8

Ocean Energy Conversion Systems

8.1 Introduction

Seventy percent of the earth's surface is covered by oceans which are a potential source of renewable energy. Mechanical energy in tides and ocean waves can be converted directly to electrical energy. The temperature difference between surface water and deep water in oceans can be used to operate heat engines for the production of electricity.

Considerable research and development has been devoted to ocean energy sources but they have not been exploited fully on a commercial scale yet. This is mainly due to practical difficulties in installing tidal and wave devices in oceans which experience extreme weather conditions, such as storms. Furthermore, transmitting power over long distances across the ocean to the shore to be fed to the local grid is expensive. In this chapter we shall consider the key conversion technologies of ocean energy. These are: (i) tidal energy [6,7], (ii) wave energy [8], and (iii) ocean thermal energy (OTEC) [9].

8.2 Tidal Energy

In this section various aspects of the tidal resource, and the two main technologies used for the exploitation of the tidal resource will be presented [5].

8.2.1 *The tidal resource*

We usually describe the motion of the earth as a rotation about the sun in an elliptical orbit. However, the earth and sun rotate around a common center of mass which is less than 500 km from the center of the sun. Similarly, the moon and earth are orbiting about a common center of mass that is about 1700 km below the surface of the earth as indicated by point C in Fig. 8.1(a). It is the rotation of the earth in this small orbit around C that is the counter part of the revolution about the sun. The above rotations generate centrifugal forces distributed over the earth [5].

The rate of change with distance of the gravitational forces acting on the earth due to the sun and the moon are inversely proportional to the cube of the distance between their centers. Based on the above observation the gravitational influence of the sun is found to be only about 46% of that due to the moon (see worked example 8.1).

The rotation of the earth-moon system creates an outward centrifugal force, which balances the gravitational force to keep the two bodies in their orbits. The centrifugal force is constant everywhere on earth, and it is directed away from the moon as indicated in Fig. 8.1(a). On the other hand, the gravitational force is always directed towards the moon.

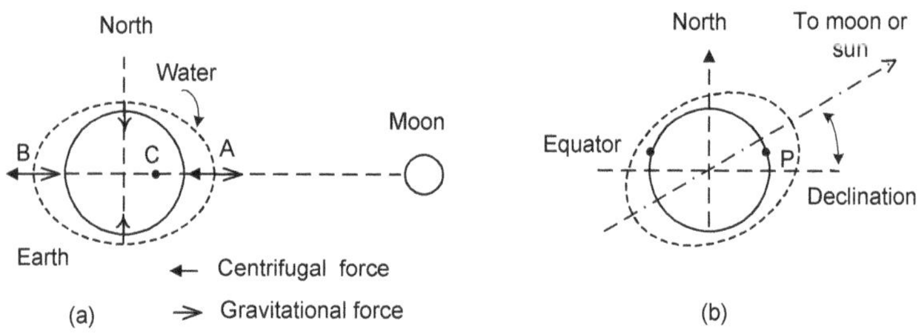

Fig. 8.1 (a) Tidal forces on the earth's surface, (b) Effect of declination

It is the difference between the gravitational and centrifugal forces that drives the tides. These forces are very small, but because they are not balanced forces they cause water movement. The distribution of these forces produce bulges of water over areas on the near and far sides of the moon where the net forces are directed outward from the earth's surface.

Figure 8.1(a) shows an idealized schematic of the earth covered with water, having no land masses, and the effect of the tidal forces. At point B on the far side of the moon, the centrifugal force exceeds the lunar attraction, whereas at point A on the near side the attractive force exceeds the centrifugal force. Depressions in the water level occur at areas where the resultant forces are directed inwards as indicated in Fig. 8.1(a).

As the earth rotates about its polar axis, each point on the surface will pass through the whole pattern of forces in one day, with two passages under the bulges and two under the depressions. This is true of all points on the earth's surface except near the poles. Hence the reason for the existence of semi-diurnal tides on diurnal rotating earth. The effect of the sun-earth system is similar to that of the moon-earth system, except that the forces involved are much smaller.

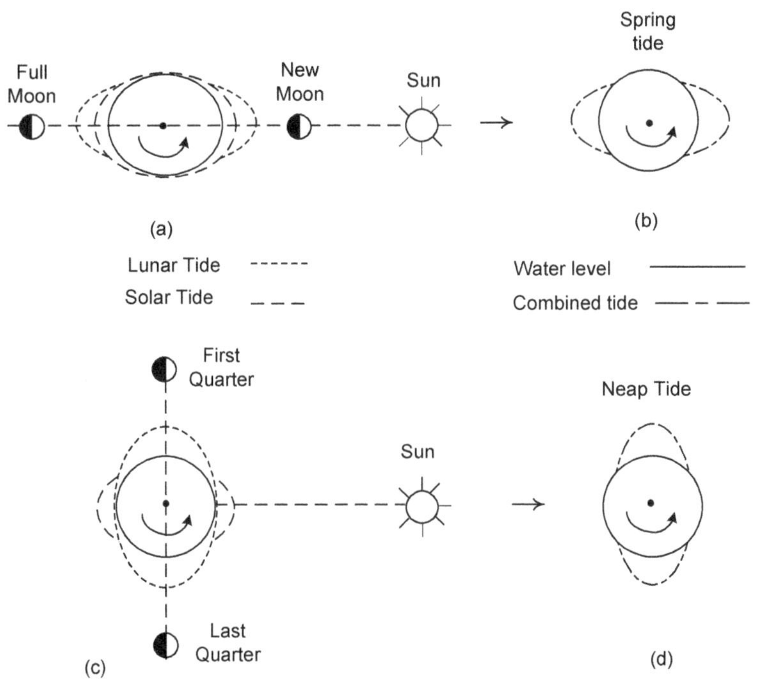

Fig. 8.2 (a) and (b) Spring tides with high ranges, (c) and (d) Neap tides with small ranges.

An important factor that effects both semi-diurnal tides and diurnal tides are the changes in solar and lunar declinations. Figure 8.1(b) shows in an exaggerated form, how declination produces an asymmetry between the two high and two low water levels at a point P as it rotates under the two bulges. The maximum monthly lunar declination north and south of the equator varies from 18.3° to 28.6° over the 18.6 year nodal period. The seasonal variation of the solar declination is from 23.5° in June to -23.5° in December.

Therefore the combined effect of the sun and the moon will increase the diurnal effect on the tides when their combined declinations are both large, either in the same or in opposite directions.

The actual tide, and the equilibrium tide discussed above, are not the same at any place on earth. However, many characteristics of the two are similar except the magnitude and time. The effective tidal forces of the moon and the sun sometimes complement one another, and sometimes act in opposition. At new and full moon, the sun and moon act together producing a large tide-raising force as depicted in Fig. 8.2(a) and (b).

During spring tide, the high tides occur near local noon and local midnight, and are higher than average due to the reinforcement of the two tide-raising forces. The two low tides are reinforced in opposite directions resulting in a tide that is lower than average.

When the moon is at its first or last quarter, its tide-raising force is partially counteracted by that of the sun. Therefore neap tides occur on the day that the sun's or moon's high waters coincides with the other's low waters. This results in a lower than average tidal range at neap tides as shown in Fig. 8.2(c) and (d).

The equilibrium tides discussed above are based on ideal conditions where the earth is covered completely with water and has no land masses or continents. In reality this does not resemble actual tides, and the rise and fall predicted by it are too small compared to observed tides.

The dynamical theory of tides represents tides as a wave driven by tide-producing forces. Its rise and fall on the coast results in the convergence and divergence of the flow. Moreover, real ocean basins have very complicated coastal and bottom topography, and therefore it is not possible to obtain exact solutions.

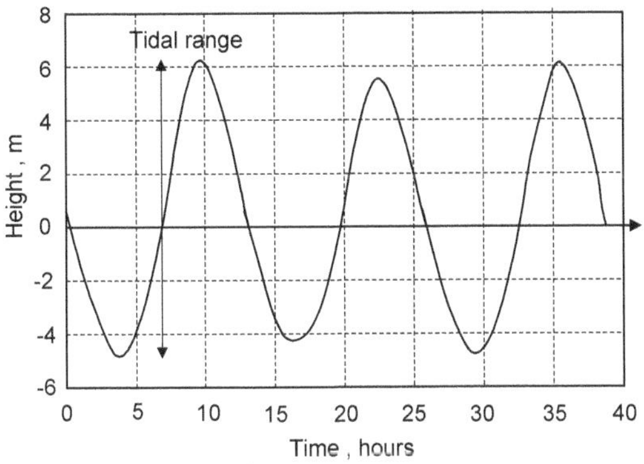

Fig. 8.3 Time variation of tidal height

8.2.2 *Analysis of tides*

The tidal range at a location depends on the combination of several complex astronomical and geographical factors. This results in a complex time series of the height of the tide at a location, that repeats every 20 years. The tidal height can be expressed in the form[1]:

$$h(t) = h_0 + \sum_1^m h_n sin(\omega_n t + \phi_n) \qquad (8.1)$$

where:

$h(t)$ is the height of the tide at time t,
h_0 is the height of mean water level,
h_n is the amplitude of the n th harmonic component,
ω_n is the angular velocity of the nth harmonic component,
ϕ_n is the phase difference of the nth harmonic component,
t is the time,
m is the number of harmonic components.

The parameters of Eq. (8.1) are determined by analyzing the measured time variation of the tidal height at a location. The mathematical technique used for this purpose is called harmonic analysis, and Fig. 8.3 shows a typical tidal height variation obtained by this procedure. Some of the harmonic components obtained are attributed to different physical factors responsible for the tides mentioned above. The *tidal range* for the variation depicted in Fig. 8.3 is about 11 m.

8.2.3 *Tidal range power generation*

The design configuration commonly used for generating power using the *tidal range method* is depicted schematically in Fig. 8.4. It is similar to systems used for hydroelectric power generation. The four main components of the tidal range plant are:

(i) Dam or barrage to form a reservoir or a basin. Tidal barrages require sites where there is a sufficiently high tidal range (see Fig. 8.3) to produce an adequate head differential between the sea and the reservoir (see Fig. 8.4).

(ii) Sluice gates to control the flow of tidal water between the sea and the reservoir. Depending on the mode of generation, tidal basins must be filled and emptied. Gates used for this purpose must be opened and closed regularly and frequently, which is not the case with conventional hydroelectric power systems. Moreover, corrosion problems associated with sea water have to addressed in the design of these gates.

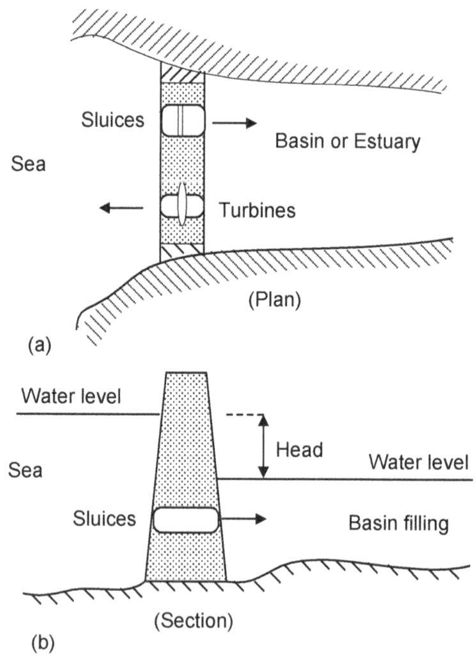

Fig. 8.4 Tidal range power generation

(iii) A set of turbines and generators to produce power using the difference in water levels on either side of the barrage. Since the water turbines used have to operate with small heads they tend to be large in size. They usually have variable pitch blades, and can function as pumps to increase the water level in the reservoir in some modes of operation.

(iv) If the barrage is across an estuary that has ports, then provision must be made for the passage of ships.

Three modes are available for the generation of power in tidal range systems. These are: (i) single-ebb cycle systems, (ii) single-tide cycle systems, and (iii) tide and ebb cycle systems.

Single-ebb cycle

The time variation of the water level in the basin and the sea during the *single-ebb cycle* is shown in the idealized diagram in Fig. 8.5. The sequence of operation is summarized in Table 8.1.

During high tide sluice gates are kept open to admit water to the tidal basin while the turbines remain shut. When the water level in the reservoir reaches the maximum level sluice gates are closed. The power generation phase begins when the water level in the sea outside the barrage is dropping while the gates remain closed. When the difference

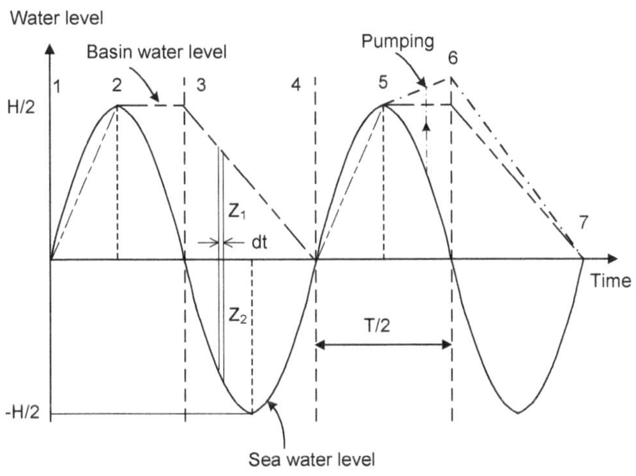

Fig. 8.5 Variation of water levels during the single-ebb cycle

Table 8.1 Sequence of operation of a single-ebb cycle

Period	Sluces	Turbines	Basin level	Sea level	Power
1-2	Open	Shut	Rising	Rising	None
2-3	Shut	Shut	Constant	Dropping	None
3-4	Shut	Open	Dropping	Drop & Rise	Variable
4-5	Open	Shut	Rising	Rising	None
5-6	Shut	Shut	Constant	Dropping	None
6-7	Shut	Open	Dropping	Drop & Rise	Variable
5-6-pump	Shut	Pumping	Rising	Drop & Rise	None
6-7-pump	Shut	Open	Dropping	Drop & Rise	Variable

in head between the two sides of the barrage is adequate, water is sent through the turbines to generate power.

This process continues until the water level in the sea starts rising in the next flood tide, and the water level in the reservoir is low. Generally, the head differential at this stage is insufficient to generate power safety and efficiently. The turbines are then shut and the gates opened to begin the next ebb cycle.

The overall performance of the plant can be improved by pumping water from the sea to the basin during 5-6 (see Fig. 8.5) to increase the effective head when generation begins at 6.

Single-tide cycle

In the *single-tide cycle*, power generation starts when the sea is at flood tide. Sluice gates are closed and sea water is admitted to the reservoir through the turbines to generate power. As the flood period ends and the water level in the sea begins to fall, the turbines are shut and the sluice gates are opened to empty the reservoir to the lower water level in the sea. This mode of power generation is generally less efficient than the ebb generation mode, and requires larger power plants due to the shorter duration of power generation.

Tide and ebb cycle

The time variation of the water level in the basin and the sea during the *tide and ebb cycle* is depicted in the idealized diagram in Fig. 8.6. The sequence of operation is summarized in Table 8.2.

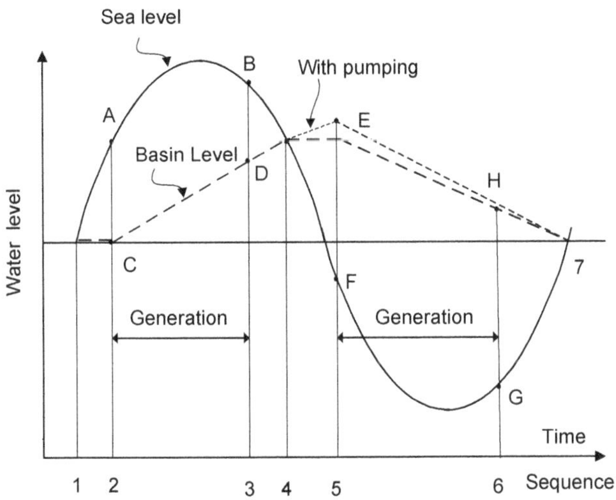

Fig. 8.6 Variation of water levels with time during ebb and tide cycle

Table 8.2 Sequence of operation of a tide and ebb cycle

Period	Sluces	Turbines	Basin level	Sea level	Power
1-2	Shut	Shut	Standing	Rising	None
2-3	Shut	Operating	Rising	Rise & Drop	Variable
3-4	Open	Shut	Filling	Dropping	None
4-5	Shut	Shut	Standing	Dropping	None
5-6	Shut	Operating	Dropping	Drop & Rise	Variable
6-7	Open	Shut	Emptying	Rising	None
4-5-pump	Shut	Pumping	Rising	Dropping	None
5-6-pump	Shut	Operating	Dropping	Drop &Rise	Variable
6-7-pump	Open	Shut	Emptying	Drop & Rise	None

The turbines are operated both during high tide and low tide, and they are usually reversible turbines. During high tide when the water level in the sea is rising the turbines receive water from the sea (C-D). At B the turbines are shut because the head is inadequate for safe power generation.

During low tide when the level of water in the sea is dropping, generation begins at EF and continues until GH. After G the head is inadequate for safe power generation.

The overall performance of the plant can be improved by pumping water from the sea to the basin during D-F to increase the effective head when generation begins at E. Though the ebb and tide mode of operation has shorter interruptions in power generation it is still not continuous.

Maximum energy available during each tidal cycle

We shall obtain an expression for the maximum energy available during each tidal cycle. If the total change in water level is H, and the area of the basin is A, then the maximum potential energy E of water is given by

$$E = \int_0^H A\rho g h dh = A\rho g \frac{H^2}{2} \tag{8.2}$$

where h is the difference in water level, ρ is the density of sea water, and g is the acceleration due to gravity.

The expression in Eq. (8.2) is applicable to each rise and fall of water level by H. There are 4 such events, that is, 2 flood tides and 2 ebb tides, during the daily tidal period. Hence the *maximum energy available* per day (24 hours) is

$$E_{dm} = 4A\rho g \frac{H^2}{2}\left(\frac{24}{24.8}\right) \tag{8.3}$$

The total period of the tides is 24.8 hours.

The maximum available power (W) is given by

$$P_{av} = 4A\rho g \frac{H^2}{2}\left(\frac{1}{24.8\times 3600}\right) \tag{8.4}$$

Analysis of the ebb cycle

The ebb cycle of power generation was described above. The time variation of the water level in the basin and the sea are shown graphically in Fig. 8.5, and the sequence of operation of the system is summarized in Table 8.1. Power is produced during two periods 3-4 and 6-7 each of which occupies a quarter of the total tidal period, $T = 24.8$ hours.

We shall obtain a simple expression for the total energy produced during one ebb cycle by making several simplifying assumptions [1]. These are: (i) the time variation of the sea water level is sinusoidal,

(ii) the water flow rate through the turbines is constant and therefore the variation of the water level in the basin is linear during generation, (iii) power generation begins and stops exactly at the mid point of the tidal range. The curves in Fig. 8.5 are based on the above assumptions.

The energy output during the period 3-4 is given by (see Fig. 8.5)

$$E_{3-4} = \int_0^{T/2} \rho g Q(z_1 + z_2)\, dt \tag{8.5}$$

The constant water flow rate, Q through the turbine is

$$Q = \frac{AH/2}{T/2} = \frac{AH}{T} \tag{8.6}$$

Substituting from Eq. (8.6) in Eq. (8.5) we have

$$E_{3-4} = \frac{A\rho g H}{T}\left(\int_0^{T/2} z_1 dt + \int_0^{T/2} z_2 dt\right) \tag{8.7}$$

The first integral in Eq. (8.7) is the area of the triangle between 3-4, and the second integral is the area under the sinusoidal curve below 3-4. Substituting appropriate expressions for the above areas we get

$$E_{3-4} = \frac{A\rho g H}{T}\left(\frac{TH}{8} + \frac{TH}{2\pi}\right) = 0.284 A\rho g H^2 \tag{8.8}$$

The total energy available during each tidal day from the two ebb tides is

$$E_d = 0.568 A\rho g H^2 \tag{8.9}$$

Comparing E_d with the maximum available energy given by Eq. (8.3) we note that ebb generation provides only about 28% of the maximum energy available. This estimate, however, is based on ideal conditions, but in practical systems we need to account for hydraulic losses, and the minimum head required to operate the turbines.

It is possible to increase the water level in the basin for ebb generation during 6-7 by reversing the turbine to pump water during 5-6 as indicated in Fig. 8.5. An important advantage of this scheme is that pumping takes place against a small head, while subsequent generation during 6-7 occurs at a higher head.

Pumped-storage with tidal generation

The inclusion of a pumped-storage system with a tidal barrage can improve the performance of the tidal plant [4]. In such a system water is pumped to the reservoir to increase the available maximum tidal range from H to $(H + h)$. The maximum available energy can be obtained from Eq. (8.2) as

$$E_{ps} = \int_0^{H+h} A\rho g h dh = A\rho g \frac{(H+h)^2}{2} \tag{8.10}$$

The energy expended to pump water to increase the level of water by h is

$$E_p = \frac{A\rho g}{\eta_p}\frac{h^2}{2} \tag{8.11}$$

where η_p is efficiency of the pumping system.

The *net increase* in energy output is given by

$$\Delta E_{net} = E_{ps} - E_p - E \tag{8.12}$$

where E is the available energy without pumping.

Substituting in Eq. (8.12) from Eqs. (8.11), (8.10) and (8.2) we get

$$\Delta E_{net} = \frac{A\rho g H^2}{2}\left[\frac{2h}{H} - \left(\frac{h}{H}\right)^2\left(\frac{1-\eta_p}{\eta_p}\right)\right] \tag{8.13}$$

Costs of tidal barrage systems

The cost of the dam or barrage across the bay is the main cost of a tidal range power system. The cost of the dam is proportional to its length L. The maximum energy from a tidal power system is proportional to the area of the basin A as seen from Eq. (8.2). To minimize the cost per unit energy output of the system the ratio L/A should be minimized [1]. The maximum value of the above ratio is found to be 80 (mkm^{-2}).

Different designs of tidal range power generation systems have been studied extensively and they are now well understood. The details of three operating tidal range plants are listed in Table 8.3. The future exploitation of this technology depends critically on the availability of suitable sites, cost-effectiveness, and environmental impact. These aspects are discussed in detail in Ref.[1].

Table 8.3 Operating tidal range power systems (Data from Ref.[1])

Location	Country	Rated power (MW)	Year of Comm.
La Rance	France	240	1966
Annapolis Royal	Canada	20	1984
Sihwa Lake	South Korea	254	2011

8.2.4 *Tidal stream power generation*

Tidal stream generation is the second scheme used to exploit the tidal resource [7]. In section 8.2 we discussed the forces responsible for the tides in oceans. The variation of water level, caused by these forces, creates a long period wave that moves across the ocean. These waves generate the tidal range as they reach land.

The speed of tidal waves are small in the open ocean, that is about 0.02 to 0.03 ms^{-1}. However, the waves accelerate as they enter narrow channels, at entrances to sea lochs, and in shallow seas. In these locations the wave speed can exceed $3ms^{-1}$. Here tidal turbines, similar to wind turbines, could be placed to extract power from tidal waves.

The frequency of tidal stream currents is the same as that of the tidal range, but in estuaries and bays they have a phase difference. The speed of the tidal wave can be represented by a series of harmonics like the variation of the tidal range.

The speed of the tidal current can be expressed in the form [1]:

$$U(t) = U_0 + \sum_1^m U_n sin(\omega_n t + \phi_n) \tag{8.14}$$

where

$U(t)$ is the speed of the tide at time t,
U_0 is the mean speed of the current,
U_n is the amplitude of the n th harmonic component of speed,
ω_n is the angular velocity of the nth harmonic component,
ϕ_n is the phase difference of the nth harmonic component,
t is the time,
m is the number of harmonic components.

The parameters of Eq. (8.14) are determined by analyzing the measured time variation of the speed of the tidal current at a location using the mathematical technique called harmonic analysis.

The speed of the current varies over a semi-diurnal cycle of 12.4 hours with a reversal of flow direction between flood flow and ebb flow. The turbulence in the stream near the water surface is caused by wave action. At greater depths obstructions and the roughness of the seabed induces turbulence. The turbulence intensity, which the ratio of standard deviation of the speed to the average speed, is about 10% compared with 15% for a good wind farm site. The variation of the speed with water depth across the tidal stream follows a power law with an exponent of $1/7$ or $1/10$.

Tidal stream generator

Tidal stream generators draw energy from water currents in much the same way as wind turbines extract energy from air currents. Therefore the energy available from the tidal stream may be expressed as

$$P = \frac{1}{2} A \rho_w C_p {U_s}^3 \tag{8.15}$$

where A is the frontal area of the tidal rotor, ρ_w is the density of sea water, C_p is the power coefficient, and U_s is speed of the stream.

The power coefficient C_p in Eq. (8.15) is subject to the Betz limit like a wind turbine. The power coefficient and thrust coefficient of a tidal turbine may be obtained by applying the expressions derived in chapter 6 using the actuator disk theory. From Eq. (8.15) we note that for the same frontal area, the energy available to a tidal turbine is 800 times that for a wind turbine because of the much higher density of water compared to air. It follows that nearly one-tenth the speed of wind provides the same power to a tidal turbine of the same size.

As with wind power systems, selection of the location is critical for tidal turbines. It is desirable to locate them in areas with fast currents where natural flows are accelerated between obstructions, such as entrances to bays and rivers, around rocky points, headlands or between islands.

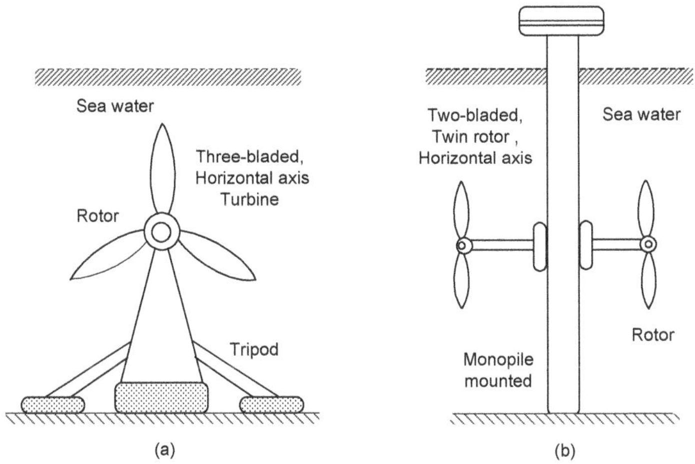

Fig. 8.7 Tidal turbine designs, (a) three-bladed horizontal axis, (b) two-bladed twin rotor, horizontal axis

Many tidal stream generators have been designed and tested. Shown schematically in Fig. 8.7(a) is a three-bladed, horizontal axis, fixed pitch, and bottom mounted tidal turbine. The two-bladed, twin rotor, horizontal axis, variable pitch turbine shown in Fig. 8.7(b), is mounted on a monopile. Several tidal devices have been operated as prototypes. Unlike wind turbines none of the tidal generators have yet emerged as the clear choice for adoption on a commercial scale [1].

8.3 Wave Power Generation

Wind blowing over ocean waters generates water waves, and the kinetic energy carried in these waves is called *wave power*. Waves in oceans are irregular but they gather, store, and transmit energy over thousands of kilometers with very little loss. Wave energy is available twenty four hours a day and round the year, and it can be considered as a concentrated form of solar energy. Its net potential as a renewable energy source is better than wind, solar and biomass energy sources [1].

Worldwide coastal wave energy has been estimated to be more than 2TW [8]. However, wave energy has not been exploited commercially to the same extent as wind energy, due partly to the more challenging conditions encountered in oceans.

Waves in oceans are classified as deep water waves, intermediate water waves, and shallow water waves. Although the available power increases significantly with depth, conveying power from deep water to the shore is more difficult and expensive.

In this section we shall discuss the physical characteristics of ocean waves including the available energy. Some devices that have been designed and tested to harness wave energy will be presented.

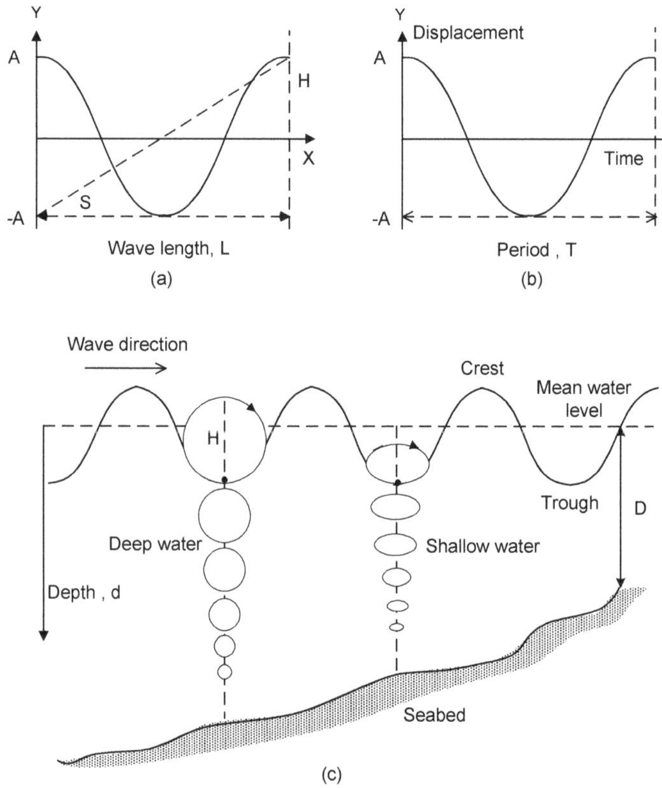

Fig 8.8 (a) Wave displacement versus distance, (b) Wave displacement versus time at a location, (c) Effect of water depth on wave characteristics.

8.3.1 *Characteristics of water waves*

Actual ocean waves are irregular both in shape and time variation. However in this section we shall use simple sinusoidal waves to obtain expressions for the various parameters that characterize a wave. Depicted in Fig. 8.8(a) is the shape of waves in the direction of propagation. The variation of displacement with time at a fixed location is shown in Fig. 8.8(b).

The height H of a wave is the total vertical distance between a crest and a trough of the wave. If A is the amplitude of the wave, then $H = 2A$. The steepness of the wave S is defined by (see Fig. 8.8a)

$$S = \frac{H}{L} = \frac{2A}{L} \tag{8.16}$$

where L is the wavelength.

For an ideal sine wave the displacement y in the direction normal to the direction of propagation is given by

$$y(t) = A\cos(\omega t) \tag{8.17}$$

where ω is the angular frequency and t is time. The angular frequency is related to the frequency f and wave period T by

$$f = \frac{\omega}{2\pi}, \quad T = \frac{2\pi}{\omega} = \frac{1}{f} \tag{8.18}$$

The wave number is defined as

$$k = \frac{2\pi}{L} \tag{8.19}$$

Speed of water waves

The speed c of a progressive wave is given by

$$c = \frac{L}{T} = \frac{\omega}{k} \tag{8.20}$$

The detailed physical modeling of water waves is beyond the scope of this book. Therefore we shall use the final expression for wave speed obtained by such an analysis. This is given by [1]

$$c = \left[\frac{gL}{2\pi} tanh\left(\frac{2\pi D}{L}\right)\right]^{1/2} \tag{8.21}$$

where D is the depth of water.

From Eqs. (8.20) and (8.21) we obtain

$$c^2 = \frac{g}{k} tanh(kD) \tag{8.22}$$

From Eqs. (8.20) and (8.22) we get

$$\omega^2 = gktanh(kD) \tag{8.23}$$

Equations (8.21) and (8.22) are known as the dispersion equations and they show that the speed of a wave increases with its wavelength.

The speed of water waves depends on the depth of water as seen from Eq. (8.21). It is possible to divide the depth into three regimes where the hyperbolic tangent function (HTF) in Eq. (8.21) behaves differently. These regimes are as follows:

(i) Deep water, where $D > \frac{L}{2}$

For this condition the argument of the HTF, $\frac{2\pi D}{L} > \pi$ and therefore

$$tanh\left(\frac{2\pi D}{L}\right) \approx 1 \tag{8.24}$$

Substituting in Eq. (8.21) we obtain the speed as

$$c = \left(\frac{gL}{2\pi}\right)^{1/2} = (1.56L)^{1/2} \tag{8.25}$$

Substituting from Eq. (8.20) in Eq. (8.25) we get

$$c = \frac{gT}{2\pi} = 1.56T \tag{8.26}$$

(ii) Shallower water, where $D < \frac{L}{20}$

For this condition the argument of the HTF, $\frac{2\pi D}{L} < 0.3$ and therefore

$$tanh\left(\frac{2\pi D}{L}\right) \approx \frac{2\pi D}{L} \tag{8.27}$$

Substituting in Eq. (8.21) we obtain

$$c = (gD)^{1/2} \tag{8.28}$$

(iii) Intermediate depths, where $D > \frac{L}{20}$ and $D < \frac{L}{2}$

For this condition, Eq. (8.21) in its complete form has to be used to find the wave speed.

Motion of water at depth

The trajectories of water particles of a wave at different depths of water are depicted schematically in Fig. 8.8(c). In deep water the particles move in circles and the net movement of water is very small. At the surface of water the diameter of the circles is equal to wave height H. The radius r_d of circular motion of water particles decreases exponentially with depth d below the surface. Hence

$$r_d = \frac{H}{2} e^{-\frac{2\pi d}{L}} \tag{8.29}$$

We note from Eq. (8.29) that when $d > \frac{L}{2}$ the radius of the circular motion is less than about 4% of the value at the surface. About 95% of the energy of the waves is contained between the surface and a depth of $\frac{L}{4}$, where L is the wavelength. Figure 8.8(c) shows that as the wave enters shallower water, the motion of the particles become elliptical, and the shape flattens as the water depth decreases.

Power of waves

The total energy of a wave consists of kinetic energy and potential energy contributing equally. The energy E [Jm^{-1}] per unit crest width averaged over a wavelength is given by [1]

$$E = \frac{1}{2}\rho g A^2 L = \frac{1}{8}\rho g H^2 L \tag{8.30}$$

where H is the wave height equal to $2A$.

The transfer of energy by waves depends on the speed of individual waves c as well as the speed of the wave as a group c_g. The rate of energy transfer is defined as the work done per unit time by the pressure in the wave over a surface normal to the direction of propagation averaged over a wave period. For regular waves with a single period in deep water

$$c_g = \frac{c}{2} \tag{8.31}$$

The power of the wave P [Wm^{-1}] is the rate of energy transfer per unit length. It may be expressed as [1]

$$P = \frac{1}{8}\rho g H^2 c_g \tag{8.32}$$

For deep water waves it follows from Eqs. (8.20) and (8.25) that

$$c = \frac{gT}{2\pi} \tag{8.33}$$

Substituting from Eqs. (8.31) and (8.33) in (8.32) we get

$$P = \frac{1}{32\pi}\rho g^2 H^2 T \tag{8.34}$$

Substituting numerical values, $g = 9.81$ ms^{-2} and $\rho = 1025$ kgm^{-3} in Eq. (8.34) we have

$$P = 981 H^2 T \tag{8.35}$$

The above equation is applicable to shallow water waves, which are of less practical importance in wave energy conversion.

Power in real ocean waves

The waves considered in the previous section were sine waves of constant frequency and amplitude. However, real ocean waves are complex and they have irregular frequencies and amplitudes that vary continuously over the day, as shown schematically in Fig. 8.9.

When manual procedures are used to measure the properties of irregular waves, the highest third of the waves, is defined as $H_{1/3}$. With modern equipment wave heights are measured automatically, and four times the root mean square of the wave height is taken as the most

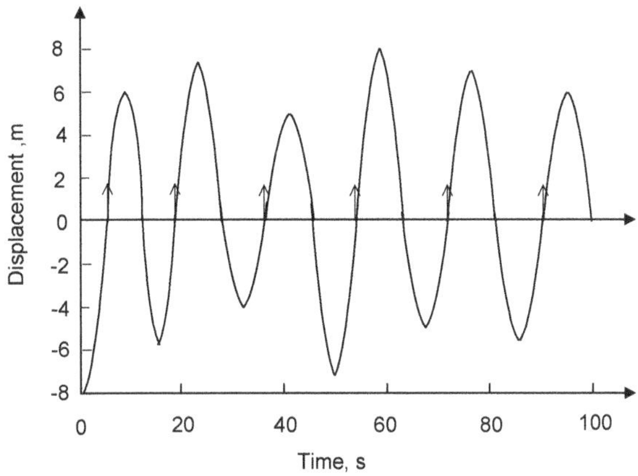

Fig 8.9 Real ocean wave – upward zero-crossings shown by arrows

significant wave height H_S. It is found that H_s is approximately equal to $H_{1/3}$.

To estimate the frequency of a real wave, the number of the upward zero crossings shown in Fig. 8.9, are observed over the measuring period. Hence the period T_z is estimated.

It is more common to define the period of an ocean wave using the concept of an energy period T_e, which is the period of a wave with the same energy as that of the measured series of waves. It is found that, $T_e = 1.12T_z$ [1].

The power of waves is given by Eq. (8.34) as:

$$P = \frac{1}{32\pi}\rho g^2 H^2 T$$

Substituting $T = T_e$ and $H = H_s/\sqrt{2}$ in the above equation the power of a real ocean wave is obtained as [1]

$$P = \frac{1}{64\pi}\rho g^2 {H_s}^2 T_e \tag{8.36}$$

Substituting numerical values, $g = 9.81$ ms^{-2} and $\rho = 1025$ kgm^{-3} in Eq. (8.36) we get

$$P = 491{H_s}^2 T_e \tag{8.37}$$

8.3.2 *Wave energy converters*

Devices used to covert the energy in waves to electricity are commonly called wave energy converters (WEC). At present there exist several technologies to capture wave energy but it is difficult to predict which technologies would be adopted for future commercialization. As with all renewable energy sources, wave energy is variable on all time scales, but less so compared to solar and wind [1,2].

Classification of WECs

Compared to other renewable energy converters, the concepts available for WEC are much larger. Despite this wide range of designs tested, WECs can be generally categorized by: (i) location, and (ii) type or mode of operation.

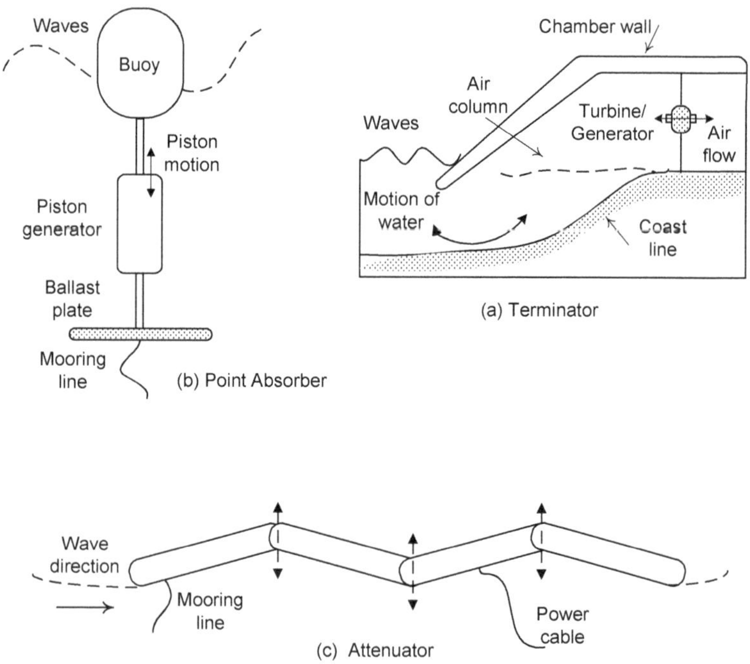

Fig. 8.10 Wave energy devices (a) terminator, (b) point absorber, (c) attenuator

1. Location

The first category based on location includes *shoreline devices*. These devices are installed at the shoreline and therefore they usually have easy access to the power grid. They are easy to maintain, and are unlikely to experience extreme conditions because waves are attenuated as they approach land. However, the wave energy available to these devices is much lower because they are located in shallow water. The sites suitable for these devices are often limited due to environmental and other constraints.

The second location based category includes *nearshore devices*. These devices are in relatively shallow water and are often attached to the seabed, which gives them a stationary base from which to execute oscillations.

The third category includes *offshore devices*, which are in deep water where the available wave energy is usually large. They must be designed to withstand extreme conditions due to the greater wave height and energy content of the waves. Therefore, these devices are more difficult to construct and maintain, and are more costly.

2. Types of devices

Despite the large variation in designs and concepts, the mode of operation of WECs can be classified under three categories. These are: terminators, point absorbers, and attenuators.

Depicted in Fig. 8.10 are examples of devices belonging the above categories that have been tested under real conditions.

(i) *Terminators*

In general, terminator devices have their main axis parallel to the wave front and physically intercept waves. As the name implies these devices terminate the progress of the wave. Such a device is shown in Fig. 8.10(a), and it is also called an oscillating water column. It consists of a concrete structure erected in shallow water at the shoreline, with a rectangular opening facing on-coming waves. As water from a wave flows into the chamber, the water surface compresses the air in the chamber like a large piston. This high pressure air flows through the turbine-generator system to produce power.

As the water in the chamber comes to rest at a higher elevation, and then recedes back to the sea, the pressure in the chamber drops. Atmospheric air is then drawn through the turbine in the opposite direction. This terminator device employs a specially designed turbine, having blades with a symmetrical airfoil cross section. The novel blade shape ensures that the turbine shaft rotates in the same direction, no matter which way air flows through it. The turbine is connected to a generator that produces electrical power continuously if there are waves impinging on the device.

Another class of devices that behave as terminators is called overtopping WECs. These devices first convert the kinetic energy of on-coming waves to potential energy. Typically, this is achieved by forcing the wave to flow up a ramp with converging side walls, and then over the top into a reservoir. The potential energy of the water collected in the reservoir is used to operate low head hydro-turbines to produce power.

(ii) *Point absorbers*

The WEC shown in Fig. 8.10(b) is an example of a point absorber. It consists of a piston attached to a floating buoy which rises and falls as waves pass under it. The cylinder in which the piston is located is held in position by a heavy ballast plate tethered to the seabed by a mooring line. The reciprocating motion of the piston is used to generate electricity directly through linear generators or indirectly through generators driven by rotating shafts.

(iii) *Attenuators*

The third category of wave energy devices, called surface attenuators, is similar in operation to point absorbers. Such a device is depicted in Fig. 8.10(c). It consists of several long floating cylindrical segments, connected to each other at one end, and the axis is oriented normal to the wave front. The device is anchored to the seabed at one end.

When the waves interact with the device the segments move relative to each other as indicated in the figure. The hinges between the segments convert this motion through a hydraulic system to produce high pressure in accumulators. These in turn operate hydraulic power systems that drive generators. Power is sent from a transformer located inside the structure, through a cable on the ocean floor to be fed into the electricity grid.

A group of WECs deployed in the sea at the same location is called a wave farm. Such commercial-scale wave farms are operating in the UK, Portugal, Australia, and the USA [8]. However, at the present time the global contribution of wave energy is small compared to solar and wind power.

8.4 Ocean Thermal Energy Conversion [OTEC]

Ocean energy resources considered thus far, that is tidal and wave energy, are available in the form of kinetic energy of water. Therefore their conversion to electricity is not subject to the Carnot limit.

Ocean thermal energy (OTEC) resource is a consequence of the difference in temperature between surface water and deeper water in oceans. The conversion efficiency of devices used to extract mechanical energy using this temperature difference is subject to the Carnot limit. However, these conversion technologies are well understood due to their numerous applications in the thermal power industry. In this section we shall present two technologies for the exploitation of ocean thermal energy for the generation of electricity.

8.4.1 *Ocean thermal energy resource*

Energy from the sun heats the surface water of oceans, and in some semitropical and tropical regions surface water can be much warmer than deep water. The denser cold water masses, formed by the interaction between the cold atmosphere and ocean water in specific areas, sink into the deep sea basins. This cold water is spread in entire deep ocean by the thermohaline (temperature-salinity) circulation[9]. Upwelling of cold water from the deep ocean is replenished by downwelling of cold surface sea water.

Among ocean sources, OTEC is one of the continuously available renewable energy resources that could contribute to base-load power supply. Up to 88,000 TWh per year of power could be generated from OTEC without affecting the ocean's thermal structure [9].

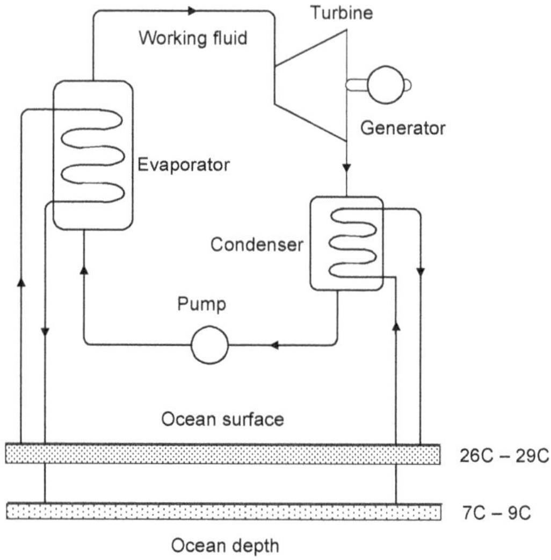

Fig. 8.11 Closed-cycle OTEC power cycle

8.4.2 *Closed-cycle thermal energy conversion*

A closed-cycle power generation system using warm ocean surface water as the heat source is depicted in Fig. 8.11. The condenser is cooled using cold water from a depth of about 1000 m. The other components of the plant, like the turbine and the feed pump are similar in operation to the Rankine cycle plants discussed in chapter 5 on solar thermal power systems.

Assuming a hot water temperature of 28°C and a cold water temperature of 6°C we obtain a Carnot efficiency of 7%, which is the maximum possible ideal efficiency. There are several irreversibilities that could decrease the actual efficiency to about 3% to 4%.

These are: (i) the temperature difference between hot water and the working fluid in the evaporator, (ii) the temperature difference between cold water and the working fluid in the condenser, (iii) frictional losses in the turbine-generator system and the pump, and (iv) parasitic losses in the system.

An experimental mini-OTEC plant using Ammonia as the working fluid was operated off of KeaholePoint, Hawaii in 1977[2]. The plant generated 50 kW (gross) and 10 kW (net) with a temperature difference of 21.2°C. The system included all the components required to build a full-scale closed-cycle OTEC plant.

8.4.3 *Open-cycle thermal energy conversion*

The open cycle OTEC system depicted in Fig. 8.12 produces power as well as fresh water. It has some features common with flash distillation used for sea-water desalination. Therefore it is ideally suited for locations that have a need for both fresh water and electric power.

Warm surface water at about 25°C is introduced into the evaporator which operates at a pressure slightly below the saturation pressure corresponding to warm sea water. A small fraction of the sea water is converted to steam (flashing) due to the pressure reduction. This steam is expanded through a turbine to produce mechanical work needed to operate the generator. Wet steam leaving the turbine is condensed using a surface condenser if fresh water is desired as an output of the plant. Otherwise a much cheaper direct-contact condenser is used. The cold water for condensation is drawn from deep sea water at a temperature of about 5°C.

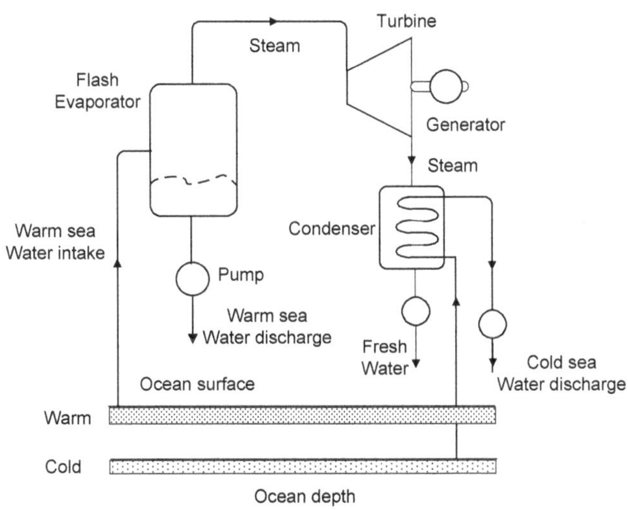

Fig. 8.12 Open-cycle OTEC power cycle

For given hot water and cold water supply temperatures, the closed-cycle plant (CC) is less efficient then the open-cycle (OC) plant. This is because the heat transfer in the evaporator and condenser in a CC plant occurs across the walls of the heat exchanger tubing. This reduces the operating temperature difference of the Rankine cycle.

A simplified analysis of an open-cycle (flash-steam) Rankine cycle power plant is presented in Chapter 9 under geothermal power systems.

Although OTEC systems have several attractive features as a renewable energy technology, they face the enormous challenge of producing net power with a relatively small temperature difference of about 20°C. Furthermore, this impacts on the size and effectiveness of the various components like the evaporator and the condenser. To produce a net power output, the required pumping power and other parasitic losses have to be kept to a minimum.

Detailed economic evaluations of OTEC systems have concluded that [2], in general, the capital cost per MW of several designs investigated decreased rapidly as the power output increased, and levelled off at an output of about 20 MW electric. The results clearly indicate that for OTEC plants to be economically viable, extremely large plants would be necessary. For such plants the initial cost is large, and this has so far discouraged investment in this renewable energy technology.

8.5 Worked Examples

Example 8.1 Calculate the gravitational force between (i) the earth and the sun, and (ii) the earth and the moon. Calculate the rate of change of the gravitational force with distance for (i) and (ii).

Solution The following are the masses of the earth, sun and the moon:

$$M_{earth} = 5.972 \times 10^{24} \text{ kg}, \quad M_{sun} = 1.989 \times 10^{30} \text{ kg},$$
$$M_{moon} = 7.347 \times 10^{22} \text{ kg}$$

The distances between the earth, the sun and the moon are:

$$R_{E-S} = 1.49 \times 10^{11} \text{ m}, \qquad R_{E-M} = 3.844 \times 10^{8} \text{ m}$$

The universal gravitational constant is

$$G - 6.674 \times 10^{-11} \text{ Nm}^2\text{kg}^{-2}$$

The gravitational force between the earth and the sun is given by

$$F_{E-S} = \frac{GM_E M_S}{R_{E-S}{}^2} = \frac{6.674\times10^{-11}\times5.972\times10^{24}\times1.989\times10^{30}}{(1.49\times10^{11})^2}$$

$$F_{E-S} = 3.57 \times 10^{22} \text{ N}$$

Similarly, the force between the earth and the moon is

$$F_{E-M} = \frac{GM_E M_M}{R_{E-M}{}^2} = \frac{6.674\times10^{-11}\times5.972\times10^{24}\times7.34\times10^{22}}{(3.8\times10^{8})^2}$$

$$F_{E-M} = 2.026 \times 10^{20} \text{ N}$$

The ratio of the forces is

$$\frac{F_{E-S}}{F_{E-M}} = \frac{3.57\times10^{22}}{2.026\times10^{20}} = 176$$

The rate of change of the gravitational force with distance is

$$F' = \frac{dF}{dR} = -\frac{2GM_E M_S}{R_{E-S}{}^3} = -F_{E-S}\left(\frac{2}{R_{E-S}}\right)$$

The required ratio is

$$\frac{F_{E-S}{}'}{F_{E-M}{}'} = \frac{F_{E-S}\times R_{E-M}}{F_{E-M}\times R_{E-S}} = 176 \times \frac{3.844\times10^{8}}{1.49\times10^{11}} = 0.45$$

We note that the effect of the force of the sun is 176 times that of the moon. However, the tidal forces are affected by the rate of change of the gravitational forces. Based on the rate of change of the gravitational force the effect of the sun is about 0.45 times that of the moon. Therefore the moon has a larger influence on the tides on earth.

Example 8.2 The average useable tidal range at a location is 7 m. The length and breadth of the basin are 6 km and 4 km respectively. Calculate the maximum daily average power available.

Solution If the total change in water level is H and the area of the basin is A, then the maximum energy E per tide is given by Eq. (8.2) as:

$$E = \int_0^H A\rho g h dh = A\rho g \frac{H^2}{2} \tag{E8.2.1}$$

where h is the difference in water level, $\rho = 1025$ kgm^{-3} is the density of sea water, and $g = 9.81$ ms^{-2} is the acceleration due to gravity.

The area of the basin is

$$A = 6 \times 4 = 24 \text{ km}^2$$

Substituting numerical values in Eq. (E8.2.1) we get

$$E = \frac{1}{2} \times 24 \times 10^6 \times 1025 \times 9.81 \times 7^2 = 5.912 \times 10^6 \text{ MJ}$$

The length of the tidal day is 24.8 h, and there are 4 tides per day.

Hence the total energy available per day is

$$E_{day} = 4 \times 5.912 \times 10^6 \times \frac{24}{24.8} = 22.88 \times 10^6 \text{MJ}$$

The daily average power is

$$P_{day} = \frac{22.88 \times 10^6}{24 \times 3600} = 264.8 \text{ MW}$$

Example 8.3 A tidal barrage built across the mouth of an estuary has created an impounded area of 18 km^2. The tidal range at the location varies between 7 m and 13 m. Estimate the maximum daily average energy that is available for (i) spring tide, and (ii) neap tide.

Solution At spring the tidal range has the highest value of 13 m and during neap it has the lowest value of 7 m. We apply the procedure in example 8.2 to the spring tide and the neap tide. Hence we have

$$E = 4 \times \frac{1}{2} A\rho g H^2 \times \frac{24}{24.8} \tag{E8.3.1}$$

The area of the tidal basin is

$$A = 18 \times 10^6 \text{ m}^2$$

For spring tide $H = 13$ m. Substititing in Eq. (E8.3.1) we get

$$E = 4 \times \frac{1}{2} \times 18 \times 10^6 \times 1025 \times 9.81 \times 13^2 \times \frac{24}{24.8} = 59.2 \times 10^6 \text{ MJ}$$

For neap tide $H = 7$m. Substititing in Eq. (E8.3.1) we get

$$E = 4 \times \frac{1}{2} \times 18 \times 10^6 \times 1025 \times 9.81 \times 7^2 \times \frac{24}{24.8} = 17.16 \times 10^6 \text{ MJ}$$

Example 8.4 A rectangular tidal lagoon has a length of 2.2 km and a breadth of 2 km. The average tidal range at the location is 8 m. Estimate the maximum daily power that could be generated using ebb generation.

Solution We shall use model developed in section 8.2.3. The ideal sequence of operation of a tidal system generating power during ebb tides is shown in Fig. 8.6. The details are listed in Table 8.1.

This idealized model assumes a sinusoidal variation of the tidal height with time. The water flow rate through the turbine during power generation is constant. The total energy extracted per ebb tide is obtained as the sum of two areas, the first is a triangle and the second has a sinusoidal shape. Thus we obtained Eq. (8.8) as:

$$E_{3-4} = \frac{A\rho gH}{T}\left(\frac{TH}{8} + \frac{TH}{2\pi}\right) = 0.284A\rho gH^2$$

The total energy available during each tidal day from the two ebb tides is

$$E_d = 0.568A\rho gH^2$$

Substituting given numerical data we get

$$E_{td} = 0.568 \times 2 \times 2.2 \times 1025 \times 9.81 \times 8^2 = 1.608 \text{ MJ}$$

The energy output per day is

$$E_d = 1.608 \times \frac{24}{24.8} = 1.556 \text{ MJ}$$

Example 8.5 A tidal power plant generates power during both the flood tide and the ebb tide. Sketch the time variation of the sea water level and the water level in the lagoon for the two tides. Indicate the effect of pumping when water in the basin is standing. Derive an expression for the energy output in terms of the above water levels.

Solution

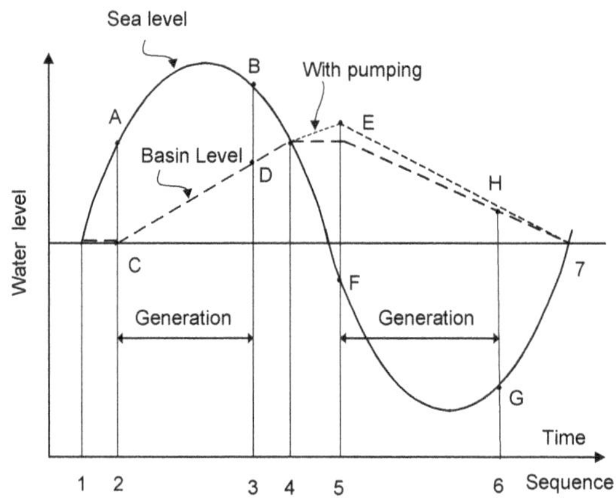

Fig. E8.5 Variation of water levels for flood and ebb generation

Table E8.5 Sequence of operation for a flood and ebb cycle

Duration	Sluces	Turbines	Basin level	Sea level	Power
1-2	Shut	Shut	Standing	Rising	None
2-3	Shut	Operating	Rising	Rise & Drop	Variable
3-4	Open	Shut	Filling	Dropping	None
4-5	Shut	Shut	Standing	Dropping	None
5-6	Shut	Operating	Dropping	Drop & Rise	Variable
6-7	Open	Shut	Emptying	Rising	None
4-5-pump	Shut	Pumping	Rising	Dropping	None
5-6-pump	Shut	Operating	Dropping	Drop & Rise	Variable
6-7-pump	Open	Shut	Emptying	Drop & Rise	None

The variation of the sea water level and the basin water level for a single flood and ebb generation is shown in Fig. E8.5. The sequence of operation is summarized in Table E8.5. The effect of pumping when the basin level is standing is also indicated.

We have assumed that the flow rates in and out of the basin are constant. If the basin area is uniform with depth, then a linear variation of its water level implies that the flow rate is constant. The variation of sea level is depicted as a sine wave.

Consider the flood generation phase from 2 to 3. The energy output is given by

$$E_f = \int_2^3 Q\rho gzdt \tag{E8.5.1}$$

where z is the difference in water level between the sea and the basin at time t. Additional details related to Eq. (E8.5.1) are indicated in Fig. 8.6. The constant flow rate Q during the period 2-3 is given by

$$Q = \frac{A(H_{3D}-H_{2C})}{t_3-t_2}\ [\mathrm{m^3s^{-1}}] \tag{E8.5.2}$$

where A is the uniform area of the basin. The heights H are between the points indicated in Fig. E8.5. The integral in Eq. (E8.5.1) may be expressed in terms of the areas of the figure. Hence we have

$$E_f = \frac{A(H_{3D}-H_{2C})\rho g[area\ ACDB]}{t_3-t_2}\ [\mathrm{J}] \tag{E8.5.3}$$

We can apply the above procedure to determine the energy required for pumping during 4-5, and the energy output during the ebb generation phase 5-6. The latter was analyzed in section 8.3.2.

Example 8.6 A tidal power system includes a pumping system. The tidal range is 8 m and water is pumped to an additional height of 1.5 m during high tide. The area of the basin is 5.4 km^2. Calculate (i) the theoretical maximum energy without pumping, (ii) the ideal increase in energy output due to pumping, and (iii) the increase in energy output if the efficiency of the pumping system is 75%.

Solution We shall use the simple model developed to obtain the additional energy produced due to pumping. The maximum energy available without pump-storage is

$$E = \int_0^H A\rho ghdh = A\rho g\frac{H^2}{2}$$

$$E = 5.4 \times 10^6 \times 1025 \times 9.81 \times \frac{8^2}{2} = 1.7375 \times 10^6\ \mathrm{MJ}$$

The energy output with pumping is

$$E_{ps} = \int_0^{H+h} A\rho g h dh = A\rho g \frac{(H+h)^2}{2}$$

The energy used to pump water to increase the level of water by h is

$$E_p = \frac{A\rho g}{\eta_p}\frac{h^2}{2}$$

where η_p is efficiency of the pumping system.

The net increase in energy output is given by

$$\Delta E_{net} = E_{ps} - E_p - E$$

Manipulating the above three equations we get

$$\Delta E_{net} = \frac{A\rho g H^2}{2}\left[\frac{2h}{H} - \left(\frac{h}{H}\right)^2\left(\frac{1-\eta_p}{\eta_p}\right)\right]$$

(i) When the efficiency of the pump is 100% we get

$$\Delta E_{net} = \frac{A\rho g H^2}{2}\left[\frac{2h}{H}\right] = \frac{5.4\times10^6\times1025\times8^2\times9.81\times2\times1.5}{2\times8} = 652 \text{ kJ}$$

(ii) When the efficiency is 75% we have

$$\Delta E_{net} = \frac{A\rho g H^2}{2}\left[\frac{2h}{H} - \left(\frac{h}{H}\right)^2\left(\frac{1-\eta_p}{\eta_p}\right)\right]$$

$$\Delta E_{net} = \frac{5.4\times10^6\times1025\times8^2\times9.81}{2}\left[\frac{2\times1.5}{8} - \left(\frac{1.5}{8}\right)^2\left(\frac{0.25}{0.75}\right)\right] = 631 \text{ kJ}$$

Example 8.7 Obtain an expression for the additional height h due to pumping that gives the maximum increase in energy for a tidal power system. The efficiency of the pumping system is η_p. Obtain the optimum pumping height if the efficiency of the pump is 50%.

Solution The energy expended to pump water to increase the level by h is

$$E_p = \frac{A\rho g}{\eta_p}\frac{h^2}{2}$$

where η_p is efficiency of the pumping system.

The net increase in energy output is given by

$$\Delta E_{net} = E_{ps} - E_p - E$$

Manipulating the above equations we get

$$\Delta E_{net} = \frac{A\rho g H^2}{2}\left[\frac{2h}{H} - \left(\frac{h}{H}\right)^2\left(\frac{1-\eta_p}{\eta_p}\right)\right]$$

To obtain the maximum value of the water height we differentiate the RHS of the above equation w.r.t to h and equate the resulting expression to zero. Hence we get

$$h_{max} = \frac{H\eta_p}{1-\eta_p}$$

If the efficiency of the pumping system is 0.5 we get, $h_{max} = H$

Example 8.8 A tidal turbine with a rotor diameter of 20 m generates 420 kW. Its hub is at a height of 16 m. The tidal stream speed is 1.8 $\mathrm{ms^{-1}}$. Calculate (i) the power coefficient, (ii) water speed at the hub, and (iii) the thrust on the rotor.

Solution The power output of the turbine is given by (see Chapter 6)

$$P_w = \frac{C_p A\rho U^3}{2} = 420 \text{ kW}$$

$$\frac{C_p \times \pi \times 10^2 \times 1025 \times 1.8^3}{2} = 420 \text{ kW}$$

The power coefficient is

$$C_p = 0.447$$

The axial induction factor is given by (see Chapter 6)

$$c_p = 4a(1-a)^2$$

Hence we have

$$4a^3 - 8a^2 + 4a - 0.447 = 0$$

The above equation has three positive roots. The physically meaningful root is given by, $a = 0.1547$

$$a = 1 - \frac{U_h}{U} = 0.1547$$

The water speed at the hub is

$$U_h = 1.8 \times 0.8453 = 1.521 \text{ ms}^{-1}$$

The thrust coefficient is (see Chapter 6)

$$c_T = 4a(1-a) = 4 \times 0.1547 \times (1 - 0.1547) = 0.523$$

The thrust on the rotor is given by

$$T = \frac{1}{2} c_T A \rho U^2$$

$$T = \frac{0.523 \times \pi \times 10^2 \times 1025 \times 1.8^2}{2} = 272.8 \text{ kN}$$

Example 8.9 A tidal turbine is located in a stream with its hub at a height of 15 m. The radius of the blades is 10 m. The velocity of the stream varies with height *z* according to the power law

$$U(z) = 1.5 \left(\frac{z}{5}\right)^{\frac{1}{10}}$$

The power coefficient of the rotor $C_p = 0.41$, and the thrust coefficient $C_t = 0.5$. Calculate (i) the power output, and (ii) the overturning moment on the support structure at the bottom.

Solution The water speed at hub height of the turbine is

$$U(15) = 1.5 \left(\frac{15}{5}\right)^{\frac{1}{10}} = 1.674 \text{ ms}^{-1}$$

The power output is given by (see Chapter 6)

$$P_w = \frac{C_p A \rho U^3}{2} = \frac{0.41 \times \pi \times 10^2 \times 1025 \times 1.674^3}{2} = 309.7 \text{ kW}$$

The thrust on the rotor is given by (see Chapter 6)

$$T_r = \frac{C_t A \rho U^2}{2} = \frac{0.5 \times \pi \times 10^2 \times 1025 \times 1.674^2}{2} = 225.6 \text{ kN}$$

The overturning moment acting at the base is

$$M = 225.6 \times 15 = 3.384 \text{ MNm}$$

Example 8.10 The rotor diameter of a tidal turbine is 15 m and it is in a tidal stream of speed 2.2 ms^{-1}. The hub height of the turbine is 16 m. At the rotor the water speed is 1.8 ms^{-1}. Calculate (i) the axial induction factor, (ii) the power and thrust of the rotor, and (iii) the overturning moment on the turbine at the base.

Solution We can apply the expressions developed for wind turbines in Chapter 6 to tidal turbines. The axial induction factor is

$$a = 1 - \frac{U_t}{U_s} = 1 - \frac{1.8}{2.2} = 0.182$$

The power coefficient is given by

$$C_p = 4a(1-a)^2$$

$$C_p = 4 \times 0.182(1 - 0.182)^2 = 0.487$$

$$P_w = \frac{C_p A \rho U^3}{2} = \frac{0.487 \times \pi \times 7.5^2 \times 1025 \times 2.2^3}{2} = 469.6 \text{ kW}$$

The thrust coefficient is given by

$$C_t = 4a(1-a)$$

$$C_t = 4 \times 0.182 \times (1 - 0.182) = 0.596$$

The thrust on the rotor is given by

$$T_r = \frac{C_t A \rho U^2}{2} = \frac{0.596 \times \pi \times 7.5^2 \times 1025 \times 2.2^2}{2} = 261.2 \text{ kN}$$

The overturning moment acting at the base is

$$M = 261.2 \times 16 = 4.18 \text{ MNm}$$

Example 8.11 The diameter of a three-bladed tidal turbine is 10 m. The moment of inertia of the rotor is 25×10^3 kgm^2. The tip speed ratio is 5 and the tidal stream flow speed is 2.4 ms^{-1}. Calculate (i) the speed of the rotor, and (ii) the rotational kinetic energy of the turbine.

Solution The tip speed ratio (TSR) is given by (see Chapter 6)

$$\lambda = \frac{R\omega}{U_w}$$

$$5 = \frac{5 \times \omega}{2.4}$$

The rotational speed of the turbine is $\omega = 2.4$ rads^{-1}

The speed of the rotor is

$$N = \frac{\omega}{2\pi} \times 60 = \frac{2.4 \times 60}{2\pi} = 22.9 \text{ rpm}$$

The rotational kinetic energy is

$$E_r = \frac{1}{2} I\omega^2 = \frac{25 \times 10^3 \times 2.4^2}{2} = 72 \text{ kJ}$$

Example 8.12 A wind turbine and a tidal turbine have the same power coefficient of 0.45, and the same thrust coefficient of 0.58. The rated wind speed and tidal stream speed are 13 ms^{-1} and 3 ms^{-1} respectively. The hub height of the tidal turbine is 13 m and the wind turbine is 50 m. The rated power of the two turbines are 1.2 MW. Calculate (i) the radii of two turbines, and (ii) the overturning moments at the base of the two turbines.

Solution The power output of the tidal turbine is

$$P_t = \frac{C_p A \rho U^3}{2} = \frac{0.45 \times \pi \times R_t^{\,2} \times 1025 \times 3^3}{2} = 1.2 \times 10^6$$

Therefore the radius of the tidal rotor is given by

$$R_t^{\,2} = \frac{2 \times 1.2 \times 10^6}{0.45 \times \pi \times 1025 \times 3^3} = 61.34$$

$$R_t = 7.83 \text{ m}$$

The power output of the wind turbine is

$$P_t = \frac{C_p A \rho U^3}{2} = \frac{0.45 \times \pi \times {R_w}^2 \times 1.125 \times 13^3}{2} = 1.2 \times 10^6$$

Therefore the radius of the wind rotor is given by

$${R_w}^2 = \frac{2 \times 1.2 \times 10^6}{0.45 \times \pi \times 1.125 \times 13^3} = 686.86$$

$$R_w = 26.2 \text{ m}$$

The thrust on the tidal rotor is

$$T_t = \frac{C_t A \rho U^2}{2} = \frac{0.58 \times \pi \times 7.83^2 \times 1025 \times 3^2}{2} = 515.27 \text{ kN}$$

The overturning moment acting at the base is

$$M_t = 515.27 \times 13 = 6.7 \text{ MNm}$$

The thrust on the wind rotor is

$$T_w = \frac{C_t A \rho U^2}{2} = \frac{0.58 \times \pi \times 26.2^2 \times 1.125 \times 13^2}{2} = 118.9 \text{ kN}$$

The overturning moment acting at the base is

$$M_w = 118.9 \times 50 = 5.9 \text{ MNm}$$

Example 8.13 A single frequency shallow water wave of 0.9 m depth has a period of 9 s. The amplitude is 1.2 m. Calculate the angular frequency, wavelength, velocity, and the steepness of the wave.

Solution For a shallow water wave the speed is given by Eq. (8.28) as:

$$c = (gD)^{1/2} = (9.81 \times 0.9)^{1/2} = 2.97 \text{ ms}^{-1}$$

The wavelength is obtained from the relation

$$c = \frac{L}{T} = \frac{L}{9} = 2.97$$

Hence the wavelength is 26.7 m.

The condition for shallow waves is

$$D < \frac{L}{20}$$

$$D < \frac{26.7}{20} = 1.335 \text{ m}$$

Therefore the condition for a shallow wave is satisfied.

The angular frequency is given by

$$\omega = \frac{2\pi}{T} = \frac{2\pi}{9} = 0.698 \text{ rads}^{-1}$$

The steepness of the wave is

$$S = \frac{2A}{L} = \frac{2\times1.2}{26.7} = 0.0898$$

Example 8.14 A deep water wave has a wavelength of 45 m and an amplitude of 1.4 m. Calculate the velocity, the period, the angular frequency, the steepness, and energy of the wave.

Solution The speed of a deep water wave is given by

$$c = \left(\frac{gL}{2\pi}\right)^{1/2} = \left(\frac{9.81\times45}{2\pi}\right)^{1/2} = 8.38 \text{ ms}^{-1}$$

The period is obtained from the relation

$$c = \frac{L}{T} = \frac{45}{T} = 8.38$$

Hence the period is 5.37 s.

The angular frequency of the wave is

$$\omega = \frac{2\pi}{T} = \frac{2\pi}{5.37} = 1.17 \text{ rads}^{-1}$$

The steepness is

$$S = \frac{2A}{L} = \frac{2\times1.4}{45} = 0.062$$

The energy of the wave is given by Eq. (8.30) as:

$$E = \frac{1}{2}\rho g A^2 L \text{ kJm}^{-1}$$

$$E = \frac{1}{2}\times 1025 \times 9.81 \times 1.4^2 \times 45 = 443.4 \text{ kJm}^{-1}$$

Example 8.15 A water wave at a depth 20 m has a period of 10 s. Calculate the wavelength and velocity of the wave.

Solution We need to check whether the deep-water condition applies. The amplitude in this case is

$$L = \frac{gT^2}{2\pi} = \frac{9.81\times 10^2}{2\pi} = 156.13 \text{ m}$$

The condition for deep water is

$$D > \frac{L}{2} = 78.06 \text{ m}$$

The given depth of 20 m is less than 78.06 m.

For shallow water the speed is

$$c = (gD)^{1/2} = (9.81 \times 20)^{1/2} = 14$$

Hence the wavelength is given by

$$L = cT = 140 \text{ m}$$

The condition for shallow water is

$$D < \frac{L}{20} = \frac{140}{20} = 7 \text{ m}$$

The given depth of 20m is larger than 7m.

Therefore the given depth is an intermediate depth. The speed of the wave is then given by Eq. (8.21) as:

$$c = \left[\left(\frac{gL}{2\pi}\right) tanh\left(\frac{2\pi D}{L}\right)\right]^{1/2} \tag{E8.15.1}$$

The speed of the wave is

$$c = \frac{L}{T} \tag{E8.15.2}$$

From Eqs. (E8.15.1) and (E18.15.2) we get

$$\frac{2\pi L}{gT^2} = tanh\left(\frac{2\pi D}{L}\right)$$

Substituting given data in the above equation we have

$$6.4048 \times 10^{-3}L = tanh\left(\frac{125.66}{L}\right) = y \qquad \text{(E8.15.3)}$$

The LHS of Eq. (8.15.3) varies linearly with L. The hyperbolic tangent function on the RHS increases smoothly from zero and approaches 1 asymptotically. The point of intersection the above graphs occur at a value of $y = 0.776$, which gives the solution of Eq. (E8.15.3). Hence the wavelength is obtained as 121 m.

The speed of the wave is

$$c = \frac{L}{T} = \frac{121}{10} = 12.1 \text{ ms}^{-1}$$

Example 8.16 A deep water wave has a wavelength of 40 m and an amplitude of 1.5 m. Calculate the wave velocity, the group velocity, the period, and the power of the wave.

Solution The speed of a deep water wave is given by Eq. (8.26) as:

$$c = \left(\frac{gL}{2\pi}\right)^{1/2} = \left(\frac{9.81\times40}{2\pi}\right)^{1/2} = 7.9 \text{ ms}^{-1}$$

The period is obtained from the relation

$$c = \frac{L}{T} = \frac{40}{T} = 7.9$$

Hence the period is 5.06 s

The group velocity for deep water waves is given by Eq. (8.31) as:

$$c_g = \frac{c}{2} = \frac{7.9}{2} = 3.95 \text{ ms}^{-1}$$

The power of the wave is given by Eq. (8.32) as:

$$P = \frac{1}{2}\rho g A^2 c_g \text{ [Wm}^{-1}\text{]}$$

$$P = \frac{1}{2} \times 1025 \times 9.81 \times 1.5^2 \times 3.95 = 44.68 \text{ kWm}^{-1}$$

Example 8.17 For a real ocean wave, the observed number of upward zero crossings during the measuring period of 160 s is 25. The most significant height is 0.85 m. Calculate power of the wave.

Solution From the number of upward zero crossings and the measuring period we obtain the wave period as

$$T_z = \frac{160}{25} = 6.4 \text{ s}$$

The most significant wave height is $H_s = 0.85$ m.

The energy period is given by [1]

$$T_e = 1.12T_z = 1.12 \times 6.4 = 7.168 \text{ s}$$

For waves in deep water, the speed is given by

$$c = \left(\frac{gL}{2\pi}\right)^{1/2} = \left(\frac{gcT}{2\pi}\right)^{1/2} \text{ Hence } c = \frac{gT}{2\pi} = 2c_g$$

Power of the wave is given by Eq. (8.32) as:

$$P = \frac{1}{8}(\rho g H^2)c_g = \frac{1}{8}(\rho g H^2)\left(\frac{gT}{4\pi}\right)$$

$$P = \frac{1}{32\pi}(\rho g^2 H^2 T) \text{ (Wm}^{-1}\text{)} \qquad \text{(E8.17.1)}$$

For real ocean waves, it can be shown that [1]:

$$H = \frac{H_s}{\sqrt{2}}$$

$$T = T_e$$

Hence Eq. (E8.17.1) maybe expressed in terms of H_s and T_e in the from

$$P = \frac{1}{64\pi}\left(\rho g^2 {H_s}^2 T_e\right) \text{ (Wm}^{-1}\text{)}$$

Substituting numerical data we have

$$P = \frac{1}{64\pi}(1025 \times 9.81^2 \times 0.85^2 \times 7.168) = 2.54 \text{ kWm}^{-1}$$

Example 8.18 An OTEC power system uses surface seawater at a temperature of 30°C as the high temperature reservoir and deep seawater at a temperature of 6°C as the cold reservoir. The plant has an output of 10 MW.

(a) Assuming ideal conditions calculate (i) the Carnot efficiency of the plant, (ii) the rate of heat input, and (iii) the rate of heat removal.

(b) If the heat transfer between the working fluid of the cycle and the reservoirs require a temperature difference of 4°C at each reservoir, calculate the thermal efficiency of the cycle.

Solution (a) The maximum efficiency of the plant is the Carnot efficiency, given by [2]

$$\eta_c = 1 - \frac{T_{cold}}{T_{hot}}$$

The hot and cold temperatures are:

$$T_h = 273 + 30 = 303 \text{ K}, \quad T_c = 273 + 6 = 279 \text{ K}$$

The Carnot efficiency is

$$\eta_c = 1 - \frac{279}{303} = 7.9\%$$

The rate of heat input is given by

$$\eta_c = \frac{W_{out}}{Q_{in}}$$

Heat input rate is

$$Q_{in} = \frac{10}{0.079} = 126.6 \text{ MW}$$

The rate of heat removal is

$$Q_{out} = Q_{in} - W_{out} = 126.6 - 10 = 116.6 \text{ MW}$$

(b) The temperature difference between the working fluid and the heat reservoirs is 4°C at each reservoir. Hence the effective temperatures of the internally reversible cycle are:

$$T_{he} = 303 - 4 = 299 \text{ K}, \quad T_{ce} = 279 + 4 = 283 \text{ K}$$

The efficiency of cycle is

$$\eta_{ce} = 1 - \frac{283}{299} = 5.4\%$$

We observe that the temperature differences decrease the efficiency significantly. Internal losses in the cycle due to frictional effects will further decrease the overall efficiency.

Example 8.19 An OTEC plant operates on the Rankine cycle using ammonia as the working fluid. Liquid ammonia enters the boiler at 10°C, and saturated ammonia vapor leaves at 20°C. The inlet and outlet temperatures of hot sea water flowing through the boiler are 25°C and 23°C respectively. Ammonia expands in the turbine to the saturation pressure at 10°C and then enters the condenser. Cold sea water at 5°C flows into the condenser and leaves at 7°C. The power output of the Rankine cycle is 10MW. Calculate (i) the work output of the turbine, (ii) the work input of the feed pump, (iii) the heat input in the boiler, (iv) the cycle efficiency, (v) heat output in the condenser, (vi) mass flow rate of ammonia, (vii) flow rate of hot water, and (viii) flow rate of cold water.

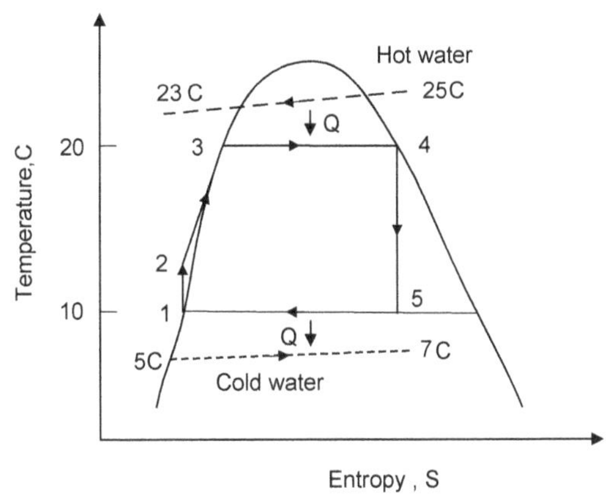

Fig. E8.19.1 Temperature- entropy diagram for the OTEC Rankine cycle

Solution A schematic of a Rankine cycle based OTEC system is shown in Fig. 18.9. The temperature-entropy diagram of the cycle is depicted in Fig. E8.19.1. The temperature variation of hot sea water and cold sea water are also shown.

We have analyzed Rankine cycle power systems in Chapter 4. The following data for ammonia are obtained from Ref.[3].

$$T_3 = T_4 = 20^{\circ}\text{C}, \quad p_3 = p_4 = 8.57 \text{ bar}$$

$$T_1 = T_5 = 10^{\circ}\text{C}, \quad p_1 = p_5 = 6.149 \text{ bar}$$

For the isentropic expansion in the turbine

$$s_4 = s_5$$

$$5.095 = 5.213x_5 + (1 - x_5) \times 0.881$$

Hence the quality $x_5 = 0.9728$

$$h_5 = 0.9728 \times 1454.3 + (1 - 0.9728) \times 227.8 = 1420.88$$

Work output per kg of ammonia is

$$w_{turbine} = h_4 - h_5 = 1462.6 - 1420.88 = 41.72 \text{ kJkg}^{-1}$$

Work input in the feed pump is

$$w_{pump} = h_2 - h_1 = v_l(p_2 - p_1)$$

$$w_{pump} = \left(\frac{1}{680}\right) \times (8.57 - 6.149) \times 10^2 = 0.356 \text{ kJkg}^{-1}$$

Net work output per kg of ammonia is

$$w_{net} = 41.72 - 0.356 = 41.36 \text{ kJkg}^{-1}$$

$$h_2 = h_1 + w_{pump} = 228.15$$

Heat input in the boiler is

$$q_{boiler} = h_4 - h_2 = 1462.6 - 228.15 = 1234.45 \text{ kJkg}^{-1}$$

The cycle efficiency is

$$\eta = \frac{w_{net}}{q_{boier}} = \frac{41.36}{1234.45} = 3.35\%$$

The heat output in the condenser is

$$q_{cond} = h_5 - h_1 = 1420.88 - 227.8 = 1193.08 \text{ kJkg}^{-1}$$

The net power output of the Rankine cycle is 10 MW.

Hence the mass flow rate of ammonia is

$$m_a = \frac{10\times10^3}{41.36} = 241.78\ \text{kgs}^{-1}$$

The heat input rate in the boiler is

$$Q_{boiler} = m_a q_{boiler} = 241.78 \times 1234.45 = 298.46\ \text{MW}$$

The heat removal rate in the condenser is

$$Q_{cond} = m_a q_{cond} = 241.78 \times 1193.08 = 288.46\ \text{MW}$$

Apply the energy balance equation to the boiler, assuming that $c_p = 3.85\ \text{kJkg}^{-1}\text{K}^{-1}$ for sea water. Hence we get

$$Q_{boiler} = m_{hw} c_{sw} (T_{hwi} - T_{hwo})$$

$$m_{hw} \times 3.85 \times (25 - 23) = 298.46 \times 10^3$$

Therefore the mass flow rate of hot water is $38.76 \times 10^3\ \text{kgs}^{-1}$.

The volume flow rate of hot water is

$$V_{hw} = \frac{38.76\times10^3}{1025} = 37.82\ \text{m}^3\text{s}^{-1}$$

The density of sea water is taken as $1025\ \text{kgm}^{-3}$.

Apply the energy balance equation to the condenser. Hence we have

$$Q_{cond} = m_{cw} c_{sw} (T_{cwo} - T_{cwi})$$

$$m_{cw} \times 3.85 \times (7 - 5) = 288.46 \times 10^3$$

The mass flow rate of cold water is $37.46 \times 10^3\ \text{kgs}^{-1}$.

The volume flow rate of cold water is

$$V_{cw} = \frac{37.46\times10^3}{1025} = 36.55\ \text{m}^3\text{s}^{-1}$$

Example 8.20 The cold water flow rate through the condenser of an OTEC plant is $36 \times 10^3\ \text{kgs}^{-1}$. Water is pumped from a depth of 1000 m through a pipe of diameter 2.8 m. Estimate the required pumping power.

Solution The previous examples show that OTEC plants in general have a relatively low efficiency of about 3-5 percent. Therefore for a given power output, the heat input rates and the heat rejection rates tend to be large. The heat exchanges needed for this purpose are also large, and so are the hot and cold water flow rates. Consequently, the pumping power expended can be a significant fraction of the power output of the system.

The volume flow rate of cold water is

$$Q = \frac{M}{\rho} = \frac{36\times10^3}{1025} = 35.12 \text{ m}^3\text{s}^{-1}$$

The water flow speed is

$$U = \frac{Q}{\pi R^2} = \frac{35.12}{\pi\times1.4^2} = 5.68 \text{ ms}^{-1}$$

The Reynolds number the flow is

$$Re_D = \frac{\rho UD}{\mu} = \frac{UD}{\nu} = \frac{5.68\times2.8}{1.32\times10^{-6}} = 12.05\times10^6$$

The flow is turbulent and the friction factor is obtained from the empirical relation given by [2]

$$f = \frac{0.184}{{Re_D}^{0.2}} = \frac{0.184}{(12.05\times10^6)^{0.2}} = 0.00705$$

The frictional head loss in the pipe is

$$\Delta H_{fr} = \frac{2fLU^2}{Dg}$$

$$\Delta H_{fr} = \frac{2\times0.00705\times1000\times5.68^2}{2.8\times9.81} = 16.56 \text{ m}$$

The pumping power is given by

$$P_{pump} = \rho g Q \Delta H_{fr}$$

$$P_{pump} = 1025\times9.81\times35.12\times16.56 = 5.85 \text{ MW}$$

The pumping power could be reduced significantly by using a larger diameter pipe. For the same water flow rate, the pumping power varies as

$1/D^5$. Therefore the pumping power could reduced by increasing the pipe diameter. For instance, if the diameter of the pipe is increased to 3.2 m, then the pumping power is about 3 MW. However, this would increase the cost of the pipe, which varies linearly with diameter.

Problems

P8.1 The area of a tidal basin is 68 km^2 and the tidal range at the location is 9.8 m. Estimate (i) the maximum daily power available, and (ii) the maximum economic length of the tidal barrage.
[*Answers*: (i) 1470 MW, (ii) 5.4 km]

P8.2 A tidal barrage is built across the mouth of an estuary creating a basin area of 15 km^2. The tidal range at the barrage varies between 5 m and 13 m. Estimate the maximum daily power that could be generated for a spring tide, and for a neap tide. [*Answers*: (i) 570 MW, (ii) 84.5 MW]

P8.3 The average tidal range at a barrage is 7.5 m and the basin area is 8 km^2. Calculate (i) the maximum average power available for ideal ebb - cycle generation, and (ii) the ratio of the power generated to the maximum daily power available at the site.
[*Answers*: (i) 28.8 MW, (ii) 28%]

P8.4 A tidal power scheme has a basin area of 6 km^2, and the average tidal range at the location is 8.5 m. The pumping system included in the scheme pumps water to an additional height of 1.5 during high tide. Calculate (a) the maximum energy available during high tide, and (b) the additional energy gained due to pumping, (i) if all losses are neglected, and (ii) if the efficiency of the pumping system is 0.75.
[*Answers*: (a) 2.18×10^6 MJ, (b) (i) 0.77×10^6 MJ, (ii) 0.74×10^6 MJ]

P8.5 A tidal turbine is located in a stream with a water speed of 2.6 ms^{-1} at the hub height of 16 m. The radius of the rotor is 5 m and the power coefficient is 0.44. Calculate (i) the power output of the turbine, and (ii) water speed at the rotor. [*Answers*: (i) 311.3 kW, (ii) 2.2 ms^{-1}]

P8.6 A tidal turbine has a rotor radius of 5 m and a hub height of 16 m. The thrust coefficient is 0.52. The tidal stream speed at hub height is 2.6 ms^{-1}. Calculate (i) the thrust on the rotor, and (ii) the overturning moment at the base of the structure.
[*Answers*: (i) 141.5 kN, (ii) 2264 kNm]

P8.7 A deep water wave has a period of 12 s and an amplitude of 1.2 m. Calculate (i) the wavelength, (ii) the speed, (iii) the steepness, and (iv) the power of the wave.
[*Answers*: (i) 224.6 m, (ii) 18.7 ms^{-1}, (iii) 0.0107, (iv) 67.8 kWm^{-1}]

P8.8 A shallow water wave at a depth of 1 m has a period of 10 s. Calculate (i) the wavelength, (ii) speed, and (iii) the power of the wave.
[*Answers*: (i) 31.3 m, (ii) 3.13 ms^{-1}, (iii) 25 kWm^{-1}]

P8.9 A water wave at a depth of 20 m has a period of 11 s. Show that the wave is an intermediate depth wave. Calculate (i) the speed, and (ii) the wavelength of the wave. [*Answers*: (i) 12.5 ms^{-1}, (ii) 137 m]

P8.10 A Rankine cycle based OTEC plant, using ammonia as the working fluid, has a power output of 8 MW. The boiler and condenser operate at 26°C and 10° C respectively. The inlet and outlet temperatures of hot water to the boiler are 30°C and 28°C. Cold water enters the condenser at 5°C and leaves at 7°C. Assuming ideal conditions, calculate (i) the efficiency, (ii) the mass flow rate of ammonia, (iii) the volume flow rate of hot water, and (iv) the volume flow rate of cold water.
[*Answers*: (i) 4.21%, (ii) 153.5 kgs^{-1}, (iii) 24 m^3s^{-1}, (iv) 23 m^3s^{-1}]

References

1. Jenkins, Nick., and Ekanayake, Janaka., *Renewable Energy Engineering*, Cambridge University Press, Cambridge, U.K. 2017.
2. Kreith, Frank and Krumdieck, Susan, *Principles of Sustainable Energy Systems*, CRC Press, Taylor & Francis Group, New York, 2015.
3. Rogers, G.F.C., and Mayhew, Y.R., *Thermodynamic and Transport Properties of fluids*, Blackwell, Oxford, U.K. 1995.

4. Shepherd, W and Shepherd, D.W., *Energy Studies,* Imperial College Press, London, U.K. 2019.
5. Webb, Paul, *Introduction to Oceanography*, Rebus Community Publishers, Open Textbook Library, 2020.
6. 'Tidal-power, wikipedia'
http://en.m.wikipedia.org/wiki/Tidal power
7. 'Tidal stream generators,wikipedia'
http://en.m.wikipedia.org/wiki/Tidal-stream generator
8. 'Wave-power, wikipedia'
http://en.m.wikipedia.org/wiki/Wave-power
9. 'Ocean thermal energy conversion, wikipedia'
http://en.m.wikipedia.org/wiki/Ocean thermal energy

Chapter 9

Hydropower and Geothermal Power

9.1 Introduction

This chapter presents two well-established renewable energy sources, namely, hydroelectric, and geothermal power. Although the physical origins of these energy sources are different, they do share several common attributes from the point of view of their utilization. These sources are capable of long term storage, are dispatchable on a MW-scale, and economically competitive. The technologies used for their conversion to electricity are well developed, with incremental improvements due to ongoing research. Several countries where hydropower and geothermal resources are abundant, have included these resources in their energy mix [4].

9.2 Hydropower

The hydro resource is based essentially on the *hydrological cycle* or the water cycle [2,5].

Solar radiation incident on surfaces of water bodies causes evaporation (see Fig. 9.1). The resulting cloud formation, and subsequent rain produces water sources, such as streams, rivers, and lakes, at high elevations. The potential energy of this water is used to operate water turbines that drive generators to produce electricity. From an energy transfer stand point the above processes constitute a heat engine cycle, with the sun as the heat source and the atmosphere (or space) as the heat sink.

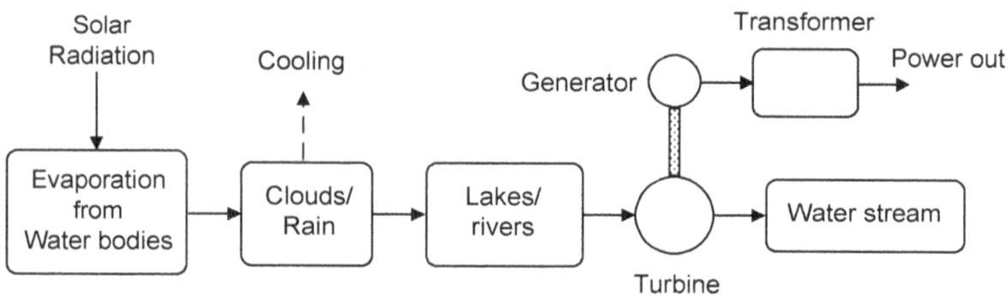

Fig. 9.1 Basic energy-flow diagram for hydroelectric energy conversion

9.3 Development of Hydropower

Hydroelectric power generation is a well-established renewable energy conversion process that contributes significantly to global electricity production. Several common aspects of hydropower were discussed in chapter 7 under pumped-hydro storage systems, and in chapter 8 under tidal range power generation.

Historically, the origin of hydropower can be traced to simple water wheels that were used during ancient times as an energy source for irrigation and the operation of mechanical devices in various industries. Typically, these devices consisted of large wooden wheels mounted on a horizontal power shaft, with blades or buckets fitted on the periphery.

The main designs are called [2]: (i) undershoot wheels, (ii) overshoot wheels, and (iii) breast wheels.

Undershoot wheels with blades on the periphery were placed in a channel diverted from a river. The momentum of water acting on the blades at the bottom of the wheel produces the torque needed to rotate the shaft. This design had a low efficiency of about 20%.

Overshoot wheels use buckets fitted on the periphery that fill with water from a stream. The water supply needs to have a head of at least equal to the diameter of the wheel. The weight of water in the bucket provides the torque to rotate the shaft. Large overshoot wheels with efficiencies up to 40% have been developed.

Water supply to breast wheels is located at the level to the power shaft and are therefore able to use both the kinetic energy and the potential energy of the water. Efficiencies up to about 70% have been achieved with breast wheels.

9.4 Operation of Modern Hydropower Schemes

Modern hydropower schemes can be broadly divided into three types based on the water head available for their operation [2]. These are:

1. High-head schemes with head between 200 m to 300 m.
2. Medium-head schemes with head from 30 m to 200 m.
3. Low-head schemes with head less than 30 m.

9.4.1 *High-head schemes*

A schematic diagram of a high-head hydroelectric scheme is shown in Fig. 9.2. The water in the reservoir is held behind a dam from which it flows through a high-pressure pipe called the *penstock*. Sometimes there are several such pipes in parallel carrying water down to the turbine house located at a lower elevation. A relief valve or a surge chamber is included in the penstock to damp out pressure fluctuations that occur during operation.

The turbines used with high-head hydropower schemes are called *impulse turbines*, a common design being the *Pelton wheel*. Other types of impulse turbines include, Turgo and Crossflow turbines.

Impulse turbines use the kinetic energy of water to rotate the turbine, which in turn spins the generator. The pressure energy of the water at turbine entry is converted to kinetic energy using a nozzle, which directs

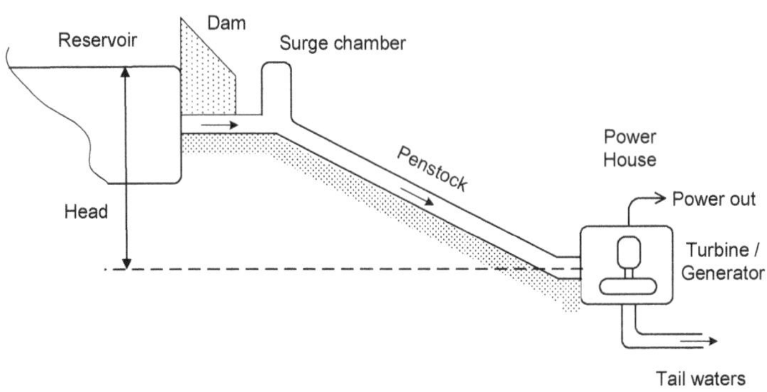

Fig. 9.2 High head hydroelectric power scheme

a high-speed water jet at buckets fitted on the periphery of the rotor, as depicted in Fig. 9.3.

The change of momentum of the water jet generates the torque needed to turn the turbine and the generator connected to it. Under steady operation the electrical torque of the generator balances the torque produced by the turbine wheel (see Fig. 9.2).

9.4.2 *Energy analysis of high-head schemes*

In this section we shall develop the energy equation required to determine the power output of water turbines. Applying Bernoulli's equation between the surface of the reservoir and the entrance to the turbine (see Fig. 9.2) we have

$$\left(z_i + \frac{p_i}{\rho g} + \frac{V_i^2}{2g}\right) = \left(z_e + \frac{p_e}{\rho g} + \frac{V_e^2}{2g}\right) + H_{loss} \tag{9.1}$$

where the subscripts i and e represent inlet and exit conditions.

The elevation is z, the pressure is p, the mean velocity is V, and H_{loss} includes all head losses in the system. Typically these losses include: friction in pipes, losses in fittings such as bends and valves, entrance and exit losses, and others. The above head losses can be expressed as a function of the velocity head using appropriate loss coefficients k_l. Thus, we get

$$\left(z_i + \frac{p_i}{\rho g} + \frac{V_i^2}{2g}\right) = \left(z_e + \frac{p_e}{\rho g} + \frac{V_e^2}{2g}\right) + \sum k_l \frac{V^2}{2g} \tag{9.2}$$

For the high-head scheme shown in Fig. 9.2, Eq. (9.2) can be simplified by making the following assumptions: (i) level of water in the reservoir is constant, (ii) the velocity at the reservoir surface is negligible, and (iii) the pressure at the exit of the turbine is atmospheric. Hence, we obtain the head available at the turbine as

$$H_{av} = \frac{V_e^2}{2g} = H - H_{loss} \tag{9.3}$$

where $H = (z_i - z_e)$ is the difference in elevation between the top of the reservoir and the entrance to the turbine as shown in Fig. 9.2.

From Eq. (9.3) we obtain the power (W) available to the turbine as

$$P_{av} = Q\rho g H_{av} = Q(H - H_{loss})\rho g \tag{9.4}$$

The power output of the generator is given by

$$P_{ele} = P_{av}\eta_t\eta_g \tag{9.5}$$

where η_t is the efficiency of the turbine, and η_g is the efficiency of the generator. Unlike wind turbines and tidal turbines, the efficiency of hydro turbines is not subject to a fundamental upper limit, like the Betz limit.

The potential energy (J) E available in the reservoir is

$$E = \rho V_w g H \tag{9.6}$$

where H is the water head in the reservoir as indicated in Fig. 9.2, and V_w is the volume of water in the reservoir. Not all the water stored in a hydro power reservoir can be used for generating electricity because some residual water must be left for flow in rivers and for environmental reasons.

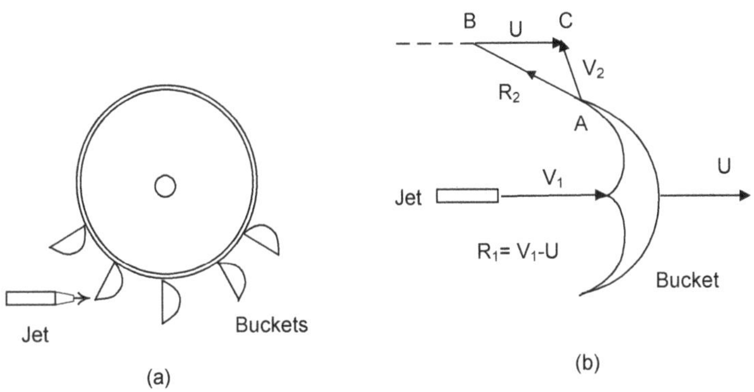

Fig. 9.3 (a) Impulse turbine – Pelton wheel, (b) Inlet and exit velocity triangles

9.4.3 *The Pelton turbine*

The Pelton turbine, shown in Fig. 9.3(a), consists of a series of buckets fitted on the periphery of a wheel. The cross-section of a bucket with the water jet striking it at the middle is depicted in Fig. 9.3(b). The plane of the section is parallel to the axis of the wheel and contains the axis of the jet.

On striking the bucket, the jet divides into two streams that flow over the bucket surface transferring a fraction of its momentum to the bucket. The change in momentum of water produces a tangential force on the bucket which provides the torque to rotate the wheel.

The absolute velocity of the jet V_1 may be expressed as

$$V_1 = C_v(2gH_n)^{1/2} \tag{9.7}$$

where the H_n is the head available at the entrance to the nozzle from penstock (see Fig. 9.2), and C_v is the loss coefficient of the nozzle.

$$H_n = H - H_{fr} \tag{9.8}$$

Friction and fitting losses in the penstock are H_{fr}, and the total gravity head between the turbine elevation and the surface of the reservoir is H.

To determine the net force on the wheel we consider a control volume surrounding the jet on one side of the bucket as show in Fig. 9.3(b). The absolute velocities of the jet at the entrance and exit are V_1 and V_2, and the corresponding velocities relative to the bucket are R_1 and R_2 respectively. The tangential velocity of the bucket at the section is U.

If the angular speed of the wheel is ω and the radius of the wheel at the section is r, then $U = \omega r$.

The velocity triangle for the jet at the exit is shown in Fig. 9.3(b). At the entrance all velocities are co-linear. From the triangle at the exit of the jet we get

$$V_2 cos\phi = R_2 cos(\pi - \theta) - U = -R_2 cos\theta - U \tag{9.9}$$

where, *angle* ABC $= \pi - \theta$, and *angle* ACB $= \pi - \phi$

The velocities of the jet at the entrance are related by

$$R_1 = V_1 - U \tag{9.10}$$

Applying the momentum theorem to the control volume, the force acting *on* the jet is obtained as

$$\overleftarrow{F}_{jet} = m(-V_2 cos\phi - V_1) \tag{9.11}$$

where m is the mass flow rate of water.

Substituting from Eqs. (9.9) and (9.10) in Eq. (9.11) we have

$$\overleftarrow{F}_{jet} = m(R_2 cos\theta + U - R_1 - U) \tag{9.12}$$

$$\overleftarrow{F}_{jet} = m(R_2 cos\theta - R_1)$$

The relative velocity of the fluid over the bucket is reduced due to friction. Therefore $R_2 = kR_1$, where k is the loss coefficient.

The force on the jet can be written as

$$\overleftarrow{F}_{jet} = m(kR_1 cos\theta - R_1) \tag{9.13}$$

By Newton's third law, the force on the bucket is in the opposite direction. Therefore

$$\vec{F}_{buc} = m(R_1 - kR_1 cos\theta) \tag{9.14}$$

Substituting from Eq. (9.10) in Eq. (9.14) we get

$$\vec{F}_{buc} = m(V_1 - U)(1 - kcos\theta) \tag{9.15}$$

The torque acting on the wheel due to the above force is

$$T = mr(V_1 - U)(1 - kcos\theta) \tag{9.16}$$

The power developed in the wheel is

$$P = mr\omega(V_1 - U)(1 - kcos\theta) \tag{9.17}$$

$$P = Q\rho U(V_1 - U)(1 - kcos\theta) \tag{9.18}$$

where Q is volume flow rate, ρ is density of fluid, and $U = \omega r$.

The speed of the jet is determined by the head and the losses, and the turbine speed U is a design variable. According to Eq. (9.18) the variation of power output with U is parabolic. The maximum power is obtained by differentiating P with respect to U. Hence, we have

$$\frac{dP}{dU} = Q\rho(1 - kcos\theta)(V_1 - 2U) = 0 \tag{9.19}$$

Therefore when the power output is a maximum, $U = \frac{1}{2}V_1$.

In practice the maximum power occurs when the turbine speed is about $0.45V_1$.

Power of the jet is

$$P_{jet} = \frac{1}{2}Q\rho V_1^{\,2} \tag{9.20}$$

Hence the efficiency of the turbine is

$$\eta_t = \frac{P}{P_{jet}} = \frac{2U(V_1-U)(1-k\cos\theta)}{V_1^{\,2}} \tag{9.21}$$

A Pelton turbine is usually mounted on the same shaft as the electrical generator that it drives. The system is operated at the condition of maximum efficiency, even when the power demand at the shaft varies. This is achieved by varying the flow rate of water through the nozzle using a spear valve, which alters the flow area while maintaining the speed of the jet.

Serious water-hammer problems could result if the flow rate through the jet is reduced suddenly in response to a change in load. This is avoided by installing a deflector plate that directs the water jet away from the bucket, while the spear valve moves to the new position.

9.4.4 *Medium-head schemes*

Medium head schemes have components like those shown in Fig. 9.2. The powerhouse is located at the base of the dam with a shorter length of penstock, and the flow rates are generally larger. The turbines used are called *reaction turbines*, a common design being the *Francis turbine*. Water from the penstock, flows through a spiral casing fitted with fixed guide vanes to direct the flow radially onto the rotor blades (see Fig. 9.5).

9.4.5 *Low-head schemes*

A low-head hydropower scheme is shown in Fig. 9.4. These usually do not include large reservoirs. The turbines used are called *axial flow turbines*, which are located at the top of a draft tube to increase the available head.

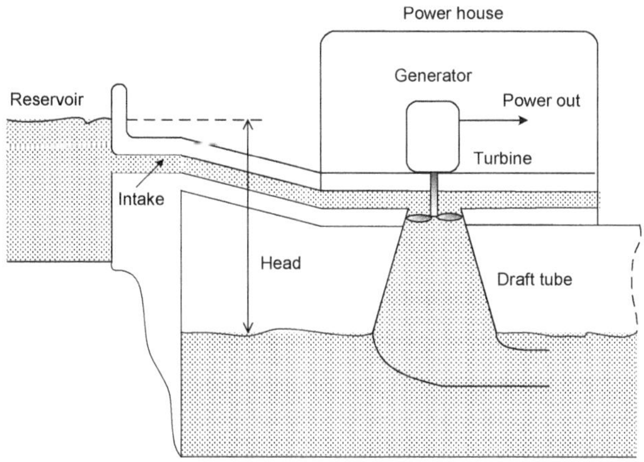

Fig. 9.4 Low-head hydropower scheme

9.4.6 *Reaction turbines*

Reaction turbines are used in medium and low head hydropower schemes. Typical designs are: the Francis wheel, the Kaplan wheel, and axial flow turbines. We shall consider the analysis of the Francis wheel in some detail.

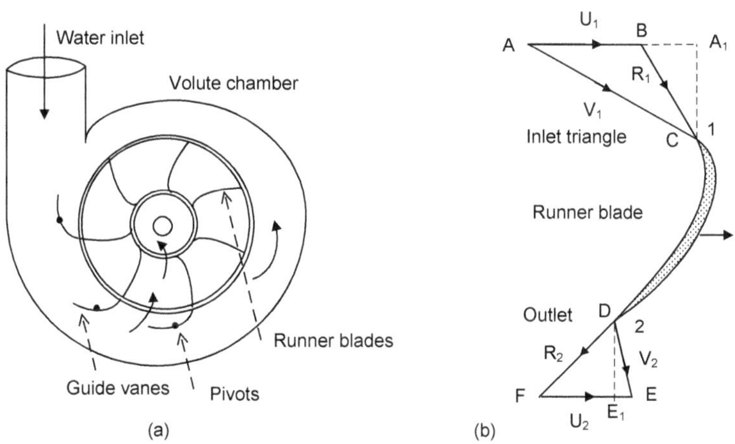

Fig. 9.5 (a) Schematic of a Francis turbine, (b) Flow over a runner blade

A schematic diagram of a Francis turbine is shown in Fig. 9.5(a). It is a reaction turbine where only a part of the available head is converted to velocity head before the turbine is reached. The runner, with moving blades fitted to its periphery, is located coaxially on a drive shaft. Guide vanes surrounding the runner are in a stationary ring called the stator. The entire assembly is housed within a volute chamber into which water flows at the periphery. The cross-sectional area of the volute chamber decreases along the flow path to keep the fluid speed constant.

Unlike a Pelton wheel, where water strikes one or two buckets, in the Francis turbine water fills all the passages in the runner. Thus, the pressure changes gradually as water passes through the runner. From the center of the runner water flows in the axial direction to the draft tube (see Fig. 9.4). The lower end of the draft tube is completely submerged below the level of water in a channel which carries used water away.

9.4.7 *Analysis of the Francis turbine*

The runner blades of the Francis turbine move only in the circumferential direction, and therefore only the components of the forces in this direction perform work. These forces are generated by the changes in momentum of the fluid in the circumferential direction, which in turn produce the torque to rotate the shaft.

We shall apply the angular momentum theorem that states the following: the torque about a fixed axis is equal to the change in angular momentum about that axis.

The velocity triangles for inlet flow and outlet flow for a typical runner blade are depicted in Fig. 9.5(b). At the inlet 1, the absolute velocity of the fluid is V_1, the blade velocity is U_1, and the fluid velocity relative to the blade is R_1. The corresponding quantities at blade exit 2 are V_2, U_2, and R_2 respectively.

Let the tangential components of the absolute fluid velocities at the inlet and the exit be V_{w1} and V_{w2} respectively. These may be obtained from the geometry of the inlet and exit velocity triangles in Fig. 9.5(b).

Applying the angular momentum theorem, we obtain the net torque T as

$$T = mV_{w1}r_1 - mV_{w2}r_2 \tag{9.22}$$

where r_1 and r_2 are the blade radii at inlet and exit, and m is the fluid flow rate. The first term on the RHS of Eq. (9.22) is the moment of momentum at 1, and the second term is its value at 2.

The power input to the shaft by the runner is given by

$$P = T\omega \tag{9.23}$$

From Eqs. (9.22) and (9.23) we get

$$P = \omega mV_{w1}r_1 - \omega mV_{w2}r_2 \tag{9.24}$$

$$P = m(V_{w1}U_1 - V_{w2}U_2) \tag{9.25}$$

The efficiency of the turbine is given by

$$\eta = \frac{P}{mgH} = \frac{m(V_{w1}U_1 - V_{w2}U_2)}{mgH} \tag{9.26}$$

where H is the head available to the turbine.

Capacity factor

The annual *capacity factor* of a hydropower system is the ratio of the actual yearly energy output to the maximum energy output based on the generator rating. It may be expressed in the form

$$ACF = \frac{Annual\ energy\ generated\ [kWh]}{Generator\ rating[kW] \times 8760} \tag{9.27}$$

The capacity factor is reduced due to the following reasons:
(i) Variability of resource, that is, the water available for power generation, (ii) breakdown of generating equipment, and (iii) lack of demand for electrical energy. The average capacity factor is between 30% to 50%.

9.5 Geothermal Energy

In this section we shall discuss the technologies used to harness geothermal energy. Geothermal energy is the thermal energy available in the uppermost crust of the earth, which is an almost inexhaustible source of energy. It is considered as a renewable source because the earth's interior will continue in a process of cooling for the indefinite future. Geothermal energy is not intermittent like solar or wind power, it has a high energy density, and is economically competitive in regions where the resource is abundant [1,5].

9.6 The Geothermal Resource

The simplified schematic diagram in Fig. 9.6 shows a typical geothermal field [1,6]. Hot molten or partially molten rock called *magma* is commonly present 24 to 40 km below the surface of the earth. However, in some locations, hot magma gets pushed up towards the earth's surface due to unusual geologic conditions. An example is an active volcano, where magma reaches the surface of the earth.

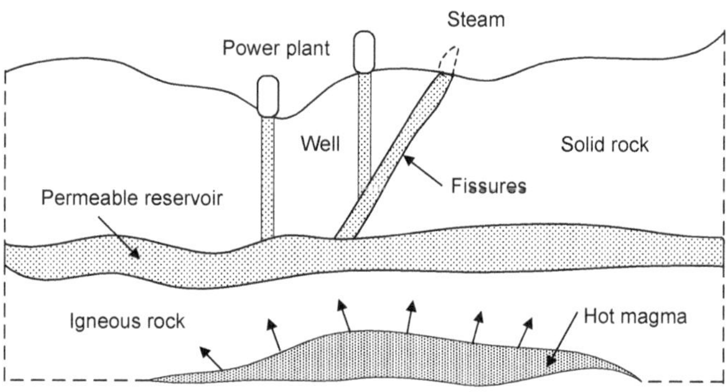

Fig. 9.6 A simplified schematic of a geothermal field

When molten magma solidifies, a layer of igneous rock is formed on the outer surface. Heat from the magma is conducted upwards to this rock layer. Ground water that finds its way to the rock through fissures in it will be heated by the rock or by hot gases emanating from the magma. The heated fluid will rise due to convection into the porous and permeable layer above the rock to form a liquid reservoir.

Hot liquid is trapped in this layer due to the presence of a solid impermeable rock layer above it. However, the solid rock has fissures that allow hot liquid to vent to the atmosphere above as geysers or hot springs. Wells drilled to the liquid reservoir or to intercept fissures are used as the heat source in geothermal power plants. The best resources for geothermal power production are called *hydrothermal convective* systems. Water in this case is heated by contact with hot rock.

In *vapor-dominated* geothermal sites water is vaporized into steam and it reaches the surface in a relatively dry condition at about 200°C. The pressure is rarely above 8 bar. Steam in this state is well suited for use in steam turbines. However, the presence of corrosive gases, and erosive materials mixed in steam is a major drawback when used directly in turbines.

In *liquid-dominated* sites hot water is trapped and circulating underground at a temperature of about 175°C to 315°C. When tapped by wells drilled at suitable locations and to the correct depths, water flows naturally to the surface, or could be drawn to the surface with the aid of pumps. When the pressure drops to about 8 bar or less, steam flashes into a two-phase mixture of low quality. Such liquid-dominated sites, are more widespread than vapor dominated sites. Hot fluid required to operate geothermal power plants or heating systems is obtained by drilling deep wells to reach these sites. Detailed presentation of these important aspects of geothermal energy utilization are available in Ref.[1,6].

9.7 Geothermal Power Generation

Geothermal power plants have many features in common with conventional Rankine-cycle thermal power systems. We analyzed the operation of such plants in chapter 5 under solar thermal power, and in chapter 8 under OTEC systems.

The hot fluid required to operate geothermal power plants are obtained directly from geothermal reservoirs in the form of steam or hot water as discussed above. No fossil fuels are used in these power plants.

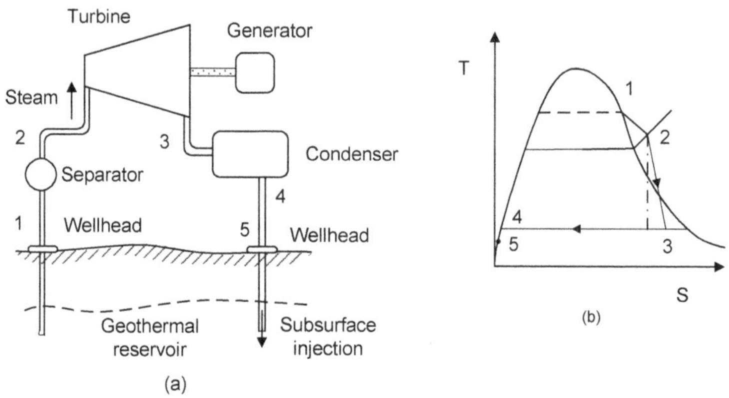

Fig. 9.7 (a) Dry steam geothermal power plant, (b) T-S diagram

9.7.1 *Dry-steam power plants*

A dry-steam power plant is shown schematically in Fig. 9.7(a). It is one of the simplest and oldest designs of geothermal plants. A simplified temperature-entropy (T-S) for the system is depicted in Fig. 9.7(b). This type of system is very efficient, but not found often because it requires a geothermal source that produces dry steam. Only about 27% of the worldwide installed capacity employ dry-steam power plants.

Steam at the bottom of the production well is nearly saturated at a high pressure of about 35 bar. As the steam flows through the well, the pressure drops and steam gets slightly superheated when it reaches the well head where the pressure is usually about 7 bar. Steam is then passed through a centrifugal separator to remove abrasive particles before it enters the turbine at 2.

After expansion through the turbine, steam enters the condenser at state 3. Unlike in a conventional Rankine cycle, condensed water is not returned to the boiler, but reinjected into a deep well after being cooled to state 5 in a cooling tower. Because the condensate is not recirculated through the system, a direct contact condenser and a cheaper dry cooling tower are commonly used in these systems.

The work output of the turbine W_{to} is given by (see Fig. 9.7b)

$$W_{to} = m_s(h_2 - h_3) \tag{9.28}$$

where m_s is the mass flow rate of steam, and h is the specific enthalpy.

The heat available in the geothermal reservoir, Q_{geo} is

$$Q_{geo} = m_s c_{pf}(T_{ext} - T_{in}) \tag{9.29}$$

where T_{ext} and T_{in} are the extraction and reinjection temperatures of the fluid and c_{pf} is the specific heat capacity of the geothermal fluid.

The gross efficiency of the system is

$$\eta_g = \frac{W_{to}}{Q_{geo}} \tag{9.30}$$

The above efficiency will be reduced due to parasitic pumping losses in the condenser cooling system.

9.7.2 *Flash-steam power plants*

A flash-steam geothermal power plant with a single flash chamber is depicted schematically in Fig. 9.8(a), and the corresponding T-S diagram is shown in Fig. 9.8(b). A double-flash system using two flash chambers in cascade is depicted in Fig. 9.9.

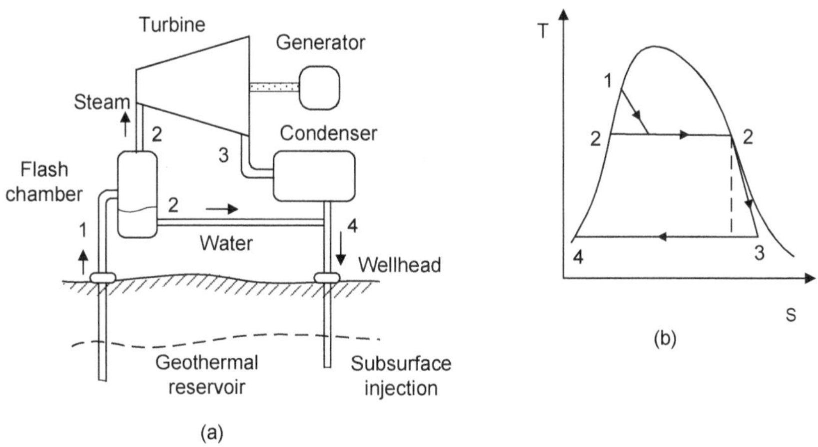

Fig. 9.8 (a) Single-flash steam power plant, (b) T-S diagram

Flash-steam power plants are used in locations with liquid-dominated geothermal sites. The worldwide installed capacity of single-flash and double-flash systems is 43% and 17% respectively. The desired water temperature for these plants is over 150°C.

Saturated liquid water reaching the wellhead at state 1 enters a flash chamber maintained at a lower pressure at state 2. A fraction of the water flashes into steam at state 2, while rest of the water is collected in the chamber from which it is discharged. Saturated steam at state 2 expands through the turbine to produce work that spins the generator. In process 3-4 exhaust steam is condensed. The water from the flash chamber and the condensate are processed and reinjected into a deep well as shown in Fig. 9.8(a).

To determine the steam flow rate to the turbine, we use a simplified equilibrium model of the flash chamber. Here the flash chamber is assumed to operate under steady adiabatic conditions (see Fig. 9.8b). Hence, we write the following conservation equations of mass and energy.

Mass balance:

$$m_1 = m_w + m_s \tag{9.31}$$

Energy balance:

$$m_1 h_1 = m_s h_s + m_w h_w \tag{9.32}$$

where m_s and m_w are the mass flow rates of steam and water, and h_s and h_w are the saturation enthalpies of vapor and liquid at state 2. The enthalpy and mass flow rate of the geothermal fluid are h_1 and m_1 respectively.

The solution of Eqs. (9.31) and (9.32) gives the steam flow rate to the turbine as

$$m_s = \frac{m_1(h_1 - h_w)}{(h_s - h_w)} \tag{9.33}$$

The work output of the turbine W_{to} is given by

$$W_{to} = m_s(h_2 - h_3) \tag{9.34}$$

The heat available from the geothermal reservoir Q_{geo} is

$$Q_{geo} = m_1 c_{pf}(T_{ext} - T_{in}) \tag{9.35}$$

where T_{ext} and T_{in} are the extraction and reinjection temperatures of the geothermal fluid, and c_{pf} is its the specific heat capacity.

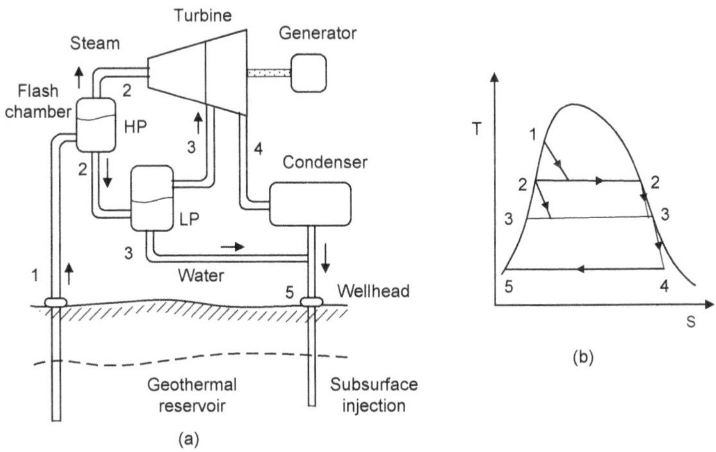

Fig. 9.9 (a) Double-flash steam power plant, (b) T-S diagram

9.7.3 *Double-flash steam power plants*

A double-flash steam geothermal power plant is depicted schematically in Fig. 9.9(a), and the T-S diagram is shown in Fig. 9.9(b). The second flash chamber operating at a lower pressure increases the mass of steam available to the turbines.

The operation of the HP flash chamber is similar to that of a single-flash system discussed above. Water leaving the HP chamber enters the LP chamber, and some water flashes into steam because of the lower pressure. The steam generated is fed to the low pressure section of the turbine as shown in Fig. 9.9(a). The expansion of this steam contributes additional power to the turbine output as seen in Fig. 9.9(b).

Equations (9.31) to (9.34) can be applied to the LP flash chamber to obtain the steam flow rate and the work output of the low-pressure turbine. The total work output of the plant is obtained by adding the outputs of the HP and LP turbines.

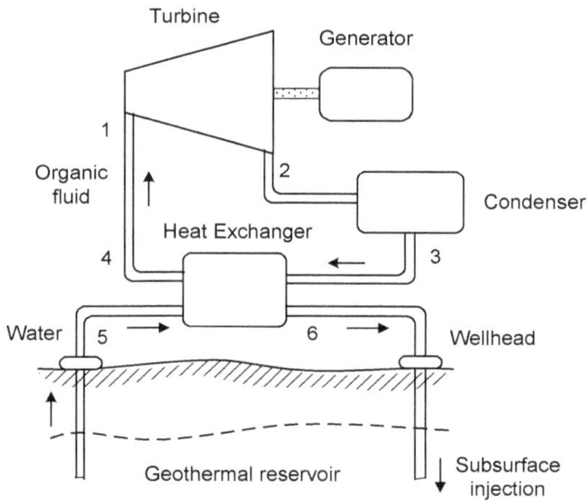

Fig. 9.10 Binary-fluid geothermal power plant

9.7.4 *Binary-fluid geothermal power plants*

We discussed binary-fluid Rankine cycle power plants in chapter 7, under OTEC energy systems. Two worked examples were also presented. Such a power plant, using a geothermal fluid as the primary fluid is shown schematically in Fig. 9.10. Binary fluid plants are used when the available geothermal temperatures are too low or the fluid quality is poor.

The boiler in a conventional Rankine power cycle is substituted with a heat exchanger where heat from the geothermal fluid is transferred to the working fluid. The working fluid is usually a low-boiling organic fluid like isobutane. The rest of the plant is a conventional Rankine power cycle.

9.8 Geothermal Heating and Cooling

Thus far we have considered power generation using geothermal energy sources. In this section heating and cooling applications of geothermal energy will be presented. There are two main types of geothermal heating applications. The older systems, known as *geothermal district heating systems* (GDHS), use geothermal fluids directly to heat commercial buildings and houses. The second system, which is a more recent application of geothermal energy, is known as a *geothermal heat pump* (GHP).

9.8.1 *Geothermal district heating*

The applications of GDHSs are limited to regions where economical geothermal sites are available. Iceland is a prime example of a country where such systems are well developed. The main applications are in space heating of houses and buildings. GDHSs may be divided into two main categories depending on whether the hot geothermal fluid is used directly in the building heating system, or indirectly by transferring geothermal heat to a secondary heating fluid using a heat exchanger.

Direct heating

If the quality of geothermal water permits, and it's below 100°C, water is passed directly through radiators or floor heating coils in buildings. Spent fluid from the radiator is discharged to waste.

Indirect heating

In indirect systems the liquid used in the district heating system is heated by the geothermal fluid in a heat exchanger. This may be necessary due to the high temperature and pressure of the geothermal fluid or its chemical characteristics. The secondary fluid, which is usually water, is distributed for district space heating. This arrangement offers more flexibility, for instance, in a situation where the geothermal system has to be shut down. An auxiliary heater could be installed to deal with such situations.

9.8.2 *Geothermal heat pumps*

Geothermal heat pumps (GHP), sometimes called ground-source, earth-coupled or water-source heat pumps, are used for heating and cooling buildings. Their operation relies on the earth as a heat reservoir whose temperature remains relatively constant despite large fluctuations in ambient air temperature.

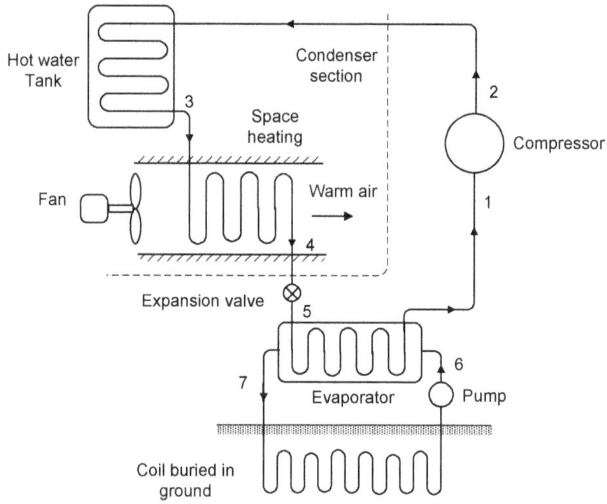

Fig. 9.11 Geothermal heat pump - heating mode

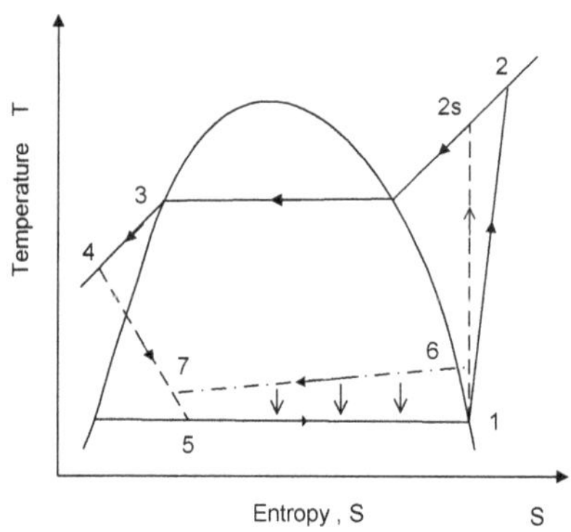

Fig. 9.12 Geothermal heat pump – T-S diagram

For cooling of buildings during summer, ordinary heat pumps, or air-source heat pumps (ASHP) as they are sometimes called, function as air conditioning systems, transferring heat from the inside of the building to the outside ambient air. The coefficient of performance (COP) of such air conditioners decreases as the ambient temperature rises.

In the winter, the same system is made to function as a heat pump, by reversing the fluid flow direction to the expansion valve, which then helps transfer heat from the ambient air to room air inside. The COP of such heat pumps decreases as the ambient temperature drops. From an energy efficiency stand point, the dependence of the COP of ASHPs on ambient temperature is a major disadvantage.

Many temperate regions experience seasonal temperature extremes, from very hot days in summer to sub-zero temperatures in winter. However, in most locations the ground temperature, a few meters below the surface, remains relatively constant. Depending on the latitude the ground temperatures range from about 7°C to 21°C. The ground is warmer than the air above in the winter and cooler than the air in the summer. GHPs take advantage of this stable ground temperature to improve the COP, and hence its energy performance.

A GHP system operating in the heating mode is depicted schematically in Fig. 9.11, and the corresponding T-S diagram is shown in Fig. 9.12. It is installed in a building where the heat pump supplies heat for domestic water heating as well as space heating. The evaporator absorbs heat from the ground-coil, which in turn, receives heat from the surrounding earth.

Ground coils are placed in vertical trenches or flat horizontal pits or a combination of the two. The design configuration, in general, depends on the available land area and the nature of the terrain. The fluid passing through the ground coil is usually a mixture of water and an antifreeze liquid.

The refrigerant flowing in the closed circuit (Fig. 9.11) is throttled from state 4 to state 5 in the expansion valve before entering the evaporator. Having absorbed heat from the surrounding earth, the warm fluid in the ground-coil enters the evaporator at 6 and leaves at 7 after transferring heat to the refrigerant. The refrigerant is compressed from state 1 to 2 before entering the heat exchanger in the hot water tank where the refrigerant condenses to state 3, thereby heating the water in the tank.

The coil in the space heating duct transfers heat from the wet vapor to the air flowing across it. The complete condenser section of the heat pump consists of the hot water tank and the space heating duct, both of which receive heat from the condensing refrigerant.

Assuming that the system is operating in a steady state and that heat losses to the ambient are negligible, we can write the following overall energy balance equations for the different components.

For the compressor:

$$W_{com} = m_r(h_2 - h_1) \tag{9.36}$$

For the hot water tank:

$$Q_{tank} = m_r(h_2 - h_3) \tag{9.37}$$

For the space heating duct:

$$Q_{air} = m_r(h_3 - h_4) \tag{9.38}$$

For the expansion valve:

$$h_4 = h_5 \tag{9.39}$$

For the evaporator:

$$Q_{evp} = m_r(h_1 - h_5) = m_{gc}c_{pg}(T_6 - T_7) \tag{9.40}$$

In Eqs. (9.36) to (9.40), Q is the heat flow rate, h is the refrigerant enthalpy, and T is the temperature. The mass flow rate and the specific heat capacity the ground-coil fluid are m_{gc} and c_{pg} respectively. The subscripts represent states of the refrigerant at the different points indicated in Figs.9.11 and 9.12.

The coefficient of performance of the heat pump is given by

$$COP_{hp} = \frac{Q_{tank}+Q_{air}}{W_{com}} \tag{9.41}$$

In addition to the above energy equations we need to write equations for the performance of individual components like the ground coil, the evaporator, the hot water tank, the space heating coil, and the compressor. These equations depend on the design of the individual component.

For the cooling mode of operation, the flow direction in the expansion valve is reversed, the space heating coil becomes the evaporator which cools the air flowing in the duct, the hot water tank and the ground-coil heat exchanger become the condenser where heat is discharged from the refrigerant. In this mode the ground coil transfers heat from the circulating fluid to the earth.

9.9 Worked Examples

Example 9.1 The head and the water flow rate of a hydropower scheme are 55 m and 4 m^3s^{-1} respectively. The frictional head loss in the penstock is 3.5 m. The efficiency of the turbine and the generator are 0.85 and 0.9 respectively. Calculate (i) power available at the reservoir elevation, (ii) head available to the turbine, (iii) the power output of the turbine, (iv) the power output of the generator, and (v) overall efficiency of the system.

Solution The power available at the reservoir level is

$$P_{res} = QH\rho g$$

$$P_{res} = 4 \times 55 \times 10^3 \times 9.81 = 2158.4 \text{ kW}$$

Head available at the turbine is

$$H_t = 55 - 3.5 = 51.5 \text{ m}$$

The power available to the turbine is

$$P_t = 4 \times 51.5 \times 10^3 \times 9.81 = 2020.86 \text{ kW}$$

Power output of the turbine is

$$P_{tur} = 2020.86 \times 0.85 = 1717.73 \text{ kW}$$

Power output of the generator is

$$P_{tur} = 1717.73 \times 0.9 = 1545.95 \text{ kW}$$

The overall efficiency is

$$\eta_{ov} = \frac{P_{tur}}{P_{res}} = \frac{1545.95}{2158.4} = 71.6\%$$

Example 9.2 The reservoir of a hydropower scheme has a capacity of 10^5 m^3. The turbine house is 40 m below the reservoir level. (a) Calculate (i) the energy available, and (ii) the power available if the water flow rate is 3.5 m^3s^{-1}. (b) If only 70% of the water in the reservoir could be used, calculate the energy available.

Solution Potential energy of water in the reservoir is

$$E_{res} = V_{res} H \rho g$$

$$E_{res} = 10^5 \times 40 \times 10^3 \times 9.81 = 392.4 \times 10^8 \text{ J}$$

The power available at the reservoir level is

$$P_r = QH\rho g$$

$$P_{res} = 3.5 \times 40 \times 10^3 \times \ 9.81 = 1373.4 \text{ kW}$$

(b) If only 70% of the water could be used for power generation, then

$$E_{res} = 0.7 \times 10^5 \times 40 \times 10^3 \times 9.81 = 274.68 \times 10^8 \text{ J}$$

Example 9.3 Obtain an expression for the maximum power output of a Pelton wheel. What is the maximum efficiency under ideal conditions?

Solution We obtained the following expression for the torque acting on the wheel in Eq. (9.16).

$$T = mr(V_1 - U)(1 - kcos\theta)$$

The power developed in the Pelton wheel is

$$P = mr\omega(V_1 - U)(1 - kcos\theta)$$

$$P = Q\rho U(V_1 - U)(1 - kcos\theta)$$

where Q is volume flow rate, ρ is density of the fluid.

The maximum power is obtained by differentiating P with respect to U. Hence we have

$$\frac{dP}{dU} = Q\rho(1 - kcos\theta)(V_1 - 2U) = 0$$

Therefore when the power output is a maximum, $U = \frac{1}{2}V_1$

Maximum power of the jet is

$$P_{max} = Q\rho\left(\frac{V_1}{2}\right)\left(V_1 - \frac{V_1}{2}\right)(1 - kcos\theta)$$

Under ideal conditions $k = 1$ and $\theta = \pi$

Substituting in the above equation we get the maximum power as

$$P_{max} = Q\rho\left(\frac{V_1}{2}\right)\left(\frac{V_1}{2}\right)(1 - \cos\pi) = \frac{1}{2}Q\rho {V_1}^2$$

$$P_{jet} = \frac{1}{2}Q\rho {V_1}^2$$

The maximum efficiency of the turbine is given by

$$\eta_t = \frac{P_{max}}{P_{jet}} = 100\%$$

With the exit angle $\theta = \pi$, the velocities at the exit of the bucket become co-linear like the inlet.

Hence the absolute flow velocity at exit becomes zero under ideal conditions. The above analysis shows that the maximum efficiency of the Pelton wheel is not subject to an upper limit like the Betz limit for wind turbines and tidal turbines.

Example 9.4 The power output and overall efficiency of a hydropower plant are 120 MW and 85% respectively. The head is 220 m and the volume of stored water is 10^8 m^3. Calculate (i) the water flow rate, (ii) the total losses in the system, (iii) total energy available, and (iv) the duration of power generation if 75% of the water could be used.

Solution Power available is given by

$$P_{av} = \frac{P_{out}}{\eta} = \frac{120}{0.85} = 141.18 \text{ MW}$$

$$P_{av} = QH\rho g$$

$$Q \times 1000 \times 220 \times 9.81 = 141.18 \times 10^6$$

Hence the water flow rate is 65.4 m^3s^{-1}

Total losses in the system are

$$P_{loss} = 141.18 - 120 = 21.18 \text{ MW}$$

The potential energy of the water in the reservoir is

$$E_{res} = VH\rho g$$

$$E_{res} = 10^8 \times 220 \times 10^3 \times 9.81 = 2158 \times 10^{11}\ \text{J}$$

Duration of generation is

$$T = \frac{0.75 \times V}{Q} = \frac{0.75 \times 10^8}{65.4} = 13.2\ \text{days}$$

Example 9.5 A hydropower scheme has a head of 750 m. The Pelton turbine used has an effective radius of 0.8 m and the jet diameter is 50 mm. Assuming ideal conditions, calculate (i) jet velocity, (ii) torque on the wheel, (iii) rotational speed, and (iv) the power developed by the turbine.

Solution We assume ideal conditions with no losses. The velocity of the jet is given by

$$V_j = (2Hg)^{1/2}$$

$$V_j = (2 \times 750 \times 9.81)^{1/2} = 121.3\ \text{ms}^{-1}$$

Volume flow rate of water is

$$Q = \pi r_j{}^2 V_j = \pi \times 0.025^2 \times 121.3 = 0.238\ \text{m}^3\text{s}^{-1}$$

For a Pelton turbine under ideal conditions, we have

$$U = \frac{V_1}{2}, \quad k = 1, \quad \theta = \pi$$

We apply these conditions to the general equations obtained.

The torque on the wheel is given by Eq. (9.16) as:

$$T = Q\rho r(V_1 - U)(1 - k\cos\theta)$$

$$T = 0.238 \times 10^3 \times 0.8\left(121.3 - \frac{121.3}{2}\right)(1 - \cos\pi)$$

$$T = 23.095\ \text{kNm}$$

The maximum power developed by the turbine is

$$P = T\omega = 23.095\omega \text{ kW}$$

The power at the jet is

$$P = QH\rho g$$

$$P = 0.238 \times 750 \times 10^3 \times 9.81 = 1751.08 \text{ kW}$$

Under ideal conditions the power output is

$$P = 23.095\omega = 1751.08 \text{ kW}$$

The rotational speed is

$$\omega = \frac{1751.08}{23.09} = 75.8 \text{ rad.s}^{-1}$$

The power output of the turbine is 1751.08 kW

Example 9.6 A hydropower scheme with a head of 800 m uses two Pelton wheels. The combined head loss due to friction in the penstock and the jet is 60 m. The diameter of each jet is 60 mm. The rotational speed of the turbine is 700 rpm and the effective bucket radius is 0.35 m. The bucket exit angle is 165°. Frictional factor for flow over the bucket is 0.94. The turbine is operating at maximum power. Calculate (i) the total power output, (ii) the rotational speed of the turbine, (iii) the runaway speed of the turbine, and (iv) the overall efficiency of the system.

Solution The head available at the turbine is

$$H_{av} = H - H_{loss} = 800 - 60 = 740 \text{ m}$$

The velocity of the jet is given by

$$V_j = (2Hg)^{1/2}$$

$$V_j = (2 \times 740 \times 9.81)^{1/2} = 120.49 \text{ ms}^{-1}$$

When the power of the turbine is a maximum

$$U = \frac{V_1}{2} = \frac{V_j}{2} = 60.25 \text{ ms}^{-1}$$

Volume flow rate of water is

$$Q = \pi r_j^2 V_j = \pi \times 0.03^2 \times 120.49 = 0.3406 \text{ m}^3\text{s}^{-1}$$

Maximum power developed by the turbine is given by Eq. (9.18) as:

$$P_{max} = Q\rho\left(\frac{V_1}{2}\right)\left(V_1 - \frac{V_1}{2}\right)(1 - k\cos\theta)$$

$$P_{max} = 0.3406 \times 10^3\left(\frac{120.49}{2}\right)\left(\frac{120.29}{2}\right)(1 - 0.94\cos 165)$$

Hence the maximum power developed is

$$P_{max} = 2.358 \text{ MW}$$

The total power output of the two Pelton wheels is 4.717 MW

The rotational speed of the turbine is given by

$$U = 60.25 = \omega \times 0.35$$

Hence the rotational speed is 172.14 rad.s^{-1}

The runaway speed of the turbine is equal to the jet speed, which is 120.49 ms^{-1}. The corresponding rotational speed is 344 rad.s^{-1}.

The total power available at the reservoir is

$$P_{res} = QH\rho g = 2 \times 0.3406 \times 800 \times 10^3 \times 9.81 = 5.346 \text{ MW}$$

The overall efficiency of the system is

$$\eta = \frac{4.717}{5.346} = 88\%$$

Example 9.7 The penstock of a hydro scheme is 160 m long and the diameter is 0.8 m. The total head is 140 m and the volume flow rate 3.2 m^{3}s^{-1}. Friction factor for the pipe is 0.032. The loss coefficients for the fittings are: intake = 0.22, valves = 0.2, bends = 0.17. Calculate (i) the head loss in the penstock, and (ii) the head available to the turbine.

Solution The flow velocity in the penstock is

$$V = \frac{Q}{A} = \frac{4Q}{\pi D^2}$$

$$V = \frac{4Q}{\pi D^2} = \frac{4\times 3.2}{\pi\times 0.8^2} = 6.37$$

The velocity head is given by

$$h_v = \frac{V^2}{2g} = \frac{6.37^2}{2\times 9.81} = 2.06$$

The head loss H_{loss} consists of pipe friction losses and fitting losses.

$$H_{loss} = \frac{fLh_v}{D} + \sum k_i h_v$$

$$H_{loss} = \frac{0.032\times 160\times h_v}{0.8} + (0.22 + 0.2 + 0.17)h_v$$

$$H_{loss} = (6.4 + 0.59) \times 2.06 = 14.4 \text{ m}$$

Head available to the turbine is

$$H_{tur} = 140 - 14.4 = 125.6 \text{ m}$$

Example 9.8 The volume flow rate through a Francis turbine 12 m^{3}s^{-1} and the net head is 140 m. The rotational speed of the turbine is 250 rpm, and the external radius of the runner is 1.2 m. Assume that the tangential speed of flow at exit is zero. Calculate (i) the velocity at inlet, (ii) the tangential flow speed at entry to the runner, (iii) the torque developed, and (iv) the power output.

Solution A schematic diagram of a Francis turbine and the velocity triangles for flow over a runner blade are shown in Fig. 9.5. The velocity is given by

$$V_j = (2gH)^{1/2}$$

$$V_j = (2 \times 9.81 \times 140)^{1/2} = 52.4 \text{ ms}^{-1}$$

The tangential velocity of the runner is

$$U_1 = \omega r_1$$

$$U_1 = 1.2 \times \frac{250\times 2\pi}{60} = 31.4 \text{ ms}^{-1}$$

Power output is given by

$$P = \rho g H Q \tag{E9.8.1}$$

$$P = 10^3 \times 9.81 \times 140 \times 12 = 16.48 \text{ MW}$$

Turbine power is given by Eq. (9.24) as:

$$P = \rho Q(V_{t1}r_1\omega - V_{t2}r_2\omega)$$

Neglecting the tangential speed at exit 2 we have

$$P = \rho Q(V_{t1}r_1\omega) \tag{E9.8.2}$$

From Eqs. (E9.8.1) and (E9.8.2) for the power output, we obtain

$$\rho g H Q = \rho Q(V_{t1}r_1\omega)$$

Hence, we have

$$V_{t1} = \frac{gH}{r_1\omega} = \frac{gH}{U_1}$$

$$V_{t1} = \frac{9.81 \times 140}{31.4} = 43.7$$

The torque is given by Eq. (9.22) as

$$T = \rho Q(V_{t1}r_1)$$

$$T = 10^3 \times 12 \times 43.7 \times 1.2 = 629.28 \text{ kNm}$$

The power output is

$$P = T\omega = 629.28 \times \frac{250 \times 2\pi}{60} = 16.47 \text{ MW}$$

Example 9.9 The following data apply to a Francis turbine: flow rate = 22 m^3s^{-1}, rotor height = 0.30 m, radius at inlet = 0.80 m, runner speed = 50 rad.s^{-1}, and stator blade angle with the radius = 72°. Calculate (i) tangential velocity at inlet, (ii) the torque on the runner, and (iii) the power output. Assume that the tangential velocity at exit is negligible.

Solution The velocity triangle ABC at inlet of the turbine is shown in Fig. 9.5(b). The radial flow velocity, V_{r1} at inlet is represented by A_1C. The flow rate is given by

$$Q = 2\pi r_1 l_1 V_{r1}$$

where l_1 and r_1 are the height the rotor and the inlet radius respectively.

Hence the radial flow velocity is

$$V_{r1} = \frac{Q}{2\pi r_1 l_1} = \frac{22}{2\pi \times 0.8 \times 0.3} = 14.59$$

Angle ACA_1 is the stator angle with the radial direction, which is equal to 72°. Hence the tangential velocity at inlet is

$$V_{t1} = V_{r1} tan(72)$$

$$V_{t1} = 14.59 tan(72) = 44.9 \text{ ms}^{-1}$$

The torque on the runner is given by Eq. (9.22) as

$$T = \rho Q (V_{t1} r_1 - V_{t2} r_2)$$

Neglecting the exit tangential velocity, we have

$$T = \rho Q V_{t1} r_1$$

$$T = 10^3 \times 22 \times 44.87 \times 0.8 = 789.7 \text{ kNm}$$

Power output is given by

$$P = T\omega = 789.7 \times 50 = 39.48 \text{ MW}$$

Example 9.10 A Francis turbine running at 250 rpm has a power output of 50 MW and the efficiency is 90%. The inlet radius of the rotor is 2.5 m and the rotor height is 1.3 m. The flow angle at inlet to the runner is 50° to the radius. Assume that the tangential velocity at exit is negligible. Calculate (i) the runner velocity at inlet, and (ii) volume flow rate of water.

Solution The runner velocity at inlet is given by

$$U_1 = \omega r_1$$

$$U_1 = 2.5 \times \left(\frac{250 \times 2 \times \pi}{60}\right) = 65.44 \text{ ms}^{-1}$$

Flow rate at the inlet is

$$Q = 2\pi r_1 l_1 V_{r1} \tag{E9.10.1}$$

From the inlet velocity triangle ABC (see Fig. 9.5), the angle ACA_1 is 50°.

Hence we get

$$V_{t1} = V_{r1} tan50 = 1.1917 V_{r1} \tag{E9.10.2}$$

From Eqs. (E9.10.1) and (E9.10.2) we have

$$Q = 2\pi r_1 l_1 V_{r1} = 2\pi r_1 l_1 \times 0.839 V_{t1} \tag{E9.10.3}$$

When the flow exits the runner radially, the power developed is obtained from Eq. (9.25) as

$$P_t = Q\rho V_{t1} U_1 \tag{E9.10.4}$$

From Eqs. (E9.10.3) and (E9.10.4) we have

$$P_t = \rho U_1 \left(\frac{Q^2}{2\pi r_1 l_1 \times 0.839}\right) \tag{E9.10.5}$$

The power developed by the turbine is

$$P_t = \frac{50 \times 10^6}{0.9} = 55.55 \times 10^6 \text{ W}$$

Substituting for P_t in Eq. (E9.10.5) we get

$$Q^2 = \frac{2\pi r_1 l_1 P_t \times 0.839}{\rho U_1}$$

$$Q^2 = \frac{2\pi \times 2.5 \times 1.3 \times 0.839 \times 55.55 \times 10^6}{10^3 \times 65.44} = 14543.4$$

Hence the flow rate is 120.6 m^3s^{-1}.

Example 9.11 The power output of a turbo-generator is 1.2 MW and the efficiency is 92%. The steady rotational speed of the shaft is

900 rpm. The moment of inertia of the turbine runner and the generator is 700 kgm^2. If the electrical load on the shaft is lost, calculate the speed of the shaft after 1.5 s.

Solution The power generated is

$$P_g = \frac{1.2\times10^6}{0.92} = 1.304 \times 10^6 \text{ W}$$

The steady torque of the turbine is given by

$$T\omega = P_g$$

$$T = \frac{1.304\times10^6\times60}{900\times2\times\pi} = 13.8 \times 10^3 \text{ Nm}$$

Under steady conditions the torque developed by the turbine is balanced by the load torque of the generator. Hence the shaft speed remains constant. When the load torque is lost the system will accelerate under the action of the turbine torque. The governing equation of motion is

$$I\frac{d\omega}{dt} = T$$

Integrating the above equation, we get

$$\omega_2 - \omega_1 = \frac{Tt}{I}$$

$$\omega_2 - \omega_1 = \frac{13.8\times10^3\times1.5}{700} = 29.57 \text{ rad.s}^{-1}$$

The increase in speed is 282.4 rpm. Hence the final speed is 1182.4 rpm.

Example 9.12 The yearly operating schedule of a hydropower scheme, rated at 50 MW, is as follows: (i) full capacity from January to March, (ii) half-capacity from April to September, and (iii) full capacity from October to December. Calculate the annual capacity factor (ACF) of the scheme. If the plant is shut down for 15 days during January and March for maintenance, calculate the new ACF.

Solution The hydro plant operation schedule is:

Full- capacity (F)	for 3 months, January, February and march
Half- capacity (0.5F)	for 6 months, April, May, June, July, August, September
Full-capacity (F)	for 3 months, October, November, December

The total yearly energy output is

$$E_{total} = \sum(N_{days}) \times capacity \times 24 \text{ MWh}$$

$E_{total} = (31 + 28 + 31)50 + (30 + 31 + 30 + 31 + 31 + 30)25 + (31 + 30 + 31)50 = 50(273.5) \times 24$ MWh

Rated yearly energy output is

$$E_{rated} = 50(365) \times 24 \text{ MWh}$$

The annual capacity factor is

$$ACF = \frac{E_{total}}{E_{rated}} = 0.749$$

The total yearly energy output with the new schedule is

$E_{total} = (16 + 28 + 16)50 + (30 + 31 + 30 + 31 + 31 + 30)25 + (31 + 30 + 31)50 = 50(243.5) \times 24$ MWh

Hence $$ACF = \frac{E_{total}}{E_{rated}} = 0.667$$

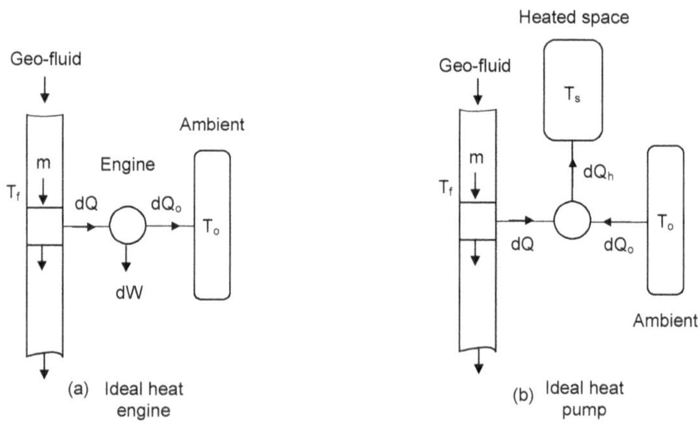

Fig. E9.13 (a) Ideal heat engine, (b) Ideal heat pump

Example 9.13 A geothermal fluid stream enters a power generation system at 95°C and leaves at 30°C. The ambient temperature is 20°C. Calculate (i) the maximum work that could be extracted from the geothermal stream using a cyclic device, and (ii) the efficiency of such a device.

Solution An ideal cyclic device extracting heat from the geothermal fluid stream and rejecting heat to a constant temperature ambient reservoir is shown schematically in Fig. E9.13(a). Because the fluid stream temperature changes with distance, we consider a series of infinitesimal Carnot heat engines, one of which is shown. The temperature of the stream at the location is T. The heat extracted by the engine is dQ, the heat rejected to the ambient is dQ_o and the work output is dW.

Energy balance for the stream gives

$$dQ = -mc_p dT \quad \text{(E9.13.1)}$$

Energy balance for the Carnot heat engine gives

$$dQ = dW + dQ_o \quad \text{(E9.13.2)}$$

Entropy balance for the reversible engine is

$$\frac{dQ}{T} = \frac{dQ_o}{T_o} \quad \text{(E9.13.3)}$$

From Eqs. (E9.13.1) to (E9.13.3) we have

$$dW = -mc_p\left(1 - \frac{T_o}{T}\right)dT \quad \text{(E9.13.4)}$$

Integrating Eq. (E9.13.4) we obtain the work output rate as

$$W_o = mc_p(T_{in} - T_{out}) - mc_p T_o ln\left(\frac{T_{in}}{T_{out}}\right)$$

where T_{in} and T_{out} are the inlet and outlet fluid temperatures.

The efficiency of the device is given by

$$\eta = \frac{W_o}{mc_p(T_{in} - T_{out})}$$

Substituting numerical data in the above equations we get

$$\frac{W_o}{m} = 4.2 \times (368 - 303) - 4.2 \times 293 ln\left(\frac{368}{303}\right) = 33.83 \text{ kJkg}^{-1}$$

The ideal efficiency is

$$\eta = \frac{33.83}{4.2(368-303)} = 12.4\%$$

The Carnot efficiency based on the mean fluid temperature and the ambient temperature is 12.7%.

Example 9.14 A dry steam plant of a vapor dominated geothermal power system has an output of 225 MW. The steam pressure at the well head is 30 bar. Steam is throttled to 10 bar before entering the turbine. The direct contact condenser is operated at 0.3bar. The temperature of the exit fluid stream from the plant is 25°C. The isentropic efficiency of the turbine is 0.85 and the efficiency of the generator is 0.9. Calculate (i) turbine work output, (ii) the quality of steam at outlet, (iii) the steam flow rate, and (iv) the overall efficiency.

Solution A schematic diagram of the dry steam plant and a T-S diagram are shown in Figs. 9.7 (a) and (b) respectively. Following data are obtained from the steam table [3].

At 30 bar

$$h_1 = h_{g1} = 2803 \text{ kJkg}^{-1}$$

For the throttling process 1-2

$$h_2 = h_1 = 2803$$

At state 2 the entropy is obtained by interpolation. The relevant superheated steam data at 10 bar are:

h_g	s_g
2778	6.586
2829	6.695

Hence by interpolation when, $h_2 = 2803$ $s_2 = 6.639$

For the isentropic expansion 2-3s

$$s_{3s} = s_2 = 6.639 \text{ kJkg}^{-1}\text{K}^{-1}$$

Steam quality at state 3s is obtained by substituting data at 0.3 bar.

$$6.639 = 7.767x_{3s} + 0.944 \times (1 - x_{3s})$$

$$x_{3s} = 0.834$$

$$h_{3s} = 2625x_{3s} + 289 \times (1 - x_{3s})$$

$$h_{3s} = 2625 \times 0.834 + 289 \times (1 - 0.834) = 2237.2$$

Isentropic work output is

$$w_{34s} = 2803 - 2237.2 = 565.8 \text{ kJkg}^{-1}$$

The actual turbine work output is

$$w_{34} = 0.85 \times 565.8 = 480.9 \text{ kJkg}^{-1}$$

Actual enthalpy at state 3 is

$$h_3 = 2803 - 480.9 = 2322.1 \text{ kJkg}^{-1}$$

Steam quality at state 3 is given by

$$2322.1 = 2625 \times x_3 + 289(1 - x_3)$$

$$x_3 = 0.87$$

The power output is given by

$$m_s \times 480.9 \times 0.9 = 225 \times 10^3$$

Hence the steam flow rate is 520 kgs^{-1}.

The energy in the geothermal fluid stream is

$$Q_{gs} = \left(h_{s1} - h_{f5}\right) \text{ kJkg}^{-1}$$

$$Q_{gs} = (2803 - 104.8) = 2698.2 \text{ kJkg}^{-1}$$

Hence the overall system efficiency is

$$\eta_0 = \frac{w_{34}}{Q_{gs}} = \frac{480.9}{2698.2} = 17.8\%$$

Example 9.15 A geothermal power plant has a single-flash chamber operating at 2.45 bar. The fluid enthalpy at inlet is 966 kJkg^{-1} and the flow rate is 150 kgs^{-1}. The condenser pressure is 0.108 bar. The isentropic efficiency of the turbine is 0.8 and the efficiency of the generator is 0.85. Calculate (i) steam flow rate to the turbine, and (ii) the power output.

Solution A single-flash geothermal power plant and the corresponding T-S diagram are shown in Figs. 9.8 (a) and (b). We apply the conservation equations of mass and energy to the flash chamber, assuming steady adiabatic conditions.

Mass balance:

$$m_1 = m_f + m_s \tag{E9.15.1}$$

Energy balance:

$$m_1 h_1 = m_s h_s + m_f h_f \tag{E9.15.2}$$

where m_s and m_w are the mass flow rates of steam and water, and h_s and h_w are the saturation enthalpies of vapor and liquid at state 2. The mass flow rate and enthalpy of the geothermal fluid are m_1 and h_1 respectively.

The relevant properties of steam are obtained from the steam tables [3].

At 2.45 bar: $h_f = 532.5$, $h_{fg} = 2183.5$, $h_g = 2716$ kJkg^{-1}

$s_g = 7.06$ kJkg^{-1}K^{-1}

From Eqs. (E9.15.1) and (E9.15.2) we get

$$m_s = 150\left(\frac{h_1 - h_f}{h_g - h_f}\right) = 150\left(\frac{966 - 532.5}{2716 - 532.5}\right) = 29.78 \text{ kgs}^{-1}$$

The steam flow rate is 29.78 kgs^{-1}.

At 0.108 bar $\quad h_f = 198, \quad h_g = 2586$ kJkg^{-1}

$$s_g = 8.123, \quad s_f = 0.668 \text{ kJkg}^{-1}\text{K}^{-1}$$

For the isentropic expansion 2-3 in thc turbinc $\quad s_2 = s_3$

Therefore

$$7.06 = (1 - x_3) \times 0.668 + 8.123 \times x_3$$

Hence $x_3 = 0.857$

$$h_3 = (1 - x_3) \times 198 + 2586 \times x_3$$

$$h_3 = (1 - 0.857) \times 198 + 2586 \times 0.857 = 2244.5$$

The isentropic work output is

$$w_{23s} = h_2 - h_3 = 2716 - 2244.5 = 471.5 \text{ kJkg}^{-1}$$

Actual work output is

$$w_{23a} = 0.8 \times 471.5 = 377.2 \text{ kJkg}^{-1}$$

Power output is

$$W_o = 29.78 \times 0.85 \times 377.2 = 9.5\ 4 \text{ MW}$$

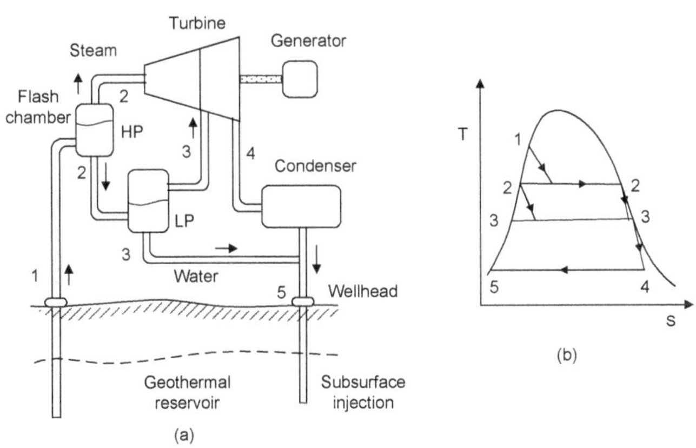

Fig. E9.16.1 (a) Double-flash steam plant, (b) T-S diagram

Example 9.16 A geothermal power plant has two flash chambers operating at 4.21 bar and 0.93 bar. The condenser pressure is 0.045 bar. The geothermal fluid enters with an enthalpy of 900 kJkg^{-1} and a flow rate of 227 kgs^{-1}. The isentropic efficiencies of the high pressure and low pressure turbines are 0.8, and the efficiency of the generator is 0.85. Calculate (i) steam flow rates from the two flash chambers, and (ii) the total power output.

Solution A double-flash geothermal power plant and the corresponding T-S diagram are shown in Figs. E9.16.1(a) and (b). We obtain the following data from the steam tables [3].

At 4.21 bar: $h_f = 612.56$, $h_g = 2741$ kJkg^{-1}

$s_g = 6.88$ kJkg^{-1}K^{-1}

From Eqs. (E9.15.1) and (E9.15.2) we get

$$m_{s1} = 227 \times \left(\frac{h_1 - h_f}{h_g - h_f}\right) \tag{E9.16.1}$$

$$m_{s1} = 227 \times \left(\frac{900 - 612.56}{2741 - 612.56}\right) = 30.65 \text{ kgs}^{-1}$$

The steam generated in the HP flash chamber is 30.65 kgs^{-1}.

Water flow rate into the LP flash chamber is

$$m_{w2} = 227 - 30.65 = 196.35 \text{ kgs}^{-1}.$$

At 0.93 bar: $h_f = 408.6$, $h_g = 2672.2$ kJkg^{-1}

$s_f = 1.280$, $s_g = 7.383$ kJkg^{-1}K^{-1}

Applying Eq. (E9.16.1) to the LP flash chamber we get the steam flow rate as

$$m_{s2} = 196.35 \times \left(\frac{h_{f1} - h_{f3}}{h_{g3} - h_{f3}}\right)$$

$$m_{s2} = 196.35 \times \left(\frac{612.56 - 408.6}{2672.2 - 408.6}\right) = 17.7 \text{ kgs}^{-1}$$

For the isentropic expansion 2-3 in the HP turbine, $s_2 = s_3$

$$6.88 = (1 - x_{3s}) \times 1.28 + 7.383 \times x_{3s}$$

Hence $$x_{3s} = 0.9176$$

$$h_{3s} = (1 - x_{3s}) \times 408.6 + 2672.2 \times x_{3s}$$

$$h_{3s} = (1 - 0.9176) \times 408.6 + 2672.2 \times 0.9176 = 2485.67$$

The isentropic enthalpy change is

$$\Delta h_{23s} = h_{2s} - h_{3s} = 2741 - 2485.67 = 255.43 \text{ kJkg}^{-1}$$

Actual enthalpy change is

$$\Delta h_{23a} = 0.8 \times 255.43 = 204.3 \text{ kJkg}^{-1}$$

Hence the actual work in the HP turbine is

$$W_{23a} = 204.3 \text{ kJkg}^{-1}.$$

Enthalpy at 3 is

$$h_3 = 2485.67 - 204.3 = 2281.37$$

Before entry to the LP turbine the expanded steam from the HP turbine is mixed with saturated steam from the LP flash chamber. Assuming adiabatic mixing we have

$$(m_{s1} + m_{s2})h_{3m} = m_{s1}h_3 + m_{s2}h_{g3}$$

$$48.35h_{3m} = 2281.37 \times 30.65 + 2672.2 \times 17.7$$

Hence $$h_{3m} = 2424.45 \text{ kJkg}^{-1}$$

The steam quality after mixing is given by

$$h_{3m} = 2424.45 = 2672 \times x_{3m} + 408.6 \times (1 - x_{3m})$$

Hence the quality after mixing is

$$x_{3m} = 0.8906$$

Entropy after mixing is

$$s_{3m} = 7.383 \times x_{3m} + 1.28 \times (1 - x_{3m}) = 6.715$$

At 0.045 bar: $h_f = 130.0,$ $h_g = 2558.0$ kJkg^{-1}

$s_f = 0.451,$ $s_g = 8.431$ kJkg^{-1}K^{-1}

For the isentropic expansion of the steam mixture in the LP turbine we have

$$6.715 = (1 - x_{4s}) \times 0.451 + 8.431 \times x_{4s}$$

Therefore $x_{4s} = 0.785$

$$h_{4s} = (1 - 0.785) \times 130 + 0.785 \times 2558 = 2035.98$$

Isentropic work in the LP turbine is

$$W_{s34} = h_{3m} - h_{4s} \text{ kJkg}^{-1}$$

$$W_{s34} = 2424.45 - 2035.98 = 388.5 \text{ kJkg}^{-1}$$

Actual work in the LP turbine is

$$W_{a34} = 388.5 \times 0.8 = 310.8 \text{ kJkg}^{-1}$$

The total work output of the HP and LP turbines is

$$W_{tot} = W_{23a}m_{s1} + W_{34a}(m_{s1} + m_{s2})$$

$$W_{tot} = 204.3 \times 30.65 + 310.8 \times (30.65 + 17.7) = 21.2 \text{ MW}$$

The total electrical power output is

$$E_{tot} = 21.28 \times 0.85 = 18.09 \text{ MW}$$

Example 9.17 A geothermal fluid stream enters a heat pump system at 90°C and exits at 20°C. An ideal absorption heat pump operates between the fluid stream and the atmosphere at 10°C and transfers heat to a space maintained at 25°C. Calculate the maximum coefficient of performance the heat pump.

Solution Depicted schematically in Fig. E9.13(b) is an ideal cyclic device extracting heat from a fluid stream, and a constant temperature ambient at T_o, and transferring it to a constant temperature space at T_S. Because the fluid stream temperature changes with distance, we consider a series of infinitesimal heat pumps, one of which is shown in the figure.

The temperature of the stream at the location is T. The heat extracted from the fluid stream is dQ, and from the ambient reservoir is dQ_o. The heat transferred to the space is dQ_S.

Energy balance for the stream gives

$$dQ = -mc_p dT \tag{E9.17.1}$$

Energy balance for the heat pump gives

$$dQ_s = dQ + dQ_o \tag{E9.17.2}$$

Since the ideal heat pump is a reversible cyclic device, its entropy remains constant under steady conditions. Entropy changes occur only in the reservoirs due to heat transfer. Therefore entropy balance gives

$$\frac{dQ_S}{T_S} - \frac{dQ}{T} - \frac{dQ_O}{T_O} = 0 \tag{E9.17.3}$$

From Eqs. (E9.17.1) to (E9.17.3) we get

$$dQ_S\left(\frac{1}{T_S} - \frac{1}{T_O}\right) = mc_p\left(\frac{1}{T_O} - \frac{1}{T}\right)dT \tag{E9.17.4}$$

Integrating Eq. (E9.17.4) we obtain the heat input to the space as

$$Q_S = \frac{\left[mc_p(T_{in}-T_{out}) - mc_pT_O ln\left(\frac{T_{in}}{T_{out}}\right)\right]}{\left(1-\frac{T_O}{T_S}\right)}$$

The total heat extracted from the fluid stream is

$$Q = mc_p(T_{in} - T_{out})$$

where T_{in} and T_{out} are the inlet and outlet fluid temperatures.

Hence the COP of the heat pump is

$$COP_{hp} = \frac{Q_S}{Q} = \frac{\left[mc_p(T_{in}-T_{out}) - mc_pT_O ln\left(\frac{T_{in}}{T_{out}}\right)\right]}{mc_p(T_{in}-T_{out})\left(1-\frac{T_O}{T_S}\right)}$$

$$COP_{hp} = \frac{\left[(T_{in}-T_{out})-T_o ln\left(\frac{T_{in}}{T_{out}}\right)\right]}{(T_{in}-T_{out})\left(1-\frac{T_o}{T_s}\right)}$$

Substituting numerical data in the above equation we get the maximum coefficient performance as

$$COP_{hp} = \frac{\left[(363-293)-283 ln\left(\frac{363}{293}\right)\right]}{(363-293)\left(1-\frac{283}{298}\right)} = 2.66$$

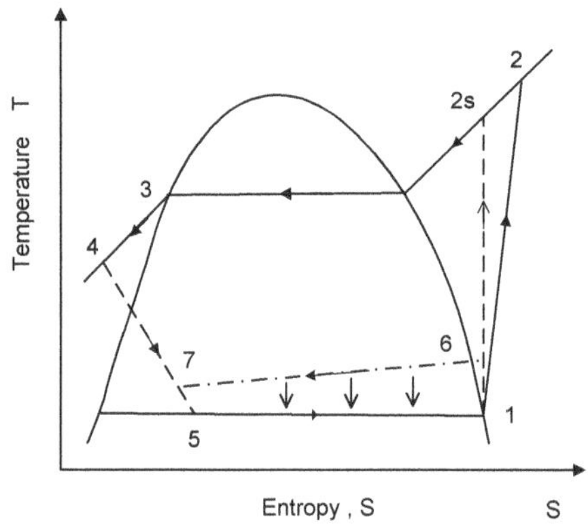

Fig. E9.18.1 Temperature-entropy diagram for heat pump in Fig. 9.11

Example 9.18 A geothermal heat pump using R-134a as the refrigerant is shown schematically in Fig. 9.11. The T-S diagram is shown in Fig. E9.18.1. Vapor enters the compressor at 1 as a dry saturated vapor at 5°C, and exits at 2 as a superheated vapor. The condensing temperature is 60°C. From 2 to 3 heat is transferred from the refrigerant to the hot water tank until it is a saturated liquid at 60°C.

From 3 to 4 liquid refrigerant transfers heat to air in the space heating duct until its temperature is 20°C. Sub-cooled liquid refrigerant undergoes throttling from 4 to 5 in the expansion valve. Heat is transferred from the ground-coil to the refrigerant during the process 5-1. The power input to the compressor is 600W and its isentropic efficiency is 0.8. Calculate (i) the refrigerant flow rate, (ii) the heat input to the water tank, (iii) the heat transferred to air in the duct, and (iv) the COP of the heat pump.

Solution The T-S diagram of the heat pump is shown in Fig. E9.18.1, where the states of the refrigerant at different points in Fig. 9.11 are indicated. We obtain the following properties of R134a from data tables [3].

At state 1: $h_{g1} = 401.3$ kJkg^{-1}, $s_{g1} = 1.7238$ kJkg^{-1}K^{-1}

For isentropic compression from 1 to 2s (see Fig. E9.18.1)

$$s_1 = s_2 = 1.7238$$

The pressure at 2 is the saturation pressure at 60°C which is 16.812 bar. Vapor at state 2s is superheated at 16.812 bar. Hence by interpolation we obtain

$$h_{2s} = 433.87$$

The isentropic enthalpy change across the compressor from 1 to 2s is

$$\Delta h_{s12} = 433.87 - 401.33 = 32.54$$

The actual enthalpy change is

$$\Delta h_{a12} = \frac{\Delta h_{s12}}{\eta_{com}} = \frac{32.54}{0.8} = 40.67 \text{ kJkg}^{-1}$$

The enthalpy at state 2 is

$$h_2 = 401.33 + 40.67 = 442.0$$

The work input to the compressor is given by

$$w_{12} = h_2 - h_1 = 442.0 - 401.33 = 40.6 \text{ kJkg}^{-1}$$

Enthalpy at state 3 is

$$h_3 = 287.51$$

Heat input to the water tank is

$$q_{23} = 442.0 - 287.51 = 154.5 \text{ kJkg}^{-1}$$

The liquid at 4 is subcooled. We ignore the effect of pressure and take the enthalpy of saturated liquid at 20°C. Therefore $h_4 = 227.45$.

The heat input to the air duct is

$$q_{34} = h_3 - h_4$$

$$q_{34} = 287.51 - 227.45 = 60.1 \text{ kJkg}^{-1}$$

The refrigerant flow rate is

$$m_{ref} = \frac{W_{com}}{w_{12}} = \frac{0.6}{40.67} = 0.01475 \text{ kgs}^{-1}$$

The heat input to the water tank is

$$Q_{wh} = 0.01475 \times 154.5 = 2278.87 \text{ W}$$

The heat input to the air duct is

$$Q_{ah} = 0.01475 \times 60.1 = 886.47 \text{ W}$$

The COP of the heat pump is

$$COP_{hp} = \frac{Q_{wh}+Q_{ah}}{W_{com}} = \frac{2278.87+886.47}{600} = 5.28$$

Example 9.19 In the geothermal heat pump described in Example 9.18, the fluid in the ground-coil transfers heat to the refrigerant in the evaporator as shown in Fig. 9.11. Its temperature drops from 14°C to 8°C. The specific heat capacity of the fluid 3.9 kJkg^{-1}K^{-1}. Calculate (i) the mass flow rate of the fluid, and (ii) the UA – product of the heat transfer coil.

Solution The heat input in the evaporator is given by (see Fig. E9.18.1)

$$Q_{evp} = m_{ref}(h_1 - h_5)$$

For the throttling process 4-5

$$h_5 = h_4 = 227.45$$

Hence we have

$$Q_{evp} = 0.01475 \times (401.33 - 227.45) = 2565 \text{ W}$$

The heat flow rate from the ground-coil to the refrigerant is equal to 2565W. From energy balance we obtain

$$Q_{gc} = m_{gc}c_{pf}\left(T_{fi} - T_{fo}\right)$$

where m_{gc} is the fluid flow rate through the ground coil and c_{pf} is specific heat capacity. Substituting numerical values in the above equation we get

$$2565 \times 10^{-3} = m_{gc} \times 3.9 \times (14 - 8)$$

The fluid mass flow rate in the ground-coil is 0.1096 kgs^{-1}.

A commonly used parameter in heat exchanger design is the (UA) - product, where U is the overall heat transfer coefficient, and A is the total heat transfer area.

In the present problem the liquid in the ground-coil is the hot fluid and the refrigerant is the cold fluid. Using the log-mean-temperature-difference (LMTD) design method we can write the following equation for the overall heat transfer.

$$Q_{gc} = UA\Delta T_{lm} \tag{E9.19.1}$$

The log-mean-temperature difference ΔT_{lm} in Eq. (E9.19.1) is given by

$$\Delta T_{lm} = \frac{(T_h-T_c)_{out}-(T_h-T_c)_{in}}{ln[(T_h-T_c)_{out}/(T_h-T_c)_{in}]} \tag{E9.19.2}$$

where T_h and T_c are the hot and cold fluid temperatures respectively. The subscripts *in* and *out* represent the inlet and outlet of the heat exchanger. Substituting numerical data in Eq. (E9.19.2) we get

$$\Delta T_{lm} = \frac{(14-5)-(8-5)}{ln\left(\frac{14-5}{8-5}\right)} = \frac{6}{ln(3)} = 5.46$$

Substituting in Eq. (E9.19.1) we have

$$2565 = 5.46 \times UA$$

Hence the (UA) – product is

$$(UA) = 469.7 \text{ WK}^{-1}.$$

A separate heat transfer analysis, based on an assumed tube diameter and the given fluid flow rate, has to be performed to determine the overall heat transfer coefficient, U. Hence the heat transfer area and the length of the required coil can be determined.

Example 9.20 A geothermal district heating systems (GDHS) uses water from a well at 80°C. The exit temperature of water is 40°C and the total water low rate is 32 litre.s^{-1}. The efficiency of the GDHS is 94% due to heat losses. Calculate (i) the total useful heat supplied during a day. (ii) If an oil heater with an efficiency of 75% is used, calculate daily fuel consumption rate. The heating value of the oil is 30 MJkg^{-1}.

Solution Heat supply rate by the geothermal water is given by

$$Q_{geo} = m_g c_g (T_{gi} - T_{go})$$

where m_g is the water flow rate, c_g specific heat capacity, T_{gi} inlet temperature and T_{go} is the outlet temperature. Substituting the given data in the above equation we get

$$Q_{geo} = 32 \times 4.19 \times (80 - 40) = 5363.2 \text{ kW}$$

The rate of useful heat supply is

$$Q_{use} = 0.94 \times 5363.2 = 5041.4 \text{ kW}$$

The total useful heat supplied during a day is

$$Q_{u,day} = 5041.4 \times 3600 \times 24 = 435.576 \text{ GJ}$$

The oil heater must deliver the above quantity of useful heat per day. Hence the required output the heater per day is

$$Q_{h,day} = \frac{Q_{u,day}}{\eta_h}$$

where η_h is the efficiency of the heater.

$$Q_{h,day} = \frac{435.576}{0.75} = 580.77 \text{ GJ}$$

The daily fuel consumption $m_{f,d}$ is given by

$$m_{f,d} = \frac{Q_{h,day}}{HVF}$$

where HVF is the heating value of the fuel.

$$m_{f,d} = \frac{580.77 \times 10^3}{30} = 19.36 \times 10^3 \text{ kg per day}.$$

Problems

P9.1 A hydropower scheme has a reservoir with a capacity of 2×10^5 m^3. The location of the turbine house is 60 m below the reservoir level. (a) Calculate (i) energy available, and (ii) the power available if the flow rate is 2.2 m^3s^{-1}. (b) If only 75% of the water in the reservoir could be used, calculate the available energy.
[*Answers*: (a) (i) 117.72 GJ, (ii) 1294.7 kW, (b) 88.29 GJ]

P9.2 The head of a hydropower scheme is 70 m and the water flow rate is 4.8 m^3s^{-1}. The penstock head loss due to friction is 4 m. The turbine efficiency is 0.85 and the generator efficiency is 0.9. Calculate (i) power available at the reservoir, (ii) head available to the turbine, (iii) the power output of the turbine, (iv) the power output of the generator, and (v) overall efficiency of the system. [*Answers*: (i) 3296.1 kW, (ii) 66 m, (iii) 2641.6 kW, (iv) 2377.4 kW, (v) 72%]

P9.3 The overall efficiency of a hydropower plant is 82% and the power output is 110 MW. The head is 200 m and the volume of stored water is 10^7 m^3. Calculate (i) the water flow rate, (ii) total power losses in the system, (iii) total energy available, and (iv) the duration of power generation if 70% of the water could be used.
[*Answers*: (i) 68.37 m^3s^{-1}, (ii) 24.15 MW, (iii) 19620 GJ, (iv) 28.4 h]

P9.4 A hydropower scheme uses a Pelton turbine with an effective radius of 0.9 m. The head at the turbine is 800 m and the jet diameter is 50 mm. Assuming ideal conditions, calculate (i) the jet velocity,

(ii) torque on the wheel, (iii) the rotational speed, and (iv) the power developed by the turbine.
[*Answers*: (i) 125.28 ms^{-1}, (ii) 27.73 kNm, (iii) 664.6 rpm, (iv) 1918.9 kW]

P9.5 A hydropower scheme using two Pelton turbines has a head of 900 m. The combined head loss due to friction in the penstock and the jet is 65 m. The diameter of each jet is 55 mm. The rotational speed of the turbine is 800 rpm and the effective bucket radius is 0.4 m. The bucket exit angle is 160°, and the friction factor for flow over the bucket is 0.95. The turbine is operating at maximum power. Calculate (i) the total power output, (ii) the rotational speed of a turbine, (iii) the runaway speed of a turbine, and (iv) the overall efficiency of the system.
[*Answers*: (i) 4.713 MW, (ii) 1527.9 rpm, (iii) 3055.8 rpm, (iv) 87.8%]

P9.6 A Francis turbine has a power output of 55 MW, a rotational speed of 225 rpm, and an efficiency of 90%. The inlet radius of the rotor is 2.5 m and the rotor height is 1.4 m. The flow angle at inlet to the runner is 50° to the radius. The tangential velocity at exit is negligible. Calculate (i) the runner velocity at inlet, and (ii) the volume flow rate of water. [*Answers*: (i) 58.9 ms^{-1}, (ii) 138.37 m^3s^{-1}]

P9.7 The power output of a turbo-generator of a hydropower scheme is 1.5 MW and the efficiency is 94%. The steady rotational speed is 950 rpm. The moment of inertia of the turbine and the generator is 750 kgm^2. (i) Calculate the steady torque produced by the turbine. (ii) If the electrical load on the turbine is lost, calculate the speed of the shaft after 2.0s. [*Answers*: (i) 16.04 kNm, (ii) 1358.4 rpm]

P9.8 The turbine of a dry-steam geothermal power plant receives saturated steam at 12 bar. The power output of the plant 250 MW. The direct contact condenser operates at 0.3 bar. The temperature of the exit fluid stream from the plant is 28°C. The isentropic efficiency of the turbine is 0.85 and the efficiency of the generator is 0.9. Calculate (i) the steam flow rate, and (ii) the overall efficiency.
[*Answers*: (i) 593.4 kgs^{-1}, (ii) 15.8%]

P9.9 A geothermal power plant has a single flash chamber operating at 4.5 bar. The fluid enthalpy at inlet is 992 $kJkg^{-1}$ and the flow rate is 810 kgs^{-1}. The condenser pressure is 0.14 bar. The isentropic efficiency of the turbine is 0.8 and the efficiency of the generator is 0.9. Calculate (i) steam flow rate to the turbine, and (ii) the power output.
[*Answers*: (i) 140.9 kgs^{-1}, (ii) 53.8 MW]

P9.10 Water from a geothermal well enters a district heating system (GDHS) at a temperature of 75°C, and a flow rate of 36 litre s^{-1}. The exit temperature of the water is 35°C. The efficiency of the GDHS is 92% due to heat losses. The specific heat capacity of the fluid is 4.16 $kJkg^{-1}K^{-1}$. Calculate (i) the total useful heat supplied during a day. (ii) If an oil heater with an efficiency of 78% is used, calculate daily fuel consumption rate. The heating value of the oil is 30 $MJkg^{-1}$.
[*Answers*: (i) 476.16 GJ per day, (ii) 20.34×10^3 kg per day]

References

1. DiPippo, Ronald, *Geothermal Power Plants*, Butterworth-Heinemann Elsevier, U.K., 2012.
2. Jenkins, Nick., and Ekanayake, Janaka., *Renewable Energy Engineering*, Cambridge University Press, Cambridge, U.K. 2017.
3. Rogers, G.F.C, and Mayhew, Y.R., *Thermodynamic and Transport Properties of fluids*, Blackwll, Oxford, U.K. 1995.
4. Shepherd, W and Shepherd, D.W., *Energy Studies,* Imperial College Press, London, U.K. 2019.
5. 'Hydropower' http://en.m.wikipedia.org/wiki/Hyropower.
6. 'Geothermalpower' http://en.m.wikipedia.org/wiki/Geothermal _power.

Chapter 10

Energy from Biomass

10.1 Introduction

Biomass is organic matter produced by plants, growing both on land and water. It includes forest crops and their residues as well as crops grown for the purpose of extracting energy from them. In industrialized countries these raw materials are used to produce transport fuels such as ethanol and biodiesel. In some developing countries, dung produced by plant-eating animals is dried and used for heating and cooking.

Unlike coal, oil, and natural gas, which take millions of years to form, biomass sources mentioned above are generally renewed in a much shorter time, typically in months and years. Therefore, they can be classified as renewable energy sources. Biomass energy is essentially stored solar energy, produced by a process known as *photosynthesis*.

In this chapter we shall consider the following aspects of biomass energy utilization: (i) the origin of the resource, (ii) the main sources of bioenergy, (iii) the methods for converting bioenergy to useful energy forms on a large scale, (iv) the processes and equipment for the manufacture of biofuels, (v) combustion of biofuels, and (vi) environmental effects.

10.2 Photosynthesis

The origin of all biomass energy is the *photosynthesis* process. However, conversion of energy in plant matter to useable forms of fuel involve several additional steps [2,4].

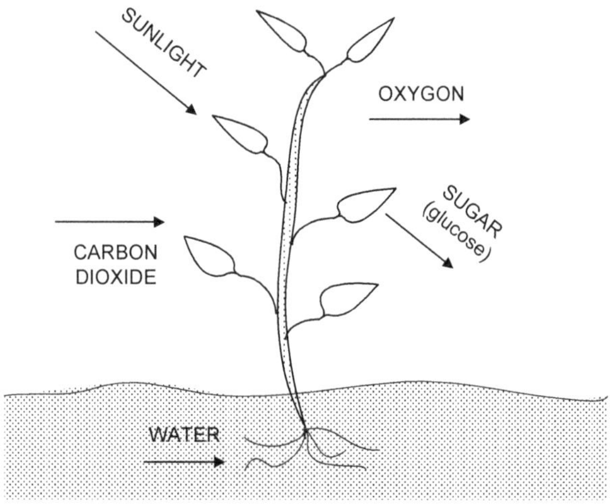

Fig. 10.1 The carbon dioxide and oxygen cycle in the atmosphere

In *photosynthesis*, green plants containing chlorophyll use solar radiation incident on the plant together with carbon dioxide in the atmosphere and water to produce carbohydrates. The process is depicted in the simplified schematic diagram in Fig. 10.1. Oxygen is created as a byproduct.

In a reverse process known as *respiration,* carbohydrates are broken down into carbon dioxide and water with the release of low grade heat. Under ideal conditions, the rate of photosynthesis in green plants is about 30 times faster than respiration. The latter two processes constitute a cycle which maintains the atmospheric balance between oxygen and carbon dioxide, and at the same time capture and store solar energy in plants.

The general chemical equation representing photosynthesis may be written as:

$$CO_2 + H_2O + solar\ energy \rightarrow [CH_2O] + O_2 \qquad (10.1)$$

where $[CH_2O]$ is a general carbohydrate.

A more specific form of Eq. (10.1) is (see Fig. 10.1)

$$6CO_2 + 6H_2O + solar\ energy \rightarrow C_6H_{12}O_6 + 6O_2 \qquad (10.2)$$

where $C_6H_{12}O_6$ is glucose.

Within the plant, glucose is transformed into cellulose, hemicellulose, and lignin that form different parts of the plant. In soft wood the proportions of cellulose, hemicellulose and lignin are 40%, 30%, and 30% respectively.

The energy of carbohydrates within a plant is more than the sum of the energies of the CO_2 and H_2O that went into creating them due to solar energy absorbed during photosynthesis. Only energy in the visible part of the solar spectrum with wavelengths from 0.40-0.70 μm is used in photosynthesis. The efficiency of conversion of solar energy to stored energy in plants is about 0.5% to 1%. However, this number can increase to about 6% to 8% in sugar cane.

The multi-step process that plants and algae use to turn carbon dioxide in air into sugar, the food needed to grow, is known as the *Calvin cycle*. It has four main steps: carbon fixation, reduction phase, carbohydrate formation, and regeneration phase. Energy to fuel chemical reactions in this sugar-generating process is provided by ATP and NADPH, which are chemical compounds containing solar energy captured by a plant.

10.3 Main Sources of Biomass

There are several sources of biomass that provide feedstock for producing bioenergy in a usable form [1,2,4]. These can be broadly categorized as wastes and dedicated energy crops. A *waste* is a material that is usually discarded because it has no apparent value or may even be an environmental pollutant. Dedicated energy crops, on the other hand, are grown specifically for use as feedstock to produce bioenergy. The following is a list of the main types of biomass feedstock.

1. *Forest residues*

These are woody materials from existing forests. Leftover wood or plant materials from logging operations, tree pruning, forest management, and land clearing are available as feedstock resources.

2. *Agricultural residues*

Agricultural residues include byproducts from agricultural crops such as wheat straw, rice husks, and sugar cane residue. Slurry and manure from animals are wastes from agricultural activities, that provide biomass feedstock.

3. *Industry residues*

These include wood industry residues, e.g., bark, sawdust, wood chips, slaps, and cutoffs from sawmills. Food industry residues that provide biomass include, cooking oils, tallow, yellow grease. Some of the industrial products are pellets from sawdust and wood shavings, bio-oil, and biofuels.

4. *Municipal waste*

These are available in the form of food and domestic waste, sewage, and other biological waste. Moreover, gases produced within landfills can be collected in various ways. They can be used in combustion processes and for thermal power generation.

5. *Food and fodder crops*

Edible parts of sugar, starch and oil plants that are usually grown as food for humans or animal fodder. At present some of these crops are being used to produce biofuels. Examples are: wheat, maize, soya, palm oil, oilseed rape, and sugar cane.

6. *Energy crops*

Energy crops are fast growing trees and perennial grasses, high in biomass production potential, grown specially for use as feedstock. They are grown in marginal land, not suitable for traditional agriculture. Examples are: miscanthus willow, oil crops such as jatropha, miscanthus, switchgrass, reed, and canary grass.

7. *Algae*

Algae are unicellular organisms capable of growing in organic salt media by fixing carbon dioxide and deriving energy from light harnessed by photosynthesis. They grow much faster than plants because they have much higher photosynthetic efficiencies compared to regular plants. An important advantage is that many strains of algae are capable of growth in saline and waste water, and therefore the need for fresh water is avoided.

10.4 Bioenergy Conversion Processes

Biomass feedstock listed in section 10.3 has to be processed and converted to a form of fuel that is suitable for different applications. Solid biomass in the form of fire wood, straw and animal dung, commonly used for cooking and heating, requires very little processing. Often in rural communities in tropical countries these resources are sun dried to remove moisture before they are burned directly.

More recently, solid biomass like woody materials from managed forests and sawmills have been converted to wood pellets for use in thermal power plants. Furthermore, solid municipal waste after careful processing is being used as fuel for power generation on a large scale.

Biomass discharged from agricultural activities are usually available as slurries and sewage. These provide feedstock for the generation of biogas. Biogas production technologies are well established, and biogas generators for domestic as well as commercial applications are being deployed in many countries.

A more advanced application of biomass is the production of biofuels for transport and power generation applications. The feedstock for these plants are wheat, maize, soya, palm oil, oilseed rape, and sugar cane.

The main processes used for the conversion of biomass into useful forms of fuel are [2,4]:

(i) *Thermochemical processes*

Thermochemical processes use heat and catalysts to convert biomass into bioenergy. Common thermochemical processes include: combustion, gasification.

(ii) *Biochemical processes*

Biochemical processes use enzymes and microorganisms to obtain biofuels from biomass. Common processes are: anaerobic digestion, alcoholic fermentation.

(iii) *Extraction of fuel from plants*

This is a more direct process where vegetable oil is extracted from plants for use as fuels. An example is the manufacture of biodiesel.

10.4.1 *Combustion of solid biomass*

Combustion is the rapid oxidation of fuel which accompanies the release of heat energy. The primary constituents of biomass fuels are carbon, hydrogen, and oxygen, and hence the main oxidation products are carbon dioxide and water. However, the presence of sulfur can produce significant amounts of sulfur oxides.

Combustion of solid fuels occurs in four main steps [2,4]:

(i) *Heating and drying*
Heating and drying of fuel particles are not accompanied by chemical reaction. As the thermal front moves into the interior of a fuel particle, water is driven out. If there is water in the fuel particle, its temperature does not attain a level to initiate pyrolysis, which is the next step in the combustion process.

(ii) *Pyrolysis*
Pyrolysis is a series of complex thermally driven chemical reactions that decompose organic compounds in a fuel. The temperatures needed for pyrolysis depends on the type of plant material used. Hemicellulose begins to pyrolyze between 150°C and 300°C, cellulose pyrolyzes between 250°C and 350°C, pyrolysis in lignite is initiated between 250°C and 500°C.

Depending on the biomass fuel and its heating value, pyrolysis produces a large variety of volatile organic and inorganic compounds. These include, carbon monoxide, carbon dioxide, methane, and tar liquid.

(iii) *Flaming combustion*
When sufficient oxygen is available, the volatile gases and the char produced during pyrolysis undergo oxidation. The products of combustion of the volatile gases are carbon dioxide and water vapor.

(iv) *Combustion of char in fuel bed*
At temperatures from 500°C to 700°C the solid carbon char is gasified in the presence of carbon monoxide, water vapor and oxygen to form carbon monoxide and hydrogen that are subsequently burnt.

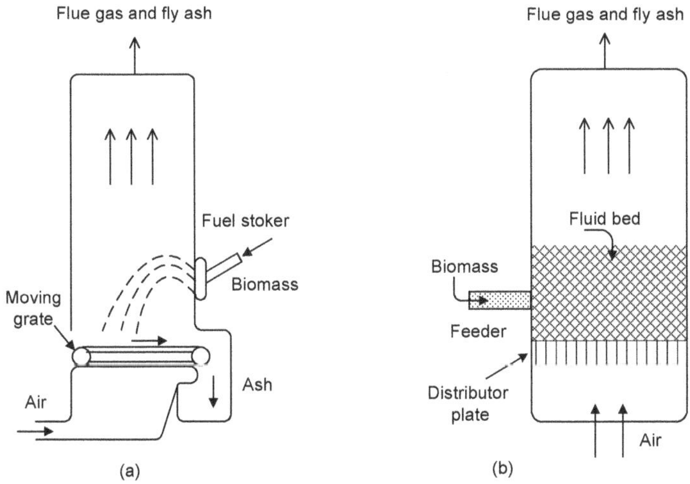

Fig. 10.2 (a) Fixed-bed combustor, (b) Fluidized – bed combustor

10.4.2 *Fixed-bed and fluidized-bed combustors*

Furnaces commonly used for biomass combustion include: (i) fixed-bed systems, and (ii) fluidized-bed systems. *Fixed-bed* combustors use static and inclined grates or moving grates with various designs to supply fuel and remove ash (see Fig. 10.2a). These include manual feed devices and screw feeders. Temperatures in fixed-bed systems vary between 850°C and 1400°C. To achieve the required temperature, large fixed-bed furnaces use two air supply streams. Primary air is supplied to the fire bed, and secondary air to the gas combustion zone above it.

Primary air oxidizes the solid carbon char in the fuel bed to provide heat to pyrolyze the newly added fuel. Secondary air ensures that the volatile gases produced undergo more complete combustion. Modern fixed-bed furnaces are designed to maximize combustion efficiency and minimize environmental impact due to harmful emissions.

Fluidized-bed combustors are a more recent development in furnace design (see Fig. 10.2b). An air stream injected from the bottom of the furnace suspends a bed of sand or granular material providing a turbulent mixture of air and solid particles. Biomass fuel to be combusted is supplied horizontally into the fluidized bed. The high heat and mass transfer rates in the combustion zone is ideal for burning a variety of fuels. The high thermal capacity of the granular bed allows the system to be operated at temperatures as low as 850°C, which lowers the emissions of nitrogen compounds.

10.4.3 *Analysis of combustion processes*

Mass balance

In a typical combustion process, fuel reacts with oxygen contained in air supplied to the combustion chamber [2,3]. During the chemical reaction between the fuel and oxygen, the mass of each element remains unchanged. Mass conservation is usually represented by the *chemical equation* for the reaction, which ensures mass balance of each element.

Consider the combustion reaction between carbon monoxide (fuel) and oxygen (oxidant). The chemical equation is

$$CO + \frac{1}{2}O_2 = CO_2 \tag{10.3}$$

In the above reaction CO and O_2 are the *reactants* and CO_2 is the *product*.

The *molar-form* of the of the mass balance of the above reaction may be stated as: one mole of carbon monoxide (CO) reacts with half a mole of oxygen O_2 to form one mole of carbon dioxide (CO_2).

The mass conservation equation may be expressed in terms of molar masses in the form: 28 kg of carbon monoxide reacts with 16 kg of oxygen to form 44 kg of carbon dioxide.

In most practical combustion chambers, the oxygen required for combustion is supplied in the form of air and not pure oxygen. Air is a mixture of oxygen (O_2), nitrogen (N_2) and argon with respective mole fractions of 21 %, 78%, and 1%. In the absence of dissociation, nitrogen does not take part in the combustion reaction. Therefore nitrogen and argon leave as constituents of the products of combustion, usually at the same temperature as the other products of the reaction.

For engineering applications, we neglect the presence of argon and assume that air is a mixture of 21% oxygen and 79% nitrogen by volume. Therefore the mole-ratio of nitrogen to oxygen in air is $79/21$ = 3.76.

When oxygen is introduced as air, the chemical reaction represented by Eq. (10.3) has to be amended to include the presence of nitrogen in the combustion system. Therefore the Eq. (10.3) is written in the form:

$$CO + \frac{1}{2}O_2 + \frac{1}{2} \times 3.76N_2 = CO_2 + \frac{1}{2} \times 3.76N_2 \tag{10.4}$$

Note that in Eq. (10.4) 3.76 moles of nitrogen per mole of oxygen are included on both sides of the chemical equation.

Similarly, for the combustion of methane with air we can write the chemical equation:

$$CH_4 + 2O_2 + 2 \times 3.76N_2 = CO_2 + 2H_2O + 2 \times 3.76N_2 \tag{10.5}$$

The minimum quantity of air required for complete oxidation of all elements of a fuel, capable of undergoing combustion, is called the '*theoretical air*' or '*stoichiometric air*'. If the quantity of air supplied to the combustion system exceeds the theoretical air required for complete combustion, then the *excess air* is included in the products of combustion. Excess air is usually supplied to ensure complete combustion of the fuel.

An important parameter of a combustion reaction is the *air-fuel ratio* (AFR), which can be determined from the chemical equation. It is defined as the ratio of the quantity of air to the quantity of fuel, which could be expressed either in the molar-form or the mass-form. For example, AFR for the stoichiometric combustion of methane can be obtained from Eq. (10.5).

The AFR on a molar-basis is:

$$(AFR)_{molar} = (2 + 2 \times 3.76)/1 = 9.52$$

moles of air per mole of fuel.

The AFR on a mass-basis is:

$$(AFR)_{mass} = \frac{(2+2\times3.76)\times29}{(1\times16)} = 17.25$$

kg of air per kg of fuel.

Air/Fuel Equivalence Ratio (AFER), λ is a parameter used to compare the *actual* AFR of a combustion reaction to the *stoichiometric* AFR. It is defined by [2]

$$\lambda = \frac{AFR}{(AFR)_{stoich}} \tag{10.6}$$

If $\lambda < 1$ the air/fuel mixture is rich in fuel
If $\lambda > 1$ the air/fuel mixture is lean in fuel

Application of the first law of thermodynamics

The application of the first law of thermodynamics, or the energy equation, to combustion processes will be presented in this section. The heat and work terms in the first-law equation are independent of the combustion process [3]. However, the evaluation of the change in internal energy and the change in enthalpy needs special care.

In the absence of chemical reactions, the internal energy and enthalpy of pure substances may be tabulated using an arbitrary selected reference state [5]. However, in chemical reactions part of the chemical binding energy of the reactants is converted to thermal energy of the products. This is called the *enthalpy of reaction*, which is either measured or computed at a *common reference state*, applicable to all substances.

The temperature and pressure of the *standard reference state* are $T^0 = 25^oC$, and $P^0 = 1\ bar$. The general approach is to determine the change in enthalpy when a given pure substance is formed from elemental substances under standard conditions.

By convention, the enthalpy of every elemental substance, which consists of only one kind of atom, is assigned a value of zero at the standard state. For example, the enthalpy of carbon is zero at the standard state and so is the enthalpy of oxygen because these are elemental substances.

The *chemical equations of formation* for methane (CH_4) and carbon dioxide (CO_2) including the *molar enthalpy of formation* $\Delta\bar{h}^0$ are listed below:

$$C(s) + 2H_2(g) = CH_4(g) + \Delta\bar{h}^0; \quad \Delta\bar{h}^0 = -74870\ \text{kJkmol}^{-1}$$

$$C(s) + O_2(g) = CO_2(g) + \Delta\bar{h}^0; \quad \Delta\bar{h}^0 = -393520\ \text{kJkmol}^{-1}$$

The relationship between the enthalpy of a substance and its enthalpy of formation may be illustrated by using the formation of water as follows:

$$\Delta\bar{h}_w{}^o = \bar{h}_{water} - \bar{h}_H - \bar{h}_O \tag{10.7}$$

Since the enthalpies of elements hydrogen and oxygen are zero, the enthalpy of water at the standard state is equal to the enthalpy of formation.

Consider the following combustion reaction between CH_4 and O_2. The chemical equation is

$$CH_4 + 2O_2 = CO_2 + 2H_2O(l) \tag{10.8}$$

The change in total molar enthalpy, $\Delta\bar{H}_R$ between the *reactants* and the *products* of the above reaction may be expressed as

$$\Delta\bar{H}_R = \bar{h}^o{}_{CO_2} + 2\bar{h}^o{}_{H_2O} - \bar{h}^o{}_{CH_4} - 2\bar{h}^o{}_{O_2} \tag{10.9}$$

Substituting the enthalpies of formation of the different substances in Eq. (10.9) we obtain the enthalpy of reaction $\Delta\bar{H}_R$ as:

$$\Delta\bar{H}_R = -393520 - 2 \times 285830 + 74870 - 2 \times 0$$

$$\Delta\bar{H}_R = -890310 \text{ kJkmol}^{-1}$$

Note that the enthalpy of formation of H_2O, CH_4 and CO_2 are -285830 kJkmol^{-1}, -74870 kJkmol^{-1}, and -393520 kJkmol^{-1} respectively.

The *heat output* $\bar{Q}_o$ during a steady flow combustion reaction is called the *heat of reaction* or the *heating value* of the fuel. If water in the products of combustion is in a liquid state then the heat output is higher due to the latent heat of condensation of water. Hence the heat output is called the *higher heating value*, HHV. When water is in the vapor state it is called the *lower heating value*, LHV. Note that heating values of fuels are also called *calorific values*, HCV and LCV.

From energy balance of the combustion process it follows that:

$$\bar{Q}_o = -\Delta\bar{H}_R$$

The use of the enthalpies of formation makes possible the calculation of the enthalpy change between reactants and products for complex reactions from data on individual constituents, as illustrated above. A similar approach is used to obtain the internal energy changes of chemical reactions.

Effect of moisture content on heating values

The moisture content of biomass effects the HHV and LHV of the fuel. The moisture content on a *wet basis* M_w and *dry basis* M_d are defined as:

$$M_w = \frac{m_w}{m_w + m_d} = \frac{m_w}{m_t} \quad (10.10)$$

$$M_d = \frac{m_w}{m_d} = \frac{m_w}{m_t - m_w} \quad (10.11)$$

where m_w, m_d, and m_t are the mass of water, the mass of dry fuel, and the total mass of fuel respectively.

When a wet hydrocarbon fuel is combusted, water in the products comes from two sources. These are: (i) moisture in the wet fuel, and (ii) combustion of hydrogen in the fuel. When water in the final products of combustion is a liquid, the heat output is the HHV, otherwise it the LHV. The combustion of one kg of hydrogen (H) produces 9 kg of water according to the equation:

$$2H + \frac{1}{2}O_2 \rightarrow H_2O \quad (10.12)$$

The relation between the HHV and LHV of a wet fuel is obtained by writing the energy balance equation:

$$m_d HHV - m_w L - m_d f_H \times 9 \times L = m_t LHV \quad (10.13)$$

where f_H is the mass fraction of hydrogen in the dry fuel, and L is the latent heat of condensation of water. Therefore

$$LHV = \left(\frac{m_d}{m_t}\right) HHV - \left(\frac{m_w}{m_t}\right) L - 9L\left(\frac{m_d}{m_t}\right) f_H$$

$$LHV = (1 - M_w)HHV - M_w L - 9Lf_H(1 - M_w) \quad (10.14)$$

Equation (10.14) shows that the LHV decreases linearly with moisture content.

The following empirical equation is recommended in Ref.[2] for the estimation of the higher heating value HHV [MJkg^{-1}] of biomass.

$$HHV = a_c X_c + a_h X_h + a_s X_s + a_o X_o + a_n X_n + a_a X_a \quad (10.15)$$

where the coefficients are given by:

$$a_c = 34.91, a_h = 117.83, a_s = 10.05, a_o = -10.34, a_n = -1.51, a_a = -2.11$$

The mass fraction of the constituents of biomass is X. The subscripts are: c – carbon, h – hydrogen, s – sulfur, o – oxygen, n – nitrogen, a – ash. The empirical model is found to over-estimate the HHV by about 2 $MJkg^{-1}$.

Some properties of solid biomass fuels are listed in Table 10.1.

Table 10.1 Properties of solid biomass fuels (Data from Ref. [2])

Biomass material	Moisture content, (wet basis) %	Ash content, (dry basis) %	Lower heating value ($MJkg^{-1}$)
Wood	10-60	0.25-1.7	8.4-17
Straw	10	4.4	12
Charcoal	1-10	0.5-6	25-32
Rice husks	13-15	20-30	14
Peat	15	1-20	9-15
Sugar cane residue	40-60	1.7-3.8	7.7-8

10.4.4 *Gasification of biomass*

The conversion of solid, carbonaceous fuels to a flammable mixture of gases consisting of carbon monoxide, hydrogen, methane, nitrogen, and carbon dioxide at a high temperature of about 750°C to 850°C is called *gasification* [4]. When the gas mixture has a high nitrogen content, it is called *producer gas*, otherwise it is known as *synthetic gas*.

The overall process is endothermic, and therefore the required heat input must be provided either by burning some of the fuel or from an external heat source. Gasification converts over 90% of biomass feed into gas with the remaining material ending as solid char and ash, or various tars in some applications.

The processes involved in gasification are: heating and drying, pyrolysis, and gas-solid reactions that determine the final composition of fuel gas. Solid particles of biomass fuel are initially heated and dried to temperatures below 100°C. When the fuel particles reach 225°C, pyrolysis commences, and becomes more intense and complete as the temperature approaches 400-500°C. The reaction produces intermediate products of char, gases, and condensable vapors, including water, methanol, acids, and heavy hydrocarbons. Pyrolysis can convert up to 80% by mass of biomass into vapors and gases. At high temperatures of over 1000°C, it is possible to convert over 95% of the carbon in the feed to gases with char-consuming reactions.

Four important reactions take place between the gases and the solid carbon char. These are [2]:

Carbon-oxygen reaction:	$C + \frac{1}{2}O_2 \leftrightarrow CO$
Boudouard reaction:	$C + CO_2 \leftrightarrow 2CO$
Carbon-water reaction:	$C + H_2O \leftrightarrow H_2 + CO$
Hydrogenation reaction:	$C + 2H_2 \leftrightarrow CH_4$

The exothermic carbon-oxygen reaction provides heat input to the endothermic Boudouard reaction and the carbon-water reaction. The hydrogenation reaction, which is mildly exothermic, produces methane.

Additional methane and hydrogen are produced from carbon monoxide in the following gas-gas reactions.

Water-gas shift reaction:	$CO + H_2O \leftrightarrow H_2 + CO_2$
Methanation reaction:	$CO + 3H_2 \leftrightarrow CH_4 + H_2O$

The main types of biomass gasifiers are [2,4]: (i) fixed-bed, updraft and downdraft, shown schematically in Figs. 10.3(a) and (b) respectively, and (ii) fluidized-bed. The gaseous fuel produced in gasifiers is called *syngas* which is used for various applications, including the operation of spark ignition and compression ignition engines.

In the updraft gasifier, depicted in Fig. 10.3(a), biomass, made into pellets or chips is fed at the top, often through a rotary valve. As it descends through the gasifier it is dried and undergoes pyrolyzes releasing hydrogen, carbon monoxide, and methane. Heat for gasification is provided by the combustion of char and tars formed above the grate. Ash and unburnt tar exit at the bottom.

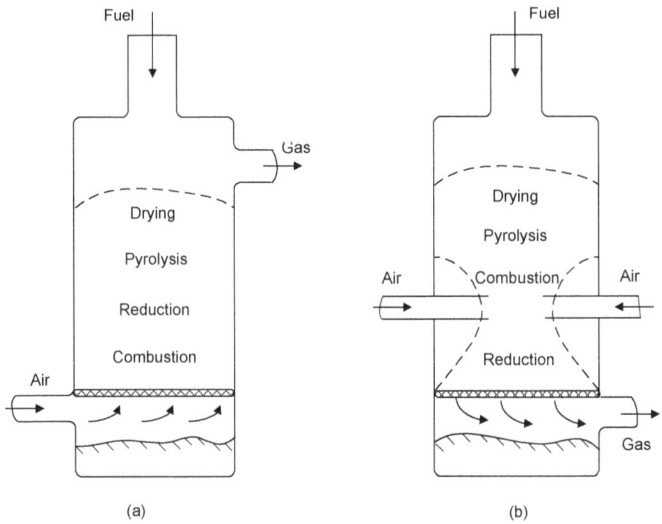

Fig. 10.3 Fixed-bed gasifiers: (a) updraft gasifier, (b) downdraft gasifiers

The temperature of the combustion zone at the bottom of the gasifier can reach up to 1200°C, but syngas leaving at the top is at about 100°C. This results in a high efficiency of conversion of biomass to gas. However, the gas leaving can contain significant quantities of tar that was not burnt in the combustion zone. Hence syngas produced in updraft gasifiers are more suitable for use in furnaces where the gas and the tar are combusted. For use in heat engines the gas needs to be filtered and cleaned to extract the tar.

In the *downdraft gasifier*, shown in Fig. 10.3(b), combustion air enters from the sides of the unit. Biomass is fed at the top and is dried and undergoes pyrolysis as it descends through the high temperature zone. The gases and tars produced during pyrolysis pass through a constriction or a throat into a reduction zone where they are broken down into carbon monoxide and hydrogen.

Fluidized-bed gasifiers and combustors (see Fig. 10.2b) have many common features. The gasification process takes place in and above a fluidized bed of sand or granular material, kept suspended by a turbulent stream of air.

The main advantages of fluidized-bed gasifiers are: (i) more compact design that achieves high reaction rates and heat transfer rates, and (ii) a wide variety of biomass feedstock can be used.

The disadvantages include: (i) the fuel has to be supplied in a small size to facilitate fluidization, typically, less than about 20 mm. (ii) fluidization requires the use of an air compressor, that consumes power, and (iii) syngas produced is at a high temperature and can contain significant quantities of tar and ash.

10.4.5 *Anaerobic digestion of biomass*

Anaerobic digestion is a process where naturally occurring micro-organisms digest biomass to produce a gas mixture of methane and carbon dioxide. It is a *biochemical process* that occurs in the absence of oxygen. Residual solid digestate from the digester can be valuable fertilizer, depending on the feedstock used. The biogas produced has numerous applications, including cooking, lighting, and power generation.

A simple batch anaerobic digestor used in rural India and China [2] is shown schematically in Fig. 10.4. It operates at constant volume and variable pressure. The feedstock, which is usually agricultural and domestic wastes, is fed into the fermentation chamber and produces biogas in a single-stage digestion process. Fresh biomass is added daily at the inlet and digestate removed through the outflow at the side. The biogas collected at the top of the chamber is piped out into the house for cooking and heating.

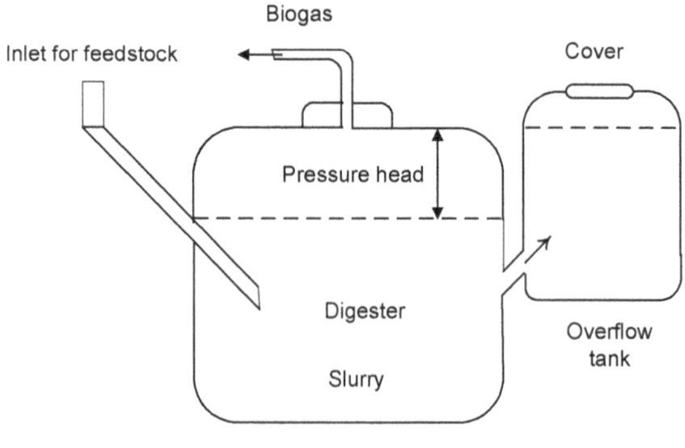

Fig. 10.4 Single-stage anaerobic digester

Anaerobic digestion is a well-developed technology for biogas production and has been used for many years.

Anaerobic digestion of cellulose can be represented by the chemical equation:

$$C_6H_{10}O_6 + H_2O \rightarrow 3CO_2 + 3CH_4$$

where $C_6H_{12}O_6$ is cellulose, CH_4 is methane, and CO_2 is carbon dioxide.

The exact composition of biogas generated by anaerobic digestion depends on the type of feedstock and the conditions of the process.

The ranges of the various constituents by volume in a typical biogas mixture are: (i) methane – 50 to 80%, (ii) carbon dioxide – 50 to 20%, (iii) trace gases, nitrogen, hydrogen, ammonia, and hydrogen sulphide < 1%.

The biochemical processes responsible for anaerobic digestion occurs in several stages in sequence, each is driven by distinct sets of bacteria [2].

(i) Hydrolysis: large molecules of carbohydrates, fats and proteins are broken down into simple sugars, lipids and amino acids which other bacteria breakdown.
(ii) Acidogenesis: products of hydrolysis are converted into alcohols and fatty acids.
(iii) Acetogenesis: alcohols and fatty acids are converted to acetic acid.
(iv) Methanogenesis: products of the above stages are converted to methane and carbon dioxide.

The main factors affecting bio-digestion or generation of gas are: temperature, pH of the contents, total solid content of feedstock, loading rate, carbon to nitrogen ratio, retention time or rate of feeding, type of feedstock, pressure, acid accumulation, diameter to depth ratio of the chamber.

The theoretical efficiency of the conversion of glucose to methane in anaerobic digestion is around 90%. However, practical digesters using agricultural and municipal feedstock have efficiencies of about 40% to 60%. Feedstock for anaerobic digestors include food wastes, farm slurry and animal manure, crops and crop residues. Typical yields of biogas from anaerobic digestion are: 25 mega-m^3 per 1000 kg of wet cow manure, and 220 mega-m^3 per 1000 kg of grass silage.

10.4.6 *Liquid biofuels from biomass*

Biomass, unlike other renewable energy sources, can be converted directly to liquid fuels, commonly called *biofuels*, that are being used as transportation fuels [6,7]. Ethanol and biodiesel are two of the widely known biofuels. Ethanol (CH_3CH_2OH) is an alcohol used as a blending agent with gasoline to improve its octane rating and reduce carbon monoxide and other smog-causing emissions.

Biodiesel is a liquid fuel produced from biomass, such as new and used vegetable oil, animal fat, or recycled cooking grease. Chemically, it consists mostly of fatty acid methyl ester. Like regular diesel fuel, biodiesel can be used in compression ignition engines, and as a blending agent in regular diesel.

Production of ethanol

The best-known alcohol-based biofuel is *ethanol*. At present, most ethanol is produced by fermenting crop sugars with pure-cultures of yeasts. The common feedstock in the USA is sugar-derived cornstarch, while in Brazil it is sugarcane. The ethanol produced using these raw materials is a highly viscous solution of up to 15% ethanol. To make it acceptable as a transport fuel with 99% or more ethanol, the viscous solution must be distilled to remove water. This is a highly energy intensive process.

Sugarcane to ethanol process

Once harvested, sugarcane stalks are transported to the ethanol plant, which is usually located close to the plantation [6]. The stalks are crushed by milling to extract sugarcane juice, which is fed to fermentation tanks containing yeast. The ethanol generated in these tanks has a concentration of about 8%.

Following fermentation, the solution in the tanks is passed through a centrifuge from which the yeast cells are extracted and recycled. The ethanol separated in the centrifuge is sent through a distillation tower to generate 99% ethanol. The solution leftover after distillation is partly recycled, and used for irrigation, sold, or used to generate methane in anaerobic digesters.

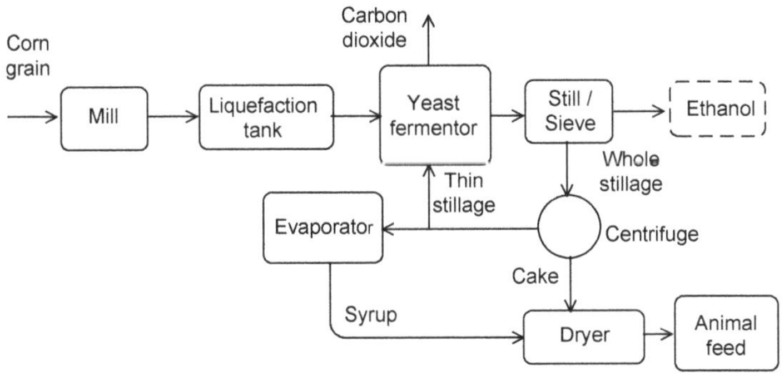

Fig. 10.5 Simplified schematic diagram of the corn grain-to-ethanol process

Corn grain to ethanol process

Two different processes, called wet milling and dry milling, are used to produce ethanol from corn grain or corn kernel [6]. In the dry milling process the entire corn kernel is milled and used as feedstock. In wet milling the corn kernel is first processed to extract purified cornstarch to be used as feedstock for ethanol production. Wet milling is more capital intensive than dry milling due the additional separation technologies required.

A simplified schematic diagram of a dry milling ethanol plant is shown in Fig. 10.5. The entire corn kernel is milled and then cooked in the presence of enzymes (α – amylases) to liquify the cornstarch in dextrins, which are polymers of glucose.

The solution is further heated with steam and then cooled to body temperature. In the fermentation tank yeast cells and enzymes are combined for simultaneous glucose formation and fermentation. After about 2 to 3 days a fermentation broth with about 15% ethanol is produced in the tank and the process is terminated.

The solution from the fermentor is passed through a distillation tower to recover ethanol with a concentration of 99%. Off gases from the fermentor are processed to recover ethanol and carbon dioxide. Liquid from the distillation tower is sent through a centrifuge to recover leftover corn grain solids and yeast cells.

The liquid flowing out of the centrifuge has large quantities of solids, and it is pumped to evaporators which produce water vapor and a syrup.

The energy input for the evaporator is usually obtained from waste heat produced in the ethanol plant. The syrup from the evaporator is mixed with the wet cake from the centrifuge and passed through a dryer to produce animal feed. This is a valuable byproduct of the ethanol production process that has an impact on the economic viability of the industry.

10.5 Energy and Environmental Costs of Biofuels

An important metric of energy or available energy in a biofuel like ethanol is the *net energy balance* (NEB) ratio[6]. It is the ratio of the energy available in biofuel produced to the energy expended, mainly from fossil fuels, for growing, harvesting, transporting, and processing biomass feedstock. If this ratio is less than one, it means that it takes more primary energy to produce the biofuel than it can deliver, and therefore unattractive from an energy utilization standpoint. The converse is true if the NEB is greater than one.

A more comprehensive measure of energy utilization is called *life-cycle-assessment* (LCA). It reflects all costs involved from extraction of the raw material to delivery of the final product, including environmental, socio-economic and other factors.

Biofuel combustion produces carbon dioxide, which is known to cause global warming, and environmental pollution. The most important of the greenhouse gases is CO_2 which contains 27% carbon.

A biofuel project is considered to be *carbon-neutral* if the carbon dioxide (CO_2) absorbed by the feedstock crop balances the greenhouse gas (GHG) emissions attributed to the project. The life cycle GHG cost of a project is the total amount of absorption and emission of CO_2 during its life cycle.

If the emissions of CO_2 during production, processing, transport, and combustion exceed what is absorbed, both above and below ground during the growth of the crop, then the GHG life cycle cost is positive. Otherwise the life cycle GHG cost is negative. The GHG life cycle cost depends both on the energy crop as well as the planting site.

The overall impact of biofuel projects on global warming must be considered before adopting them on a large scale [1]. However, an advantage of locally produced biofuels is that they give some measure of fuel security for countries that import a large fraction of fuel used from foreign sources.

10.6 Worked Examples

Example 10.1 Estimate the land area that would be needed to generate 0.2 GW of electricity steadily over a year using biomass fuel. The annual insolation at the location is 900 $kWhm^{-2}$. Assume a photosynthetic efficiency 0.5%, and power conversion efficiency of 25%.

Solution The yearly energy required is

$$E_e = 0.2 \times 10^6 \times 24 \times 365 \text{ kWh}$$

Energy available from biomass is

$$E_{bm} = A \times 900 \times \frac{0.5}{100} \times \frac{25}{100} \text{ kWh}$$

where A is the cultivated area.

When the energy from biomass is adequate to generate the electricity needed

$$E_e = E_{bm}$$

Substituting numerical data we have

$$A = \frac{0.2 \times 10^6 \times 24 \times 365}{900 \times 0.005 \times 0.25} = 1557.3 \times 10^6 \text{ m}^2$$

Hence the required area is 1557.3 km^2.

Example 10.2 The moisture content and ash content of a biomass are 22% and 4.5% on a wet basis. Calculate its moisture content on a dry basis and ash-free basis.

Solution The moisture content on a wet basis is given by

$$M_w = \frac{m_w}{m_t} = 0.22$$

$$m_t = m_w + m_f + m_a$$

where m is mass, and the subscripts t, f, w and a represent total, fuel, water, and ash respectively. The ash content on a wet basis is

$$M_a = \frac{m_a}{m_t} = 0.045$$

The moisture content on a dry basis is

$$M_d = \frac{m_w}{m_t - m_w} = \frac{0.22m_t}{m_t - 0.22m_t} = 0.282$$

The moisture content on an ash-free basis is

$$M_{af} = \frac{m_w}{m_t - m_w - m_a} = \frac{0.22m_t}{m_t - 0.22m_t - 0.045m_t} = 0.299$$

Example 10.3 An oven-dried biomass material has a higher heating value (HHV) of 14 MJkg^{-1}. The latent heat of vaporization of water is 2.3 MJkg^{-1}.
(i) If the moisture content of the material is 38%, calculate the higher heating value (HHV).
(ii) If the hydrogen content by mass in the material is 6%, calculate the lower heating value (LHV).

Solution The moisture content is given by

$$M_w = \frac{m_w}{m_f + m_w} = 0.38$$

Hence we have $$m_w = 0.613m_f$$

The mass of moist biomass fuel is

$$m_{mf} = (1 + 0.613)m_f = 1.613m_f$$

Therefore the HHV of moist fuel is

$$HHV = \frac{14 \times m_f}{1.613m_f} = 8.68 \text{ MJkg}^{-1}$$

We use Eq. (10.14) to determine the lower heating value.

$$LCV = (1 - M_w)HCV - M_w L - 9Lf_H(1 - M_w)$$

$$LCV = (1 - 0.38) \times 8.68 - 0.38 \times 2.3 - 9 \times 2.3 \times 0.06(1 - 0.38)$$

Hence the LHV is 3.73 $MJkg^{-1}$.

Example 10.4 Octane, C_8H_{18} undergoes stoichiometric combustion with air. Calculate (i) the air fuel ratio, and (ii) the kg of carbon dioxide produced per kg of fuel.

Solution The chemical equation for the combustion reaction is

$$C_8H_{18} + \frac{25}{2}O_2 + \frac{25}{2}(3.76)N_2 \rightarrow 8CO_2 + 9H_2O(v) + \frac{25}{2}(3.76)N_2$$

Mass of fuel is

$$m_f = (8 \times 12) + (1 \times 18) = 114 \text{ kg}$$

Mass of air is

$$m_a = \frac{25}{2} \times (2 \times 16) + \frac{25}{2}(3.76 \times 28) = 1716 \text{ kg}$$

The air-fuel ratio is

$$AFR = \frac{m_a}{m_f} = \frac{1716}{114} = 15.05 \text{ kg per kg}$$

Mass of carbon dioxide is

$$m_{co_2} = 8(12 + 32) = 352 \text{ kg}$$

Mass of carbon dioxide per unit mass of fuel is

$$\frac{m_{co_2}}{m_f} = \frac{352}{114} = 3.088$$

Example 10.5 Liquid ethanol (C_2H_5OH) is burned with 150% theoretical air. Calculate (i) the mass of air required per kg of fuel, and (ii) the mass of carbon dioxide produced per kg of fuel.

Solution The stoichiometric chemical equation is

$$C_2H_5OH + 3O_2 + 3 \times 3.76N_2 \rightarrow 2CO_2 + 3H_2O + 3 \times 3.76N_2$$

Mass of fuel is

$$m_f = 24 + 5 + 16 + 1 = 46 \text{ kg}$$

Mass of oxygen is

$$m_o = 3 \times 32 = 96 \text{ kg}$$

Mass of nitrogen is

$$m_n = (3 \times 3.76) \times 28 = 315.84 \text{ kg}$$

Mass of air is

$$m_a = 96 + 315.84 = 411.8 \text{ kg}$$

Mass of air with 50% excess air is

$$m_{ae} = 411.8 \times 1.5 = 617.7 \text{ kg}$$

Air fuel ratio is

$$AFR = \frac{617.7}{46} = 13.42 \text{ kg per kg}$$

Carbon dioxide produced per kg of fuel is

$$m_{co_2} = \frac{2\times(12+32)}{46} = 1.913 \text{ kg}$$

Example 10.6 A solid fuel has the following ultimate analysis in percent: $C = 70$, $H = 5$, $S = 1$, $O = 12$, $N = 2$, $ash = 10$. For complete combustion with 20 excess air at 1 bar, determine (i) mass of air required per kg of fuel, and (ii) the mass of carbon dioxide produced.

Solution The combustible constituents of the fuel are carbon, hydrogen and sulfur. Consider the constituents of one kg of fuel undergoing the following stoichiometric reactions:

For carbon

$$C + O_2 \rightarrow CO_2$$

Mass of oxygen: $m_c = 0.7 \times \frac{32}{12} = 1.867$ kg

For hydrogen

$$H_2 + \frac{1}{2}O_2 \rightarrow H_2O$$

Mass of oxygen: $m_h = 0.05 \times \frac{16}{2} = 0.4$ kg

For sulfur

$$S + O_2 \rightarrow SO_2$$

Mass of oxygen: $m_s = 0.01 \times \frac{32}{32} = 0.01$ kg

Contribution of oxygen in the fuel is: $m_{fo} = -0.12$ kg.

Note that the oxygen present in the fuel reduces the external oxygen required for combustion. Nitrogen and ash do not participate in combustion.

Total mass of oxygen required

$$m_{ot} = 1.867 + 0.4 + 0.01 - 0.12 = 2.157 \text{ kg}$$

Mass of air required is

$$m_a = 2.157 \times \left(1 + \frac{3.76 \times 28}{32}\right) = 9.253 \text{ kg}$$

Total mass of air required with 20% excess air is

$$m_{ae} = 1.2 \times 9.253 = 11.1 \text{ kg}$$

Mass of carbon dioxide produced is

$$m_{co_2} = \frac{44}{12} \times 0.7 = 2.57 \text{ kg}$$

Example 10.7 The fuel in worked example 10.6, undergoes complete combustion with 20% excess air at 1 bar. Calculate the volumetric analysis of the dry products of combustion.

Solution Consider one kg of fuel.

Mass of CO_2 in products is

$$m_{co_2} = \frac{44}{12} \times 0.7 = 2.57 \text{ kg}$$

Mass of SO_2 in products is

$$m_{so_2} = \frac{64}{32} \times 0.01 = 0.02 \text{ kg}$$

Mass of H_2O in products

$$m_{H_2O} = \frac{18}{2} \times 0.05 = 0.45 \text{ kg}$$

Mass of stoichiometric oxygen required is 2.157 kg.

Mass of nitrogen supplied with stoichiometric oxygen is

$$m_{N_2} = 2.157 \times \frac{3.76 \times 28}{32} = 7.096 \text{ kg}$$

Excess oxygen supplied

$$m_{eo} = 0.2 \times 2.157 = 0.431 \text{kg}$$

Total mass of nitrogen in air is

$$m_{N_2} = (7.096 + 3.29 \times 0.431) = 8.51 \text{ kg}$$

Composition of dry products by mass is

$$CO_2 = 2.57 \text{ kg}, \quad SO_2 = 0.02 \text{ kg}, \quad O_2 = 0.431 \text{ kg}, \quad N_2 = 8.51 \text{ kg}$$

Composition by moles is obtained by dividing by the individual molar masses. Hence we get

$$CO_2 = 0.0584, \quad SO_2 = 0.0003, \quad O_2 = 0.0134, \quad N_2 = 0.304$$

Total moles of dry products is 0.376.

The volumetric analysis (percent) is given by the mole fractions of the constituents in the products. Hence we get the volumetric analysis of the products as:

$$CO_2 = 0.155, \quad SO_2 = 0.0008, \quad O_2 = 0.0356, \quad N_2 = 0.809$$

Example 10.8 The ultimate analysis of a solid biomass fuel is: carbon 53%, hydrogen 17%, oxygen 10%, nitrogen 2%, ash 4%, and moisture 14%. Calculate (i) the mass of air required for complete combustion with 25% excess air, (ii) mass of carbon dioxide produced, and (ii) the higher heating value of biomass.

Solution The combustible constituents of biomass fuel are carbon and hydrogen. Consider 1 kg of biomass undergoing stoichiometric combustion:

For carbon

$$C + O_2 \rightarrow CO_2$$

Mass of oxygen: $m_c = 0.53 \times \frac{32}{12} = 1.413 \text{ kg}$

For hydrogen

$$H_2 + \frac{1}{2}O_2 \rightarrow H_2O$$

Mass of oxygen: $m_h = 0.17 \times \frac{16}{2} = 1.36 \text{ kg}$

Contribution of oxygen in the fuel is: $m_{fo} = -0.1 \text{ kg}$

The oxygen present in biomass reduces the oxygen required for combustion. Nitrogen and ash do not participate in combustion.

Total mass of oxygen required

$$m_{ot} = 1.413 + 1.36 - 0.1 = 2.673 \text{ kg}$$

Mass of air required is

$$m_a = 2.673 \times \left(1 + \frac{3.76 \times 28}{32}\right) = 11.467 \text{ kg}$$

Air required with 25% excess air is

$$1.25 \times 11.467 = 14.33 \text{ kg}$$

Mass of carbon dioxide produced is

$$m_{co_2} = \frac{44}{12} \times 0.53 = 1.943 \text{ kg}$$

The following empirical equation (Eq.10.15) is recommended in Ref.[2] for the estimation of the higher heating value HHV [$MJkg^{-1}$] of biomass.

$$HHV = a_c X_c + a_h X_h + a_s X_s + a_o X_o + a_n X_n + a_a X_a$$

where the coefficients are given by:

$$a_c = 34.91,\ a_h = 117.83,\ a_s = 10.05,\ a_o = -10.34,\ a_n = -1.51,\ a_a = -2.11.$$

$$HHV = 34.91 \times 0.53 + 117.83 \times 0.17 - 10.34 \times 0.1 - 1.51 \times 0.02 - 2.11 \times 0.04$$

The higher heating value is 37.38 $MJkg^{-1}$

Example 10.9 The general chemical formula for a biomass fuel is $CH_{1.4}O_{0.6}$. Calculate the mass of air required for the complete combustion of 1 kg of biomass with 35% excess air.

Solution The general chemical equation for stoichiometric combustion is:

$$CH_xO_y + aO_2 + 3.76aN_2 \rightarrow bCO_2 + cH_2O + 3.76aN_2$$

where a, b, and c are coefficients to be determined.

We write the following mass balance equations:

Oxygen: $y + 2a = 2b + c$
Hydrogen: $x = 2c$
Carbon: $1 = b$

Solving the above equations we have

$$a = \frac{1}{2}\left(2 + \frac{x}{2} - y\right)$$

Substituting the given values of x and y we have

$$a = 1.05\ c = 0.7$$

Hence the chemical reaction equation is:

$$CH_{1.4}O_{0.6} + 1.05O_2 + 3.76 \times 1.05N_2 \rightarrow$$
$$CO_2 + 0.7H_2O + 3.76 \times 1.05N_2$$

Mass of fuel: $m_f = 12 + 1.4 \times 1 + 0.6 \times 16 = 23$ kg

Mass of oxygen: $m_o = 1.05 \times 32 = 33.6$ kg

Mass of air: $m_a = 33.6 + 33.6 \times \frac{3.76 \times 28}{32} = 144.2$ kg

Mass of air per kg of fuel is 6.27 kg

Mass of air with 35% excess air is 8.46 kg

Example 10.10 Calculate the higher and lower heating values of propane (C_3H_8) at 1 bar and 25°C. The molar enthalpy of formation of CO_2, $H_2O(g)$, and C_3H_8 are respectively -393510 kJkmol^{-1}, -241820 kJkmol^{-1}, and -104680 kJkmol^{-1}. The latent heat of water is 43952.4 kJkmol^{-1}.

Solution The chemical equation for the combustion process is

$$C_3H_8 + 5O_2 = 3CO_2 + 4H_2O$$

The enthalpy of reaction per kmol of C_3H_8 is given by

$$\Delta\overline{H}_R = 3\bar{h}^o{}_{CO_2} + 4\bar{h}^o{}_{H_2O} - \bar{h}^o{}_{C_3H_8} - 5\bar{h}^o{}_{O_2}$$

$$\Delta\overline{H}_R = -3 \times 393510 - 4 \times 241820 + 104680 - 5 \times 0$$

$$\Delta\overline{H}_R = -2043130 \text{ kJkmol}^{-1}$$

Molar mass of C_3H_8 is 44 kgkmol^{-1}.

The lower heating value under standard conditions is

$$LHV = \frac{2043130}{44} = 46434.8 \text{ kJkg}^{-1}$$

When the 4 moles of water vapor in the products condense to water the enthalpy of reaction becomes

$$\Delta\overline{H}_R(l) = -2043130 - 4 \times 43952.4 = 2218939.6 \text{ kJkmol}^{-1}$$

The higher heating value under standard conditions is

$$HHV = \frac{2218939.6}{44} = 50430.4 \text{ kJkg}^{-1}$$

Example 10.11 The ultimate analysis of a biomass fuel is: carbon 39%, oxygen 36%, nitrogen 1%, hydrogen 6%, sulfur 1%, ash 17%. The moisture content on a wet basis is 28%. Calculate the HHV and the LHV of the fuel.

Solution The following empirical equation (Eq.10.15) is recommended in Ref.[2] for the estimation of the higher heating value HHV [MJkg^{-1}] of biomass.

$$HHV = a_c X_c + a_h X_h + a_s X_s + a_o X_o + a_n X_n + a_a X_a$$

where the coefficients are given by:

$a_c = 34.91$, $a_h = 117.83$, $a_s = 10.05$, $a_o = -10.34$, $a_n = -1.51$, $a_a = -2.11$.

$$HHV = 34.91 \times 0.39 + 117.83 \times 0.06 + 10.05 \times 0.01 - 10.34 \times 0.36 - 1.51 \times 0.01 - 2.11 \times 0.17$$

The higher heating value is 16.69 MJkg^{-1}.

The moisture content on a wet basis is

$$M_w = 0.28$$

The lower heating value is given by Eq. (10.14) as

$$LCV = (1 - M_w)HCV - M_w L - 9Lf_H(1 - M_w)$$

$$LCV = (1 - 0.28) \times 16.69 - 0.28 \times 2.3 - 9 \times 2.3 \times 0.06(1 - 0.28)$$

Hence the lower heating value is 10.48 MJkg^{-1}.

Example 10.12 The ultimate analysis of a biomass fuel is: carbon 39%, oxygen 36%, nitrogen 1%, hydrogen 6%, sulfur 1%, ash 17%. The moisture content on a wet basis is 28%. The fuel undergoes combustion with 35% excess air. Calculate the mass of air required for the combustion of 1 kg of biomass.

Solution Consider one kg of dry biomass undergoing stoichiometric combustion.

For carbon

$$C + O_2 \rightarrow CO_2$$

Mass of oxygen: $m_c = 0.39 \times \frac{32}{12} = 1.04$ kg

For hydrogen

$$H_2 + \frac{1}{2}O_2 \rightarrow H_2O$$

Mass of oxygen: $m_h = 0.06 \times \frac{16}{2} = 0.48$ kg

For sulfur

$$S + O_2 \rightarrow SO_2$$

Mass of oxygen: $m_s = 0.01 \times \frac{32}{32} = 0.01$ kg

Contribution of oxygen in the fuel is: $m_{fo} = -0.36$ kg

Mass of nitrogen in fuel is: $m_{n_2} = 0.01$ kg

Oxygen in biomass reduces the oxygen required for combustion. Nitrogen and ash do not participate in combustion.

Total mass of oxygen required

$$m_{ot} = 1.04 + 0.48 + 0.01 - 0.36 = 1.17 \text{ kg}$$

Mass of air required is

$$m_a = 1.17 \times \left(1 + \frac{3.76 \times 28}{32}\right) = 5.02 \text{ kg}$$

Air required with 35% excess air is

$$m_{ae} = 1.35 \times 5.02 = 6.78 \text{ kg}$$

The moisture content of biomass is 28%. Hence the mass of dry biomass fuel is 0.72 kg per kg of wet biomass.

The air required for the combustion of 1 kg of (wet) biomass is

$$m_a = 0.72 \times 6.78 = 4.88 \text{ kg}$$

Example 10.13 A biomass project requires the following energy inputs: (i) 42 GJ per hectare of fossil fuel energy for cultivation and fertilizer, (ii) 10.5 GJ per tonne (10^3 kg) for harvesting and processing. The land yields 18 dry tonne of biomass per hectare with an energy content of 20 MJkg^{-1}. Calculate the net energy balance (NEB) of the project.

Solution The total fossil fuel energy consumed in the project is

$$E_{ff} = 42 + 10.5 \times 18 = 231 \text{ GJ per hectare}$$

The total energy available from biomass is

$$E_{bm} = 18 \times 10^3 \times 20 = 360 \text{ GJ per hectare}$$

The ratio of the energy available in biomass to the fossil fuel energy consumed is the NEB.

$$NEB = \frac{360}{231} = 1.55$$

Example 10.14 The following data apply to a corn ethanol plant.
Feedstock corn required per liter of ethanol: 2.58 kg
The energy required per hectare of land to grow corn:
Machinery – 4.78 GJ, Fossil energy – 5.9 GJ, Nitrogen – 10.15 GJ, Additional energy inputs – 12.5 GJ.
Energy inputs to the ethanol production process per 1000 liters of ethanol:
Transportation – 1.3 GJ, Steam for distillation – 10.7 GJ, Electricity – 4.8 GJ, Additional energy inputs – 0.61 GJ.
The corn yield per hectare of land: 9400 kg.
The net energy available in a liter of ethanol produced: 21.4 MJ.
Calculate the net energy balance (NEB) of the project.

Solution Consider the production of 1000 liters of ethanol.

The total energy required in the ethanol production plant:

$$E_{plant} = 1.3 + 10.7 + 4.8 + 0.61 = 17.41 \text{ GJ}$$

Corn feedstock required to produce 1000 liters of ethanol:

$$m_{corn} = 2.58 \times 1000 \text{ kg}$$

Required farm area is

$$A_{corn} = \frac{m_{corn}}{9400} = \frac{2.58\times1000}{9400} = 0.2745 \text{ hectare}$$

Energy required to produce corn is

$$E_{corn} = (4.78 + 5.9 + 10.15 + 12.5) \times 0.2745 = 9.149 \text{ GJ}$$

Total energy required is

$$E_{input} = E_{plant} + E_{corn} = 17.41 + 9.149 = 26.56 \text{ GJ}$$

Total energy available in 1000 liters of ethanol is

$$E_{av} = 21.4 \times 1000 = 21.4 \text{ GJ}$$

Hence the net energy balance ratio is

$$NEB = \frac{E_{av}}{E_{input}} = \frac{21.4}{26.56} = 0.806$$

Note that the ethanol plant produces only about 80% of the energy input.

Example 10.15 The following data apply to a corn ethanol plant.
Distance of ethanol plant from corn farm: 42 km
The price of corn: \$ 305 per tonne (10^3 kg)
The capacity of truck used for transportation: 11 tonne of corn
Weight of empty truck: 10.2 tonne
The equivalent CO_2 emission of the truck: 210 gm per tonne per km
The corn feedstock required by the plant: 132 tonne per year
Calculate (i) the cost of feed stock, and (ii) the equivalent greenhouse emissions during a year.

Solution Cost of corn feedstock per year

$$C_{corn} = 305 \times 132 = \$\,40{,}260$$

Number of truck loads of corn required per year

$$N_{tr} = \frac{132}{11} = 12$$

One-way distance travelled to transport corn

$$L_{tr} = 42 \times 12 = 504 \text{ km}$$

Equivalent CO_2 emission form truck carrying corn per year

$$m_{ft} = 210 \times (11 + 10.2) \times 504 = 2.244 \text{ Mg of } CO_2 \text{ - equivalent}$$

Equivalent CO_2 emission form returning empty truck per year

$$m_{et} = 210 \times 10.2 \times 504 = 1.079 \text{ Mg of } CO_2 \text{ - equivalent}$$

Total equivalent CO_2 emission per year

$$m_{total} = 2.244 + 1.079 = 3.323 \text{ Mg}$$

In choosing the source (farm) for purchasing corn for the ethanol plant both the cost of corn as well as the equivalent CO_2 emissions for transporting corn need to be considered.

Problems

P10.1 The annual insolation at a location 960 kWhm^{-2}. Estimate the average electrical energy that could be generated from 25 km^2 of land producing biomass. The efficiency of power generation is 30%.
[*Answer*: 36 GWh]

P10.2 The contents of 1 kg of biomass are as follows: 0.78 kg of dry, ash-free matter, 0.16 kg of water and 0.06 kg of ash. Calculate the moisture content on a wet, dry and ash-free basis.
[*Answers*: 0.16, 0.19, 0.205]

P10.3 Methane (CH_4) undergoes combustion with 30 % excess air. Calculate per kg of fuel (i) the mass of air required, (ii) the mass of carbon dioxide produced, and (iii) the mass of nitrogen in the products of combustion. [*Answers*: (i) 22.3 kg, (ii) 2.75 kg, (iii)17.1 kg]

P10.4 Cellulose used for anaerobic digestion has the chemical formula $C_6H_{10}O_5$. The mixture of cellulose and water in a gas generator produces methane (CH_4) and carbon dioxide. The density of methane under standard conditions is 0.657 kgm^{-3}. Calculate the volume of methane produced per kg of cellulose under ideal conditions.
[*Answer*: 0.45 m^3]

P10.5 The higher heating value (HHV) of a dry biomass is 11.8 MJkg^{-1}. If the moisture content is 16%, and the hydrogen content by mass in the material is 6%, calculate the HHV, and the LHV of moist biomass. [*Answers*: 9.916 MJkg^{-1}, 6.918 MJkg^{-1}]

P10.6 A biomass fuel undergoes combustion with 30% excess air. The ultimate analysis of the fuel is: $C = 52$ %, $H = 7\%$, $S = 0\%$, $O = 37\%$, $N = 1\%$, $ash = 3\%$. Calculate (i) the mass of air required per kg of biomass, and (ii) the volumetric analysis of the dry products of combustion. [*Answers*: (i) 8.75 kg, (ii) $CO_2 = 14.6\%, O_2 = 4.96\%$, $N_2 = 80.44\%$]

P10.7 A biomass fuel has the following ultimate analysis: $C = 52\%$, $H = 7\%$, $S = 0\%$, $O = 37\%$, $N = 1\%$, $ash = 3\%$. Calculate the higher heating value of the fuel. [*Answer*: 22.49 MJkg^{-1}]

P10.8 The volumetric composition of a gaseous fuel is:

$$H_2O = 26\%, CH_4 = 38\%, CO = 26\%, N_2 = 6\%, CO_2 = 4\%$$

Calculate (i) the air-fuel ratio of a stoichiometric mixture of air and fuel, and (ii) the volumetric composition of the products of combustion when water is in the vapor state. [*Answers*: (i) 8.13 kg per kg, (ii) $CO_2 = 12.1\%, H_2O = 18.3\%$, $N_2 = 69.6\%$]

P10.9 A biomass project requires the following energy inputs: (i) 10.7 GJ per tonne for harvesting and processing, and (ii) 44 GJ per hectare for transportation, cultivation and fertilizer. The land yields 16 dry tonne of biomass per hectare with an energy content of 20 $MJkg^{-1}$. Calculate the net energy balance (NEB) of the project. [*Answer*: 1.486]

P10.10 The following data applies to a corn ethanol plant.
Distance of ethanol plant from corn farm: 45 km
The price of corn: $ 290 per tonne
The capacity of truck used for transportation: 9.0 tonne of corn
Weight of empty truck: 9.8 tonne
The equivalent CO_2 emission of the truck: 209 gm per tonne per km
The corn feedstock required by the plant: 135 tonne per year
Calculate (i) the cost of corn feedstock, and (ii) the equivalent greenhouse emissions per year. [*Answers*: (i) $ 39150, (ii) 4.03 Mg]

References

1. Jacobson, Mark Z, 100% *Clean, Renewable Energy and Storage for Everything*, Cambridge University Press, Cambridge, U.K., 2021.
2. Jenkins, Nick., and Ekanayake, Janaka., *Renewable Energy Engineering*, Cambridge University Press, Cambridge, U.K. 2017.
3. Jones, J.B., and Hawkins, G.A., *Engineering Thermodynamics: An Introductory Textbook*, John Wiley & Sons, Inc. New Jersey, 1986.
4. Kreith, Frank and Krumdieck, Susan, *Principles of Sustainable Energy Systems*, CRC Press, Taylor & Francis Group, New York, 2015.
5. Rogers, G.F.C., and Mayhew, Y.R., *Thermodynamic and Transport Properties of fluids*, Blackwell, Oxford, U.K. 1995.
6. Vanek, Francis M, Albright, Louis D, and Angenent,Largus T, *Energy Systems Engineering: Evaluation and Implementation*, McGraw-Hill Book Company, New York, 2012.
7. Biofuels - ' http:/en.m.wikipedia.org/wiki/Biofuel'

Appendix A1

Transmission and Absorption of Solar Radiation

A1.1 Fundamental Equations

The laws of refraction relate the incident angle θ_1 and the refracted angle θ_2, both measured from the normal to the interface, by the equation [1]:

$$n_1 sin\theta_1 = n_2 sin\theta_2 \tag{A1.1}$$

where the refractive indices of the two media are n_1 and n_2 respectively.

Fresnel's expressions for the reflection of unpolarized radiation at the interface between medium 1 and medium 2 are:

$$r_p = \frac{sin^2(\theta_2-\theta_1)}{sin^2(\theta_2+\theta_1)} \tag{A1.2}$$

$$r_a = \frac{tan^2(\theta_2-\theta_1)}{tan^2(\theta_2+\theta_1)} \tag{A1.3}$$

$$r = r_p + r_a \tag{A1.4}$$

where r_p, r_a and r are the perpendicular component, parallel component, and the total reflectivity respectively of the incident unpolarized radiation. For radiation incident normal to the interface, the reflectivity r is given by

$$r = \frac{(n_1-n_2)^2}{(n_1+n_2)^2} \tag{A1.5}$$

A1.1.1 *Beer-Lambert law*

When radiation passes through a slab of material with a wavelength-dependent absorption coefficient $\alpha(\lambda)$, the intensity decreases exponentially according to *Beer-Lambert* law. Hence the intensity at a thickness L is given by

$$I(L) = I(0)exp(-\alpha L/cos\theta_2) \tag{A1.6}$$

The radiation absorbed in the slab is given by

$$A(L) = I(0)[1 - exp(-\alpha L/cos\theta_2)] \tag{A1.7}$$

where $I(0)$ is the intensity at the interface.

A1.2 Solar Radiation Transmission and Absorption

We shall apply the net radiation method [2] to obtain expressions for the transmittance of solar radiation through a multi-layered cover system, and the overall reflectivity.

A1.2.1 *Effective properties of a single slab*

Thc slab shown in Fig. A1.1(a) is a representative component of the multi-layered cover system depicted in Fig. A1.1(b). The flux of solar radiation incident on surface 1 of the layer is, Q_i and the reflected and transmitted net-radiation fluxes are Q_r and Q_1 respectively. At surface 2, the respective reflected and transmitted net-radiation fluxes are Q_2 and Q_t. The net radiation flux is the algebraic sum of all the different radiation currents that result from multiple reflections at the two faces of the slab.

The reflectivities r_i and r_o of the two surfaces 1 and 2 of the layer are assumed different to allow for the possible presence of surface coatings and thin reflecting films on these surfaces [1]. The transmittance τ of the layer can be computed knowing the angle of incidence of the radiation beam, the refractive index, the extinction coefficient of the material, and the layer thickness using Eq. (A1.6).

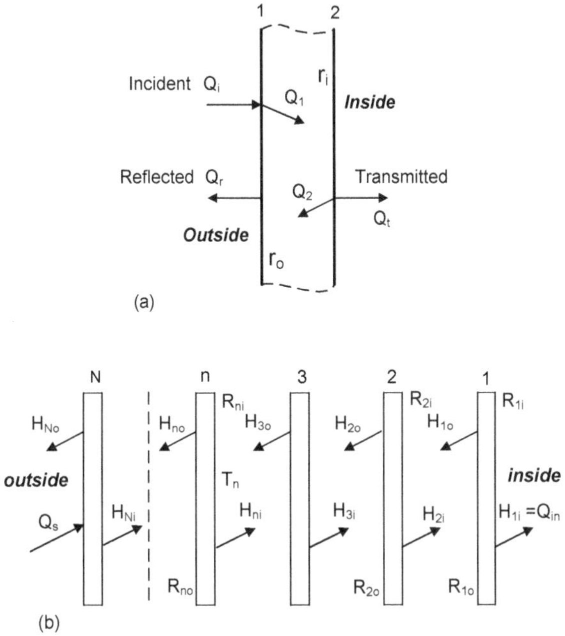

Fig. A1.1 (a) Net radiation fluxes for one layer (b) Multi-layered system

For most cover materials, the reflectivity and the transmittance vary with the wavelength of incident radiation. Therefore these spectral properties have to be averaged over the solar radiation spectrum before they are used in the radiation balance equations given below.

The following radiation balance equations relate the various net radiation fluxes for the layer depicted in Fig. A1.1(a).

$$Q_1 = (1 - r_o)Q_i + \tau r_o Q_2 \tag{A1.8}$$

$$Q_2 = \tau r_i Q_1 \tag{A1.9}$$

$$Q_t = \tau(1 - r_i)Q_1 \tag{A1.10}$$

$$Q_r = r_o Q_i + \tau(1 - r_o)Q_2 \tag{A1.11}$$

Eliminating Q_1 and Q_2 in Eqs. (A1.8) to (A1.11) we obtain the following expressions for the *effective transmittance, T* and the *effective reflectivity* R_o, used to characterize the layer:

$$T = \frac{Q_t}{Q_i} = \frac{\tau(1-r_o)(1-r_i)}{1-\tau^2 r_o r_i} \tag{A1.12}$$

$$R_o = \frac{Q_r}{Q_i} = r_o + \frac{\tau r_i (1-r_o)^2}{1-\tau^2 r_o r_i} \tag{A1.13}$$

Similarly, for radiation incident on surface 2 of the slab, the effective reflectivity, R_i is

$$R_i = \frac{Q_r}{Q_i} = r_i + \frac{\tau r_o (1-r_i)^2}{1-\tau^2 r_o r_i} \tag{A1.14}$$

A1.2.2 *Transmittance of a multi-layered cover system*

In this section we shall develop a series of general relationships applicable to cover systems consisting of different types of semi-transparent materials like glass and plastic.

Consider the multi-layered cover system, consisting of N partially-transparent layers separated by non-absorbing gas spaces, as shown schematically in Fig. A1.1(b). The layer, n ($n = 1, N$) is characterized by the three independent properties T_n, R_{ni} and R_{no} given by Eqs. (A1.12), (A1.13) and (A1.14) respectively.

The radiation flux incident on the outer surface of layer N is Q_s and the radiation flux entering the inside through inner most layer 1 is Q_{in}. The net radiation fluxes entering and leaving the different layers are indicated in Fig. A1.1(b). Applying the net radiation balance equation to the different layers we have

$$Q_{in} = H_{1i} = T_1 H_{2i} \tag{A1.15}$$

$$H_{1o} = R_{1o} H_{2i} \tag{A1.16}$$

$$H_{2i} = T_2 H_{3i} + R_{2i} H_{1o} \tag{A1.17}$$

$$H_{2o} = R_{2o} H_{3i} + T_2 H_{1o} \tag{A1.18}$$

$$H_{3i} = T_3 H_{4i} + R_{3i} H_{2o} \tag{A1.19}$$

$$H_{3o} = R_{3o} H_{4i} + T_3 H_{2o} \tag{A1.20}$$

$$H_{ni} = T_n H_{(n+1)i} + R_{ni} H_{(n-1)o} \tag{A1.21}$$

$$H_{no} = R_{no} H_{(n+1)i} + T_n H_{(n-1)o} \tag{A1.22}$$

$$H_{(N+1)i} = Q_s \tag{A1.23}$$

For a cover system consisting of N layers, the *overall transmittance* is defined as the ratio,

$$T_{on} = \frac{H_{1i}}{H_{(n+1)i}} \tag{A1.24}$$

Regular pattern of the above set of equations enables us to develop the following set of expressions for the overall transmittance.

For 1 layer

$$\frac{1}{T_{o1}} = \frac{1}{T_1} \tag{A1.25}$$

For 2 layers

$$\frac{1}{T_{o2}} = \frac{(1-R_{2i}R_{1o})}{T_2}\left(\frac{1}{T_{o1}}\right) \tag{A1.26}$$

For 3 layers

$$\frac{1}{T_{o3}} = \frac{(1-R_{3i}R_{2o})}{T_3}\left(\frac{1}{T_{o2}}\right) - \frac{T_2 R_{3i} R_{1o}}{T_3}\left(\frac{1}{T_{o1}}\right) \tag{A1.27}$$

For larger number of layers the expressions for the overall transmittance could be written down by observing the form of the above expressions. Alternatively, a step-by-step solution procedure using Eqs. (A1.15) to (A1.23) may be developed to compute the overall transmittance.

A1.2.3 *Reflectance of multiple layers*

The *overall reflectance* of n layers is defined as,

$$\bar{R}_n = \frac{H_{no}}{H_{(n+1)i}} \tag{A1.28}$$

By manipulating Eqs. (A1.15) to (A1.23) we obtain the following expressions for the *overall reflectance* of different number of layers.

For 1 layer

$$\bar{R}_1 = \left(\frac{H_{1o}}{H_{2i}}\right) = R_{1o} \tag{A1.29}$$

For 2 layers

$$\bar{R}_2 = \left(\frac{H_{2o}}{H_{3i}}\right) = R_{2o} + \frac{\bar{R}_1\, T_2{}^2}{(1-R_{2i}\bar{R}_1)} \tag{A1.30}$$

For 3 layer

$$\bar{R}_3 = \left(\frac{H_{3o}}{H_{4i}}\right) = R_{3o} + \frac{\bar{R}_2\, T_3{}^2}{(1-R_{3i}\bar{R}_2)} \tag{A1.31}$$

References

1. Duffie John A., Beckman William A., with Blair Nathan Blair, *Solar Engineering of Thermal Processes, Photovoltaics and Wind*, John Wiley & Sons, New Jersey, USA (2020).
2. Wijeysundera, N.E., 'A net radiation method for the transmittence and absorptivity of series of parallel regions'. *Solar Energy*, 17, 1975, 75-77.

Appendix A2

Heat Transfer Correlations

A2.1 Heat Transfer Correlations for Solar Collectors

The analysis of energy transfer processes in solar collector systems requires the values of several convective heat transfer coefficients. These are usually obtained using various heat transfer correlations. A few important and relevant correlations are listed below.

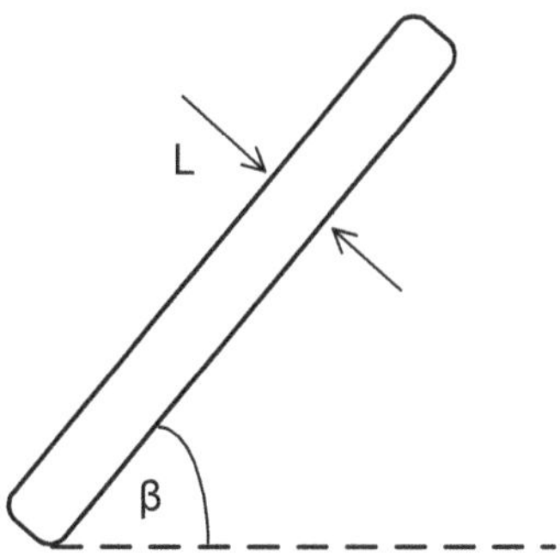

Fig. A2.1 Heat transfer in an air-filled rectangular cavity

A2.1.1 *Heat transfer in an inclined rectangular cavity*

The free convection heat transfer coefficient for an inclined rectangular cavity, shown in Fig. A2.1, is available in most standard text books on solar energy [1,2]. It is expressed in the form:

$$Nu = 1 + 1.44\left[1 - \frac{1708}{Racos\beta}\right]^{+}\left(1 - \frac{1708(sin1.8\beta)^{1.6}}{Racos\beta}\right) + \left[\left(\frac{Racos\beta}{5830}\right)^{1/3} - 1\right]^{+} \tag{A2.1}$$

where the meaning of the + exponent is that only *positive* values of the terms in the square brackets are to be used. (if the term is negative the value is taken as zero). The tilt angle β varies from 0 to 75°.
The meaning of the symbols are:

$Nu = \frac{hL}{k}$, is the Nusselt number

$Ra = \frac{g\beta'\Delta TL^3}{\alpha v}$, is the Rayleigh number

$Pr = \frac{v}{\alpha}$, is the Prandtl number

h = heat transfer coefficient, L = plate spacing, k = thermal conductivity, g = acceleration due to gravity, $\beta' = 1/T$ = volumetric coefficient of expansion, ΔT = temperature difference between plates, v = kinematic viscosity, α = thermal diffusivity.

A2.1.2 *Heat transfer due to wind*

Solar collectors mounted outdoors experience cooling due to wind blowing over the outer surface. Several heat transfer correlations have been developed to estimate wind heat transfer coefficients. These are listed below.

Data from wind tunnel tests on rectangular flat plates at various orientations have been correlated to obtain the following expression [1].

$$Nu = 0.86Re^{1/2}Pr^{1/3} \tag{A2.2}$$

The applicable Reynolds number range is:

$$2 \times 10^4 < Re < 9 \times 10^4$$

The characteristic length is four times the area divided by the perimeter of the plate.

There are two simpler equations that relate the heat transfer coefficient h ($Wm^{-2}C^{-1}$) directly to the wind speed V (ms^{-1}).

These are

$$h = 5.7 + 3.8V \tag{A2.3}$$

In Eq. (A2.3) the heat transfer coefficient may also include the effects of free convection and thermal radiation. A correlation that includes only wind effects is

$$h = 2.8 + 3.0V \tag{A2.4}$$

For very low wind speeds, the dominant mode of heat loss is free convection. Correlations for this situation are available in standard text books on heat transfer.

A2.1.3 *Forced convection across a cylinder*

For *forced convection* across a single cylinder the following correlations are recommended [1].

$$Nu = 0.40 + 0.52Re^{0.52} \tag{A2.5}$$

for $0.1 < Re < 1000$

$$Nu = 0.24Re^{0.6} \tag{A2.6}$$

for $10^3 < Re < 50 \times 10^3$

A2.1.4 *Forced convection inside tubes and rectangular ducts*

The absorbers of water heating collectors have circular or oval-shapes tube through which water flows. The following heat transfer correlations are available for such flow situations [2].

For *laminar flow* in tubes or ducts the following equation can be used to determine the average heat transfer coefficient, $\bar{h}_c$ for $\frac{RePrD_H}{L} > 10$.

$$\overline{Nu}_c = \frac{\bar{h}_c D_H}{k} = 1.86(RePrD/L)^{1/3}(\mu_b/\mu_w)^{0.14} \tag{A2.7}$$

where

D_H = the hydraulic mean diameter
L = length of tube or duct
k = fluid thermal conductivity at bulk fluid temperature
μ_b = viscosity of the fluid at the bulk temperature at the middle of the duct length
μ_w = viscosity of the fluid at the wall temperature.

When the duct is *short*, and $\frac{RePrD_H}{L} > 100$, the following equation applies [2]:

$$\overline{Nu}_c = \frac{\bar{h}_c D_H}{k} = \frac{RePrD_H}{4L} ln\left[\frac{1}{1-2.654/Pr^{0.167}(RePrD_H/L)^{0.5}}\right] \quad \text{(A2.8)}$$

For turbulent flow inside tubes and ducts the following equations may be used to obtain the heat transfer coefficient [2].

For Prandtl numbers between 0.7 and 700, and $\frac{L}{D_H} > 60$

$$\overline{Nu}_c = \frac{\bar{h}_c D_H}{k} = 0.023Re^{0.8}Pr^{1/3} \quad \text{(A2.9)}$$

When the duct is short such that $\frac{L}{D_H} < 60$, the heat transfer coefficient calculated using Eq. (A2.9) is multiplied by $\left(1 + \frac{D_H}{L}\right)^{0.7}$.

A2.1.5 *Forced convection between parallel plates*

For turbulent flow between two parallel plates with one side heated, the following equation may be used to calculate the heat transfer coefficient [2].

$$\overline{Nu}_c = \frac{\bar{h}_c D_H}{k} = 0.0196Re^{0.8}Pr^{1/3} \quad \text{(A2.10)}$$

References

1. Duffie John A., Beckman William A., with Blair Nathan Blair, *Solar Engineering of Thermal Processes, Photovoltaics and Wind*, John Wiley & Sons, New Jersey, 2020.
2. Goswami, Y., Kreith Frank,and Kreider Jan F., *Principles of Solar Energy Engineering*, 2nd Edition, Taylor & Francis, Philadelphia, PA, 2000.

Index

www.ingramcontent.com/pod-product-compliance
Lightning Source LLC
LaVergne TN
LVHW010338220926
848484LV00004BB/928